ADVANCED CONCEPTS IN OCEAN MEASUREMENTS
FOR MARINE BIOLOGY

THE BELLE W. BARUCH LIBRARY IN MARINE SCIENCE

THE BELLE W. BARUCH LIBRARY IN MARINE SCIENCE NUMBER 10

Advanced Concepts in Ocean Measurements for Marine Biology

Edited by Ferdinand P. Diemer

F. John Vernberg

Donna Z. Mirkes

Published for the Belle W. Baruch Institute for Marine Biology and

Coastal Research by the

UNIVERSITY OF SOUTH CAROLINA PRESS

Library of Congress Cataloging in Publication Data
Main entry under title:

Advanced concepts in ocean measurements for marine
  biology.

  (The Belle W. Baruch library in marine science; no. 10)
  Papers presented at a symposium co-sponsored by the U.S.
Office of Naval Research and the Belle W. Baruch Institute
for Marine Biology and Coastal Research, and held Oct.
24-28, 1978, at Hobcaw Barony, Georgetown, S.C.
  Includes index
  1. Marine biology—Technique—Congresses. 2. Population
biology—Congresses. 4. Mensuration—Congresses.
I. Diemer, Ferdinand P.   II. Vernberg, F. John,
1925— III. Mirkes, Donna Z. IV. Belle W. Baruch Institute
for Marine Biology and Coastal Research.
V.   United States. Office of Naval Research
VI.   Title: Ocean measurements for marine biology.
VII.   Series: Belle W. Baruch library in marine sciences;
no. 10.
QH91.57.A1A3   574.92'028   79-24801
ISBN 0-87249-388-1

PREFACE

The complexity of the world's oceans is reflected in the diverse nature of the disciplines involved in studies of oceanic processes. Frequently there is an information exchange gap between these disciplines. An attempt to bridge this gap resulted in a symposium held October 24-28, 1978, at Hobcaw Barony, Georgetown, South Carolina. The symposium design allowed for tutorial input from signal physicists, an insight into problems from the bioscience and fisheries communities, and perspective from these marine-related disciplines. The established aims of the symposium were to: (1) identify problems in the measurement of biotic and abiotic factors in the sea; (2) review recent advances in measurement and analysis of various oceanic parameters and to assess their applicability to problems in biological oceanography; and (3) recommend future research and development direction. The major topics included: Perspectives for Marine Biological Measurements, Past, Present and Future; Sampling, Experimental Design, and Signal Physics; Satellite and Aircraft-Based Remote Sensing; and Hydroacoustics.

To maintain the high quality of the Belle W. Baruch Library in Marine Science series, all papers have been externally peer-reviewed. We extend special thanks to the various scientists throughout the country who served in this capacity.

We gratefully acknowledge the Department of Navy, Office of Naval Research and the Belle W. Baruch Institute for Marine Biology

and Coastal Research, University of South Carolina, who co-sponsored
and financed this symposium. We also thank the Belle W.
Baruch Foundation for providing conference facilities at Hobcaw
Barony.   The efforts of Ms. Sue Mushock, Dr. Dennis Allen, and the
staff at Hobcaw Barony are especially appreciated.

F. P. D.
F. J. V.
D. Z. M.

CONTENTS

PERSPECTIVES FOR MARINE BIOLOGICAL MEASUREMENTS: RESEARCH AND PROBLEMS

SAMPLING, EXPERIMENTAL DESIGN, SIGNAL PHYSICS

SATELLITE AND AIRCRAFT-BASED REMOTE SENSING

MARINE BIOLOGICAL MEASUREMENT AND INSTRUMENTATION

HYDROACOUSTICS AND MARINE BIOLOGY

STATISTICAL PROBLEMS IN MARINE BIOLOGY

WORKING GROUP REPORTS

ADVANCED CONCEPTS IN OCEAN MEASUREMENTS FOR MARINE BIOLOGY

Introductory Comments

Robert L. Edwards

Mensuration, the topic to be addressed at this workshop, is so obvious and so critical and so pervasive that one wonders why it is seldom addressed in its own right. When Ferd Diemer asked me to give some introductory remarks, I agreed without thinking very deeply about the significance of the topic. It was some fifteen minutes later that I fully realized that in fact we were to be directly concerned with the ultimate, vital substance of science, namely measurement. Our work is only really useful when it may be replicated and/or compared with similar work. Our data is useful only when there is a universally recognized basis for communicating it.

I have been in my office only two days in the last thirty. While admittedly that was an unusually busy period, nowadays the average marine research administrator spends much time at meetings, conferences, planning sessions, and so forth. It is easy to regard the various topics being dealt with in descriptive terms, e.g., 'the early life history of bottom fish' or 'temperature and salinity fronts in the Gulf of Maine'. The following is a list of topics that occupied my time during this 30-day period--not the entire list--just those topics that received a significant amount of attention:

1. fish population assessments;

2. productivity research, especially primary and secondary
    production;
3. ecosystem modeling, with an emphasis on the specification
    of required, but lacking, areas of understanding;
4. hydroacoustic theory and practice;
5. chromosomal aberrations in the eggs of marine organisms;
6. pollutants--distribution, flux, and abundance;
7. ocean variability, in this instance the shelf and the evi-
    dence of variability as detected by remote sensors;
8. statistical theory, especially that pertinent to sampling
    design for programs that monitor biological aspects (dis-
    tribution and abundance).

One can scan this list, and depending upon his bent, find one
or more topics of particular interest. But all items have one
common thread--that is nothing more nor less than the problem of
measuring things. Some comments about a few of these items are in
order and relevant with respect to our meeting.

The topic 'primary production' came up as a noteworthy item
at the International Commission for the Exploration of the Seas
(ICES) meeting as a consequence of a mini-symposium concerned with
comparing the energy flow in the North Sea with that of the region
off the Middle Atlantic and New England. Much to the surprise of
many people, it developed that there was a wider variation in the
use of ^{14}C than had been apparent previously, and that biologists
generally had been woefully reluctant to standardize their procedures to date. Further, there were some disturbing semantic problems associated with the interpretive use of the production values
obtained. Biologists generally have lagged far behind the physical
oceanographers and chemists in getting on with the task of standardization and intercalibration of techniques and the matter became
painfully apparent in this instance. One could justifiably suggest
that the next five years be dedicated to just that--the standardization and intercalibration of all the presumed conservative techniques used by marine biologists. The only area where I have seen
much progress in this matter is in the recognition that certain
collecting gear, for example plankton nets and trawls for fish,
need standardization in themselves, as well as the vessels hauling
the nets. The attachment of different experts to different types
of gear, all too often personally designed, has rendered much of
the data collected to date less than useless when it comes to comparing areas and seasons. While it serves no useful purpose to be
overly critical about this state of affairs, it is time we frankly
recognized the consequences of our actions and do more than give
lip service to solutions. Incidentally, ICES quickly established
a working group to deal with standardizing the use of ^{14}C and at
least has scratched the surface of the general problem.

The issues in the hydroacoustic arena are not very different.
Echo sounders have a variety of applications, both practical
(searching for fish to capture) and scientific (estimating abundance). Its tactical use is one thing and its strategic use another. The fisherman uses it tactically and it can serve him well.
Its usefulness to the fisherman is determined to a large degree by

his own internal data bank concerning fish behavior and distribution and his ability to integrate this data with what is, or is not, revealed by the echo sounder. The scientist, if he is to be objective, cannot use the instrument for his strategic purposes on this basis, at least until he can quantify his interpretive inputs. These two uses, tactical and strategic, have been shamelessly confused. This confusion is now being dealt with and a useful series of international workshops and symposia have been decided upon to come to grips with the many problems. Surprisingly, even though many of those working in the field were engineers and not biologists, it took several years to get some individuals to recognize the virtue of calibration, both within and between.

Chromosomal aberrations were mentioned. This area is very exciting scientifically. Time doesn't allow a review of the specifics, but suffice it to say that the marine physiologist and specialists in genetics are developing some very promising quantitative techniques for assessing some of the environmental insults we read so much about, as for example AMOCO CADIZ and ARGO MERCHANT, pollutants such as sewage, heavy metals, and sundry chlorinated hydrocarbons. There is no need to deal further with the list. Every item has a problem associated with it and in each case it is concerned either with the act of measurement or the interpretation and comparability (archivability) of measurements, or both. From my point of view there is one overriding message, and that is that the time has come for marine biologists to address directly and formally the need for standardization and calibration. It is past time that we 'hardened' up that science (made it less an 'art'). The physical oceanographers and marine chemists have long since recognized the problem and continue to deal with it as a significant and serious part of their scientific responsibility. They have shown the way and the biologists can well afford to take a page or two from their book.

I will now deal briefly with a second topic, also related to mensuration, but somewhat further afield and a bit more philosophic. That topic is the question of 'roles'. What is the principal role of the federal scientist, the academic scientist, other governmental (e.g., State) scientists, and the scientist in a consulting firm? All too often some politician or private citizen gives vent to his or her frustrations about the inadequacy of the scientific community when it comes to dealing with the pressing problems of society. The criticism may address the 'ivory tower attitude' (regarding the inability to respond to the 'real world'), their competitiveness, or their inability to communicate. Even though I sometimes fully agree with such charges, I tend to resent them because of the implication that there is a deliberate intention on the part of scientists not to do what they should. Virtually every scientist that I know will make every effort to explain his work, in Esperanto if necessary, to any interested person. When contentious issues are involved, the problem is all too often the narrow band width of the receiver. With regard to the charge of nonresponsiveness, here the critic has usually imbued the scientist with responsibilities and attributes he doesn't have, particularly a unique right and competence to make value judgments for society. When it comes to the concern expressed

about the 'competitiveness' of scientists, here the critic frequently
has my ear and usually my concurrence, almost regardless of how
shallow or misguided a particular criticism may be. And now, we are
back to mensuration.

For convenience's sake I categorize work in the marine environment as either process or time series oriented from one point of
view, or as micro-, meso-, or macro-scale from another point of
view. Each of us here probably has his own somewhat unique taxonomic frame to express the same set of ideas, and each scheme is no
more than an artificial intellectual construct to assist in the
communication of ideas. Bear with me as I proceed and remember
that it is necessary to generalize. I fully recognize that nuances
exist, but we cannot deal with them here and now.

The scientist in the public sector, such as I am, serves no
particular group, but the public at large. He has no authority
to make decisions for society--there are appointed officials to
do that as well as elected officials. The scientists in the
commercial sector, an individual in a consulting firm for example,
often does work to serve a particular interest or segment of society, hopefully without losing the objectivity normally considered one of the mandatory qualities of a scientist. A scientist
in the private sector works in his chosen field on topics of his
own choosing. He is expected to subject himself continuously to
his peers to maintain his credibility. The other (first) two
categories must do the same, but are also subject to review by
nonscientists who have quite different standards of judgment,
some of which are useful and appropriate while others, at times,
can only be described as venal. Under the circumstances government scientists will tend to be less daring--more conservative--
than scientists in private institutions.

In general terms, governmental institutions can provide longer
term and greater logistic support for programs considered worthy in
practical terms than can private institutions. The academic scientist frequently resides in a university and is constrained to deal
with problems that can be chewed piecemeal by graduate students.
The contract scientist tends to be in between these two extremes.

Ignoring obvious exceptions, it seems clear to me that the
principal responsibility of the government scientist lies more in
the macro-scale, long-time series arena, that where considerable
logistic support and dedication to the development of long-time
series data bases are necessary. The basis for such studies will
tend to be scientifically conservative and the results designed
to serve the general need of the political unit involved--city,
county, state, or country.

The academic scientist will generally be more involved in
process-oriented studies, on a micro- or meso-scale, and will
generally be either unable or unwilling to take on a further
responsibility for the maintenance of a data base to service
other needs. To serve that last purpose, data centers and museums
exist. The academic scientist is constrained to be conservative
only to the degree his peers find him straying beyond their standards for objectivity, etc.

Each of these groups supports the other. They are not, or at

least should not be, competitive. The macro-scale studies provide a framework within which process-oriented or micro-scale
studies fit and evolve. The credibility of the longer-term time
series, macro-scale studies should be preceded by the other studies
which were initially judged on the basis of scientific merit alone.

   We all know that there are elements of competitiveness that
don't really stand close examination. But that is not my point.
We aren't paying enough attention to the fact that there is a
great deal of merit in and need for these different approaches,
that they are interactive, mutually supportive, and that neither·
is viable without the other. We should be paying more attention
to the interactive potentials, doing a better job recognizing and
defining the different roles, and just generally putting more emphasis on working together toward common objectives, regardless
of our particular situation or our unique responsibilities. Let's
measure ourselves better than we have in the past.

   To summarize, my prescription for the next five years is--

1. To deal with the need for the standardization and calibra-
    tion of instruments and techniques, especially those used
    by biologists. And, related to this, it is time to address
    seriously the difficulties associated with archiving biolog-
    ical data.

2. To define more clearly the different roles and responsibil-
    ities of the various types of institutions and individuals
    doing research in the ocean. Both to the end that the full
    potential for interaction with consequent better results by
    all parties is achieved, and to the end that the present
    confusion on the part of the public about our different
    roles and responsibilities is significantly reduced if not
    eliminated, and lastly to the end that nonconstructive
    competitiveness is reduced at least to the level that it
    doesn't sully the credibility of each of our research
    efforts.

PERSPECTIVES FOR MARINE BIOLOGICAL MEASUREMENTS: RESEARCH AND PROBLEMS

MARMAP, a Fisheries Ecosystem Study in the Northwest Atlantic: Fluctuations in Ichthyoplankton-Zooplankton Components and Their Potential for Impact on the System

Kenneth Sherman

ABSTRACT: Initial results of a fishery ecosystems
study based on time-series surveys of fish, plankton,
benthos, hydrography, fish catch data, and ancillary
process-oriented projects are described for the continental shelf area off the northeast coast of the
United States. The study is conducted from the Gulf
of Maine to Cape Hatteras as part of the MARMAP
(Marine Resources Monitoring, Assessment and Prediction) Program of the National Marine Fisheries Service. The region has been subjected to heavy fishing mortality over the past decade. During the
1968 to 1975 period, the biomass of principal fish
species declined by 50% over this region of the
continental shelf. The full impact of the removal
of several million metric tons of predators from
the continental shelf ecosystem is not known. Two
major questions are addressed. 1) Does the reduction in the stocks of important predator species,
herring, mackerel, cod, haddock, hake, and others,
release secondary production to be consumed by
short-lived, fast growing, smaller, less desirable
species? 2) What are the probabilities associated
with the return of over-exploited species to their

former abundance levels? Ecosystem studies now
underway by the Northeast Fisheries Center focus
on the biological and environmental factors controlling mortality, recruitment, survival, and
productivity of the fish stocks on the continental
shelf. Changes in population levels of ichthyoplankton and zooplankton detected on MARMAP surveys
are discussed in relation to potential impact on
the ecosystem. Preliminary evidence suggests that
the recent (1974-1977) increase in sand lance
(*Ammodytes* spp.) abundance may be indicative of a
shift in dominance among the fish species in the
Northwest Atlantic. Emphasis is given to the importance of time-series studies at all levels in
the ecosystem in relation to emerging theory, which
holds that stressed ecosystems favor the development
of fast growing, short-lived, and generally less
desirable commercial species, like sand lance. Comparisons are made between primary productivity and
fish biomass in the North Sea, Mid-Atlantic Bight,
and Georges Bank. The higher fish biomass of
Georges Bank is attributed to the higher primary
productivity of the area. The importance for
accelerating studies of secondary production in
both the benthos and zooplankton in order to understand better fish-plankton-benthos relationships is
stressed.

FISHERIES ECOSYSTEM STUDIES OFF THE NORTHEAST COAST OF THE UNITED
STATES

The fishery resources off the northeast coast of the United
States support a fish-catching and processing industry contributing
one billion dollars annually to the economy of the coastal states
from Maine to North Carolina. These resources are now, under the
terms of the recently passed Fisheries Conservation and Management
Act of 1976, (U.S.A. [FCMA] Public Law No. 94-265, 94th Congress H. R.
200, April 13, 1976) subject to management by the New England and Mid-
Atlantic Regional Fisheries Management Councils. The Councils are
required to develop management plans for the resources under their
jurisdiction that ensure optimal sustained yields based on ecological, economic, and social considerations. Input for the ecological
decisions are to be based on the "best scientific information available".

The best and most sought after scientific information from a
fisheries management point of view is the accurate prediction of
future stock sizes and the effects of different levels of fishing
on the continued production of economically viable resource populations. This need has not changed since the early days of whaling,
when the U. S. Wilkes Expedition of 1838 was supported by Congress
to improve our knowledge of Pacific whaling areas. Henry Bigelow
was supported, in part, by Federal funds when in the 1920's he in-

vestigated the fish, plankton, and oceanography of the Gulf of Maine
for the U. S. Fish Commission with an end to improving the fishing
industry. Subsequent studies on both sides of the Atlantic focused
on the yields of single species, not from any lack of intellectual
awareness of the interaction and interdependence of species, but
rather from the constraints of meager budgets provided to support
fishery research organizations. Those days have passed. But the
early orientation to single species assessments has not been easily
shed. "Special" interests will continue to demand information on
particular species, and we will need to continue providing single
species estimates of abundance levels.

Under the FCMA some 2.2 million square miles of contiguous
ocean water falls under the jurisdiction of the United States as
a Fisheries Management Zone (FMZ). At present only 150,000 square
miles of the zone, most of which is off the northeast coast, is
being systematically monitored for seasonal, areal, and annual
changes in plankton, fish, benthos, and hydrography. There are
no shortcuts to obtain the comprehensive population and environmental information required to improve forecasts of fish abundance
within the Fisheries Management Zone. A balanced approach is needed
that allows for: 1) a time-series of observations in the form of
routine multispecies fish, plankton, benthos, and hydrographic monitoring surveys; 2) a systematic collection of fish-catch data; and
3) process-oriented studies dealing with biological and environmental linkages among plankton production, benthos production, and
fish production. The three level approach allows for the monitoring
of changes, ensures that the appropriate critical events are being
measured, and advances the basic understanding of biological energy
transfer mechanisms in marine ecosystems. This fisheries ecosystem
program, called MARMAP (Marine Resources Monitoring, Assessment, and
Prediction) is conducted by the National Marine Fisheries Service on
the continental shelf from the Gulf of Maine to Cape Hatteras.
During the past decade, this region of the continental shelf, including the Gulf of Maine, Georges Bank, Southern New England, and
the Mid-Atlantic Bight, has been subjected to extreme fishing pressure. From 1968 to 1975 the biomass of the principal fish species
declined approximately 50%; much of the decrease in biomass correlates with increased fishing effort, indicating overfishing (Clark
and Brown, 1977). Environmental conditions, coastal pollution, inter-
and intra-specific competition may also have contributed to the decline, but no quantitative estimate of this mortality is now available.

The full impact of the removal of several million metric tons of
predators from the continental shelf ecosystem is not known. Significant questions remain unanswered. Does the reduction in the stocks
of important predator species, herring, mackerel, cod, haddock, hake,
and others, release secondary production to be consumed by short-
lived, fast growing, smaller, less desirable species? What are the
probabilities associated with the return of over-exploited species
to former abundance levels? Studies are now underway by the Northeast Fisheries Center (NEFC) to address these questions. These
studies focus on the critical linkages between the principal food
species of fish and the recruitment, survival, and productivity of

the fish stocks on the continental shelf from the Gulf of Maine to
Cape Hatteras.

   Multispecies Assessments

   Studies of single species do not provide the assessment information required for effective management of multispecies fisheries
operating at different trophic levels. While it is important to continue these studies, they are now being pursued within a broader matrix that measures interactions in changing abundances among the
species in the ecosystem. Single-species yield models have recently
been augmented with multispecies models that are ecologically sensitive (Regier and Henderson, 1973; Parrish, 1975; Laevastu et al.,
1976; Anderson and Ursin, 1977). These models deal with multispecies fishery interactions at different trophic levels. They
are important approximations of the consequences of predator-prey
dynamics based on fishery imposed selective mortality, and hold
promise for providing a basis for the management of marine ecosystems. If ecosystem models are to assume an appropriate role
in the management of marine resources, it will be necessary to
overcome present deficiencies in: 1) understanding relationships
between stock-size and recruitment; 2) identifying the linkages
between primary, secondary, and fish production; and 3) quantifying
predator-prey dynamics.

   Predator-Prey Interactions

   Predator-prey interactions are complex. They reflect a series
of interrelationships that can change significantly the abundance
of important fish stocks. A schematic representation of the predator-prey interactions for eight of the more abundant species of fish
and squid off the northeast coast of the United States is given in
Figure 1. It is presented as a qualitative example of the complexity of the known interactions between fish and their prey. In
addition, the significant changes in the size of prey consumed as a
fish moves through larval, juvenile, and adult stages compound the
difficulty in sorting out predator-prey relationships. The feeding
habits of codfish illustrate the problem. Codfish larvae feed principally on microzooplanktonic copepods, crustacean eggs, pteropods,
and larvae of meroplankton. As juveniles they feed on macrozoo-
plankton including euphausids and amphipods; as adults, fish become
a principal food.
   It has recently been suggested that shifts in the abundances
of predators in stressed marine ecosystems can lead to significant
changes in the species composition and size structure of the prey
populations (Steele and Frost, 1977). The degree to which species
shifts in abundance can result in changes of fish production in
the continental shelf ecosystem is now the subject of an expanded
research effort by the Northeast Fisheries Center. For example,
it appears that the predatory consumption levels of a single species, the silver hake, is sufficient to consume 30% of all fish
produced on the continental shelf of the northeast coast (Edwards
and Bowman, 1978). Aspects of this research, including recent

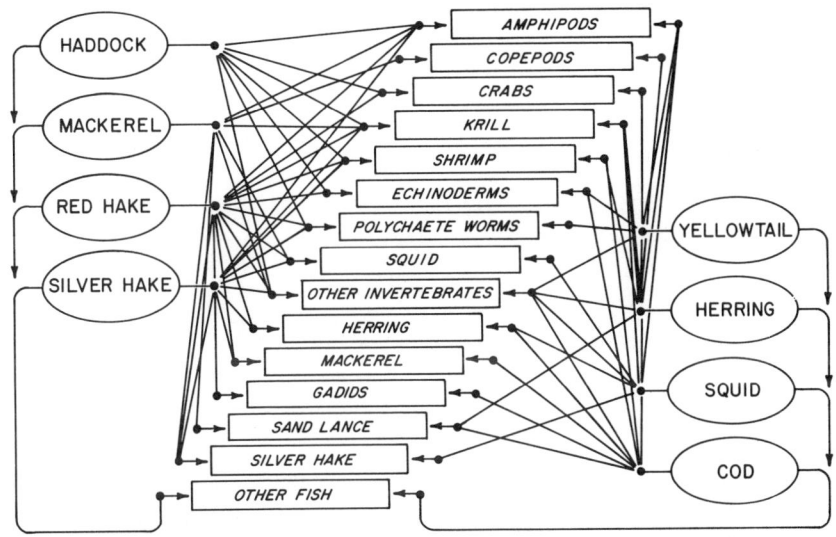

FIGURE 1: Schematic predator-prey interactions
for the more important species of fish and squid
off the northeast coast of the United States.
Predator names are enclosed in ovals; prey are
shown in rectangles. From Langton and Bowman
(1977).

evidence of changes in abundance among ichthyoplankton and zooplankton species, are given in the present report.

MARMAP SURVEYS

     Fishery science is undergoing major changes in the approach
to improving assessments of the abundance levels and forecasting
potential yields of fish stocks. The new approach represents a
balance between the more traditional studies of biological and
physical processes as they relate to productivity of coastal waters,
and the requirement for committing ships and personnel to fisheries-
independent time-series surveys of annual changes in the productivity levels of plankton, fish, and benthos populations. Time-
series surveys are dull and routine, but necessary for measuring
population, environmental, and pollution changes over time and space
and for sorting out the causes of these changes with respect to
fishing mortality, natural mortality, or pollution mortality. In
addition to the surveys, controlled ecosystem experiments are now
being conducted in large enclosures containing the smaller population components of the pelagic ecosystem. The best example of this
effort is in the Controlled Ecosystem Experiments (CEPEX) sponsored
by the National Science Foundation. The CEPEX operation is conducted as a multidisciplinary study of the interactions of primary,

secondary, and tertiary trophic levels of a northwest temperate
deep-water embayment in British Columbia under the effects of physical-chemical changes. Findings of this study apply to our investigations of fishery ecosystems by providing more insight into the
critical functions that need to be measured on time-series surveys.
Other technical advances in hydroacoustics, remote sensing, and
electronic data processing, when applied to the time-series approach,
will undoubtedly contribute significantly to increased efficiencies
and reduced costs of the MARMAP surveys.

In 1971 a systematic macroscale sampling of zooplankton and
ichthyoplankton was initiated by the Northeast Fisheries Center
(NEFC) in the Gulf of Maine, Georges Bank, Southern New England,
and Mid-Atlantic Bight during bottom trawl surveys in autumn and
spring. In 1976 the zooplankton-ichthyoplankton surveys were expanded to bimonthly coverage of the areas to examine the zooplank-
ton-fish linkages including 1) changes in the distributions, abundance, and growth of juvenile and adult fish, and 2) larval fish
growth and mortality in relation to their planktonic prey and predators. MARMAP surveys are conducted systematically at stations
selected from a stratified random design for fish, shellfish, benthos, phytoplankton, ichthyoplankton, and zooplankton. Bottom
trawl surveys for fish are conducted in spring and autumn, and
since 1977, in summer. Two shellfish surveys are made annually.
Benthic sampling is limited, contingent on the analyses of 25
years of collections now being completed at the Woods Hole Laboratory of NEFC. Surveys of zooplankton-ichthyoplankton, phytoplankton, primary productivity, and hydrography are conducted on
a bimonthly basis for a total of six surveys per year. The survey
data are augmented with a comprehensive system for obtaining catch
data at each of the major fishing ports from Cape Hatteras to the
Gulf of Maine. As required, special surveys are conducted to deal
with specific problems (e.g., tagging, feeding, current meter deployment and retrieval, vertical distribution studies of ichthyoplankton, and sampling for sharks and other large predators). Mesoscale studies of larval herring mortality have been done jointly
with other countries. In addition, microscale "patch-studies" have
been conducted to investigate factors controlling larval survival
by examination of larval predator-prey relationships within the
water column. The area under investigation is extensive, requiring
a heavy logistic commitment. The MARMAP studies are being conducted jointly with scientists and ships of the Federal Republic
of Germany, German Democratic Republic, Poland, and the USSR.

Ichthyoplankton-Zooplankton Monitoring

All ichthyoplankton-zooplankton collections are made with 61 cm
diameter bongo nets towed obliquely through the water column between 1.5 and 3.5 kts. At each location nets with 0.505 mm and
0.333 mm mesh apertures are used; at selected stations a 20 cm bongo
sampler is fitted with 0.165 mm and 0.253 mm mesh nets and positioned
above the larger bongo frame to sample the microzooplankton. Sample
processing is done at the Narragansett Laboratory of NMFS under the
supervision of Mrs. Ruth Byron, and at the Polish Plankton Sorting
Center, Szczecin, under the direction of Dr. Leonard Ejsymont.

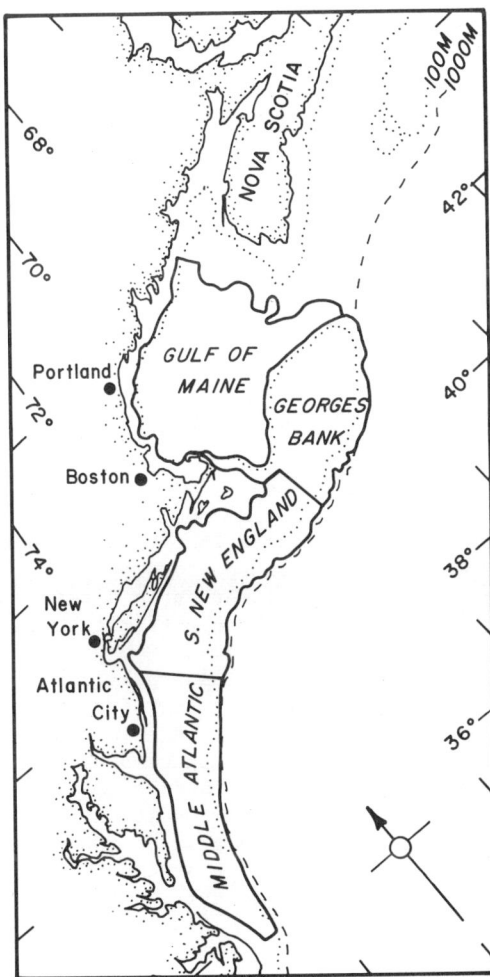

FIGURE 2: The four geo-
graphic areas of the Northwest Atlantic surveyed from
1971 through 1977 during
MARMAP operations of the
Northeast Fisheries Center,
Woods Hole, Massachusetts.

Observational Protocol

    The four areas on the continental shelf surveyed, Gulf of Maine,
Georges Bank, Southern New England, and Mid-Atlantic Bight, are
shown in Figure 2. At each station tows are made for zooplankton
and ichthyoplankton; chlorophyll, nutrients, salinity, and temperature are measured. During each survey of the continental shelf ^{14}C
measurements of primary production are made daily. Although the
1971 through 1977 time-series of ichthyoplankton-zooplankton collections is not yet fully sorted and identified, several findings are
particularly noteworthy and will be reported here.

ICHTHYOPLANKTON

Shifts in the abundance of fish species may be established in
their egg and larval stages. Reduction in abundance can be caused
by: 1) a decline in the size of the spawning biomass of traditionally
abundant resource species; 2) shifts in spawning-time causing a mis-
match between newly hatched larvae and optimal densities of their
planktonic prey; 3) changes in environmental and climatic events
leading to mass mortalities from shifts in circulation away from
nursery areas; and 4) increased embryonic and larval mortality
levels imposed by chronic releases of hazardous substances over
principal spawning areas.

Ichthyoplankton Pulses in Abundance

On the average, peak spawning of the principal species follows
a temporal and spatial pattern that may serve as a critical adaptive
mechanism for survival. Based on recent observations reported by
Colton et al. (1978) spawning peaks of larvae tend to be successive
rather than simultaneous. For example, among the 11 most common
gadids, only two species overlap with respect to both spawning area
and time, cod, *Gadus morhua*, and haddock, *Melanogrammus aeglefinus*,
on Georges Bank in March. It should be mentioned, however, that
the peak of haddock spawning is more prolonged beginning in February
and extending through March. Among the pelagic species, mackerel,
Scomber scombrus, reach peak spawning in the Mid Atlantic Bight in
May, and in June in the western Gulf of Maine. The butterfish,
Peprilus triacanthus, is at peak spawning in southwest Georges Bank
in June and July. The peak spawnings of the remaining pelagics also
differ. Menhaden, *Brevoortia tyrannus*, reach their peak from September through November in the Mid-Atlantic Bight, and Atlantic
herring, *Clupea harengus*, in October-November on Georges Bank and
western Nova Scotia, with less intensive spawning in the Gulf of
Maine. It is, however, the deviation from average conditions that
will have an important influence on the size of a year-class.
Measures of these deviations cannot be made without a time-series
of observations over the continental shelf. This data-base is just
now being developed.

Shifts in Abundance of *Ammodytes* Larvae

In the southern half of the survey area, Southern New England
and the Mid-Atlantic Bight and Georges Bank, significant changes in
the species composition of ichthyoplankton have been detected
during the 1974-1977 period (Smith et al., 1978). Since 1974, the
increase in abundance of sand lance has been sharp and consistent,
increasing from below 50% of the ichthyoplankton constituents in
1974 to a peak of over 90% of the larval fish assemblage in early
spring, 1977. The taxonomic status of the sand lance, *Ammodytes*
spp., has not been resolved. More than one species may be in the
collections. Abundance of cod, haddock, pollock, and herring larvae for the same period declined from a level of approximately 50%
of the ichthyoplankton to less than 10% in 1977 (Fig. 3). The in-

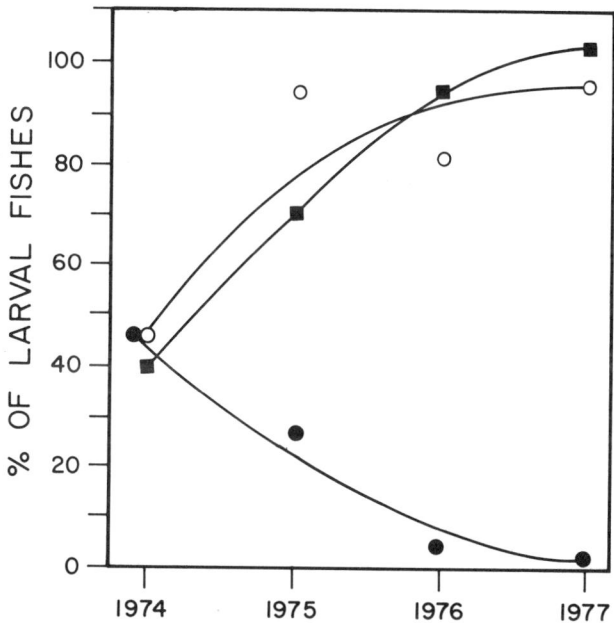

FIGURE 3: Percentage composition of *Ammodytes*
spp. larvae in the Georges Bank, Southern New
England, and Mid-Atlantic Bight sub-areas of
the coastal ecosystem 1974-1977. Changes in
the percentage composition are also given for
cod, haddock, pollock, and herring in the
Southern New England and Georges Bank sub-areas.
Southern New England, Georges Bank:
Ammodytes spp. ;   Cod, Haddock,
Pollock, Herring; Middle Atlantic Bight:
Ammodytes sp. From Smith et al. (1978).

crease has been widespread along the shelf with principal centers
of abundance spreading from off the Maryland and Delaware coasts
in 1974 to encompass most of the southern half of the survey area
in 1977. The increase in abundance, although of lower magnitude,
was also evident in the Gulf of Maine. On Georges Bank, the increase in numbers of sand lance larvae followed a similar upward
trend in 1974, 1975, and 1976, but showed a decrease in 1977.

The sand lance preys on zooplankton in larval, juvenile, and
adult stages (Bigelow and Schroeder, 1954). With respect to
feeding, it can be considered a competitor of two other important
pelagic species, herring and mackerel. Sand lance and herring co-
occur as larvae and adults. In the North Sea herring have been
reported to prey heavily on larval sand lance. Mackerel reaches
peak spawning in May-June, limiting any potential competition between the larvae of both species; they could be food competitors
in the juvenile and adult stages. The extent of mackerel preda-

tion on sand lance is not known. With the decline in the biomass
of herring and mackerel, feeding conditions for sand lance would
have improved greatly and the probability of high mortalities
through predation would be reduced. The sand lance is also abundant in the nursery areas of cod and haddock and may be a serious
competitor for zooplankton. The full impact of an apparent increase in abundance of a zooplankton predator in relation to other
pelagic and demersal species is not yet clear. Studies on inter-
specific competition among sand lance, herring, mackerel, cod, and
haddock are continuing.

ZOOPLANKTON AND FISHERIES

     Since the early work of Bigelow fifty years ago, little atten-
tion has been given to a systematic assessment of the changes in
the biomass and species composition of the zooplankton off the
northeast coast of the United States and how the changes relate to
fish distribution and production. Studies over the past five
decades have generally been limited to observations over a one or
two year cycle and over a limited geographic area. Reports of
these studies have been summarized by Colton (1963). As recently
as 1976, information on seasonal changes in zooplankton standing
stock was limited to a rather confused series of data compiled
from the available literature in which both spring and autumn were
depicted by different authors as annual peak periods of zooplankton
abundance (Cohen, 1976).

     Zooplankton Pulses in Biomass Abundance, 1977

     The present analysis includes a summary of the standing stock
of biomass and species composition of the most abundant zooplankters
in 1977 in the Gulf of Maine, Georges Bank, and Southern New England.
Biomass values are based on displacement volumes of the 0.333 mm mesh
samples. Species composition reflects the relative abundance of only
the larger zooplankton constituents (adults and copepodites III, IV,
V) retained by the relatively coarse meshes of the 0.333 mm bongo net.
     Statistical profiles of the biomass data were prepared for the
1977 data including calculations of the mean, median, range, standard
deviation, variance, and coefficient of variation. The variances of
the biomass values exceeded the means in each of the surveys. In recognition of the skewness in the data, median values were used by depict trends in biomass changes.
     During 1977, six surveys were completed. The standing stock of
zooplankton was estimated from surveys made in early, mid-, and late
spring, in summer, and in mid- and late autumn. Seasonal pulses in
zooplankton abundances in Southern New England and Georges Bank were
similar. Median values increased from early to mid-spring. The
greatest seasonal pulse was in late spring, followed by a decline
in summer, and a continuing decline through autumn. In the Gulf
of Maine median zooplankton volumes increased from an early spring
low to a late spring high. The zooplankton level is apparently relatively constant through summer, followed by a second pulse that

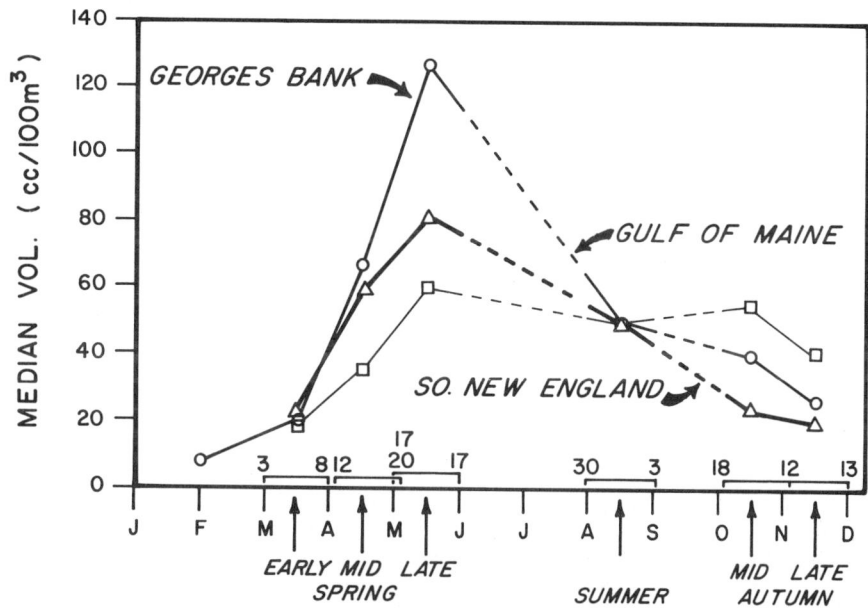

FIGURE 4: Median zooplankton volumes in the Gulf
of Maine, Georges Bank, and Southern New England,
March-December, 1977; time periods covered during
each of the six surveys are bracketed; mid-points
of the survey periods are shown by arrows; dashed
lines represent interpolated values for inter-
survey periods.

reached an annual high in mid-autumn. By late autumn, zooplankton
levels had declined (Fig. 4).
     Changes in the magnitude of the seasonal pulses in abundance
of zooplankton were compared among seasons for each of the areas
by means of the Kruskal-Wallis Analysis of Variance; between-season differences in abundance were also compared. Zooplankton standing stock median values were significantly different for the volumes
in each of the three areas (P<0.0001). Between-season tests for significance were made for each of the areas. Of the 15 pairs of comparisons made, four were not significantly different (P>0.05): mid-
autumn and late autumn in Southern New England; mid-autumn and late
autumn on Georges Bank; and between summer-mid-autumn and mid-
autumn-late autumn in the Gulf of Maine.
     The among-area differences in zooplankton standing-stock were
tested for significance. Of the six comparisons made, the early
spring and summer zooplankton levels were not significantly different.
     Seasonal differences in biomass are most significant in late
spring. The peak spring pulse in abundance on Georges Bank is 1.6 x
greater than in the Gulf of Maine and 1.4 x greater than in Southern
New England. By summer the biomass along the entire area off New

England is at a similar level of moderate abundance (50 cc/100 m^3).
By mid-autumn, areal similarity changes and the lowest biomass is
in Southern New England and the highest in the Gulf of Maine with
values in Georges Bank at an intermediate level.

Zooplankton Species Composition

Among the 51 taxa in the samples, copepods were the most abundant, constituting 75% or more of the zooplankton. Our sampling may
not have been sufficiently frequent to detect any major increase in
cohorts of copepod populations produced during the inter-survey
periods. The zooplankton standing-stock values for summer and
autumn in the Georges Bank and Southern New England areas, therefore, should be considered minimal. Other zooplankters including
chaetognaths, euphausids, coelenterates, larval cirripeds, cladocerans, salps, pelecypods, appendicularians, and ostracods occurred
over a wide area but in low densities ($\leq$20% of the zooplankton/sur-
vey/area). Larval cirripeds were swarming on Georges Bank in early
spring and in Southern New England waters in mid-spring. Chaetognaths were most numerous on Georges Bank in summer and late autumn;
pulses of chaetognath abundance occurred in late spring and summer
in Southern New England. Cladoceran pulses were particularly high
in summer and autumn on Georges Bank and pelecypods swarmed in late
fall on Georges Bank.

Copepod Species Composition

The total zooplankton in each of the three areas surveyed was
dominated by three species: *Calanus finmarchicus, Pseudocalanus
minutus,* and *Centropages typicus.* Indices of species abundance
were based on changes in population densities (e.g., median numbers/
100 m^3/survey), percentage composition of all copepod species, and
a modified Fager and McGowan index of dominance:

$$D = \frac{S(100)}{N}$$

where: D = % dominance
        S = number of stations where the species represent $\geq$50% of
            the total copepod fauna
        N = number of stations

In addition to the three dominant species, the copepod *Metridia
lucens* persisted in relatively low numbers in all samples, particularly in the deeper waters of the Gulf of Maine. Other species were
dominant at only one or two locations. In this regard, the status
of *Oithona similis* is not clear. This small copepod was undersampled
in the 0.333 mm mesh net, along with the naupliar and early copepodite
stages of other species. Based on the frequency of *Oithona* in larval
fish stomachs, it is likely to be among the dominant species in fine-
mesh samples (0.153 mm and 0.253 mm) presently archived at Narragansett.
Seasonal abundances for the *C. finmarchicus, P. minutus,* and *C.
typicus* are plotted in Figures 5a-5c. Cumulative population increases
are considered "pulses". They represent continuous production for

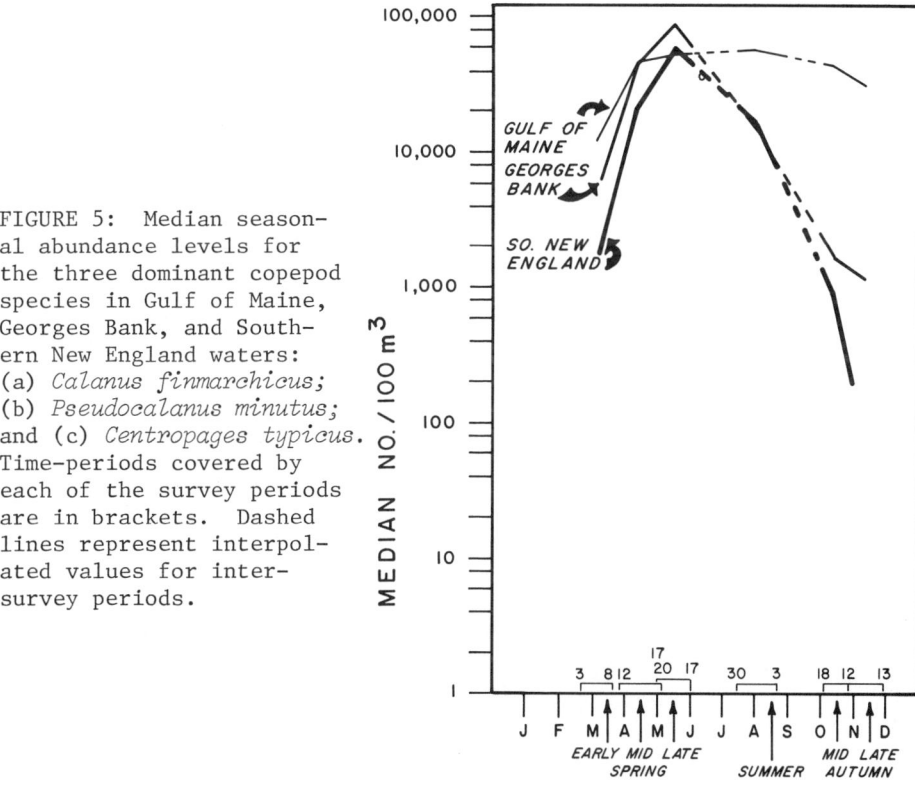

(a) **_Calanus finmarchicus_**

FIGURE 5: Median season-
al abundance levels for
the three dominant copepod
species in Gulf of Maine,
Georges Bank, and South-
ern New England waters:
(a) *Calanus finmarchicus;*
(b) *Pseudocalanus minutus;*
and (c) *Centropages typicus.*
Time-periods covered by
each of the survey periods
are in brackets. Dashed
lines represent interpol-
ated values for inter-
survey periods.

several cohorts and do not represent an index of turnover rate.

Calanus finmarchicus Pulses

In the Gulf of Maine, *C. finmarchicus* is at a high level of
abundance throughout most of the year. Median values ranged between 30,000 and 60,000/100 m3/season from mid-spring through autumn. On Georges Bank, *C. finmarchicus* pulses to an annual maximum
in late spring (*ca.*100,000/m3); it declines sharply in abundance
from summer through late autumn (1,000/100 m³). In Southern New
England waters, *C. finmarchicus* undergoes an initial pulse in late
spring (*ca.*150,000/100 m³) and an abrupt decline continuing to an
annual low in late autumn (<200/100 m³). Among-area values in *C.
finmarchicus* abundance for each of the seasons were significantly
different (P<0.0001); comparisons of abundance levels between the
sampling periods were different in Southern New England and Georges
Bank except for the mid-autumn and late-autumn period. In the Gulf
of Maine seasonal differences were significant between early and
mid-spring and late spring. Standing stocks among the three areas
were significantly different in early spring among all three areas.
The other differences reflected highest abundance levels of *C.
finmarchicus* in the Gulf of Maine in summer and autumn (P<0.0001).

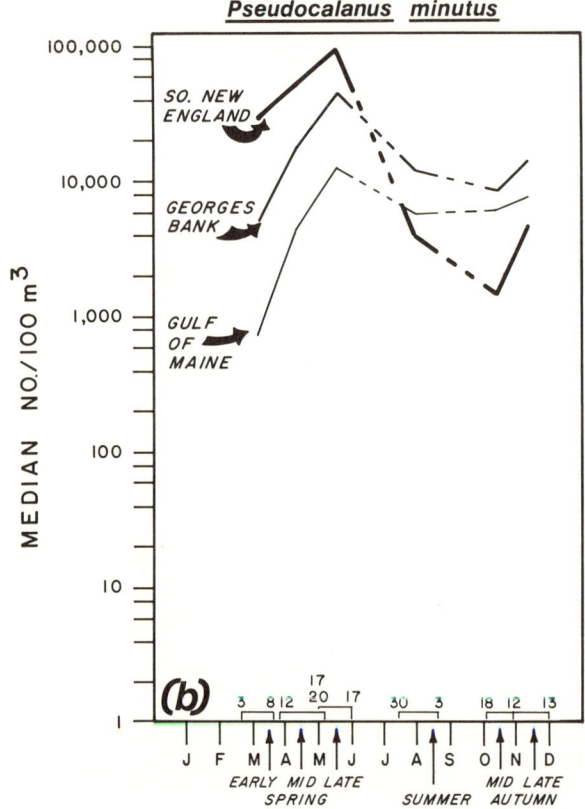

Pseudocalanus minutus Pulses

This small calanoid undergoes an annual pattern of abundance
similar to *C. finmarchicus* in all areas (Fig. 5b). On Georges Bank
and in the Gulf of Maine numbers of *P. minutus* increase from an
early spring annual low to a late spring high; abundance is reduced
from late spring through summer. A secondary weaker pulse in abundance is reached between mid- and late autumn.. Abundance levels for
each of the areas were different among the seasons (P<0.001); in
early and mid-spring standing-stocks of *P. minutus* were different in
each of the areas. In Southern New England the summer to early autumn decline was not significant. However, the increase in abundance
shown between mid- and late autumn was significant (P<0.05).   On
Georges Bank the pulses in abundance increasing from an early spring
low to a late spring high were significant; however, the level of
abundance from summer through late autumn was not significantly different.   In the Gulf of Maine early seasonal differences were different with increasing densities of *P. minutus* through late spring;
autumn levels were similar.   The among-area differences shown in Figure 5 were significant in each of the survey periods, except for
late autumn.

Centropages typicus

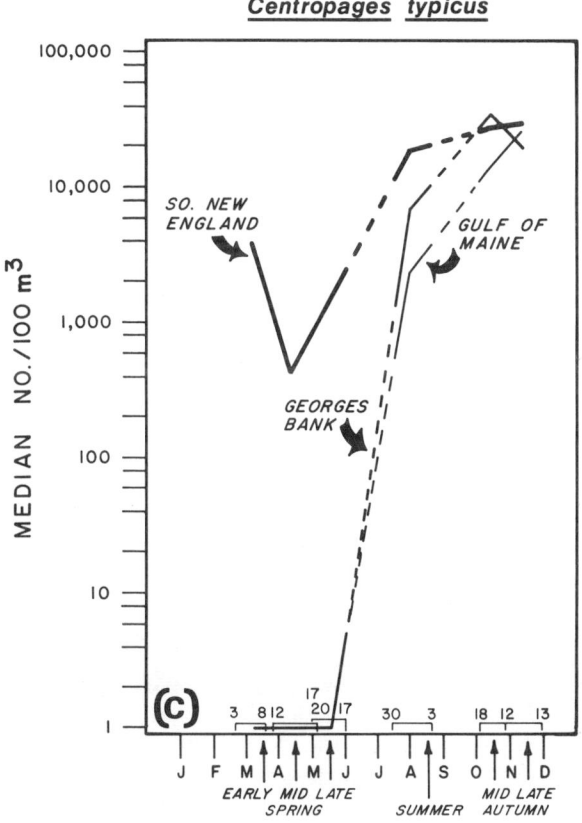

Centropages typicus Pulses

The center of *C. typicus* abundance is in Southern New England. In each of the areas sampled *C. typicus* increased steadily in abundance from an early spring low to a mid- to late autumn pulse of approximately 25,000/100 m^3 (Fig. 5c).

Differences in *C. typicus* abundance among seasons were significant for each of the areas (P<0.0001). In Southern New England the early to mid-spring decline was significant, as were the increases from late spring through summer. On Georges Bank, abundance levels increased significantly from late spring through late autumn. The general increase shown for the Gulf of Maine was based on too few occurrences to test for significance. The among-area differences for each of the seasons were significant with the exception of late autumn values.

Zooplankton Pulses in Abundance and Larval Fish Survival

The probability for successful larval growth is enhanced from spring through autumn by the combined and nearly continuous production of eggs, nauplii, and copepodites of the three dominant zooplankton species--*C. finmarchicus*, *P. minutus*, and *C. typicus*. The standing stock of copepods is reduced during summer.

Survival and growth of larvae produced in successive spring and autumn spawnings is likely dependent on the match/mismatch in time and space of the production of *C. finmarchicus* and *P. minutus* in spring, and *P. minutus* and *C. typicus* in autumn. Cod, haddock, mackerel, and flatfish larvae spawned in spring optimize survival potential by spawning on the ascending limb of the spring copepod production curve. Herring spawnings are supported by *P. minutus* production in autumn (Sherman and Honey, 1971). In summer it would appear that growth and survival of hake and other first-feeding larvae may be dependent on the dynamics of *C. typicus* production.

In spring, successive pulses in abundance from early spring through late spring of *P. minutus* and *C. finmarchicus* are indicative of accelerated production of copepod eggs, nauplii, and copepodites; the presence of early developmental stages (copepodites) in our samples is noted, although no actual counts are made as they are undersampled in the 0.333 mm mesh used in the bongo sampler. Early developmental stages of *C. typicus* occur in our summer and mid-autumn samples, and *P. minutus* copepodites occur in the mid-autumn samples.

In the Gulf of Maine, the initial spring pulse is provided by increases in the abundance of *C. finmarchicus* and *P. minutus*, followed by a summer through mid-autumn pulse of *C. typicus* abundance and a third pulse from mid- to late autumn of *C. typicus*. On Georges Bank *C. finmarchicus* and *P. minutus* generate the initial pulse in copepod abundance. This is followed by a second pulse extending from late summer through mid-autumn of *C. typicus*. In Southern New England waters *C. finmarchicus* and *P. minutus* generate an early to late spring pulse. The annual *C. typicus* pulse is advanced over Georges Bank and the Gulf of Maine areas, beginning in late spring and continuing to a late autumn maxima for the second major seasonal pulse in copepod abundance in Southern New England. *P. minutus* undergoes a slight increase in abundance in late autumn that represents a third, but minor, copepod pulse.

1971-1975 Zooplankton Time-Series

Zooplankton Biomass. An initial examination of the zooplankton sampled during the fall and spring bottom trawl surveys has been completed. Samples were analyzed from the 0.333 mm mesh standard MARMAP bongo sampler collected at 39 or more locations on Georges Bank during each of the seasons. Biomass values were determined by displacement volumes and oven-dry weights. Good agreement was found between the two sets of values in each of the years ($r \geq 0.717$), except for spring 1971 ($r = 0.548$).

Statistical profiles of the biomass data were prepared, including calculations of the mean, range, median, standard deviation, variance, standard error, and coefficient of variation. The variance, standard error, and coefficient of variation. The variances of the biomass values exceeded the mean in each of the seasons. In recognition of the skewness in the data, plots were made using median values to depict trends in biomass changes. In autumn, the biomass increased from moderate levels in 1971-1972 (26 and 29 cc/100 m^3) to a high in 1973 (42 cc/100 m^3) and declined in 1974 to 25 cc/100 m^3 and a low of 19 cc/100 m^3 in 1975. The 1973 values were the highest among the five years of the time-series. The Mann-Whitney U test was used to examine between-year differences in biomass for both seasons. Volumes in 1973 were significantly higher than those in each of the other years and the 1975 volumes were significantly lower (P≤0.05).

In spring the volumes were lowest in 1971-1972, and were not significantly different (P>0.05) between years. Volumes in 1973 and 1974 were similar (P>0.05). Both were significantly greater than those in 1971-1972 (P<0.05). In 1975 the biomass was significantly higher than that in other years (P<0.05) (Sherman et al., 1978).

Species Composition   The zooplankton sampled on the 1971-1975 bottom trawl surveys were dominated by a few species. Copepods were the predominant taxa, ranging from 88% of the zooplankton in 1972 to 94% in 1971. In spring, *Calanus finmarchicus* and *Pseudocalanus minutus* were dominant in each of the years. Their abundance varied among the years; they were most numerous in the springs of 1973, 1974, and 1975. Population densities of *C. finmarchicus* were lowest during the 1971 and 1972 surveys, when percentages of adults to copepodites were highest, ≥41%. In the succeeding years, the surveys were later in the "biological" sense for *C. finmarchicus*. Large-scale cohort production was well underway with the percentages of adults to copepodites reduced to 3%, 8%, and 2% respectively, for 1973, 1974, and 1975. The percentage difference of adults to copepodites did not exceed 64% for the smaller *P. minutus*. It is likely that the low densities of copepods in 1971 and 1972 were the result of a delay in the onset of spring swarming characteristic of *C. finmarchicus* on Georges Bank (Bigelow, 1926), rather than a reflection of a change in levels of secondary production.

In autumn copepods were the principal zooplankters. The species *Centropages typicus* and *Pseudocalanus minutus* were dominant in each of the years (Sherman et al., 1978). Median densities of *C. typicus* populations were highest in 1972 and 1973. Highest volumes for the five-year period occurred in 1973. The abundance of the mysid *Neomysis americana* (median 7,000/100 m^3) in addition to the high numbers of *C. typicus* combined to increase the biomass of zooplankton in 1973 over values observed during the other years of the survey. The hydrographic and biological factors controlling the size of the zooplankton populations are not fully understood. Research on these factors is continuing. A significant factor that could have contributed to the low numbers of *C. typicus* in autumn, 1975, was the abundance of the chaetognath, *Sagitta elegans*, and the coelenterate, *Nanomia cara* on Georges Bank, two important copepod predators. The presence of unusually dense concentrations of *N. cara* on Georges Bank in the fall and winter of 1975 caused loss of fishing time and income to coastal fishermen operating in the Gulf of Maine and Georges Bank; evidence of *N. cara* feeding on copepods during this period has been reported recently (Rogers et al., 1978).

Annual Zooplankton Pulses

Observations of monthly changes in zooplankton abundances in
the offshore waters of the United States coast are limited. Based
on earlier studies, Cohen (1976) attempted to depict an annual cycle
for the Gulf of Maine-Georges Bank area. The trend of displacement
volumes indicated a spring peak in May-June. In addition, Cohen
shows high annual values in September. Recent MARMAP findings are
consistent with the spring peak; no evidence of an autumn peak was
found in the 1977 volume data. However, increases were observed
in our 1977 data for the copepods, *C. typicus,* on Georges Bank and
in the Gulf of Maine, and *P. minutus* in Southern New England in autumn.

In coastal waters of the Gulf of Maine, the monthly trends in
zooplankton displacement volumes show an April minimum, followed by
an increase in late spring to a July peak, a decline in August and
a secondary peak in October followed by a decline in December. The
microzooplankton, e.g., copepod eggs, crustacean nauplii, and copepodites, were most abundant in summer and autumn; in winter and early
spring, the zooplankters were predominately adults (Sherman et al.,
1976).

The fluctuations in abundance of zooplankton along the inner
margin of the Gulf of Maine coast are likely the result of local
environmental conditions rather than any large-scale advective
processes. In early spring, onset of stratification triggers the
spring phytoplankton bloom followed by rapid production of copepod
cohorts. Meroplankton can swarm and assume dominance over the background abundance of copepods for limited periods. Over the entire
annual cycle, copepods are the dominant zooplankters in coastal waters. Of the 19 species common to the Central Gulf of Maine coastal
area, nine are numerous, but only one, *Pseudocalanus minutus,* is
abundant in all seasons, and is an important food of larval herring.
Survival of larval herring may be related to the synchronous development of *P. minutus* cohorts and the first feeding of larval herring.
Pseudocalanus minutus is the most abundant species of the proper size
to be ingested by small herring larvae in Gulf of Maine coastal waters.
This is the only small copepod species that is undergoing high rates
of cohort production in synchrony with the abundance of first feeding
larvae (Fig. 6). *Temora longicornis* also increases in abundance in
autumn, but is too large to be eaten by young larvae. The other
abundant species, *Acartia clausi* and *C. finmarchicus,* are declining
in abundance in autumn and are predominately in the late copepodite
and adult stages (Sherman and Honey, 1971; Sherman et al., 1976).

In the coastal band from Cape Ann to Machias Bay, 36 species of
copepods are found, but only seven are numerous: *Acartia clausi, A.
longiremis, Centropages typicus, Calanus finmarchicus,* harpacticoid
spp., *Pseudocalanus minutus,* and *Temora longicornis.* The most important environmental factor shaping the distribution of the species is
depth rather than any particular range of salinity or temperature.
Four species are abundant in shoal areas of the coast: *A. longiremis,
A. clausi,* harpacticoid spp., and *T. longicornis.* The other three
species, *P. minutus, C. finmarchicus,* and *C. typicus,* decline in
abundance from offshore to inshore. Densities of copepods are

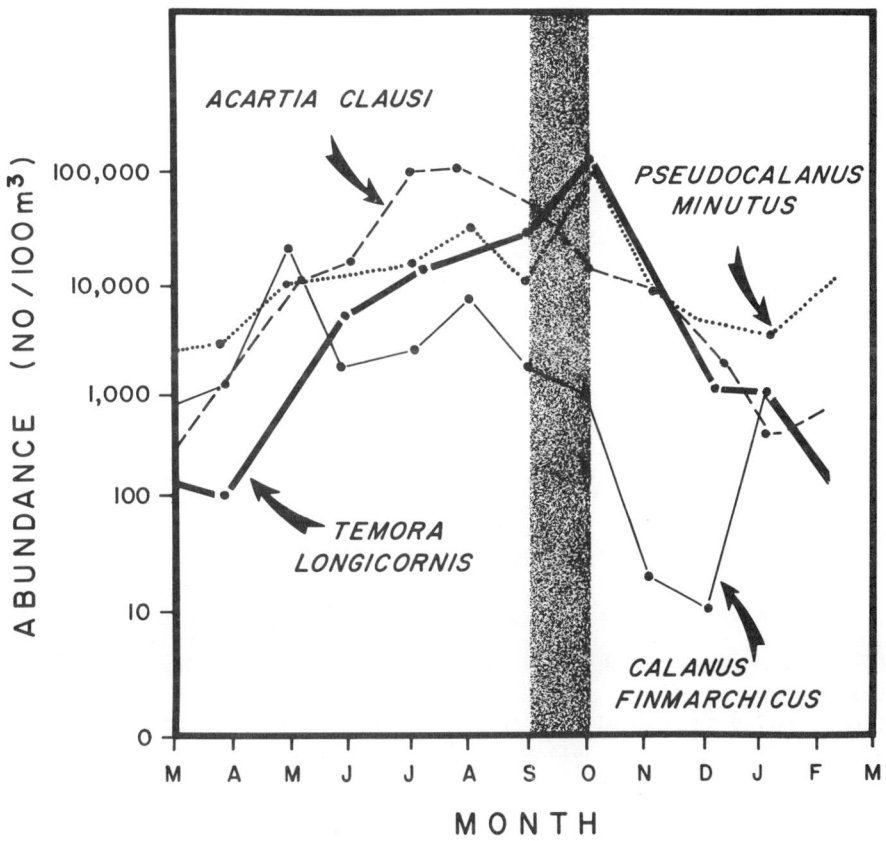

FIGURE 6: Monthly changes in the abundance of
the dominant copepod species in coastal waters
of central Maine, October 1968 through February
1970. The shaded area depicts the peak spawning period of herring along the coast of the
Gulf of Maine. From Sherman et al. (1976).

greatest in the western area where conditions for feeding and growth
are better in the stratified waters than in the turbulent vertically
mixed waters in the eastern area (Sherman, 1970)

Zooplankton and Fish Distributions

In addition to playing a critical role in the survival, growth,
recruitment, and mortality of larval fish, zooplankton can be a
dominant influence on migrations of adult populations. Pavshtics
(1963) and Zinkevitch (1967) have reported that movements of herring
between the Mid-Atlantic Bight and Georges Bank follow progressive
seasonal swarming of zooplankters as the vernal bloom moves from
west to east along the coast in the frontal zones between coastal,

slope, and shoal waters around the periphery of Georges Bank. This
apparent relationship may hold promise for forecasting herring and
mackerel movements along the coast.

Information on the movements of pelagic fish in coastal waters
is scanty. Adult herring are fished on Jeffreys Ledge and other
shoals within the Gulf in autumn. Herring feeding in these areas
appear to favor larger zooplankton particles including the copepods
Calanus finmarchicus and *Centropages typicus*. It is unlikely that
adults feeding on large calanoids, which are most numerous beyond
the headlands, would move into the coastal embayments.

Larval Physiology and Feeding Studies

Physiological and predator-prey studies of fish species off the
northeast coast are now underway by NEFC on the quality and densities
of food required for optimal survival and growth of larval fish under
controlled laboratory and field conditions (Laurence, 1974; 1977a).
Experiments have been conducted with larval cod, haddock, scup, tautog,
and winter flounder. Recently, growth and mortality models have been
developed for winter flounder, *Pseudopleuronectes americanus*, that
consider temperature, prey density, and larval size (Laurence, 1977b;
Beyer and Laurence, 1978).

Recent hypotheses developed from models describing the quantity
and quality of food required by fish larvae for survival and growth
have been tested in the sea. Off the California coast, Lasker (1975)
has demonstrated that dinoflagellates of the proper size and nutritional value for rapid growth of larval anchovy were found to co-
occur with anchovy in horizontal and vertical patches. However, if
upwelling is early, it appears that the synchronous relationship can
be unbalanced. Patches of dinoflagellates can be dispersed in the
early upwelling and replaced by copepodites which may be too large
for the anchovy to ingest, thereby reducing the survival potential
of first-feeding larvae. Laboratory studies of winter flounder
growth based on empirical models indicate that densities of food
required for rapid growth usually exceed values reported in the
literature (Laurence, 1975). Patches of larval fish food of the
proper density (approximately 200 copepodites/liter) have recently
been observed in estuarine and embayment areas in Southern New
England (Laurence et al., 1979). Both observations confirm the importance
of microdistributions in the sea. If we are to understand the relationships between food availability and larval fish survival, our
approach to sampling needs to include systems for measuring microdistributions of larvae, their predators, and food, by utilizing
pumps, particle counters, and other "real-time" methods.

Secondary Production Estimates

The relationship of zooplankton numbers to biomass and productivity rates is under investigation. Available information, unfortunately, is fragmentary. Secondary production based on older stages
of *C. finmarchicus* in spring was recently estimated at 79.46 mg
carbon/m^2/day by Green et al. (1977). The instantaneous growth
method was used in the calculation from the formula:

$$P = \frac{W_B - W_A}{t} \ \frac{(N_1 + N_2)}{2}$$

where: W_A = weight at the beginning of a stage
W_B = weight at the end of a stage
t = stage duration in days
N_1 = count of organisms in a stage
N_2 = count of organisms in succeeding stage

The production rates for each of five surveys made during the spring generations of *C. finmarchicus* populations were used to estimate total production for the 100 day period of the surveys. As was pointed out by Green et al. (1977) the values were of the same order of magnitude, 46 mg C/m^2/day, estimated for *C. finmarchicus* in the North Sea by Mullin (1969) and 77 mg C/m^2/day reported for *Acartia tonsa* by Heinle (1966). Studies are now underway in the Northeast Fisheries Center to refine the estimates of secondary production by enlarging the number of species examined and extending the analyses to all of the growing seasons.

Biochemical Studies

Studies of larval fish viability and growth from samples collected at sea have been traditionally based on morphological differences. The objective of these studies has been to identify areas where the physiological condition of larvae indicated optimal food and environmental conditions for feeding, growth, and survival. The procedure is time-consuming and subject to bias from preservative distortion. A rapid biochemical method has recently been developed that provides an index of physiological condition based on RNA/DNA ratios (Buckley, 1977).

EMERGING ECOSYSTEM THEORY IN RELATION TO PLANKTON STUDIES IN THE NORTHWEST ATLANTIC

Trophodynamic studies of larval survival will need to consider the early modelling of the production cycle developed by Riley (1941, 1946); Riley and Bumpus (1946); Riley (1947); and Riley et al. (1949). They have not been used effectively in the Gulf to forecast conditions, principally for lack of systematic input parameters that would allow for prediction in annual, seasonal, and/or more frequent time scales. Since the early studies by the Woods Hole and Bingham groups of primary production, we have experienced a hiatus in making systematic observations. Some beginnings are underway. Matrices of information are being developed now for key elements in the ecosystem. Our goal is to develop a food budget for Georges Bank, and to continue satisfying our most critical information voids by expanding our effort from Georges Bank to the inner Gulf of Maine and Mid-Atlantic area. Progress is being made on defining the benthic communities on the shelf (Williams and Wigley, 1977). Changes in fish distribution and abundance have been monitored with bottom trawl surveys conducted over

the shelf for the past 15 years (Grosslein, 1969; Grosslein and
Bowman, 1973). The MARMAP oceanographic effort includes the monitoring of current systems, temperature, and salinity changes and the
effects of the movement of water masses on the fish stocks in the
area. Surface temperature and salinity charts have been prepared
for the bottom trawl surveys. In addition, new information on water
movements is now becoming available from moored current meters monitoring flow in the northeast channel of Georges Bank. High velocity
in flow from the Scotian Shelf into the Gulf of Maine and Georges
Bank areas has been observed directly for the first time.

Primary Productivity Studies and Fish

Significant activity is now underway in the New York Bight. A
group headed by John Walsh of Brookhaven National Laboratory is now
investigating the energy flux of the Bight and principal driving
forces of the system. Nutrient enrichment through storm mixing
appears to be the principal mechanism for recycling nutrients within
the ecosystem. The pulses of phytoplankton growth associated with
wind events are thought to be the cause of the low fish yield of the
Bight (Walsh et al., 1978).

In contrast to the New York Bight area and North Sea, fish
yields of the Georges Bank area are significantly greater. These
differences appear related to the different levels of primary production among the three areas. The estimated levels of primary
productivity range from 100 gC/m^2/yr for the North Sea to 150-200 gC/
m^2/yr for the New York Bight, and 400-500 gC/m^2/yr for Georges Bank.
The latter value has recently been reported by Cohen et al. (1978),
based on ^{14}C measurements made during MARMAP surveys in 1975. The
estimated average production of fish biomass for the three areas increases from a low of 10 t/km^2/yr for the North Sea to 15 t/km^2/yr
in the New York Bight, and a high of 19 t/km^2/yr for Georges Bank.
The extremely high value for Georges Bank primary production is now
being reexamined critically. A series of ^{14}C measurements were made
in the area in 1976 and preliminary evidence supports the high value.

Observations of Stressed Ecosystems in the Northwest Atlantic

Recently Steele and Frost (1977) described a model of a stressed
ecosystem, wherein overall levels of primary production are not affected by stress; however, significant shifts in species composition
were in evidence. In a stressed system the production of small,
fast growing species is favored over larger species. The theoretical
basis for the model is supported by field data obtained with a Continuous Plankton Recorder describing long-term changes in the plankton of
the North Sea. In addition, recent experimentation in large plastic
enclosures have shown a shift from essentially large-celled diatom
populations to smaller dinoflagellates following dosings with copper
and petroleum hydrocarbons (Steele and Frost, 1977; Thomas and Seibert,
1977).

In addition to plankton changes in the North Sea described by the
Institute of Marine Environmental Research (IMER) group in Plymouth,
Ursin (1977) reported a decline in mackerel and herring stocks coinci-

dent with an increase in the biomass of fast growing and short-lived species, sprat, Norway pout, and sandeel. Although much of the data is preliminary in nature, the need for comprehensive fisheries ecosystem studies is clear. In a summary statement describing the results of the symposium on the "Changes in the North Sea Fish Stocks and Their Causes", Hempel (1978) reported that

> ...direct and indirect effects of changes in the fisheries as well as climatic changes in their consequences for the biotic environment caused the recent changes in the fish stocks of the North Sea. It is not possible to quantify the effects of man-made and natural factors separately because of the complexity of interactions between the various fish stocks and the stages of their early life history. A new approach to the problem has been developed which emphasizes ecosystem modelling, detailed studies of the early life history and the feeding relationships of the various fish stocks.
> The studies of the changes in the North Sea fish stocks face a problem which is common to many ecosystem studies. They require monitoring over large areas and decades in order to describe correlations. On the other hand the events are small-scale in space and time which determine the year-to-year differences in recruitment and hence in the build up and decline of stocks.

Implications of Biomass Changes in the Northwest Atlantic

In the Northwest Atlantic off the United States coast, the changes in size of the fish biomass over the past eight years have been dramatic. Recent estimates of primary production levels indicate that Georges Bank is far more productive than previously estimated. Considerable effort is now being directed to verify the higher values. Given that the values are reasonable, we will look for evidence of changes at trophic levels below primary production. Significant shifts have occurred among the fish species. Mackerel and herring stocks have declined, and coincidentally, the abundance of sand lance, a short-lived, fast growing species with high-ecological efficiency, has increased dramatically, particularly in the Southern New England and Mid-Atlantic Bight areas. Some between-year and among-area differences in zooplankton abundance have been detected. However, the causes for the changes in zooplankton densities are not clear. It will be important in the future to partition zooplankton mortalities into environmental and predator compartments, a difficult but important task.

The implications of these kinds of species shifts are clear. Resource managers, now, more than ever before, need to evaluate the consequences of multispecies interactions and to sort out the impacts resulting from the removals of presently unfished "ecological species" (e.g., four-bearded rockling, and/or sandeels) in favor of "commercial" species (e.g., cod, haddock, herring, flounders, and mackerels). Proper evaluation is dependent on the best scientific advice available, and to this end it is necessary to study marine ecosystems with a fishery management perspective.

Available evidence from the northeast Atlantic indicates that the large-scale fluctuations in fish biomass were measurable with the present methods of combining, and analyzing data from fish catches and independently conducted research surveys to monitor changes in the population levels of species, their predators and prey, and their environments. In the northeast Atlantic, largely through the framework of ICES (International Council for the Exploration of the Sea) catch data is systematically reported. Joint international surveys for demersal and pelagic adult, juvenile, and larval fishes are conducted, usually for target species. Measurements of pollutants, primary production, hydrography and zooplankton biomass, species composition and productivity are generally studies of limited areas and/or relatively short-duration. The exceptions are the Continuous Plankton Recorder surveys of the North Atlantic underway for nearly three decades.

The approach in the northwest Atlantic is largely an evolution of ICNAF (International Commission for Northwest Atlantic Fisheries) joint international studies to support the total fish biomass management regime adopted in 1973. Standardized MARMAP surveys are underway for monitoring population changes of fish, plankton, shellfish and benthos, and hydrography by the Northeast Fisheries Center.

Principal focus in the MARMAP ecosystem study is on the early life stages of fish (Fig. 7). Estimates are being made of the magnitude of the spawning biomass of a stock based on abundance and mortality estimates of eggs and early stage larvae. To support the activity, a major commitment has been made to conduct a "patch" study in 1979 in cooperation with other countries. The operations will be conducted on Georges Bank on a microscale level using new sampling strategies, including pumps, to observe the relationships among fish larvae and their prey and predators in relation to growth and survival. Macroscale ichthyoplankton-zooplankton surveys of up to six times a year are continuing to monitor temporal and spatial changes in: 1) fish spawning and estimates of larval fish production; 2) changes in zooplankton abundance and species composition; and 3) changes in hydrography and the effects on fish. The common denominator to the studies is energy flow among the principal species. An initial energy budget has been developed by Cohen et al. (1978) for Georges Bank. Serious deficiencies exist in several of the key components. One of the most significant lies in the fragmentary nature of the secondary production component, both in the zooplankton and in the benthos. The initial stimulus for developing the budget was in fact to identify the weakest components and develop the necessary research initiatives to overcome them.

The principal focus over the next several years in the MARMAP program will be on predator-prey interactions at all trophic levels in both mesoscale and microscale studies. To date approximately 70,000 fish stomachs have been examined to frame the most critical questions. Preliminary evidence suggests that silver hake, *Merluccius bilinearis,* play a principal predation role in the ecosystem (Edwards and Bowman, 1978) off the northeast coast. The estimates of turnover rate of benthos and plankton will need to be refined.

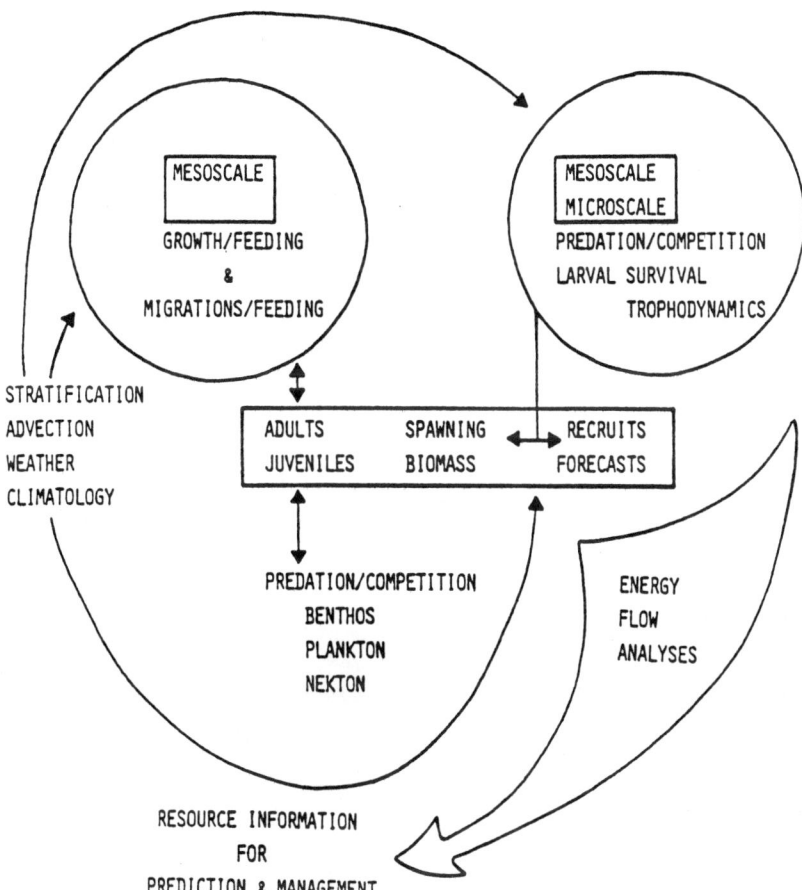

FIGURE 7: Principal focus of the MARMAP ecosystem study of the Northeast Fisheries Center.
The rectangle depicts the interactions under
investigation to obtain a better understanding
of the relationship between the size of a spawning biomass of fish and subsequent year-class
recruitment. Studies are underway on the larval, juvenile, and adult fishes within the context of measuring energy flow through the system, and the effects of fishing, pollution and
environmental changes on the flow. Macroscale
surveys are made up to six times a year to monitor changes of fish, plankton, and hydrography.
Mesoscale surveys are conducted from the onset
of larval hatching up to juvenile development on
target species. Herring sp. has been the target
species since 1971 in studies of recruitment processes off the northeast coast. Microscale
studies of larval herring growth and predator-
prey studies are underway for 1979.

The degree to which changes in primary production levels serve to notify the species abundance and composition of predators in the ecosystem now needs to be examined carefully. Studies of primary production will be accelerated in an effort to confirm or modify the extremely high values attributed to Georges Bank, and the impact of these values on the fish stocks. The MARMAP studies are serving to improve primary production estimates, and provide a better understanding of the links between plankton and fish production.

REFERENCES

Anderson, K. P. and E. Ursin, 1977. A multispecies extension to the Beverton and Holt theory of fishing, with accounts of phosphorous, circulation, and primary production. *Medd fra Danmarks Fiskeri-og Havundersogelser N. S.* 7: 319-435.

Beyer, J. E. and G. C. Laurence, 1978. A stochastic model of larval fish growth. *International Council for the Exploration of the Sea, 66th Annual Meeting, Copenhagen, Denmark, October 2-11, 1978.* 1978/L: 28. 28 pp.

Bigelow, H. B., 1926. Plankton of the offshore waters of the Gulf of Maine. *Bull. U. S. Bureau Fish.* 40: 1-509.

Bigelow, H. B. and W. C. Schroeder, 1954. Fishes of the Gulf of Maine. *U. S. Fish. Bull.,*(No. 74) 53. 577 pp.

Buckley, L., 1977. Biochemical changes during ontogenesis of winter flounder (*Pseudopleuronectes americanus*) and the effect of starvation. *International Council for the Exploration of the Sea, 65th Annual Meeting, Reykjavik, Iceland, September 25-October 2, 1977. 1977/L: 29.* 9 pp.

Clark, S. H. and B. E. Brown, 1977. Changes of biomass of finfishes and squids from the Gulf of Maine to Cape Hatteras, 1963-1974, as determined from research vessel survey data. *U. S. Fish. Bull.* 75: 1-21.

Cohen, E. B., 1976. An overview of the plankton communities of the Gulf of Maine. *Int. Comm. Northw. Atlant. Selected Papers, Number 1,Dartmouth, Canada.* 89-105.

Cohen, E., M. Grosslein, M. Sissenwine, and F. Steimle, 1978. Production studies at the Northeast Fisheries Center. *Northeast Fisheries Center, Woods Hole, Massachusetts, Laboratory Reference Report No. 78-14.* 35 pp.

Colton, J. B., Jr., 1963. History of oceanography in the offshore waters of the Gulf of Maine. *U. S. Fish. Wildl. Serv. Spec. Sci. Rep. Fish. No. 496.* 18 pp.

Colton, J. B., Jr., W. Smith, A. Kendall, P. Berrien, and M. Fahay, 1978. Principal spawning areas and times of marine fishes, Cape Sable to Cape Hatteras. *U. S. Fish. Bull.* 76: 911-915.

Edwards, R. L., 1976. Middle Atlantic fisheries:recent changes in population outlook. *Amer. Soc. Limnol. Oceanogr. Special Symposium 2.* pp. 302-311.

Edwards, R. L. and R. E. Bowman, in press. An estimate of the food consumed by continental shelf fishes in the region between New Jersey and Nova Scotia. *Proceedings of an International Symposium on Predator-Prey Systems in Fish Communities and Their Role*

in Fisheries Management, Atlanta, Georgia, July 23-27, 1978.
Sport Fishery Institute.

Green, J., J. B. Colton, Jr., and D. T. Bearse, 1977. Preliminary
estimates of secondary production on Georges Bank. *International
Council for the Exploration of the Sea, 65th Annual Meeting,
Reykjavik, Iceland, September 25-October 2, 1977.* 1977/L:30.
8 pp.

Grosslein, M. D., 1969. Groundfish survey program of Bureau of
Commercial Fisheries Woods Hole. *Comm. Fish. Rev.* 31: 22-35.

Grosslein, M. D. and E. Bowman, 1973. Mixture of species in Sub-
areas 5 and 6. *Int. Comm. Northw. Atlant. Fish Redbook 1973
Part III:* 163-208.

Heinle, D. R., 1966. Production of a calanoid copepod *Acartia tonsa*
in the Patuxent River estuary. *Chesapeake Sci.* 7: 59-74.

Hempel, G., ed., 1978. North Sea fisheries and fish stocks -- a
review of recent changes. *Rapp. et P.-v Réun. Cons. int. Explor.
Mer.* 173: 145-167.

Laevastu, T., F. Favorite, and W. B. McAlister, 1976. A dynamic
numerical marine ecosystem model for evaluation of marine resources in eastern Bering Sea. *NOAA Outer Continental Shelf
Environmental Assessment Program Report RU 77.* 69 pp.

Langton, W. and R. E. Bowman, 1977. An abridged account of predator-prey interactions for some northwest Atlantic species of
fish and squid. *Northeast Fisheries Center, Woods Hole, Massachusetts, Laboratory Reference Report No. 77-17.* 34 pp.

Lasker, R., 1975. Field criteria for survival of anchovy larvae:
the relationship between inshore chlorophyll maximum layers and
successful first feeding. *U. S. Fish. Bull.* 73: 453-462.

Laurence, G. C., 1974. Growth and survival of haddock (*Melanogrammus
aeglefinus*) larvae in relation to the planktonic prey concentrations. *J. Fish. Res. Bd. Can.* 31: 1415-1419.

Laurence, G. C., 1975. Laboratory growth and metabolism of the winter flounder *Pseudopleuronectes americanus* from hatching through
metamorphosis at three temperatures. *Mar. Biol.* 32: 223-239.

Laurence, G. C., 1977a. Comparative growth, respiration, and delayed
feeding abilities of larval cod (*Gadus morhua*) and haddock (*Melanogrammus aeglefinus*) as influenced by temperature during laboratory studies. *International Council for the Exploration of the
Sea, 65th Annual Meeting, Reykjavik, Iceland, September 25-Octo-
ber 2, 1977.* 1977/L:31. 16 pp.

Laurence, G. C., 1977b. A bioenergetic model for the analysis of
feeding and survival potential of winter flounder, *Pseudopleuronectes americanus,* larvae during the period from hatching to metamorphosis. *U. S. Fish. Bull.* 75: 529-546.

Laurence, G. C., T. Halavik, B. Burns, and A. Smigielski, 1979. An
environmental chamber for monitoring "in situ" growth and survival
of larval fishes. *Trans. Amer. Fish. Soc.* 108: 197-203.

Mullin, M. M., 1969. Production of zooplankton in the ocean: the
present status and problems. *Oceanogr. and Mar. Biol. Ann. Rep.*
7: 293-314.

Parrish, J. D., 1975. Marine trophic interactions by dynamic stimulation of fish species. *U. S. Fish. Bull.* 73: 695-716.

Pavshtics, Ye. A., 1963. Distribution of plankton and summer feeding of herring in the Norwegian seas and on Georges Bank. *Spec. Publ. Int. Comm. Northw. Atl. ant. Fish. Res. Bull.* 6: 583-590.

Regier, H. A. and H. F. Henderson, 1973. Towards a broad ecological model of fish communities and fisheries. *Trans. Amer. Fish. Soc.* 102: 56-72.

Riley, G. A., 1941. Plankton studies. IV. Georges Bank. *Bull. Bingham Oceanogr. Coll.* 7: 1-73.

Riley, G. A., 1946. Factors controlling phytoplankton populations on Georges Bank. *J. Mar. Res.* 6: 54-73.

Riley, G. A., 1947. A theoretical analysis of the zooplankton population on Georges Bank. *J. Mar. Res.* 6: 104-113.

Riley, G. A. and D. F. Bumpus, 1946. Phytoplankton-zooplankton relationships on Georges Bank. *J. Mar. Res.* 6: 33-47.

Riley, G. A., H. Stommel, and D. F. Bumpus, 1949. Quantitative ecology of the plankton of the western North Atlantic. *Bull. Bingham Oceanogr. Coll.* 12: 1-169.

Rogers, C. A., D. C. Biggs, and R. A. Cooper, 1978. Aggregation of the siphonophore *Nanomia cara* in the Gulf of Maine: observations from a submersible. *U. S. Fish. Bull.* 76: 281-284.

Sherman, K., 1970. Seasonal and areal distribution of zooplankton in coastal waters of the Gulf of Maine, 1967 and 1968. *U. S. Fish and Wildl. Serv. Spec. Sci. Rep. Fish. No. 594.* 8 pp.

Sherman, K. and K. A. Honey, 1971. Seasonal variations in the food of larval herring in the coastal waters of Central Maine. *Rapp. et P.-v. Réun. Cons. int. Explor. Mer.* 160: 121-124.

Sherman, K., L. Sullivan, and R. Byron, 1978. Pulses in the abundance of zooplankton prey of fish on the continental shelf off New England. *International Council for the Exploration of the Sea, 66th Annual Meeting, Copenhagen, Denmark, October 2-11, 1978.* 1978/L: 25. 42 pp.

Sherman, K., L. Sullivan, K. Honey, and D. Busch, 1976. Changes in the availability of food of larval herring in Maine coastal waters. *International Council for the Exploration of the Sea, 64th Annual Meeting, Copenhagen, Denmark, October 4-13, 1976.* 1976/L:38. 22 pp.

Smith, W. G., L. Sullivan, and P. Berrien, 1978. Fluctuations in production of sand lance larvae in coastal waters off the Northeastern United States, 1974 to 1977. *International Council for the Exploration of the Sea, 66th Annual Meeting, Copenhagen, Denmark, October 2-11, 1978.* 1978/L:30. 16 pp.

Steele, J. H. and B. W. Frost, 1977. The structure of plankton communities. *Phil. Trans. of the Royal Soc. London. B. Biol. Sci.* 280: 485-534.

Thomas, W. and D. Seibert, 1977. Effects of copper on the dominance and diversity of algae. *Bull. Mar. Sci.* 27: 23-33.

U. S. 94th Congress, 1976. Fisheries Conservation and Management Act of 1976. Public Law No. 94-265, 94th Congress H. R. 200, April 13, 1976. U. S. Government Printing Office, Washington, D. C.

Ursin, E., 1977. Multispecies fish stock assessment for the North Sea. *International Council for the Exploration of the Sea, 65th Annual Meeting, Reykjavik, Iceland, September 25-October 2, 1977.* 1977/F:42. 19 pp.

Walsh, J. J., T. E. Whitledge, F. W. Barvenik, C. D. Wirick, S. O. Howe, W. E. Esaias, and J. T. Scott, 1978. Wind events and food chain dynamics within the New York Bight. *Limnol. Oceanogr.* 23: 659-683.

Williams, A. B. and R. L. Wigley, 1977. Distribution of decapod crustacea off northeastern United States based on specimens at the Northeast Fisheries Center, Woods Hole, Massachusetts. *NOAA Tech. Rept. NMFS Circ. 407*. 44 pp.

Zinkevitch, V. N., 1967. Observations on the distribution of herring, *Clupea harengus* L. on Georges Bank and in adjacent waters in 1962-1965. *Int. Comm. Northw. Atlant. Fish. Res. Bull.* 4: 101-115.

The Need to Improve Data Acquisition and Data Processing in Biological Oceanography

T. T. Packard

ABSTRACT: A comparison is made of data acquisition and data processing in biological and physical oceanography. The comparison suggests that research in plankton biomass and physiological rates would benefit from the application of fluorescence, enzyme, and particle counting techniques to improve data acquisition and would benefit from the utilization of microprocessors to improve data processing.

The solution to complex oceanographic problems requires multidisciplinary research. Without this type of research and the ensuing input from biological, chemical, and physical oceanographers, phenomena such as the oxygen minimum zone, El Niño, oceanic gyres, and upwelling cannot be understood. Furthermore, unless this input is coordinated and is rapidly made available to all the participants of the multidisciplinary program, the synthesis of the data and a comprehensive understanding of the phenomenon cannot be achieved within the lifetime of the program. This coordination and rapid data exchange is not being accomplished in current multidisciplinary programs. This situation exists, not because there is a lack of enthusiasm or willingness on the part of the participants, but because there is a large data acquisition and data processing gap between the physicists and geochemists on one side and the chemists, biologists, microbiologists, and biochemists on the other side. If a physical oceanographer wants to know the density structure of a seawater column, he can lower a Salinity/Temperature/Depth (STD) probe (e.g., Foster and Carmack, 1976) and within minutes he can locate, with the aid of a plotter, the major pycnoclines and water masses in the seawater column. If a biological oceanographer wants to determine the biomass profile of the seawater column, he can lower a fluorometer (Whitledge, 1978; Whitledge, in press), or he can acoustically survey for fish schools (e.g., Thorne et al., 1977), but those measurements give him the distribution of only two parts of the biomass: the phytoplankton and the nekton. To determine the bacterial and zooplankton components of the plankton, the biologist must lower water bottles (e.g., Devol et al., 1976) and nets (e.g., Shulenberger, 1978) to discrete depths to collect samples. Then, after lengthy sample preparation, he must measure carbon, nitrogen, or dry weight in each sample and calculate and plot the data before he can locate the major features of the depth distribution. Where the physicist waits only an hour, the biologist may wait as long as several months before he sees his results. Consequently, collaboration between the two scientists is severely retarded. As another example, if we compare the physical oceanographer's ability to measure particle profiles by beam transmittance (e.g., Kitchen et al., 1978) or volume scattering (Kullenberg, 1978) with the chemist's ability to measure profiles of carbohydrate (Hitchcock, 1977) or protein (e.g., Packard and Dortch, 1975) by wet chemistry, we find again that the physicist can review his results at the completion of the oceanographic station while the chemist must wait at least until the end of the day.

The above is the situation with static properties; with dynamic properties, such as ocean currents and biological rates, the situation is just as bad. The physical oceanographers can moor an array of current meters and return months later to recover the taped data (Enfield et al., 1978).

For the biological oceanographer to measure rates of photosynthesis or nitrate uptake over this same period, the biologist must sample the seawater column daily, incubate his radioactively labelled samples in deck incubators onboard ship (e.g., Huntsman and Barber, 1977), run the filtered samples through a scintillation counter (e.g., Barber et al., 1978) or mass spectrometer (Dugdale and Goering, 1967), record the data by hand, and finally enter the data into a computer. In place of this hand recording and entering, the physical oceanographer plays the current meter tapes on a computer's tape reader (Enfield et al., 1978). Once in the computer the data can be machine filtered, calculated, plotted, and statistically analyzed. Within a short time, the physicist is ready to synthesize his data with the biological and biochemical data, but those data, unfortunately, are still on filters in a drying oven or in a freezer.

With regard to mapping biological or chemical properties along horizontal surfaces, biological oceanographers do not have any instrumentation comparable to long-range side-scan sonar (e.g., Whitmarsh and Laughton, 1976) that the geophysicists use in bottom mapping. Nevertheless, mapping has proved to be useful to biological oceanographers in developing optimum sampling schemes while on field sur-

veys (Walsh et al., 1971; Walsh, 1972; and Kelley and Rix, 1974).
On recent cruises to upwelling areas, biologists conducted mapping
surveys at night to assess the movement of phytoplankton patches,
fish schools, and boluses of upwelled water (Kelley et al., 1975;
Thorne et al., 1977). Similar maps have been useful for experimental physiologists who needed seawater with different combinations
of chlorophyll and nutrients (MacIsac and Dugdale, 1972; Walsh,
1972; Packard, 1979). They have also been useful for ecologically oriented research that was focused on understanding the changes
in the biomass level and species composition at fixed positions in an
upwelling area (Blasco, 1977). For this research maps helped distinguish *in situ* biological changes from advective ones. At the
current level of development, biological mapping is a useful tool,
but it would be even more useful if other biological properties of
the seawater could be mapped. The most important properties for
plankton research, would be phytoplankton productivity, zooplankton
biomass, and secondarily, respiration of either plankton group.
For deep ocean research, microplankton biomass and metabolism would
be key variables to map. This mapping is not done because reliable
indices of these variables have not been developed.

Thus, regardless of the dimensionality of the research, the
only biological variables which can be continuously assessed in
real-time are phytoplankton and nekton biomass. How can this situation be improved? One of the first steps is to employ microprocessors (Ratzlaff, 1978) to minimize the human interaction between
the primary instrument (i.e., mass spectrometer, scintillation counter, spectrophotometer, etc.) and the central data file. Microprocessors can be programmed to acquire raw data and calculate results,
thereby eliminating the data recording, manipulating, and entering
that are routinely done by hand. In doing so, time could be saved
and human errors avoided. Alone, this step will not yield automated
real-time assessment of biological variables, but by reducing the
time gap between data acquisition and data reporting it will enable
biologists to collaborate in interdisciplinary research projects more
effectively.

The second step towards improving the situation is to focus
attention on the primary sensor end and develop indicies of biomass
and physiological rates. We could adapt existing technology about
particle counters to develop continuous flow and *in situ* microplankton and zooplankton counters as has been explored by Boyd
(Boyd, 1973; Mackas and Boyd, 1978). Existing knowledge about
fluorescent dyes could be adapted to develop continuous flow and
in situ assessment of particulate protein (Udenfriend et al., 1972;
Torres et al., 1976; Schiltz et al., 1977) and other chemical constituents of plankton. It may even be possible now to use immunofluorescent techniques to develop real time assessments of plankton
species (Richards and Cowland, 1967; Anonymous, 1973). These approaches or the adaptation of particle counters will advance "real-
time" assessment of the static aspects of biology, the stock assessment. But to advance the real-time assessment of rate measurements
(i.e., productivity, respiration, excretion, etc.), another approach
must be taken.

In plankton research there are four difficult problems inherent

in the measurement of physiological rates: 1) the rates are low and
challenge the sensitivity limits of conventional methodology; 2) the
organisms responsible for the rates require a simulated natural
environment for the rate measurements to be representative; 3) the
organisms are widely dispersed and require concentration at some
stage of the rate measurement; and 4) many of the rates of the bacterial, zooplankton, and phytoplankton components of the plankton
cannot now be separated.

One approach to developing continuous real-time assessment of
physiological rates in plankton is to focus attention on the chemical
basis of these rates. This basis involves enzyme chemistry since all
of the physiological rates (i.e., photosynthesis, respiration, nitrogen assimilation, and nitrogen excretion) are controlled by a series
of enzymatic reactions (Lehninger, 1977). If the activity level of
the concentration of the controlling enzymes could be measured, then
the maximum rate of the physiological process would be known. Furthermore, if the regulatory mechanism of the rate-limiting enzymatic reaction were understood, then the *in situ* rate of the physiological process could be calculated from the maximum rate. In marine research,
progress has been made in identifying and measuring the rate-control-
ling enzymatic reactions associated with nitrogen assimilation (Eppley
et al., 1969; Falkowski and Rivkin, 1977), respiration (Curl and
Sandberg, 1961; Packard 1969, 1971), and photosynthesis (Mukerji and
Morris, 1976), but little progress has been made in understanding the
regulatory mechanisms. When more progress has been made in understanding enzyme regulation in marine organisms and when enzyme analyses are more fully automated, physiological rates could be mapped and
monitored in the same way nutrient salts and fluorometric chlorophyll
are today (Kelley et al., 1975).

In summary, improvement in the assessment of biological fields
in the ocean, and conversion to real-time data processing will enable
biologists to interact more effectively in interdisciplinary research
projects. The first step should be made in data processing by utilizing more microprocessors. The second step should be made in field
assessment by developing and adapting fluorescence techniques to
measure static biochemical properties, enzyme techniques to measure
dynamic physiological rates, and particle counting technology to
measure stocks of microplankton and zooplankton.

ACKNOWLEDGMENTS

I thank R. C. Dugdale, J. O'Brien, P. Sherman, B. Jones, and N.
Breitner for helpful discussions about this paper, and V. Jones and
B. Royal for help in preparing it. This effort was supported by
ONR contract No. 00014-76-C-0271 and the State of Maine. The manu-
script is contribution number 78034 from the Bigelow Laboratory for
Ocean Sciences.

REFERENCES

Anonymous, 1973. Immunofluorescent staining technique for the
    detection of wild *Saccharomyces* contaminant in *Saccharomyces
    cerevisiae*. *J. Institute of Brewing* (London) 79: 134-136.
Barber, R. T., S. A. Huntsman, J. E. Kogelschatz, W. D. Smith,
    B. H. Jones, and J. C. Paul, 1978. Carbon, chlorophyll and
    light extinction from JOINT II 1976 and 1977. *Coastal Up-
    welling Ecosystems Analysis Data Report No. 49.* 476 pp.
Blasco, D., 1977. Red tide in the upwelling region of Baja, Cali-
    fornia. *Limnol. Oceanogr.* 22: 255-263.
Boyd, C. M., 1973. Small scale spatial patterns of marine zoo-
    plankton examined by an electronic *in situ* zooplankton detect-
    ing device. *Neth. J. Sea Res.* 7: 103-111.
Curl, H., Jr., and J. Sandberg, 1961. The measurement of dehydro-
    genase activity in marine organisms. *J. Mar. Res.* 19: 123-
    138.
Devol, A. H., T. T. Packard, and O. Holm-Hansen, 1976. Respiratory
    electron transport activity and adenosine triphosphate in the
    oxygen minimum of the eastern tropical North Pacific. *Deep-
    Sea Res.* 23: 963-975.
Dugdale, R. C. and J. J. Goering, 1967. Uptake of new and regenerated
    forms of nitrogen in primary productivity. *Limnol. Oceanogr.* 12:
    196-206.
Enfield, D. B., R. L. Smith, and A. Huyer, 1978. A compilation of
    observations from moored current meters, Vol. XII: Wind, currents
    and temperature over the continental shelf and slope off Peru
    during JOINT II. *Data Report No. 70*, Oregon State University.
    343 pp.
Eppley, R. W., J. L. Coatsworth, and L. Solórzano, 1969. Studies of
    nitrate reductase in marine phytoplankton. *Limnol. Oceanogr.* 14:
    194-205.
Falkowski, P. G. and R. B. Rivkin, 1977. The role of glutamine syn-
    thetase in the incorporation of ammonium in *Skeletonema costatum*
    (Bacillariophyceae). *J. Phycol.* 12: 448-450.
Foster, T. D. and E. C. Carmack, 1976. Frontal zone mixing and an-
    tarctic bottom water formation in the southern Weddell Sea.
    Deep-Sea Res. 23: 301-319.
Garfield, P., T. T. Packard, and L. A. Codispoti (in press). Parti-
    culate protein in the Peru upwelling system. *Deep-Sea Res.*
Hitchcock, G. L., 1977. The concentration of particulate carbohy-
    drate in a region of the West Africa upwelling zone during
    March, 1974. *Deep-Sea Res.* 24: 83-94.
Huntsman, S. A. and R. T. Barber, 1977. Primary production off
    Northwest Africa: the relationship to wind and nutrient condi-
    tions. *Deep-Sea Res.* 24: 25-35.
Kelley, J. C. and J. L. Rix, 1974. An automated contouring system
    for the interactive real-time information system (IRIS). *Tech.
    Rep. Dept. Oceanogr., Univ. Wash. (CUEA) No. 4, Ref. No. M74-05.*
Kelley, J. C., T. E. Whitledge, and R. C. Dugdale, 1975. Results of
    sea surface mapping in the Peru upwelling system. *Limnol.
    Oceanogr.* 25: 784-794.
Kitchen, J. C., J. Ronald, V. Zaneveld, and H. Pak, 1978. The ver-

tical structure and size distribution of suspended particles off Oregon during the upwelling season. *Deep-Sea Res.* 25: 453-469.

Kullenberg, G., 1978. Light scattering observations in the north-west African upwelling region. *Deep-Sea Res.* 25: 525-543.

Lehninger, A. L., 1977. *Biochemistry.* Worth Publishers, Inc., New York. 1104 pp.

MacIsac, J.. and R. C. Dugdale, 1972. Interaction of light and in-organic nitrogen in controlling uptake in the sea. *Deep-Sea Res.* 19: 521-524.

Mackas, D. L. and C. M. Boyd, 1978. Spectral analysis of zooplankton spatial heterogeneity. *Science* 204: 62-64.

Mukerji, D. and I. Morris, 1976. Photosynthetic carboxylating en-zymes in *Phaeodactylum tricornutum:* assay methods and properties. *Mar. Biol.* 36: 199-206.

Packard, T. T., 1969. The estimation of oxygen utilization rate in seawater from the activity of the respiratory electron transport system in plankton. Ph.D. Thesis, Univ. Wash., Seattle. 115 pp.

Packard, T. T., 1971. The measurement of respiratory electron trans-port activity in marine phytoplankton. *J. Mar. Res.* 29: 235-244.

Packard, T. T. and Q. Dortch, 1975. Particulate protein-nitrogen in North Atlantic surface waters. *Mar. Biol.* 33: 347-354.

Packard, T. T., 1979. Half-saturation constants for nitrate re-ductase and nitrate translocation in marine phytoplankton. *Deep-Sea Res.* 26A: 321-326.

Ratzlaff, K. L., 1978. Microprocessors in minicomputer applications. *American Laboratory* 10: 17-27.

Richards, M. and T. W. Cowland, 1967. The rapid detection of brewery contaminants belonging to the genus *Saccharomyces* by a serologi-cal technique. *J. Institute of Brewing* (London) 73: 552.

Schiltz, E., K. D. Schnackerz, and R. W. Gracy, 1977. Comparison of ninhydrin, fluorescamine, and o-phthaldialdehyde for the detec-tion of amino acids and peptides and their effects on the reco-very and composition of peptides from thin-layer fingerprints. *Anal. Biochem.* 79: 33-41.

Shulenberger, E., 1978. Vertical distributions, diurnal migrations, and sampling problems of hyperiid amphipods in the North Pacific central gyre. *Deep-Sea Res.* 25: 605-625.

Thorne, R. E., O. A. Mathisen, R. J. Trumble, and M. Blackburn, 1977. Distribution and abundance of pelagic fish off Spanish Sahara during CUEA Expedition JOINT 1. *Deep-Sea Res.* 24: 65-75.

Torres, A. R., V. L. Alvarez, and L. B. Sandberg, 1976. The use of o-phthaldialdehyde in the detection of proteins and peptides. *Biochim.Biophys.Acta* 434: 209-214.

Udenfriend, S., S. Stein, P. Bohlen, W. Dairman, W. Leimgruber, and M. Weigele, 1972. Fluorescamine: a reagent for assay of amino acids, peptides, proteins, and primary amines in the picomole range. *Science* 178: 871-872.

Walsh, J. J., J. C. Kelley, R. C. Dugdale, and B. W. Frost, 1971. Gross features of the Peruvian upwelling system with special reference to possible diel variation. *Invest. Pesq.* 35: 25-42.

Walsh, J. J., 1972. Implications of a systems approach to oceano-graphy. *Science* 176: 969-975.

Whitledge, T. E., 1978.   The biological current meter – a moored
     fluorometer.   *Abstract Amer. Soc. Limnol. Oceanogr. Victoria,*
     B. C., June, 1978.
Whitledge, T. E., in press.   Probes 1978; Processes and resources
     of the Bering Sea shelf.   *Data Report R/V T. G. THOMPSON cruise*
     131, leg II.   Institute of Marine Science, Univ. Alaska, Fair-
     banks.
Whitmarsh, R. B. and A. S. Laughton, 1976.   A long-range sonar
     study of the mid-Atlantic Ridge crest near 37°N (FAMOUS area)
     and its tectonic implications.   *Deep-Sea Res.* 23:   1005–1024.

SAMPLING, EXPERIMENTAL DESIGN, SIGNAL PHYSICS

Statistical Methods in Optimal Curve Fitting

Howard L. Weinert

ABSTRACT: Many optimal curve fitting and approximation problems have the same structure as certain estimation problems involving random processes. This structural correspondence has many useful consequences for curve fitting problems, including recursive algorithms and computable error bounds. The basic facts of this correspondence are reviewed and some new results on error bounds and optimal sampling are presented.

INTRODUCTION

Consider the basic problem of approximating $f(t)$ given (possibly inaccurate) measurements of $\lambda_1 f, \lambda_2 f, \ldots, \lambda_N f$, where $\{\lambda_j\}_1^N$ are linear functionals. The approximation can be carried out by fitting a curve $s(\cdot)$ to the data, and then approximating $f(t)$ by $s(t)$. If the data are error free, it makes sense to require $\lambda_j s = \lambda_j f$, in which case we have an *interpolation* problem. If the data contain errors, we have a *smoothing* problem. In both cases, the data values will be denoted by $\{r_j\}_1^N$.

49

Before anything can be said about the choise of $s(\cdot)$, something must be said about the underlying function $f(\cdot)$. I will assume that $f(\cdot)$ belongs to the following Hilbert space (see Remark 1):

$$H = \{g(t), \ t \ \varepsilon \ [0,T]: \ g^{(n)} \text{exists a.e. and} \int_0^T [g^{(n)}]^2 < \infty\}. \quad (1)$$

Here, $g^{(n)}$ denotes the nth derivative of g, and n is a fixed integer satisfying $1 \leq n \leq N$. Since dependent or unbounded functionals provide no useful information about $f(\cdot)$, I will also assume that the measurement functionals $\{\lambda_j\}_1^N$ are linearly independent and bounded on H. For future reference, a constant-coefficient (see Remark 2) differential operator L is defined as

$$L = \frac{d^n}{dt^n} + a_{n-1} \frac{d^{n-1}}{dt^{n-1}} + \ldots + a_1 \frac{d}{dt} + a_0. \quad (2)$$

Now if the measurements are error-free, they localize $f(\cdot)$ to a linear variety U in H, where

$$U = \{g \ \varepsilon \ H: \ \lambda_j g = r_j \ , \ 1 \leq j \leq N\}. \quad (3)$$

As the interpolating function $s_0(\cdot)$ choose a function in U that satisfies

$$\int_0^T (Ls_0)^2 = \min_{g \varepsilon U} \int_0^T (Lg)^2. \quad (4)$$

If the measurements contain errors, the smoothing function $s_1(\cdot)$ is chosen as a solution to the optimization problem

$$\min_{g \varepsilon H} \ \{\int_0^T (Lg)^2 + \sum_{j=1}^N \rho_j^{-1} (r_j - \lambda_j g)^2\}. \quad (5)$$

Here, the weights $\{\rho_j\}_1^N$ are strictly positive real numbers.

Both problems (4) and (5) have unique solutions if $\{\lambda_j\}_1^n$ are linearly independent on the null space of L. This condition will be assumed in all that follows. The solution to (4) is called the *Lg-interpolating spline*. The solution to (5) is called the *Lg-smoothing spline*. We can see that as all the weights ρ_j approach zero, the spline smoothing problem (5) reduces to the spline interpolation problem (4).

Because of their optimality properties, splines are superior to ordinary polynomials in curve fitting. At the same time, the functional form of a spline can be almost as simple as a polynomial (see Remark 2). Splines have been applied to many problems in numerical analysis, statistics, and engineering. For further information on spline theory, history, and applications, see Greville (1969); Jerome and Schumaker (1969); Schoenberg (1969); Bellman and Roth

(1971); Schultz (1973); Caprihan (1975); de Montricher et al. (1975); de Figueiredo and Netravali (1976); Larsen et al. (1977).

Remark 1

Functions in H can be differentiated at least n times without producing a delta function. Similar curve fitting problems can be formulated in other types of Hilbert spaces; for example, the space of bandlimited functions as in Yao (1967).

Remark 2

All the results discussed in this paper extend in the obvious way to variable-coefficient differential operators provided the co-efficients are sufficiently differentiable. As discussed later on, the operator L determines the functional form of the spline.

INTERPOLATING AND SMOOTHING SPLINES AS PROJECTIONS

Subsequent results require that the interpolation and smoothing problems (4) and (5) be reformulated as minimum-norm problems with inner product constraints, after which they can be solved using the Projection Theorem. The interpolation problem will be considered first (see Remark 3).

The space H is a reproducing kernel Hilbert space (see Aronszain, 1950) with norm

$$||g||^2 = \sum_{j=1}^{n} (\lambda_j g)^2 + \int_0^T (Lg)^2 , \quad g \varepsilon H \qquad (6)$$

and reproducing kernel

$$K_0(t,\tau) = \sum_{j=1}^{n} z_j(t) z_j(\tau) + \int_0^T G(t,\xi) G(\tau,\xi) d\xi, t,\tau \varepsilon [0,T] \qquad (7)$$

where $\{z_i(\cdot)\}_1^n$ is the basis for the null space of L that is dual to $\{\lambda_i\}_1^n$ and $G(\cdot,\cdot)$ is the Greens function for L that satisfies $\lambda_j G(\cdot,\tau)=0$ for $1 \leq j \leq n$ and $\tau \varepsilon [0,T]$. In terms of the inner product determined by (6), we have the reproducing property

$$<g(\cdot), K_0(\cdot,t)>=g(t), \quad g \varepsilon H, \quad t \varepsilon [0,T] \qquad (8)$$

and thus equation (3) can be restated as

$$U = \{g \varepsilon H: \quad <g(\cdot), h_j(\cdot)>=r_j, \quad 1 \leq j \leq N\} \qquad (9)$$

where

$$h_j(t) = \lambda_j K_0(\cdot,t), \quad t \varepsilon [0,T] \tag{10}$$

Since the first term in (6) is fixed for all $g \varepsilon U$, it is clear from (4) that the interpolating spline $s_0(\cdot)$ is the function in U that satisfies

$$||s_0||^2 = \min_{g\varepsilon U} ||g||^2. \tag{11}$$

In the light of (9) and (11), we see that $s_0(\cdot)$ is the projection of any function in U onto the span of $\{h_j(\cdot)\}_1^{N0}$.

   In order to get a similar formulation of the spline smoothing problem, we must work in the following space H^+ of order pairs:

$$H^+ = \{(g(\cdot),\theta): \quad g \varepsilon H \text{ and } \theta = [\theta_1,\theta_2,\ldots, \theta_N]' \varepsilon E^N\} \tag{12}$$

It can be shown (see Remark 4) that H^+ is a Hilbert space with norm

$$||(g,\theta)||^2 = \int_0^T (Lg)^2 + \sum_{j=1}^N \rho_j^{-1} \theta_j^2 + \sum_{j=1}^n (\lambda_j g + \theta_j)^2. \tag{13}$$

Now define two index sets by

$$I = [0,T] \tag{14a}$$

$$J = \{1,2,\ldots, N\} \tag{14b}$$

and let

$$K_1(t,\tau) = \sum_{j=1}^n (1+\rho_j)z_j(t)z_j(\tau) + \int_0^T G(t,\xi)G(\tau,\xi)d\xi, \tag{15}$$

   t, $\tau\varepsilon I$.

Also let e_j, $j \varepsilon J$, be the jth unit N-vector, let $z(\cdot)$ be the N-vector

$$z(\cdot) = [z_1(\cdot), z_2(\cdot),\ldots, z_n(\cdot),0,\ldots, 0]' \tag{16}$$

and let Q be the diagonal NxN matrix

$$Q = \text{diag } (\rho_1,\rho_2,\ldots, \rho_N). \tag{17}$$

Then H^+ with norm given by (13) has the reproducing kernel

$$K_t^+ = \begin{cases} (K_1(\cdot,t), \ - \underset{\sim}{Q} \, z(t)), \ t \ \varepsilon \ I \\ \\ (-\underset{\sim}{z}'(\cdot)\underset{\sim}{Q} \, \underset{\sim}{e}_t, \ \underset{\sim}{Q} \, \underset{\sim}{e}_t), \ t \ \varepsilon \ J. \end{cases} \qquad (18)$$

In terms of the inner product determined by (13), the reproducing property of H^+ is

$$<(g(\cdot),\underset{\sim}{\theta}), \ K_t^+> = \begin{cases} g(t), \ t \ \varepsilon \ I \\ \\ \theta_t, \ t \ \varepsilon \ J \end{cases}, \ (g(\cdot),\underset{\sim}{\theta}) \ \varepsilon \ H^+ . \qquad (19)$$

Now since $\{r_j\}_1^N$ are fixed, it is clear that the optimization problem in (5) is equivalent to minimizing $||(g,\underset{\sim}{\theta})||^2$ subject to the contraints $\lambda_j g + \theta_j = r_j$, $j \ \varepsilon \ J$. These constraints can be written as inner product constraints:

$$<(g,\underset{\sim}{\theta}), \ (d_j,\underset{\sim}{\omega}_j)> = \lambda_j g + \theta_j = r_j, \ j \ \varepsilon \ J \qquad (20)$$

where

$$d_j(t) = \lambda_j K_1(\cdot,t) - \underset{\sim}{e}_j' \underset{\sim}{Q} \, z(t), \ t \ \varepsilon \ I \ , \ j \ \varepsilon \ J \qquad (21)$$

$$\underset{\sim}{\omega}_j = \underset{\sim}{Q}(\underset{\sim}{e}_j - \lambda_j \, z), \ j \ \varepsilon \ J. \qquad (22)$$

Let

$$U^+ = \{(g,\underset{\sim}{\theta}) \ \varepsilon \ H^+ : <(g,\underset{\sim}{\theta}),(d_j,\underset{\sim}{\omega}_j)> = r_j, \ j \ \varepsilon \ J\}. \qquad (23)$$

Then if $(\hat{g},\hat{\underset{\sim}{\theta}})$ solves

$$\begin{array}{c} \text{minimize} \quad ||(g,\underset{\sim}{\theta})||^2 \ , \\ (g,\underset{\sim}{\theta})\varepsilon U^+ \end{array} \qquad (24)$$

we can see that the smoothing spline $s_1(t) = \hat{g}(t)$, $t \ \varepsilon \ I$. Therefore, $s_1(\cdot)$ is the component in H of the projection of any element of U^+ onto the span of $\{(d_j, \ \underset{\sim}{\omega}_j)\}_1^N$.

Remark 3

Nothing in the interpolation problem (4) forces a particular choice of norm and reproducing kernel for H. However, the choice in (6) and (7) seems to be the most convenient one for developing recursive spline algorithms and computable error bounds. This particular framework for interpolation was developed, to varying degrees, by Golomb and Weinberger (1959), de Boor and Lynch (1966), and Weinert and Kailath (1974).

Remark 4

Details of this new approach to the smoothing problem can be found in Weinert et al (in press). It is clear that spline smoothing is nothing more than spline interpolation in an augmented space. Note that because of the cross-term in (13), H^+ is not the usual product of H and E^N; hence the more complicated reproducing kernel and reproducing property. For other approaches to spline smoothing, see Kimeldorf and Wahba (1970 a, b, 1971), Reinsch (1971), Schumaker (1973), Wahba and Wold (1975), de Figueiredo and Caprihan (1977). Anselone and Laurent (1968) characterize Ls_1 as a component of the solution to a projection problem in $LHxE^N$.

SPLINES AND STOCHASTIC ESTIMATES

In this section the stochastic estimation problems that are structurally equivalent to spline interpolation and smoothing are described (see Remark 5). Let $\{y_0(t), t \varepsilon I\}$ be a zero-mean random process with covariance function $K_0(\cdot,\cdot)$. It can be shown (see Parzen, 1961) that H is congruent (isometrically isomorphic) to the space of zero-mean random variables spanned by $\{y_0(t), t \varepsilon I\}$. As a result, if $\hat{y}_0(t)$ is the linear least-squares estimate of $y_0(t)$ given random variables $\{\lambda_j y_0\}_1^N$, and if $\hat{\hat{y}}_0(t)$ is the sample value of $\hat{y}_0(t)$ obtained by setting $\lambda_j y_0 = r_j$, $j \varepsilon J$, then

$$s_0(t) = \hat{\hat{y}}_0(t), t \varepsilon I. \tag{25}$$

For the spline smoothing case, let $\{y_1(t), t \varepsilon I\}$ be a zero-mean random process with covariance function $K_1(\cdot,\cdot)$, and let $\underset{\sim}{v} = [v_1, v_2, \ldots, v_N]'$ be a zero-mean random vector such that

$$E[y_1(t)\underset{\sim}{v}] = -\underset{\sim}{Q} \underset{\sim}{z}(t), t \varepsilon I \tag{26}$$

$$E[\underset{\sim}{v} \underset{\sim}{v}'] = \underset{\sim}{Q}. \tag{27}$$

Now H^+ is congruent to the space of zero-mean random variables spanned by $\{y_1(t), t \varepsilon I; v_j, j \varepsilon J\}$. If $\hat{y}_1(t)$ is the linear least-squares estimate of $y_1(t)$ given noisy data $\{\lambda_j y_1 + v_j\}_1^N$, and if $\hat{\hat{y}}_1(t)$ is the sample value of $\hat{y}_1(t)$ obtained by setting $\lambda_j y_1 + v_j = r_j$, $j \varepsilon J$, then

$$s_1(t) = \hat{\hat{y}}_1(t), t \varepsilon I. \tag{28}$$

Thus the interpolating spline $s_0(\cdot)$ and smoothing spline $s_1(\cdot)$ are sample functions of the random processes $\hat{y}_0(\cdot)$ and $\hat{y}_1(\cdot)$, respectively. As a result recursive algorithms for $\hat{y}_0(\cdot)$ and $\hat{y}_1(\cdot)$ serve as recursive algorithms (see Remark 6) for $s_0(\cdot)$ and $s_1(\cdot)$. Of course, $\hat{y}_0(\cdot)$ and $\hat{y}_1(\cdot)$ can be found using just the covariance information given above. However, computation of the covariances $K_0(\cdot,\cdot)$ and $K_1(\cdot,\cdot)$ can be avoided by using the dynamical linear models that generate the processes $y_0(\cdot)$ and $y_1(\cdot)$ in response to white noise. The dynamics of these models are immediately determined by the coefficients of the operator L, without intermediate computations. Before giving the models, we will restrict attention to a broad class of constraint functionals $\{\lambda_j\}_1^N$ called extended Hermite-Birkhoff functionals, which have the form

$$\lambda_j g = \sum_{k=1}^{n} \alpha_{jk} g^{(k-1)}(t_j) \ , \ j \ \varepsilon \ J \tag{29}$$

where $0 \le t_1 \le t_2 \le \ldots \le t_N \le T$ and the $\{\alpha_{jk}\}$ are known real numbers. For future reference, let

$$\underset{\sim}{c}_j = [\alpha_{j1}, \alpha_{j2}, \ldots, \alpha_{jn}] \ , \ j \ \varepsilon \ J. \tag{30}$$

For the spline interpolation case, it can be shown that if $u(\cdot)$ is a zero-mean, unit intensity white noise process, $y_0(\cdot)$ is generated by the model

$$\frac{d}{dt} \underset{\sim}{x}_0(t) = \underset{\sim}{A} \underset{\sim}{x}_0(t) + \underset{\sim}{b} \ u(t) \tag{31a}$$

$$y_0(t) = \underset{\sim}{c} \underset{\sim}{x}_0(t) \tag{31b}$$

where

$$\underset{\sim}{b} = [0 \ . \ . \ . \ 0 \ 1]' \tag{32}$$

$$\underset{\sim}{c} = [1 \ 0 \ . \ . \ . \ 0] \tag{33}$$

$$\underset{\sim}{A} = \begin{bmatrix} \underset{\sim}{0} & \vdots & \underset{\sim}{I} \\ \text{---} & \text{---------------} \\ -a_0 & -a_1 \ldots & -a_{n-1} \end{bmatrix} \tag{34}$$

and $\underset{\sim}{x}_0(\cdot)$ is the n-dimensional state vector. All that remains is to specify the unique initial conditions that guarantee that $y_0(\cdot)$ has covariance $K_0(\cdot,\cdot)$. The most convenient initialization point is t_n. Let $\theta(\cdot,\cdot)$ be the fundamental matrix for $\underset{\sim}{A}$; i.e.,

$$\theta(t,\tau) = \exp \underset{\sim}{A}(t-\tau). \tag{35}$$

If $\underset{\sim}{M}$ is the nxn matrix with i^{th} row $\underset{\sim}{M'_i}$:

$$\underset{\sim}{M'_i} = \underset{\sim}{c_i} \; \underset{\sim}{\theta}(t_i, t_n) \tag{36}$$

and if $\underset{\sim}{\Delta}(\cdot)$ is the nxn matrix with i^{th} row $\underset{\sim}{\Delta'_i}(\cdot)$:

$$\underset{\sim}{\Delta'_i}(t) = \begin{cases} \underset{\sim}{c_i} \; \underset{\sim}{\theta}(t_i, t), & t \; \varepsilon \; [t_i, t_n] \\ \\ \underset{\sim}{0}, & \text{otherwise} \end{cases} \tag{37}$$

then the initial conditions for (31) are

$$E[\underset{\sim}{x_0}(t_n)] = \underset{\sim}{0}$$

$$E[\underset{\sim}{x_0}(t_n)\underset{\sim}{x'_0}(t_n)] = \underset{\sim}{M}^{-1}[\underset{\sim}{I} + \int_{t_1}^{t_n} \underset{\sim}{\Delta}(\tau)\underset{\sim}{bb'}\underset{\sim}{\Delta'}(\tau)d\tau]\underset{\sim}{M}^{-T} \tag{38}$$

$$E[\underset{\sim}{x_0}(t_n) \; u(t)] = \begin{cases} \underset{\sim}{M}^{-1} \; \underset{\sim}{\Delta}(t)\underset{\sim}{b}, & t \; \varepsilon \; [t_1, t_n] \\ \\ \underset{\sim}{0}, & \text{otherwise}. \end{cases} \tag{39}$$

For the spline smoothing case, the model for $y_1(\cdot)$ is identical to (31) – (39) except that the state vector, now denoted by $x_1(\cdot)$, has initial covariance

$$E[\underset{\sim}{x_1}(t_n)\underset{\sim}{x'_1}(t_n)] = \underset{\sim}{M}^{-1}[\underset{\sim}{I} + \underset{\sim}{Q_n} + \int_{t_1}^{t_n} \underset{\sim}{\Delta}(\tau)\underset{\sim}{bb'}\underset{\sim}{\Delta'}(\tau)d\tau]\underset{\sim}{M}^{-T} \tag{40}$$

where $\underset{\sim}{Q_n} = \text{diag}(\rho_1, \rho_2, \ldots, \rho_n)$. In addition, the correlation between $y_1(\cdot)$ and $\underset{\sim}{v}$ expressed by (26) can be accounted for by the alternate relations

$$E[\underset{\sim}{x_1}(t_n)\underset{\sim}{v'}] = -\underset{\sim}{M}^{-1}[\underset{\sim}{Q_n}|\underset{\sim}{0}] \tag{41}$$

$$E[u(t)\underset{\sim}{v'}] = \underset{\sim}{0}, \; t \; \varepsilon \; I. \tag{42}$$

Remark 5

The stochastic correspondence (25) for interpolation was proved by Weinert and Kailath (1974) and, in a different way, by Weinert and Sidhu (1978). The correspondence (28) for smoothing was proved by Weinert et al. (in press). Different choices of norm and re-

producing kernel for H and H⁺ lead to different stochastic correspondences. In fact, our early work in this area was motivated by
that of Kimeldorf and Wahba (1970a, 1970b, 1971) who did make a
different choice. Their stochastic correspondences for interpolation
and smoothing involve random processes with unknown mean values, and
do not seem to be as useful for developing recursive spline algorithms. Note that every reproducing kernel is a covariance function
and vice-versa.

Remark 6

With a recursive algorithm, the spline can be computed using
one data point at a time. Recursive algorithms are therefore useful
in real-time applications and those in which data storage is a
limiting factor. A recursive interpolation algorithm based on the
stochastic realization (31) - (39) was developed by Weinert and
Sidhu (1978). See also Sidhu and Weinert (1979). Similar
results for smoothing were established by Weinert et al. (in press).
Proofs of (31) - (42) are also given in those papers. The generalization to vector-valued splines is treated by Sidhu and Weinert (in
press). The uniqueness assumption stated in section 1 guarantees
the invertibility of the matrix $\tilde{M}$. This uniqueness condition is
equivalent to a strong type of observability for the model (31);
see Weinert and Sidhu (1977) for details.

Remark 7

Note that in going from the estimation problem corresponding
to interpolation to that corresponding to smoothing, one does not
simply add noise to the previous observations. The *signal* covariance $K_0(\cdot,\cdot)$ must be changed to $K_1(\cdot,\cdot)$. Then, noise that is
correlated with the signal is added. The change in signal co-
variance is easily accomplished here by changing the covariance
matrix of the initial state of the realization of $K_0(\cdot,\cdot)$. An
attempt to generalize the results of Weinert and Sidhu (1978) to
the smoothing case was made by de Figueiredo and Caprihan (1977).
However, I believe that the stochastic correspondence and resulting
recursive algorithm reported there are incorrect, though a valid
nonrecursive formula for smoothing splines is given.

ERROR BOUNDS FOR SPLINE INTERPOLATION AND SMOOTHING

Returning to the case of general constraint functionals $\{\lambda_j\}_1^N$,
we will first examine what, if anything, can be said about the
interpolation error $|f(t)-s_0(t)|$ when all that is known about $f(\cdot)$
is that f ε H. Much work has been done on upper-bounding this
error (e.g., Golomb and Weinberger, 1959 and Schultz, 1973) and
all these bounds have the general form

$$|f(t)-s_0(t)| \leq dq(f) \tag{43}$$

for a fixed t ε I, where d is a known constant and q(f) is some non-

linear functional of $f(\cdot)$. Obviously, in order to evaluate this
bound, one must have additional *a priori* information about $f(\cdot)$ in
the form of $q(f)$. In most cases, this information is either completely lacking or is too imprecise to give a tight bound. The best
one can do in such cases is bound a normalized version of the interpolation error. In fact, using the Schwartz inequality, the reproducing property (8), and the stochastic correspondence discussed in
section 3, it can be shown that

$$\max_{f \varepsilon H} \frac{|f(t)-s_0(t)|}{||f||} = e_0(t) \tag{44}$$

where $e_0^2(t)$ is the minimum mean-square error in estimating $y_0(t)$
from $\{\lambda_j y_0\}_1^N$; i.e.,

$$e_0^2(t) = E[(y_0(t)-\hat{y}_0(t))^2]. \tag{45}$$

Note that $e_0(t)$ can be computed once L is chosen, without additional
a priori information about $f(\cdot)$. In fact, recursions for computing
$e_0(t)$ can easily be added to the spline interpolation algorithm of
Weinert and Sidhu (1978), thus avoiding explicit use of the covariance $K_0(\cdot,\cdot)$. Note also that $e_0(t)$ is independent of the spline
interpolation data $\{r_j\}_1^N$ and can thus be computed before any measurements are made. As discussed in the next section, this fact permits
the development of optimal sampling schemes. The corresponding result for spline smoothing is

$$\max_{(f,\theta) \varepsilon H^+} \frac{|f(t)-s_1(t)|}{||(f,\theta)||} = e_1(t)$$

where

$$e_1^2(t) = E[(y_1(t)-\hat{y}_1(t))^2].$$

The above remarks on computing $e_0(t)$ apply also to $e_1(t)$.

Remark 8

Larkin (1972) took a different approach to the problem of lack
of *a priori* information. He imposed a Gaussian probability distribution on H, effectively converting every element of H into a
random process. His interpolation bounds are then of the form

$$|f(t)-s_0(t)| \leq m(\alpha) \text{ with probability } \alpha \tag{46}$$

where t is fixed and $m(\alpha)$ depends on α, N, and $\{\lambda_j f\}_1^N$. The problem
with this approach is that there is no natural probability distri-

bution associated with H. A nonlinear constraint is thus being added to the interpolation problem, equivalent to that of guessing some $q(f)$ in bounds of the type (43). As a result, $m(\alpha)$ has no relation to the actual interpolation error. Sacks and Ylvisaker (1970) obtained results analogous to (44) - (45) for the problem of optimal quadrature (approximate integration). Wahba (1969) gives partially computable error bounds for some general interpolation and smoothing problems.

OPTIMAL SAMPLING FOR SPLINE INTERPOLATION

In this section we shall restrict attention to the problem of interpolating from samples $\{f(t_j)\}_1^N$. Sampling functionals are of course a special case of (29). It is known that the interpolating spline $s_0(\cdot)$ satisfies the following conditions:

$$L^a L \, s_0(t) = 0 \quad t_j < t < t_{j+1} \, , \, 1 \le j \le N-1 \tag{47a}$$

$$L \, s_0(t) = 0 \, , \, 0 < t < t_1 \text{ and } t_N < t < T \tag{47b}$$

where L^a is the formal adjoint of L. In addition $s(\cdot)$ has $2n-2$ continuous derivatives on $[0,T]$. Obviously, L determines the functional form of the spline.

The following experimental design problem is of interest here. Suppose we can choose N points in $[0,T]$ at which to sample an unknown function $f \, \varepsilon \, H$, after which the samples will be interpolated with the spline $s_0(\cdot)$. If the spline is to give the best approximation to $f(\cdot)$ over the entire interval $[0,T]$, it makes sense in light of the bound (44) to choose the sampling times $\{t_j\}_1^N$ to minimize $\int_0^T e_0^2(t)dt$. This is a difficult nonlinear optimization problem, and results for an arbitrary L are not yet available, but if $L = d/dt$, in which case the spline is piecewise linear, then it is easy to show that

$$\int_0^T e_0^2(t)dt = 1/2 \, t_1^2 + 1/2 \, (T-t_N)^2 + 1/6 \sum_{j=1}^{N-1} (t_{j+1} - t_j)^2.$$

As a result, the optimal sample times are

$$t_1^* = \frac{T}{3N-1} \, , \, t_{j+1}^* = t_j^* + \frac{3T}{3N-1} \, , \, 1 \le j \le N-1. \tag{48}$$

Remark 9 .

Results on optimal sampling for other types of approximation problems, including approximate integration, can be found in Karlin (1969); Larkin (1970); Sacks and Ylvisaker (1970); Richter-Dyn (1971); Wahba (1971); Hajek and Kimeldorf (1974); Wahba (1974); Karlin et al. (1976); Samaniego (1976); Wahba (1976); Bojanov (1977); Bojanov and Chernogorov (1977). The result (48) appears to be new.

ACKNOWLEDGMENTS

I wish to acknowledge the valuable collaboration of T. Kailath, G. S. Sidhu, and R. H. Byrd at various stages of the research described in this paper. I am also indebted to the referee for valuable suggestions.

REFERENCES

Anselone, P. M. and P. J. Laurent, 1968. A general method for the construction of interpolating or smoothing spline functions. *Num. Math.* 12: 66–82.

Aronzajn, N., 1950. Theory of reproducing kernels. *Trans. Amer. Math. Soc.* 68: 337–404.

Bellman, R. and R. S. Roth, 1971. The use of splines with unknown end points in the identification of systems. *J. Math. Anal. Appl.* 34: 26–33.

Bojanov, B. D., 1977. Characterization and existence of optimal quadrature formulas for a class of differentiable functions. *Soviet Math. Dokl.* 18: 215–218.

Bojanov, B. D. and V. G. Chernogorov, 1977. An optimal interpolation formula. *J. Approx. Theory* 20: 264–274.

Caprihan, A., 1975. Finite-duration digital filter design by use of cubic splines. *IEEE Trans. Circuits and Systems* 22: 204–207.

deBoor, C. and R. Lynch, 1966. On splines and their minimum properties. *J. Math. Mech.* 15: 953–969.

de Figueiredo, R. J. P. and A. Caprihan, 1977. An algorithm for the construction of the generalized smoothing spline with application to system identification. *Proc. Conf. Inf. Sci. Systems*, pp. 494–500. Barton Press, Baltimore.

deFigueiredo, R. J. P. and A. N. Netravali, 1976. On a class of minimum energy controls related to spline functions. *IEEE Trans. Auto. Control* 21: 725–727.

de Montricher, G. F., R. A. Tapia, and J. R. Thompson, 1975. Nonparametric maximum likelihood estimation of probability densities by penalty function methods. *Ann. Statis.* 3: 1329–1345.

Golomb, M. and H. Weinberger, 1959. Optimal approximation and error bounds. In: *On Numerical Approximation*, (R. Langer, ed.), pp. 117–190. Univ. Wisconsin Press, Madison.

Greville, T. N. E., ed., 1969. *Theory and Applications of Spline Functions.* Academic Press, New York. 212 pp.

Hajek, J. and G. Kimeldorf, 1974. Regression designs in autoregressive stochastic processes. *Ann. Statis.* 2: 520–527.

Jerome, J. and L. L. Schumaker, 1969. On Lg-splines. *J. Approx. Theory* 2: 29–49.

Karlin, S., 1969. The fundamental theorem of algebra for monosplines satisfying certain boundary conditions and applications to optimal quadrature formulas. In: *Approximations with Special Emphasis on Spline Functions,* (I. J. Schoenberg, ed.), pp. 467–484. Academic Press, New York.

Karlin, S., C. A. Micchelli, A. Pinkus, and I. J. Schoenberg, 1976.
    Studies in Spline Functions and Approximation Theory. Academic
    Press, New York. 500 pp.
Kimeldorf, G. and G. Wahba, 1970a. A correspondence between Bayesian
    estimation of stochastic processes and smoothing by splines.
    Ann. Math. Statist. 41: 495–502.
Kimeldorf, G. and G. Wahba, 1970b. Spline functions and stochastic
    processes. *Sankhya* 132: 173–180.
Kimeldorf, G. and G. Wahba, 1971. Some results of Tchebycheffian
    spline functions. *J. Math. Anal. Appl*. 33: 82–95.
Larkin, F. M., 1970. Optimal approximation in Hilbert spaces with
    reproducing kernel functions. *Math. Comput*. 24: 911–921.
Larkin, F. M., 1972. Gaussian measure in Hilbert space and applica-
    tions in numerical analysis. *Rocky Mt. J. Math*. 2: 379–421.
Larsen, R. D., E. F. Crawford, and P. W. Smith, 1977. Reduced
    spline representations for EEG signals. *Proc. IEEE*. 65: 804–
    807.
Lyche, T. and L. L. Schumaker, 1973. Computation of smoothing and
    interpolating natural splines via local bases. *SIAM J. Numer.
    Anal*. 10: 1027–1038.
Parzen, E., 1961. An approach to time series analysis. *Ann. Math.
    Statist*. 32: 951–989.
Reinsch, C., 1971. Smoothing by spline functions, II. *Num. Math*.
    16: 451–454.
Richter–Dyn, N., 1971. Minimal interpolation and approximation in
    Hilbert spaces. *SIAM J. Numer. Anal*. 8: 583–597.
Sacks, J. and D. Ylvisaker, 1970. Statistical designs and integral
    approximations. In: *Proc. 12th Biennial Sem. Canadian Math.
    Congress*, (R. Pyke, ed.), pp. 115–136. Canadian Math. Congress,
    Montreal.
Samaniego, F. J., 1976. The optimal sampling design for estimating
    the integral of a process with stationary independent incre-
    ments. *IEEE Trans. Inf. Theory* 22: 375–376.
Schoenberg, I. J., ed., 1969. *Approximations with Special Emphasis
    on Spline Functions*. Academic Press, New York. 488 pp.
Schultz, M. H., 1973. *Spline Analysis*. Prentice-Hall, Inc., Engle-
    wood Cliffs, New Jersey.
Sidhu, G. S. and H. L. Weinert, 1979. Dynamical recursive algo-
    rithms for Lg-spline interpolation of EHB data. *Applied Math.
    Computation* 5: 157–185.
Sidhu, G. S. and H. L. Weinert, in press. Vector-valued Lg-splines.
    J. Math. Anal. Appl.
Wahba, G., 1969. On the numerical solution of Fredholm integral
    equations of the first kind. *Tech. Report 217, Dept. of Sta-
    tistics*. Univ. of Wisconsin, Madison. 60 pp.
Wahba, G., 1971. On the regression design problem of Sacks and
    Ylvisaker. *Ann. Math. Statist*. 42: 1035–1053.
Wahba, G., 1974. Regression design for some equivalence classes
    of kernels. *Ann. Statist*. 2: 925–934.
Wahba, G., 1976. On the optimal choice of nodes in the collocation-
    projection method for solving linear operator equations. *J.
    Approx. Theory* 16: 175–186.
Wahba, G. and S. Wold, 1975. A completely automatic French curve;

Fitting spline functions by cross validation. *Commun. Statist.* 4: 1-17.

Weinert, H. L., R. H. Byrd, and G. S. Sidhu, in press. A stochastic framework for recursive computation of spline functions - Part II. Smoothing splines. *J. Optimization Theory Appl.* 30.

Weinert, H. L. and T. Kailath, 1974. Stochastic interpretations and recursive algorithms for spline functions. *Ann. Statist.* 2: 787-794.

Weinert, H. L. and T. Kailath, 1976. A spline-theoretic approach to minimum energy control. *IEEE Trans. Auto. Control* 21: 391-393.

Weinert, H. L. and G. S. Sidhu, 1977. On uniqueness conditions for optimal curve fitting. *J. Optimization Theory Appl.* 23: 211-216.

Weinert, H. L. and G. S. Sidhu, 1978. A stochastic framework for recursive computation of spline functions - Part I. Interpolating splines. *IEEE Trans. Inf. Theory* 24: 45-50.

Yao, K., 1967. Applications of reproducing kernel Hilbert spaces - bandlimited signal models. *Inf. and Control* 11: 429-444.

Linear Prediction for Signal Modelling and Spectrum Analysis

Bradley W. Dickinson

ABSTRACT: This paper presents a description of
linear prediction and two of its main uses in
signal processing applications: spectrum analysis and signal modelling. Briefly, linear
prediction denotes the process of fitting a
linear difference equation model to an observed
time series. The model may be used to give an
estimate of the signal power spectrum or to aid
in digital encoding of the signal for data compression or communication purposes. The presentation is discursive with an emphasis on
general ideas rather than particular details.
A number of techniques used for linear prediction are described, and finally some areas of
current research are discussed. The main goal
of the paper is to present linear prediction in
such a way that nonspecialists in signal processing can appreciate its principles and principal uses, with enough references to serve as
an introduction to the literature.

63

INTRODUCTION

Ever-increasing amounts of digital computing power available
for signal processing applications have greatly altered the nature
of research efforts on signal analysis. A major shift has been the
change in emphasis from continuous-time signals to discrete-time
signals or *time series*. However, there remain many common ties between these two topics, including the use of linear dynamical systems for modelling signals. The aim in this survey paper is to
give a fairly detailed overview of one aspect of time series analysis that has attracted a considerable amount of theoretical and
applied research interest, called *linear prediction* (LP).

Fortunately, this task is made a bit easier by the recent
survey of Makhoul (1975). I hope to give an exposition of the
many facets of LP, leaving details to the references, with particular emphasis on work that is not discussed by Makhoul, either
because it is more recent work or because my own perspective is
inevitably different from his.

A main goal of the paper will be to provide enough of the
basic ideas of LP so that several general applications areas can
be discussed and so that the more active areas of research can be
explained. It is my hope that such a general overview, rather
than a wealth of details, will be most valuable to the scientist
who may have a particular signal processing application in mind.
If linear prediction should turn out to be a potentially useful
tool for such a reader, there will be no problem in finding details in the literature and applying the "fine tuning" that is
appropriate for the application. Of course, this last step is
not necessarily trivial and may lead to new research on linear
prediction itself.

In view of the above goal, the presentation in following
sections will not be totally self-contained. A background in
elementary signals and systems will be presumed, and no extensive mathematical derivations will be included. An ample, but
not exhaustive, reference list will be included. After starting
with some basics of signal analysis and linear prediction, the
main body of the paper will present and compare a variety of
general areas to which linear prediction has been applied. A
discussion and comparison of linear prediction techniques follows.
A final section will cover current research efforts to extend the
range of applicability of LP and some of the limitations and problems encountered.

SIGNALS, SAMPLING, SPECTRA, AND SYSTEMS

The signals of interest in this paper are of the form $\underline{s} = \{i;$
$0 \leq i \leq N\}$ or $I = \{i; -\infty < i < \infty\}$. We think of $\underline{s}$ as a sequence of *measure-*
ments, that is a sequence of sample values of some function $s(t)$:
$s_i = s(t_i)$, $i \epsilon I$. In this paper $s(t)$ is regarded as a function of
time and $\underline{s}$ is called a *time series*; the sampling instants will be
integer multiples of a fixed sampling interval T_s : $t_i = iT_s$, $i \epsilon I$.

However, the measurements could just as well be functions of position in some coordinate system, of depth, or of any other independent variable.

It is crucial to understand that not every function s(t) can
be reconstructed from its samples taken T_s seconds apart. By
thinking in terms of simple sine wave functions (Steiglitz, 1974)
we are led to an elementary version of the *sampling theorem*:

> Let s(t) be a function composed of a sum of sinu-
> soidal signals whose highest frequency is W Hz (i.e.,
> whose shortest period is 1/W sec). Then s(t) is unique-
> ly determined by its sample values, $\{s(iT_s), -\infty < i$
> $< \infty)$, if and only if $T_s < \frac{1}{2W}$.

Using more sophisticated ideas from Fourier analysis and probability theory, the sampling theorem can be generalized to cover
all functions s(t) of practical importance. Intuitively, the
theorem states that the uniformly spaced sampling instants must be
closely enough spaced so that the function s(t) changes slowly between sampling instants. When the sampling rate is lower than
twice the highest frequency component in the function being sampled,
the sample values obtained may give no information at all about the
function. For example the function s(t) = A cos (4πt/5) + B cos
(6πt/5) contains frequencies 2/5 Hz and 3/5 Hz. Suppose the sampling
rate is 1 Hz (T_s = 1 sec). The sample values are the same as those
obtained for the function (A+B) cos (4πt/5). When B = -A, all the
samples are zero! No matter what A and B are, it is impossible to
determine anything but A+B from the samples. In general, the contribution of any frequency component above one-half the sampling
frequency to the sample values can be duplicated by a lower frequency term called its alias. In practice, as long as a sufficiently
large fraction of the signal power lies at frequencies below half of
the sampling frequency, little distortion is introduced by aliasing
and we may confidently deal only with the uniformly spaced samples.

It is customary to scale the time axis into units of T_s, the
sampling period. Thus a time series will be regarded as samples of
a signal whose highest frequency is ½ Hz. A fairly general class of
discrete signals can be analyzed by the z-transform (Steiglitz, 1974)
defined for signal s as

$$S(z) = \sum_{i \epsilon I} s_i z^{-i} \tag{1}$$

The values of S(z) at the points $z = e^{j\omega}$ ($j = \sqrt{-1}$) then represent
the frequency content, in a sense that can be made precise using
the inverse z-transform, of the signal s.

Processing of signals by linear systems extends naturally to
the discrete-time case. Here an operational definition of a linear
time-invariant discrete-time system will suffice: it is regarded

as an operator mapping an input series $\underline{u}$ into an output series $\underline{y}$ by the z-transform equation

$$Y(z) = H(z)U(z) \tag{2}$$

where $H(z)$ is the *transfer function* of the system. The linear systems treated here all have the form

$$H(z) = \sum_{k=0}^{q} b_k z^{-k} / (1 + \sum_{k=1}^{p} a_k z^{-k}) \tag{3}$$

and correspond to linear difference equations of the form

$$y_n = \sum_{k=0}^{q} b_k u_{n-k} - \sum_{k=1}^{p} a_k y_{n-k} \tag{4}$$

Following the statistical literature, such systems are called *autoregressive-moving average* (ARMA) models. Two special cases are worth noting: the moving average (MA) case, in which $p = 0$ so the second sum in (4) is missing; the autoregressive (AR) case, in which $q = 0$ and $b_0 \neq 0$.

Transfer functions are used to describe conveniently the input-output properties of systems. For example, a unit amplitude (sampled) sinusoidal input at frequency ω is transformed by linear system $H(z)$ into a sinusoidal output of the same frequency with amplitude $|H(e^{j\omega})|$ and relative phase arg $(H(e^{j\omega}))$.

Random signals will also be used as inputs to linear systems and the autocorrelation function, defined by

$$R_x(k) = E(x_i x_{i+k}) \qquad -\infty < k < \infty \tag{5}$$

where E denotes statistical expectation, will be a useful partial characterization of the random signal $\underline{x}$. Let $G_x(z)$ denote the z-transform of the series $\{R_x(k)\}$. Then $G_x(e^{j\omega}) = Sx(\omega) = S_x(\omega)$ is the power spectral density function of $\underline{x}$, that is the amount of power in the signal $\underline{x}$ per unit frequency, as a function of frequency. The power spectral density of the output of a linear system $H(z)$ driven by input $\underline{u}$ with power spectral density $S_u(\omega)$ is

$$S_y(\omega) = |H(e^{j\omega})|^2 S_u(\omega) \tag{6}$$

The simple form of $H(z)$ for the ARMA models makes the evaluation of such quantities fairly easy.

Having recalled these basic ideas, we turn in the next section to the subject of interest: linear prediction.

LINEAR PREDICTION AND ITS USES

From equation (4) we note a fundamental property of ARMA models: the output at time t, y_t, can be computed (*predicted!*) as a *linear*

combination of previous outputs and inputs. *Linear prediction* (LP)
is thus naturally defined as the fitting of an ARMA model to an observed time series. We draw a distinction between LP and *system
identification* (SI), reserving the latter term for the fitting of
an ARMA model when both the input and output time series are observed. SI has some clear similarities to LP; its main application
area, however, has been in identifying dynamical models from operating records for the purpose of developing controllers or regulators
that appropriately modify the behavior of the overall system-con-
troller configurations. Applications to complex technological systems have been a primary driving force behind research in this field.
Some recent summaries are available in Eykhoff (1974), Mehra and
Lainiotis (1976), and a special issue of the IEEE Transactions on
Automatic Control, edited by Kailath (1974b).

The classic work on ARMA modelling in statistical time series
analysis is that of Box and Jenkins (1976). However, a statistical
approach to linear prediction is not always satisfactory because the
assumption that the observed time series is a sample from a stochastic process, an ensemble of time series with statistical regularity,
is often inappropriate. Rather than characterizing the observed data
as the output of a linear system excited by a random input, it is
often more realistic to make no assumptions at all about the input or
to view the input as a deterministic time series. Often an admissible
class of deterministic input time series can be hypothesized because
a partial characterization of the input series can be inferred from
physical considerations.

A good example of the latter is the modelling of speech sounds
by linear prediction; this is the subject of a recent book by Markel
and Gray, 1976) and is an area of active research for which the main
archival source is the IEEE Transactions on Acoustics, Speech, and
Signal Processing. From physiology, it is known that voiced speech
sounds, like long vowel sounds, are generated by a glottal pulse
train exciting the vocal tract which serves as an acoustical filter
(Flanagan, 1972). Thus pulse-train inputs are the most appropriate
class of concern in modelling voiced speech signals. A stochastic
model is more appropriate, however, for fricative sounds, like the
s in glass, where the turbulent excitation of the vocal tract closely
resembles a white noise sequence.

It is customary to preserve much of the terminology of the statistical approach to LP; even more importantly, the statistical approach suggests techniques for model fitting that may be used even
when a stochastic model is not assumed. In what follows, this use
of mixed metaphor will be continued. For example, properties of the
autocorrelation function of a stochastic process will be considered
as a basis for a model fitting procedure. In practice, whether the
observed time series is random or not makes little difference because estimates of the autocorrelation function based on observed
data are used. It remains to examine the quality of the LP models
in a particular application to judge the effectiveness of any model
fitting method.

The first application of linear prediction to be considered is
spectrum estimation -- or, equivalently, autocorrelation function
estimation, since the two are connected by a transform relationship.

Spectrum estimation by LP is based on equation (6). The spectrum of
the observed signal can be computed from the spectrum of the input
signal and the magnitude-squared of the fitted model transfer function evaluated on the unit circle in the complex plane, $z = e^{j\omega}$.

Because most physical signals are characterized by a few peaks
in their (short-term) power spectrum, it is quite appropriate to
consider modelling them with AR or ARMA models, assuming a flat
power spectrum for the input (e.g., white noise). This is because
these models have poles, values of the complex variable z where the
transfer function H(z) becomes infinite. If these poles are located
close to the unit circle, $z = e^{j\omega}$, the magnitude of $H(e^{j\omega})$ will have
sharp peaks. Pure MA models, which lack poles, can only approximate
such behavior using a very high model order. The AR model is determined, up to scale, by the pole locations, since the poles completely
determine the denominator polynomial of H(z) and the numerator is a
constant. ARMA models have more flexibility for matching spectral
shapes. They possess zeros, values of z for which H(z) vanishes, and
in combination with the poles, the magnitude of H(z) in this case can
demonstrate deep nulls and a wider variety of peak shapes. Such
features are demonstrated by AR models only of very high order.

There is, unfortunately, a price to be paid for the added flexibility of the ARMA models. The techniques for fitting these models
generally lead to sets of nonlinear simultaneous equations, so that
iterative or approximate computational methods must be used. In
contrast, a wide range of AR fitting methods that are linear or stage-
wise linear are available. Fortunately many spectra are adequately
characterized using AR models. However, given the increasing amount
of computer power available to the signal processor, there are ARMA
fitting methods available that would probably be more widely used if
they could be shown to be advantageous in more applications.

First, some general references on spectrum estimation will be
given. The classic reference on the topic is the book by Blackman
and Tukey (1959). A treatment of ARMA methods and classical Fourier
methods is given in Jenkins and Watts (1968). Two fine surveys of
the more recent work on AR spectrum estimation are those by Ulrych
and Bishop (1975) and Lacoss (1977). Many of the other entries in
our bibliography contain material pertinent to spectrum estimation
applications of LP.

Classical spectrum estimation techniques are based on two
roughly equivalent approaches. Either a sample autocorrelation
function is tapered and Fourier transformed or a smoothed periodo-
gram of the data is computed. There is an inherent tradeoff between
variability of the estimate, which is related to inaccurate estimates of the longer lags of the autocorrelation estimate caused by
the limited data, and high resolution, which requires that the
longer lags be used. Classical methods assume that the lags beyond
those actually used are zero, and this unrealistic assumption can
be viewed as the source of decreased resolving power.

Spectrum estimation techniques based on LP can be thought of
as a way of avoiding this unnatural assumption, resulting in improved resolution. For example, an ARMA model can be chosen to
fit, exactly or approximately, the estimated autocorrelation function lags; the spectrum of the model then specifies, indirectly as
a result of the transform relationship, a nonzero extension of the

autocorrelation function that is usually more reasonable than an all-zero extension. There are a number of criteria and associated techniques for fitting the model, including *least squares, maximum likelihood* (a misnomer that has become entrenched by common usage) and the so-called *maximum-entropy methods*.

This last method, due to Burg (1975 and his earlier work), takes the following approach. If the initial $p + 1$ lags of the autocorrelation function, $\{R_i, 0 \leq i \leq p\}$, are known exactly, then maximization of the functional

$$\underline{H} = \int_{-\pi}^{\pi} \hat{S}_y(\omega) d\omega \qquad , \tag{7}$$

which is the entropy of a Gaussian random process with power spectrum $S_y(\omega)$, subject to the transform constraints

$$R_i = \int_{-\pi}^{\pi} \hat{S}_y(\omega) e^{j\omega i} \, d\omega/2\pi \tag{8}$$

leads to an AR model fitting problem. The fitted model, say

$$y_t + \hat{a}_1 y_{t-1} + \ldots + \hat{a}_p y_{t-p} = u_t \qquad , \tag{9}$$

is determined by the linear equations

$$\sum_{k=1}^{p} R_{|i-k|} \, \hat{a}_k = -R_i \qquad , \qquad 1 \leq i \leq p \qquad . \tag{10}$$

The correlation function extension can be found from

$$R_i = \sum_{k=1}^{p} \hat{a}_k R_{k-i} \qquad , \qquad i > p \tag{11}$$

but of course the spectrum estimate is more easily computed using the AR model (ignoring scaling)

$$\hat{S}_y(\omega) = |H(e^{j\omega})|^2 \tag{12}$$

where

$$H(z) = (1 + \sum_{k=1}^{p} \hat{a}_k z^{-k})^{-1} \qquad . \tag{13}$$

In practice, of course, the "true" autocorrelation lags are not known (even if the Gaussian assumption does hold!) and must be estimated. Estimates may be substituted for the $\{R_i\}$; an extension that accounts for the uncertainty of these estimates has been suggested by Newman (1977a). Alternately, equations (13) and (14) may be used to obtain a spectrum estimate from other AR estimates that do not require autocorrelation function estimates explicitly.

In this context, it is important to require that the estimated

AR model have all of its poles located within the unit circle of the
complex plane so that a true autocorrelation function estimate could,
if necessary, be obtained. (The autocorrelation function must be
positive definite). While this requirement imposes rather complicated constraints in the "reference frame" of the AR coefficients,
an alternative filter structure, known as a *lattice filter*, provides
a convenient way to incorporate it. The use of lattice filters was
pioneered by Burg (1975) and by Itakura and Saito (1971). The term
maximum entropy spectrum estimate usually refers to the particular
estimate of Burg (1975). More recently, Makhoul (1977) and Dickinson
(1978b) have developed LP methods based on the lattice filter which
are computationally efficient and which lead to admissible autocorrelation function estimates, hence also to reasonable spectrum estimates. A short description of lattice filters is given in the
appendix.

AR spectrum estimates of modest order can be used in combination
with classical methods to advantage in some situations. An AR model
can be fit to greatly reduce the dynamic range of the sample spectrum.
Having fit the model, the input signal is calculated by "inverse filtering" using equation (9) and classical spectrum estimates for the
input signal can be applied more successfully because of the reduced
dynamic range. An application of this approach is discussed by
Thomson (1977). Other uses of LP for spectrum analysis include the
classic work of Yule (1927) on sunspot periodicities, extensive work
on seismic petroleum prospecting techniques (Claerbout, 1976), mechanical vibration analysis (Gersch et al., 1973), EEG analysis (Gersch,
1970) doppler spectrum estimation for distributed radar targets
(Cooper and McGillem, 1978), and the speech analysis applications
mentioned briefly above.

Other application, where line spectra are involved, has been
treated with LP techniques. We prefer to use the term *frequency
estimation* rather than spectrum estimation to describe problems involving one or more sine waves in additive noise. A good survey of
recent work on this problem is given by Frost (1977). Such spectra
can be viewed as the narrow bandwidth limit of a form of ARMA spectra. For high signal to noise ratios, the maximum entropy methods
give good resolution and frequency estimates (Lacoss, 1971; Marple,
1978). When the noise grows larger, a more general ARMA model
should be used. A reasonable linear approximation was proposed by
Pisarenko (1973). An estimate of the noise power is subtracted from
the autocorrelation function and then an AR model is fit; the procedure may be iterated for improvement. Applications of frequency estimation have been reported in a number of geophysical and seismolog-
ical areas (Chen and Stegun, 1974; Ulrych and Bishop, 1975; Landers
and Lacoss, 1977) and in radar imaging (Bowling, 1977).

The second major use of linear prediction is *signal modelling*.
The areas of application of signal modelling include waveform coding,
data compression, and signal synthesis. Much of the work in this
area has been motivated by audio signal processing needs, especially
the desire to develop efficient encoding of speech signals for transmission on digital communications links.

A key requirement for efficiency in the coding context is the
removal of redundancy which, in speech, is characterized by the signal

correlation or short term power spectrum. It is natural to try to de-correlate or "whiten" a signal by removing the contribution that can be predicted from the past; hence linear prediction using an AR model! Let the estimate of the signal y_t, based on p previous observations, be denoted $\hat{y}_t$, where

$$\hat{y}_t = - \sum_{k=1}^{p} a_k y_{t-k} \tag{14}$$

A natural criterion for choosing the parameters is least squares, suggested for this application by Atal and Hanauer (1971); the parameter estimates are chosen by minimizing the cost function

$$J = \sum_t (y_t - \hat{y}_t)^2 \tag{15}$$

leading to the normal equations

$$\sum_{j=1}^{p} r_{ij} a_j = -r_{i0} \quad , \qquad 1 \le i \le p \tag{16}$$

where

$$r_{ij} = \sum_t y_{t-i} y_{t-j} \quad . \tag{17}$$

Noting the similarity of (16) and (10), or viewing (17) as an auto-correlation estimate, has led to using the apparently simpler form (10) to approximate (16), (for example, Markel and Gray, 1976). The estimates defined by (16) may lead to poles outside the unit circle, so some care must be taken when least squares is applied to spectrum estimation.

Given any estimate of the AR model parameters, the input or residual series $\{u_t\}$ can be computed by inverse filtering (equation (9)). For any reasonable observed signal, the residuals have less inherent information and can be more efficiently encoded in digital form using any one of a number of waveform coding techniques. The original waveform can be recovered from the encoded version and, using the AR parameters in (9), the original observations can be computed. Such a signal synthesis system for speech signals is called a residual-excited vocoder.

For speech signals, more efficiency yet is obtained by ignoring details of the residual series and characterizing it as a periodic pulse train, in the case of voiced sounds, or as a noise input, for unvoiced sounds. Only the input power and the period of the pulse train is needed by the synthesizer in order to create a signal with the same audible characteristics as the original speech. Details of such analysis-synthesis systems are described in Markel and Gray (1976).

In order to track the time-varying character of speech, such systems need an adaptive component. The simplest approach is to segment the speech into 20-30 ms intervals over which the analysis

is performed. Viswanathan et al. (1978) discuss this approach using
lattice filter methods. Morf et al. (1977a) have proposed the lattice filter be continuously adapted, eliminating segmentation.
Turner (1978) proposed a lattice-based method for variable-length
segmentation of the speech based on phonetic similarity. The ability
to track the changing spectral characteristics of a signal is crucial,
and work continues on this problem.

In principle, the LP techniques for signal modelling or spectrum
analysis can be extended to give ARMA models. The availability of
increased computer power makes the solution of the inherently non-
linear estimation problem tractable in many cases. However problems
of sensitivity and of initial conditions for signal synthesis arise
and must be carefully considered.

As an example of the ARMA modelling problem, least squares will
be discussed. This method is equivalent, for long data records, to
the maximum likelihood method proposed by statisticians (Box and
Jenkins, 1976). In a somewhat overlooked paper on system identification (Steiglitz and McBride, 1965) an iterative method of fitting
the least squares ARMA model to input-output series was described.
For signal modelling, somehow an appropriate input-output series
must be determined. For noise excited systems, a high order AR
model is fitted and the resulting residuals are used as the input
series to accompany the observed output series. This technique
(Mayne, 1977) has demonstrable convergence properties under suitable
assumptions.

When the "true" input sequence is well-modelled as a pulse
train, as in voiced speech, an alternate method can be applied. A
nonparametric estimate of the system impulse response is obtained by
homomorphic processing or *cepstral analysis* (Kopec et al., 1977).
Then the iterative ARMA fitting technique may be applied to the
impulse-impulse response input-output pair (Steiglitz, 1977). In
order to reduce the sensitivity of the impulse response estimate to
the phase of the pulse train, an effect caused by the finite observation record, prefiltering of the speech with an all-pass digital
filter has been proposed (Steiglitz and Dickinson, 1978).

A real understanding of the ARMA fitting problem remains to be
found through further research. One conclusion is fairly certain:
it is essential that the AR and MA coefficients be fit simultaneously.
Methods based on fitting the AR parameters first (Gersch and Sharp,
1973; Kopec et al., 1977) do not produce spectral features such as
broad peaks with sharp skirts that are observed in real physical
signals such as speech.

Naturally, the signal modelling or spectrum analysis functions
of LP are often used as "front-end" processing in more complicated
information systems such as speech understanding systems and waveform classification or pattern recognition systems. Two such examples are the speech recognition system of Gupta et al. (1978) and
the EEG classification work of Gersch and Yonemoto (1977). Many
future applications of LP can be envisioned including, hopefully,
some in marine biology.

RESEARCH PROBLEMS IN LINEAR PREDICTION

Having completed an overall survey of linear prediction and
its uses, a few comments on unfinished business from the author's
perspective are in order.

There has been a considerable amount of recent research on
utilizing the structure of the linear equations that must be solved
to obtain AR estimates. Morf et al. (1977a) demonstrated that the
least squares estimates in equation (17) may be computed using
$O(p^2)$ arithmetic operations, the number being only about three times
that required when the matrix is Toeplitz as in equation (10).
Friedlander et al. (1979) showed that both of these matrices may be
expressed as a sum of products of lower and upper triangular Toeplitz matrices, and they developed a general inversion formula for
this class. A small generalization (Dickinson, 1978a) allows for
an approximate maximum likelihood estimate to be similarly obtained.
These results show that the computational burden of fitting high
order AR models by a number of different methods is not excessive.

Perhaps more important is the fact that the $O(p^2)$ methods may
be arranged to give intermediate results which lead to a *minimum
phase* estimate, one whose poles all lie within the unit circle
(Dickinson, 1977). Further work on this so-called *residual energy
ratios estimate* has led to a computational algorithm based on
Cholesky factorization which is more efficient for low order fits,
p<20, and which has the excellent numerical stability associated
with Cholesky factorization (Dickinson, 1978b). In other work
Makhoul (1977) has shown that the computational burden of the lattice methods can be reduced to roughly the same level as solving
linear equations; recall that these estimates are also minimum
phase. However, no comparable work for the ARMA fitting methods
described here has been reported. Several workers in the field are
approaching such problems from a number of viewpoints. The answer
will be closely related to the question of extending lattice methods
to give ARMA models. This is a major outstanding problem of linear
prediction.

Another problem receiving much recent attention is that of model order determination. Several statistical approaches to this
question have been devised; see Lacoss (1977) for some discussion.
Even in the noise excited case, there is no clear consensus, only
a number of reasonable approaches. Extensions to ARMA fitting will
be the aim of further research. For signal modelling applications,
where accuracy of parameters takes a back seat to fidelity of reproduction, often measured in a subjective way, modest overfitting
is not a major problem, at least in the AR case. The ARMA fitting
problem is much more sensitive, due to its nonlinear nature, and the
analyst must use all information available about the signal in order
to make a sensible choice; automatic techniques need to be more fully
developed.

The extension of LP methods to multivariate (also called multichannel) signals has been vigorously pursued because of major applications in seismic and biomedical signal processing applications.
All of the major approaches in the univariate case have now successfully translated to multivariate versions. Partial summaries are

given by Lacoss (1977) and by Jones (1978). These predated the crucial work of Morf et al. (1978a, b) where the properly normalized
multivariate lattice filter structure was first introduced. Before
this work, several authors had proposed rather ad hoc extensions of
the univariate lattice filter and of related LP methods but were not
always able to prove or disprove that the resulting estimate was
minimum phase. The work of Morf et al. (1978a, b) clarified this
situation and developed one maximum entropy lattice-based estimate.
Recently (Dickinson, 1979) the residual energy ratios estimate has
been extended to the multivariate case based on this work. Of
course, model order selection problems are worse than in the univariate case (Jones, 1978) and ARMA extensions are hardly developed.

There is also ongoing research aimed at extending LP methods,
especially in the spectrum analysis context, to deal with signals
defined on a two dimensional grid rather than on a one dimensional
time or space axis. Applications include picture processing and
array processing. While the work on this problem is beyond the
scope of the current paper, it is interesting to note how difficult
the problem becomes when the signal points are not associated by a
linear ordering. Interested readers may refer to the work of Woods
(1976) and Newman (1977b) for some approaches to two dimensional
spectrum estimation.

More relevant to the work discussed here is an interest in
robust and nonparametric statistical methods that may be applied
to LP and used for signal modelling or spectrum analysis. In the
latter case, Thomson (1977) has already demonstrated the utility of
robust methods in practice. More work can be expected in the future
because the world is far from linear and Gaussian.

In conclusion, we expect the interest in linear prediction to
remain high in the coming few years as more applications become recognized. The current research in methods, techniques and theory
is motivated by practical problems encountered in real world signal
processing problems. These range from a desire for real-time processing capability to the need to handle situations where the least
squares fit AR models are inadequate. Researchers in the field will
especially welcome suggestions from people who try LP and find it
almost the tool they expected!

ACKNOWLEDGMENTS

Besides the survey of Makhoul (1975), the work of a number of
people has greatly influenced this presentation. The ideas of Ken
Steiglitz and his friendship have been invaluable. Martin Morf,
whom I have had the pleasure of working with from time to time, has
been responsible for a wealth of new ideas. For my background in
systems and estimation, I owe Thomas Kailath many thanks; his survey of linear estimation theory (Kailath, 1974a) has been a continual source of inspiration and further reference. Finally I want
to thank John Turner for providing me with numerous insights through
conversations and his thesis (Turner, 1978).

The financial support of the National Science Foundation under
grant ENG77-28523 during the preparation of this paper is gratefully
acknowledged.

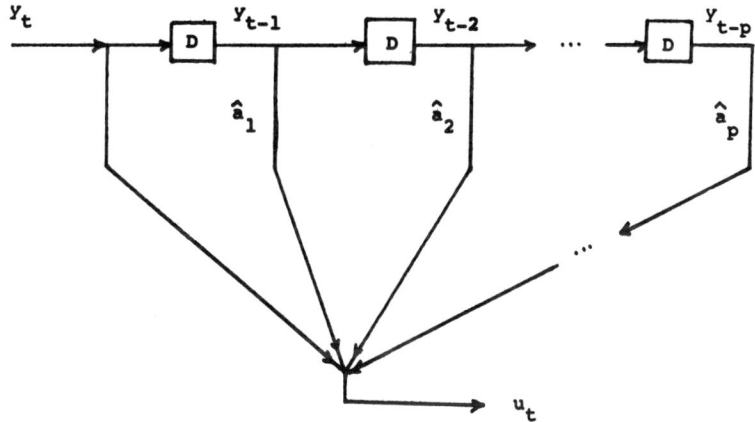

FIGURE 1: A direct form filter.

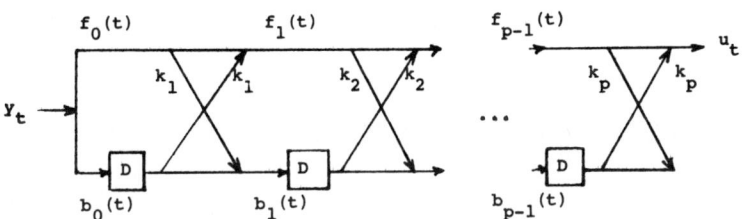

FIGURE 2: An equivalent lattice filter.

APPENDIX: LATTICE FILTERS

As noted in the text, equation (9) can be used to compute the residual series $\{u_t\}$ given the observations $\{y_t\}$ and the AR parameters. A block diagram for this computation is shown in Figure 1. We use the symbol D to denote a unit delay element. By introducing some intermediate variables into the computation, the structure in Figure 1 may be transformed to the alternate lattice structure shown in Figure 2. The details of this transformation will be described shortly; at this point the main advantage of the lattice structure will be noted.

The condition that the polynomial $a(z) = 1 + \hat{a}_1 z^{-1} + \ldots + \hat{a}_p z^{-p}$ have all its roots inside the unit circle of the complex plane is equivalent to the condition that $|k_i| < 1$, $1 \leq i \leq p$, where the k_i coefficients appear in the lattice structure (Fig. 2) derived from the so-called direct form (Fig. 1) as shown in what follows. This condition, essential for the minimum phase property of the AR estimate which is in turn essential for the fitting of an admissible autocorrelation function, is thus easily checked once the transformation to lattice form is made. On the other hand, we may as well base our

estimation method on the lattice form where the minimum phase property is readily controlled, obtaining the equivalent direct form
only if it is really needed. This approach was pioneered by Burg
(1975, summarizing earlier work) and Itakura and Saito (1971). More
recent expositions appear in Markel and Gray (1976) and Makhoul
(1977).
        To obtain the lattice structure, we define two sets of signals,
$\{f_j(t),\ 0 \leq j \leq p\}$ and $\{b_j(t),\ 0 \leq j \leq p\}$, taking $f_0(t) = b_0(t) = y_t$, the
observed signal. Then we define recursively

$$f_{i+1}(t) = f_i(t) + k_{i+1}b_i(t-1)$$

$$\qquad\qquad\qquad\qquad\qquad\qquad\qquad\qquad (A1)$$

$$b_{i+1}(t) = b_i(t-1) + k_{i+1}\, f_i(t)$$

We find by induction that

$$f_i(t) = y_t + \sum_{j=1}^{i} c_{ij}y_{t-j} \qquad\qquad\qquad (A2)$$

where the coefficients are given by

$$c_{ij} = c_{i-1,j} + k_i c_{i-1,i-j}\ , \qquad 1 \leq j \leq i-1$$

$$\qquad\qquad\qquad\qquad\qquad\qquad\qquad 1 \leq i \leq p \quad . \quad (A3)$$

$$c_{ii} = k_i$$

In order to identify this result with equation (9) we may take
$f_p(t) = u_t$ and $c_{pj} = \hat{a}_j$, $1 \leq j \leq p$. Now the recursions in (A3) may
be solved to obtain expressions for the desired k_i coefficients.
We have

$$k_i = c_{ii}$$

$$\qquad\qquad\qquad\qquad\qquad\qquad\qquad p \leq i \leq 1 \quad . \qquad (A4)$$

$$c_{i-1,j} = (c_{i,j} - k_i c_{i-j})/(1-k_i^2), \qquad 1 \leq j \leq i-1$$

The lattice structure follows (of course!) from (A1).
        In order to compute $\{y_t\}$ from $\{u_t\}$, it is easy to rearrange the
calculations in (A1) to derive the lattice filter shown in Figure 3.
This figure also shows the intuitive reason for the choice of notation. The signals $f_i(t)$ and $b_i(t)$ represent forwards and backwards
moving signal (waves!) at the i-th layer of the system. The equation
relating behavior at the boundary of two layers (A1) shows that the
coefficient k_{i+1} plays the role of a *reflection coefficient*. The
minimum phase condition, that $|k_i| < 1$ for $1 \leq i \leq p$, is interpreted as a

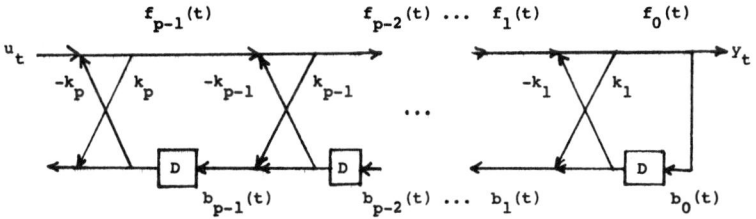

FIGURE 3:   The inverse lattice filter.

passivity condition: energy flowing into a layer is no greater than
that flowing out.   Burg (1975) exploited this analogy with seismic
signal propagation through a layered earth to obtain the lattice
structure in the first place.

REFERENCES

Atal, B. S. and S. Hanauer, 1971.   Speech analysis and synthesis by
    linear prediction of the speech wave.   *J. Acoust. Soc. Amer.* 50:
    637-655.
Blackman, R. B. and J. W. Tukey, 1959.   *The Measurement of Power
    Spectra.*   Dover, New York.   190 pp.
Bowling, S. B., 1977.   Linear prediction and maximum entropy spectral
    analysis for radar applications.   *MIT Lincoln Laboratory, Proj.
    Rep. RMP-122.*   74 pp.
Box, G. E. P. and G. M. Jenkins, 1976.   *Time Series Analysis: Fore-
    casting and Control*, 2nd Ed.   Holden Day, San Francisco.   553 pp.
Burg, J. P., 1975.   Maximum entropy spectral analysis.   Ph.D. Thesis,
    Stanford University, Geophys. Dept.   123 pp.
Chen, W. Y. and G. R. Stegun, 1974.   Experiments with maximum entropy
    power spectra of sinusoids.   *J. Geophys. Res.* 79:   3019-3022.
Claerbout, J. F., 1976.   *Fundamentals of Geophysical Data Processing
    With Applications to Petroleum Prospecting.*   McGraw-Hill, New
    York.   274 pp.
Cooper, G. R. and C. D. McGillem, 1978.   Doppler spectrum estimation
    for continuously distributed radar targets.   *Proc. RADC Spectrum
    Estimation Workshop, Rome, N.Y.:*   273-283.
Dickinson, B. W., 1977.   An approach to stationary autoregressive
    estimation.   *Proc. 1977 Conf. on Inform. Sciences and Systems,
    Johns Hopkins Univ.:*   507-511.
Dickinson, B. W., 1978a.   Two recursive estimation schemes for auto-
    regressive models based on maximum likelihood,   *J. Stat. Comp.
    and Simul.* 7:   85-92.
Dickinson, B. W., 1978b.   Autoregressive estimation using residual
    energy ratios.   *IEEE Trans. Inform. Theory, IT-24:*   503-506.
Dickinson, B. W., 1979.   Estimation of partial correlation matrices
    using Cholesky decomposition.   *IEEE Trans. Automat. Contr.* AC-
    24:   302-305.

Eykhoff, P., 1974. *System Identification: Parameter and State Estimation.* Wiley, New York. 555 pp.

Flanagan, J. L., 1972. *Speech Analysis, Synthesis and Perception,* 2nd Ed. Springer-Verlag, New York. 317 pp.

Friedlander, B., M. Morf, T. Kailath, and L. Ljung, 1979. New inversion formulas for matrices classified in terms of their distance from Toeplitz matrices. *J. Lin. Alg. Appl.*

Frost, O. L., 1977. Power-spectrum estimation. In: *Aspects of Signal Processing, Part 1,* (G. Tacconi, ed.), pp. 125-162. D. Reidel, Dordreht-Holland.

Gersch, W., 1970. Spectral analysis of EEG's by autoregressive decomposition of time series. *Math. Biosci.* 7: 205-222.

Gersch, W., N. N. Neilsen, and H. Akaike, 1973. Maximum likelihood estimation of structural parameters from random vibration data. *J. Sound and Vibration* 31: 295-308.

Gersch, W. and D. R. Sharpe, 1973. Estimation of power spectra with finite order autoregressive models. *IEEE Trans. Automat. Contr.,* AC-18: 367-369.

Gersch, W. and J. Yonemoto, 1977. Automatic classification of multivariate EEGs using an amount of information measure and the eigenvalues of parametric time series model features. *Comp. Biomed. Res.* 10: 297-318.

Gupta, V. N., J. K. Bryan, and J. N. Gowdy, 1978. A speaker-independent speech-recognition system based on linear prediction. *IEEE Trans. Acoust., Speech, Signal Processing,* ASSP-26:27-33.

Itakura, F. and S. Saito, 1971. Digital filtering techniques for speech analysis and synthesis. *Conf. Rec., 7th Int. Congr. Acoustics,* Paper 25C1.

Jenkins, G. M. and D. G. Watts, 1968. *Spectral Analysis and Its Applications.* Holden Day, San Francisco. 525 pp.

Jones, R. H., 1978. Multivariate autoregression estimation using residuals. In: *Applied Time Series Analysis,* (D. Findley, ed.), pp. 139-162. Academic Press, New York.

Kailath, T., 1974a. A view of three decades of linear filtering theory. *IEEE Trans. Inform. Theory,* IT-20: 146-181.

Kailath, T., ed., 1974b. Special issue on system identification and time series analysis. *IEEE Trans. Automat. Contr.,* AC-19: 637-951.

Kopec, G. E., A. V. Oppenheim, and J. M. Tribolet, 1977. Speech analysis by homomorphic prediction. *IEEE Trans. Acoust., Speech, Signal Processing,* ASSP-25: 40-49.

Lacoss, R. T., 1971. Data adaptive spectral analysis methods. *Geophys.* 36: 661-675.

Lacoss, R. T., 1977. Autoregressive and maximum likelihood spectral analysis methods. In: *Aspects of Signal Processing, Part 2,* pp. 591-615. D. Reidel, Dordreht-Holland.

Landers, T. C. and R. T. Lacoss, 1977. Some geophysical applications of autoregressive spectral estimates. *IEE Trans. Geosci. Elect.,* GE-15: 26-31.

Makhoul, J., 1975. Linear prediction: a tutorial review. *Proc. IEEE* 63: 561-580.

Makhoul, J., 1977. Stable and efficient lattice methods for linear prediction. *IEEE Trans. Acoust., Speech, Signal Processing,* ASSP-25: 423-428.

Markel, J. D. and A. H. Gray, 1976. *Linear Prediction of Speech.*
   Springer-Verlag, New York. 288 pp.

Marple, L., 1978. Frequency resolution of high-resolution spectrum
   analysis techniques. *Proc. RADC Spectrum Estimation Workshop,
   Rome, N. Y.*: 19-35.

Mayne, D. Q., 1977. An efficient, multistage, linear identification
   method for ARMA processes. *Proc. 1977 IEEE Conf. Dec. Contr.,
   New Orleans* 435-438.

Mehra, R. K. and D. G. Lainiotis, eds., 1976. *System Identification:
   Advances and Case Studies.* Academic Press, New York. 593 pp.

Morf, M., B. W. Dickinson, T. Kailath, and A. Vieira, 1977a. Efficient solution of covariance equations for linear prediction.
   IEEE Trans. Acoust., Speech, Signal Processing, ASSP-25: 429-
   433.

Morf, M., A. Vieira, and T. Kailath, 1978a. Covariance characterization by partial autocorrelation matrices. *Ann. Stat.* 6: 643-648.

Morf, M., A. Vieira, and T. Kailath, 1978b. Recursive multi-channel
   maximum entropy method. *IEEE Trans. Geosci. Elect., GE-18:* 85-
   94.

Morf, M., A. Vieira, and D. T. Lee, 1977b. Ladder forms for identification and speech processing. *Proc. 1977 IEEE Conf. Dec.
   Contr., New Orleans* 1074-1078.

Newman, W. I., 1977a. Extension to the maximum entropy method.
   IEEE Trans. Inform. Theory, IT-23: 89-93.

Newman, W. I., 1977b. A new method of multidimensional power spectral analysis. *Astron. Astrophys.* 54: 369-380.

Pisarenko, V. F., 1973. The retrieval of harmonics from a covariance
   function. *Geophys. J. Roy. Astron. Soc.* 33: 347-366.

Steiglitz, K., 1974. *An Introduction to Discrete Systems.* Wiley,
   New York. 318 pp.

Steiglitz, K., 1977. On the simultaneous estimation of poles and
   zeros in speech analysis. *IEEE Trans. Acoust., Speech, Signal
   Processing, ASSP-25:* 229-234.

Steiglitz, K. and B. W. Dickinson, 1978. Conditioning problems in
   pole-zero modelling. *Proc. 1978 IEEE Int. Symp. Circuits and
   Systems, New York*: 463-464.

Steiglitz, K. and L. E. McBride, 1965. A technique for the identification of linear systems. *IEEE Trans. Automat. Contr., AC-
   10:* 461-464.

Thomson, D. J., 1977. Spectrum estimation techniques for characterization and development of WT4 waveguide. *Bell System. Tech. J.*
   56: 1769-1815.

Turner, J. M., 1978. Linear predictive modelling and efficient
   speech encoding. Ph.D. Thesis, Princeton Univ., Dept. of Elect.
   Eng. Comp. Sci. 173 pp.

Ulrych, T. J. and T. N. Bishop, 1975. Maximum entropy spectral
   analysis and autoregressive decomposition. *Rev. Geophys. and
   Space Physics* 13: 183-200.

Viswanathan, R., J. Makhoul, and A. W. F. Huggins, 1978. *Speech
   Compression and Evaluation.* Bolt Beranek and Newman Inc., BBN
   Rep. 3794. 228 pp.

Woods, J. W., 1976. Two dimentional Markov spectral estimation.
   IEEE Trans. Inform. Theory, IT-22: 552-559.

Yule, G. U., 1927. On a method of investigating periodicities in

disturbed series, with special reference to Wolfer's sunspot
numbers. *Phil. Trans. Roy. Soc., 226-A:* 267-298.

Optimal Objective Mapping: A Technique for Fitting Surfaces to Scattered Data

Michael Karweit

ABSTRACT: In oceanography it is frequently useful
to estimate scalar, two-dimensional, time dependent
fields (e.g., temperature field) from collected
data. Usually, the data cannot be collected synoptically and so some temporal and spatial interpolation scheme must be used to produce an estimated instantaneous field. Although many mechanical procedures exist for doing this, they
suffer from not being able to take into account
the underlying physical processes which produce
the fields.

This paper discusses an interpolation/mapping procedure proposed by Gandin (1965) which depends on
the underlying covariance functions of the field.
This objective mapping procedure (actually, a two-
dimensional extension of ideas developed by Wiener,
1949, and Kolmogorov, 1941) gives a "best," in the
least-squared error sense, estimate of the field.
The principal feature is that the statistics of
the field, not the data individually, are used to
calculate the necessary coefficients. The procedure also provides for a variety of random measure-

81

ment errors, and, as a by-product of the calculations, can produce an expected-error map. The
technique is derived in some detail for fields
having simplifying assumptions. For the more complex situations, such as vector fields, the technique is merely outlined.

What is presented here is not new, and in fact
closely parallels Bretherton et al., 1976. The
purpose of this paper is to bring this relatively
new idea in physical oceanography to the attention
of other members of the oceanographic community.

INTRODUCTION

In many areas of science, one is often confronted with the task
of constructing an accurate map of a field from which scattered data
have been obtained. Figure 1 exemplifies the situation. The contours represent the actual field, the crosses the data samples, and
the intersection of the grid lines the points at which the field is
to be estimated.
In the simplest case, this problem is easy: the field does not
change with time, or at least changes little over the duration of data
collection; an arbitrarily large number of data points might be collected; the locations of each data point can be regulated; and, if in
the course of producing a map from existing data one finds that in
certain areas more information is needed, one can always return to
those locations and obtain more. An example of such a simple case
might be the construction of a geologic or topographic map of some
region.
If the field to be mapped changes with time so that the mapping
one would be interested in would be kind of a snapshot of the field
at one instant of time, the problem becomes more difficult. Unless
all the necessary information can be collected simultaneously, the
data are actually from a sequence of different fields. An analogous
difficulty exists if the field is constant with time, but one is simply unable to sample at all the needed locations.
These more complicated types of fields are the ones encountered
by oceanographers all the time, ones that require mapping for purposes
of display or input to numerical models: fields of temperature, velocity, fish density, and nutrient concentration. All are time dependent, and, for a variety of reasons, poorly sampled. The question becomes how can one assemble scattered information sampled from a time-
varying field and come up with a composite estimate of that field?
It is this problem, a problem of interpolation, on which the present
paper will focus.

FORMULATION OF THE PROBLEM

In a mathematical formulation, we are concerned with the field
$\theta(x,y,z,t)$. As suggested by the arrow, the field may be a vector

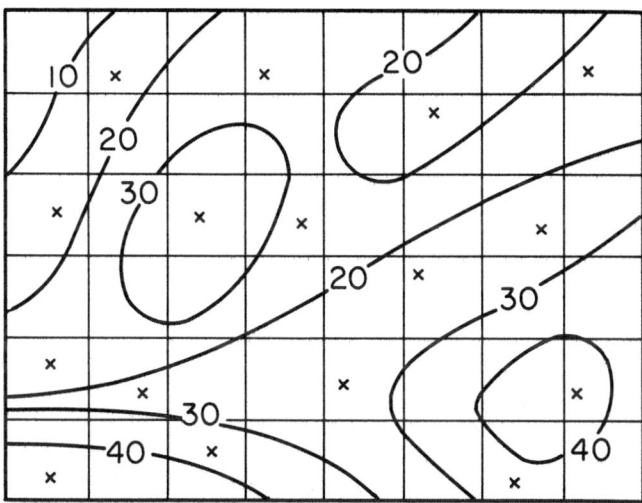

FIGURE 1: A contour map of a hypothetical 2-dimensional scalar field. The field is to be sampled at location x and approximated at the intersection of the superposed grid. All units are arbitrary.

field such as a velocity field. It may also be three-dimensional and time varying. We want to take N sample points $\phi_i(x_i,y_i,z_i,t_i)$

$i = 1,N$ where $\phi_i = \theta_i + \epsilon_i$ the actual value of the field plus some

measurement error; on the basis of these ϕ's we try to estimate the actual field θ at M other points x_j,y_j,z_j,t_j , $j = 1,M$, perhaps imposed grid points. The general form of estimator that we want to consider is

$$\tilde{\theta}_j = \sum_{i=1}^{N} \sum_{k=1}^{P} \alpha_{ijk} f_k(\phi_i) \tag{1}$$

where $\tilde{\theta}_j$ is the estimate of θ at the point (x_j,y_j,z_j,t_j); f_k are potentially P different functions of the sampled data, and are the coefficients to be determined.

To proceed through some of the concepts involved, we will at first observe only the simplest formulation. Mathematically, suppose we are concerned with the fluctuations $\theta(x,y)$ of a single-valued, time-independent, scalar field. (The assumption of zero mean will be elaborated upon later.) On the basis of N measured values $\phi_i(x_k,y_i)$, we want to estimate θ as $\tilde{\theta}$ at specified points

(x_j, y_j) j = 1,M. The ϕ's that we measure are the true θ's, but with some measurement error, so $\phi_i = \theta_i + \varepsilon_i$. Our estimate for $\tilde{\theta}_j(x_j, y_j)$ will be assumed to be simply a linear combination of our observations ϕ_i. So,

$$\tilde{\theta}_j = \sum_{i=1}^{N} \alpha_{ij} \phi_i \tag{2}$$

Our job is to find the coefficients α_{ij} which will produce a suitable set of $\tilde{\theta}$'s.

The simplest methods for picking the α's are strictly mechanical and make no use of any prior knowledge of the field. They may be local or global, i.e., they may make use of only those data nearest the point to be estimated, or they may make use of all of the data.

Our approach is that, for a given estimate $\tilde{\theta}_j(x_j, y_j)$, the α's yield relative contributions from the sampled data and perhaps should reflect some measure of the distance from the point to be estimated. So $\alpha_{ij} = f(x_i - x_j, y_i - y_j)$; for example

$$\alpha_{ij} \propto \frac{1}{(x_i - x_j)^2 + (y_i - y_j)^2} \quad \text{or} \quad \alpha_{ij} \propto \frac{1}{K + |x_i - x_j| + |y_i - y_j|}$$

the constant of proportionality being determined by $\sum_{i=1}^{N} \alpha_{ij} = 1$ for each j. Such procedures weight each observation in a prescribed way for every desired estimate, and in fact can become global or local schemes depending on how quickly α_{ij} becomes zero as $(x_i - x_j)^2 + (y_i - y_j)^2$ becomes large. Notice that in this type of scheme the α's depend only on the coordinates of the data, vis-a-vis the coordinate of the point to be estimated, and not on the data values themselves. Implied in the use of such an algorithm is that the true field is spatially correlated and that correlation decreases with increasing separation, certainly a reasonable assumption for most physical fields.

Another approach which might be construed as a mechanical method for picking the α's is one that involves fitting a surface to nearby data points (in our simplified exposition this is limited to a plane surface) and using the resulting equation to deduce local estimates. If three non-colinear points are used to calculate the equation of the plane, the whole field can be mapped out deterministically as triangular segments, the vertices of which are the sample points. This maps the entire field (interior to the data points) as a segmented planar surface. Algorithms for connecting scattered points by triangular segments are nontrivial and only recently have been derived (Cavendish, 1974). A lot of attention has been given to this problem because of recent interest in finite element analysis.

A similar approach is to allow more than three data points to

contribute to the definition of the local surface. In this case, the
problem is overspecified and other conditions must be imposed. The
most common condition is minimum squared differences between the calculated surface and the data points. This so-called least squares
technique, while producing a local planar estimate of the field as in
the above case, uses the data themselves in the estimation, unlike the
above case.

Compared to the method discussed earlier, this last one is decidedly different. Its solution will not necessarily reduce to the
same value as the data at the coordinates of the data. Thus, whereas
the previous methods were strictly schemes of interpolation, this method is a scheme of approximation. The rationale for which method to
choose is based largely on potential measurement error, something
which, so far, we have ignored. If measurement error can be ruled out,
the mapping schemes which ensure surfaces passing through the data are
ordinarily used. If measurement error plays a role, the approximation
schemes are usually more appropriate. In the case of the least squares
technique, if one is confident that the field is approximately locally
planar then the difference between the calculated surface and the data
can be construed as an indicator of measurement error.

The procedures for choosing the α's so far have been quite mechanical. That is, they have depended on only some simple mathematical
assumptions not necessarily discriminant of any of a variety of fields.
In many of the natural sciences, certainly in oceanography, much more
than that about the fields is usually known. Insofar as the fields we
observe are the result of physical processes, we can often make valid
inferences from the governing equations and obtain characteristics specific to the particular field in question. This, coupled with prior
knowledge of similar fields, can be used to provide much less *ad hoc*
procedures for finding the α's. Such procedures will be commented on
and serve as the principal focus of this paper.

Many of the fields under investigation in oceanography, while
different from one another in detail, are similar statistically. For
example, two one-kilometer square areas of the ocean will each exhibit certain changing patterns of temperature at, for example, a depth
of 500 m, which will not be identical. But if you can assume that
these two areas are similarly driven and are subject to approximately
the same conditions, then you can expect some of the average features
to be the same, maybe the average distance between temperature maxima
or the mean-square temperature variation within the one kilometer
square.

Three hypothetical temperature fields are illustrated in Figure
2. The first two (2a,b) might well have been produced under comparable circumstances and which would exhibit similar statistics. The
third (2c), with a relatively pronounced up-down pattern, most likely
was produced differently. We will proceed with deducing an interpolation scheme for statistically similar fields.

AN OPTIMAL INTERPOLATION TECHNIQUE

A scheme of "objective analysis" was introduced by Gandin (1965)
as a technique for interpolating meteorological fields. We will

assume that the field we are investigating is only one of a large
ensemble of fields which was produced by the same set of governing
equations, all with similar boundary conditions. As before, $\phi_i =$
$\theta_i + \varepsilon_i$ and $\tilde{\theta}_j = \sum\limits_{i=1}^{N} \alpha_{ij} \phi_i$ We are interested in minimizing the

squared error of our estimate $\tilde{\theta}_j$, over the whole ensemble. Define
the square error as

$$\overline{E_j} = \overline{(\theta_j - \tilde{\theta}_j)} = \overline{(\theta_j - \sum\limits_{i=1}^{N} \alpha_{ij} \phi_i)^2} \tag{3}$$

$$= \overline{\theta_j \theta_j} - 2\sum\limits_{i=1}^{N} \alpha_{ij} \overline{\theta_j \phi_i} + \sum\limits_{i=1}^{N} \sum\limits_{k=1}^{N} \alpha_{ij} \alpha_{kj} \overline{\phi_i \phi_k}$$

the overbars indicating averages taken over the entire ensemble.
This can be construed as an equation in which the α's are arbitrary. However, we want to choose them so that $\overline{E_j}$ is a minimum.
The Gauss-Markov theorem is used to obtain an optimum linear estimate
for θ_j,

$$\frac{\partial \overline{E_j}}{\partial \alpha_{kj}} = 2\overline{(\theta_j - \sum\limits_{i=1}^{N} \alpha_{ij} \phi_i)(-\phi_k)} = -2\overline{\theta_j \phi_k} + 2\sum\limits_{i=1}^{N} \alpha_{ij} \overline{\phi_i \phi_k} \tag{4}$$

$$\frac{\partial \overline{E_j}}{\partial \alpha_{kj}} = 0 \quad \blacktriangleright \quad \overline{\theta_j \phi_k} = \sum\limits_{i=1}^{N} \alpha_{ij} \overline{\phi_i \phi_k} \tag{5}$$

This final expression yields an N x N matrix equation for finding
the N α_{ij}'s necessary to deduce an estimate of θ at (x_j , y_j)

$$\begin{bmatrix} \overline{\phi_1 \phi_1} & \overline{\phi_1 \phi_2} & \cdots & \overline{\phi_1 \phi_N} \\ \overline{\phi_2 \phi_1} & & \cdots & \\ \cdot & \cdot & & \cdot \\ \cdot & \cdot & \cdots & \cdot \\ \cdot & \cdot & \cdots & \cdot \\ \overline{\phi_N \phi_1} & \cdot & \cdots & \overline{\phi_N \phi_N} \end{bmatrix} \begin{bmatrix} \alpha_{1j} \\ \alpha_{2j} \\ \cdot \\ \cdot \\ \cdot \\ \alpha_{Nj} \end{bmatrix} = \begin{bmatrix} \overline{\theta_j \phi_1} \\ \overline{\theta_j \phi_2} \\ \cdot \\ \cdot \\ \cdot \\ \overline{\theta_j \phi_N} \end{bmatrix} \quad \blacktriangleright$$

$$\tag{6}$$

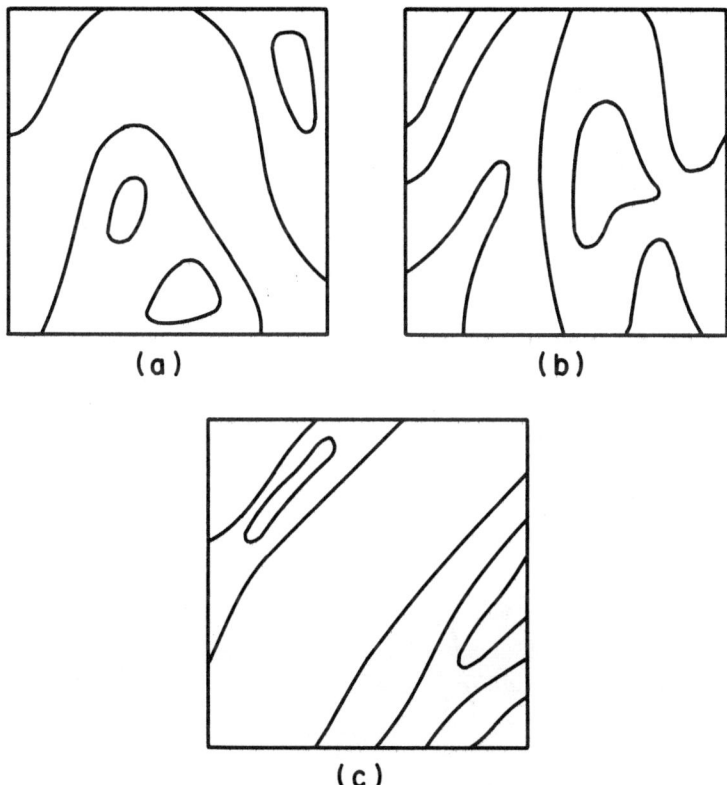

FIGURE 2: Qualitative contour maps of 2-dimensional scalar fields. All three maps are different in detail but [a] and [b] are similar in a statistical sense.

$$\alpha_{ij} = \sum_{k=1}^{N} [\overline{\phi_i \phi_k}]^{-1} \overline{\theta_j \phi_k}$$

If the α_{ij}'s are chosen as above, then $\overline{E}_j$, the mean square error, is

$$\overline{E}_j = \overline{\theta_j \theta_j} - 2 \sum_{i=1}^{N} \alpha_{ij} \overline{\theta_j \phi_i} + \sum_{i=1}^{N} \alpha_{ij} \left(\sum_{k=1}^{N} \alpha_{jk} \overline{\phi_k \phi_i} \right) \qquad (7)$$

$$\angle \overline{\theta_j \phi_i}$$

$\overline{E}_j$ can then be written as comprising two terms,

$$\overline{E}_j = \overline{\theta_j \theta_j} - \sum_{i=1}^{N} \sum_{i=1}^{N} [\overline{\phi_i \phi_k}]^{-1} \overline{\theta_j \phi_i} \; \overline{\theta_j \phi_k} \tag{8}$$

true total variance — variance accounted for by estimation scheme

We must now interpret these ensemble averages. Assume, for convenience, that the measurement error ε_i in $\phi_i = \theta_i + \varepsilon_i$ is uncorrelated with θ or with other ε_k's, and it has an ensemble mean of zero and a single mean-square value common to all measurements, i.e.,

$$\overline{\varepsilon_i \theta_j} = 0 \quad , \quad \overline{\varepsilon_i \varepsilon_k} = \sigma^2 \delta_{ik} \quad , \quad \overline{\varepsilon_i} = 0 \; .$$

Deduce the following:

$$\overline{\phi_i \theta_j} = \overline{\theta_i \theta_j} + \underset{= 0}{\overline{\varepsilon_i \theta_j}} = \overline{\theta_i \theta_j}$$

$$\overline{\phi_i \phi_k} = \overline{(\theta_i + \varepsilon_i)(\theta_k + \varepsilon_k)} = \overline{\theta_i \theta_k} + \underset{= 0}{\overline{\varepsilon_i \theta_k}} + \underset{= 0}{\overline{\varepsilon_k \theta_i}} + \overline{\varepsilon_i \varepsilon_k} =$$

$$\overline{\theta_i \theta_k} + \sigma^2 \delta_{ik}$$

Now

$$\overline{E}_j = \overline{\theta_j \theta_j} - \sum_{i=1}^{N} \sum_{k=1}^{N} \left[\overline{\theta_i \theta_k} + \sigma^2 \delta_{ik} \right]^{-1} \overline{\theta_i \theta_j} \; \overline{\theta_j \theta_k} \tag{9}$$

where

$$\alpha_{ij} = \sum_{k=1}^{N} \left[\overline{\theta_i \theta_k} + \sigma^2 \delta_{ik} \right]^{-1} \overline{\theta_j \theta_k} \tag{10}$$

What we now have is an interpolation scheme, which tells us how to weight our observations to get optimally estimated values of the true field θ. The α_{ij}'s are calculated in terms of the statistical characteristics and could in fact be obtained from prior knowledge. Unlike virtually all of the mechanical schemes for the interpolation of fields, this objective mapping procedure incorporates the inherent physical processes which produce them.

If the covariance statistics are in fact known, then estimates
of θ will have different properties depending on our assumption of
the measurement error σ^2. If in the sampling of a particular field
$\sigma^2 = 0$, the estimated field at point of sampling will coincide with
the sample value, i.e., the objective mapping scheme is a scheme
strictly of interpolation. If $\sigma^2 \neq 0$, the estimated field will not
necessarily coincide with a sampled value, because the sample value
would be noisy and nearby sample values would temper its effect. In
this case, the scheme is one of surface fitting.

  Our problem is solved if we know the spatial covariance function
$\overline{\theta_i \theta_j}$, of the ensemble of fields. But usually we do not and it must
be estimated. Actually, in some simple oceanographic problems,
enough is known of the physics of the field that covariance functions
have been theoretically deduced. Using our notation, we can write the
necessary covariance function C as,

$$\overline{\theta_i \theta_j} = \overline{\theta(x_i, y_i)\theta(x_j, y_j)} = C(x_i, y_i, x_j, y_j)$$

Since estimating a four parameter function is hopeless, we want to see
if some simplifications might be possible. In oceanography, fields
away from boundaries or particularly persistent features (like the
Gulf Stream) are often statistically homogeneous; that is, the statistics in one part of the field are the same as another part of the
field. In terms of the covariance functions, we could write

$$C(x_i, y_i, x_j, y_j) = C[x_i, y_i, x_i + (x_j - x_i), y_i + (y_j - y_i)]$$

If this function were spatially homogeneous, then it could not be a
function of any specific location $x_i y_i$ by itself; so

$$C = C[(x_j - x_i), (y_j - y_i)] = C[\Delta x, \Delta y]$$

a function of the distance between the points.

  A further simplification is also usually appropriate, that of
assuming that the field is statistically isotropic. In such a case,
C is invariant with respect to direction and

$$C[\Delta x, \Delta y] \implies C\left[\{(\Delta x)^2 + (\Delta y)^2\}^{\frac{1}{2}}\right] = C[r]$$

a function of undirected separation only.

  With no *a priori* information of the covariance function, we have
to estimate it from the sample data themselves. If assumptions of
homogeneity and isotropy are made, we can obtain N! (not all independent) estimates of the covariance function by taking all possible
pairs of data and calculating

$$C(r_{ij}) = \theta(x_i, y_i)\theta(x_j, y_j)$$

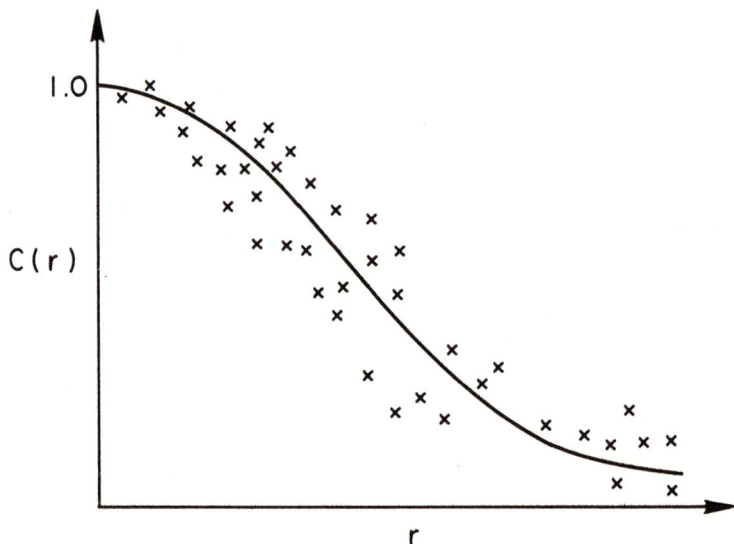

FIGURE 3: An approximation of an isotropic
correlation function C(r), where r is distance.
The x's are typical data points from which C(r)
is estimated.

Figure 3 shows a possible outcome of such a calculation and indicates
a smoothed estimate of C(r). With this function C(r) and for our data
set, we can then calculate the α's

$$\alpha_{ij} = \sum_{k=1}^{N} \left[C(r_{ik}) + \sigma^2 \delta_{ik} \right]^{-1} C(r_{jk}) \tag{11}$$

and produce $\tilde{\theta}_j = \sum_{i=1}^{N} \alpha_{ij} \phi_j$ and for the point x_j, y_j.

The C(r) which we have just approximated should be a function
reflecting an ensemble of fields. Since it was deduced from only
one, it is not necessarily a good estimate. Depending on the ensemble in question, and the area over which data were taken, either
of two extremes could occur. Figure 4 depicts the two extremes.
The bold line shows the true ensemble covariance function and the
thin lines indicate estimates of C(r) from three realizations of
the ensemble. In each case averaging the three realizations gives
the correct ensemble function. However, to estimate C(r) from only
one realization gives dramatically different results for [a] and [b].
In [a] the scatter of the data is uncorrelated along C(r) and can, in
part, be smoothed to give a reasonable estimate. In [b] the scatter
is not uncorrelated along C(r) and we can see that any estimate we
make will be in error. Fortunately, the optimal interpolation

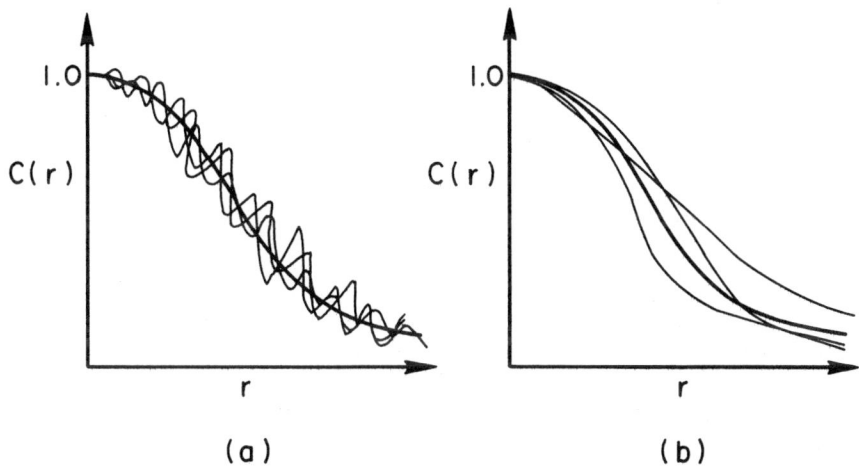

(a)   (b)

FIGURE 4:   Both [a] and [b] estimate the same
correlation function C(r) (heavy line) as the
average of three sets of data (light lines).
In [a] variations in the data are uncorrelated
along r; in [b] variation in the data are correlated along r.

scheme is relatively insensitive to the details of $C(r)$ and so not-
perfect estimates are acceptable.

In fact, for non-periodic phenomena, $C(r) \to 0$ as $r \to \infty$; and on

physical grounds, $C(0)$ is a maximum and $\frac{dC}{dr} = 0$ at $r = 0$. Conse-

quently, users often simply assume $C(r) \propto \exp[-(r/R)^2]$, the sample
data being used to determine R, the width of the function.

An initial goal of this whole procedure is to produce an algorithm which depends more on the statistics of the true field and
less on the data themselves. The hope is that greater certainty can
be assumed for statistical properties than for individual measurement. The interpolation/smoothing scheme which we have just outlined aids field estimation in two ways: 1) assuming we have non-
zero ε_i's, we can improve on individual measurements by allowing
surrounding measurements to temper the error at a sample point. As
was pointed out earlier, if $\varepsilon_i = 0$, then the interpolation method
gives the sample value at the sample point; 2) we can interpolate for
other points in the field in a way which is consistent with the ensemble statistics of the field, i.e., we do not produce estimated
fields whose statistics are different from the true field. This
last feature is usually not true for the *ad hoc* interpolation methods.

COMPUTATIONAL CONSIDERATIONS

For the jth point on the field we estimate θ, using (2) and (6) to get

$$\tilde{\theta}_j = \sum_{i=1}^{N} \sum_{k=1}^{N} \left[\overline{\theta_i \theta_k} + \sigma^2 \delta_{ik} \right]^{-1} \overline{\theta_j \theta_k} \, \phi_i \qquad (12)$$

Thus for each estimate of θ it is necessary to invert an N x N matrix and do a double sum. For even moderate values of N, this calculation would be terribly expensive. However, note that the matrix to be inverted depends only on the positions of the sample data, i.e., indices i and k. This means that for any or all interpolation positions j we need only the same inverted matrix. Rearrange (12) to get

$$\tilde{\theta}_j = \sum_{k=1}^{N} \left\{ \sum_{i=1}^{N} A_{ik}^{-1} \, \phi_i \right\} \overline{\theta_j \theta_k} \qquad (13)$$

where

$$A_{ik} = \overline{\theta_i \theta_k} + \sigma^2 \delta_{ik}$$

With this rearrangement we have everything which depends solely on the N data points within the squiggly bracket and can perform that calculation once. So let

$$\eta_k = \sum_{i=1}^{N} A_{ik}^{-1} \, \phi_i \qquad (14)$$

$$\tilde{\theta}_j = \sum_{k=1}^{N} \eta_k \, \overline{\theta_j \theta_k} \qquad (15)$$

Once the η's are calculated only N calculations are required for each estimate of θ. In all these calculations $\overline{\theta_j \theta_k}$ is, of course, just a value picked off the covariance function $C(\overline{r})$.

Even with this calculation procedure, computing costs may be significant. Although A_{ik} is a symmetric, positive-definitive matrix, its inversion for large N is expensive. The only way to reduce the problem is to consider doing the mapping on only a part of the field at a time using only the closer data points. Since matrix inversion requires $O(N^3)$ operations, any reduction of N at the expense of having to run more parts is worthwhile. Especially in those cases where the size of the field of interest is large compared to the length over which the covariance function is significantly different from zero, a piecewise mapping is desirable.

ADDITIONAL COMMENTS ON THE SIMPLIFIED MAPPING PROCEDURE

At the onset of our discussion we assumed that the errors of measurement ε_i were uncorrelated with one another and had a common mean-square value, $\overline{\varepsilon_i \varepsilon_k} = \sigma^2 \delta_{ik}$. What happens to the calculation if these assumptions are not met? Such conditions can easily occur: for example, several instruments, each with different probable errors, might have been used in sampling the field, or a spatially correlated disturbance affects the measurement taking (in an oceanographic application, maybe sea-surface conditions). Carrying out the calculation without the simplifying assumption is not difficult. The only change is that the matrix A is modified.

$$A_{ik} = \overline{\theta_i \theta_k} + \overline{\varepsilon_i \varepsilon_k} \qquad\qquad (16)$$

where some form for the error covariance $\overline{\varepsilon_i \varepsilon_k}$ would have to be assumed. The effect of this non-simplified error covariance, however, might be undesirable depending on the relation between $\overline{\theta_i \theta_k}$ and $\overline{\varepsilon_i \varepsilon_k}$. This relation is best described through the introduction of the concept of "length scale", which was already alluded to above. For our purposes, the length scale of a field is defined as that distance over which the field is more or less coherent, i.e., its values are correlated. If the field is expressed as a covariance function $C(r)$, then the length scale might be defined as the distance r at which C is 0.5 or 0.05 of its maximum value; or as the distance r at which C has a maximum second derivation. The exact definition is unimportant, because the same definition will be used for each of the fields to be compared, and whatever we choose will qualitatively differentiate between the fields.

The matrix A is comprised of two fields, the true field identified by its covariance function $\overline{\theta_i \theta_k}$, and an error field identified by its covariance function $\overline{\varepsilon_i \varepsilon_k}$. Each of the fields can be characterized by a length scale, say λ_θ and λ_ε. The mapping procedure we have been describing is a smoothing procedure as well as an interpolation procedure; and smoothing is equivalent to attenuating short length scale differences. Differences over large length scales remain. In electrical engineering parlance, the procedure is a low pass filter, whose characteristics are determined by the covariance functions $\overline{\theta_i \theta_k}$ and $\overline{\varepsilon_i \varepsilon_k}$. If $\lambda_\theta \gg \lambda_\varepsilon$, then our mapping scheme will do what we hope, extract only that part of the measured value of θ due to probable instrument error, and not that part due to the actual field. If, on the other hand, $\lambda_\theta \lesssim \lambda_\varepsilon$, the algorithm cannot distinguish between the error field and true field and any smoothing must result in attenuation of the estimated value of the actual field. This complication is well-described for a continuous two-dimensional scalar field by Thompson (1956).

It should be noted that the expected error field is immediately available from the calculation. From (3) $\overline{E}_j = \overline{(\theta_j - \tilde{\theta}_j)^2}$ is the expected error at the jth point on the field. If $\tilde{\theta}_j$ is calculated using the "best" α_{ij}'s from (11), then $\overline{E}$ gives the variation in the measured field accounted for by the optimal mapping technique. A normalized plot of this field gives a two-dimensional "confidence" map.

COMPLICATIONS

Up to now we have assumed that the fields we have been dealing with have had zero means or at least known means which could then be subtracted out. Further, we have also assumed that we knew or could estimate the covariance function of the field. Often we will come across problems in which neither of the assumptions is met.

If the expected value of every point on the field is the same, then under certain conditions we can estimate the field mean from the measured data. Again, we introduce another length scale L_M, the length over which measurements are taken and over which the mapping is to be performed. Although L_M is not defined in the same way as λ_θ and λ_ϵ, a comparison must be made. Simply, L_M must be much greater than $\lambda_\theta, \lambda_\epsilon$ or else a biased estimate of the mean of the field will be obtained. The importance of dealing with the fluctuation about the mean field rather than the field itself will not be elaborated upon except to say that the parallel analysis for non-zero mean fields is non-optimal. Whereas in our optimal interpolation scheme we are guaranteed a minimum error variance, in the parallel scheme the weighting factors α_{ij} are additionally subjected to the constraint $\sum_{i=1}^{N} \alpha_{ij} = 1$. This point is well elaborated upon by Gandin (1965).

A related problem occurs if we must estimate the covariance function. Again unless L_M is considerably bigger than λ_θ, the covariance function cannot be well determined, and a substitute must be found. The difficulty with the estimation of the covariance function is that if L_M is not large enough with respect to λ_θ, we can never get measurements far enough apart to estimate the total variance of the field. This usually results in a distorted covariance function estimate and a poor estimate of the mean. A better defined function over short distances is the structure function $S(r)$.

$$S(r) = S_{ik} = \overline{(\theta_i - \theta_k)^2} = 2\left[\overline{\theta_i\theta_i} - \overline{\theta_i\theta_k}\right] = 2[C(0) - C(r)]$$

Again isotropy is assumed and $r = \left[(x_i - x_k)^2 + (y_i - y_k)^2\right]^{\frac{1}{2}}$. Interpolation can be performed in the usual way by using $-\frac{1}{2}S(r)$ instead of $C(r)$. But again, since the saturating value $S(\infty)$ (equivalent to $C(0)$ is unobtainable for $L_M < \lambda_\theta$, a scheme which does not assume a zero-mean must be used.

We should point out that a "best" interpolation technique constraining the weights α_{ij} so that $\sum_{i=1}^{N} \alpha_{ij} = 1$ can be calculated, but is slightly tedious to implement. An acceptable compromise is to calculate the α_{ij}'s in the regular way and normalize them as $\alpha'_{ij} =$

$$\alpha_{ij} \Big/ \sum_{i=1}^{N} \alpha_{ij}. \quad \text{This is not optimum, but satisfies} \quad \sum_{i=1}^{N} \alpha'_{ij} = 1 \quad \text{and}$$

has been found to be satisfactory for most meteorological fields.

EXTENSIONS OF THE BASIC THEORY

For convenience, we have considered only time-dependent fields. Fortunately, this is not a necessary restriction, because most fields which oceanographers would be interested in are time varying. One of the most important applications of our optimal mapping technique is the estimation of a time-varying field at a single point in time which was sampled non-synoptically. That is, suppose several ships were measuring a field of isotherm depths over the course of several weeks, is there a way to get a best guess of what the field actually looked like in the middle of that period? With extensions to our mapping scheme, the answer is yes.

The problem now becomes one of estimating $\theta_j(x_j, y_j, t_j)$ as

$$\tilde{\theta}_j(x_j, y_j, t_j) = \sum_{i=1}^{N} \alpha_{ij} \, \phi_i(x_i, y_i, t_i)$$

As before, optimal interpolation will be accomplished when

$$\alpha_{ij} = \sum_{k=1}^{N} \overline{[\phi_i \phi_k]}^{-1} \, \overline{\theta_j \phi_k}$$

Recall that in the time-independent problem, and with assumption of isotropy and homogeneity, the averaged products $\overline{\phi_i \phi_k}$ were obtained from a covariance function $C(r)$, a measure of the field's spatial coherence. Now this function must be restipulated to portray the field's temporal coherence as well. Assuming that the temporal statistics do not depend on time explicitly, but only on time difference (temporal homogeneity), a two-parameter covariance function $C(r, \Delta t)$ is required. So

$$\overline{\phi_i \phi_k} \quad \Longrightarrow \quad C[r, \Delta t] \text{ where } r = \left[(x_i - x_k)^2 + (y_i - y_k)^2 \right]^{\frac{1}{2}},$$

$$\Delta t = t_i - t_k$$

The function $C[r, \Delta t]$ at $\Delta t = 0$ will be identical to $C(r)$.

Of course, the level of complexity of the covariance function

of the field is dictated by the field itself. Even if the details
of the field have a systematic drift u in, say, the x-direction,
we can deduce a covariance function parameterized as $C(\Delta x - u\Delta t,$
$\Delta y, \Delta t)$ and perform an optimal mapping. This more complex technique
was used by Karweit (1976) to produce a computer-generated movie of
the time history of isotherm fields during the MODE-I experiment.
The point is that our technique is not limited to only those fields
with the simplest statistics.

        Up to now we have considered only scalar fields as candidates
for optimal interpolation. We will not describe how our scheme can
be utilized for vector fields. To this end, we will use as an example the velocity field as measured by current meters and drogues.
From the data we wish to estimate the actual velocity field $u(x, y)$,
$v(x, y)$ in the same area. Because all but the smallest scales of
motion are horizontal, we know that the velocity field is largely
nondivergent and consequently is derivable from a stream function
$\psi(x, y)$.

$$u = - \frac{\partial \psi}{\partial y} \qquad v = \frac{\partial \psi}{\partial x} \tag{17}$$

If we can map out the function $\psi(x, y)$, we will have our desired
field, the velocity vectors following lines of constant ψ. Our
technique will proceed in the ordinary way except that now we interpolate ψ from

$$\tilde{\psi}_j = \sum_{i=1}^{N} \left\{ \alpha_{ij} u_i + \beta_{ij} v_i \right\} \tag{18}$$

By carrying out the same sort of analysis as before we will find that
the following cross-variances are required:

$$A_{2i-1, 2k-1} = \overline{u_i u_k} + \sigma^2 \delta_{ik} \; ,$$

$$A_{2i, 2k-1} = \overline{u_i v_k} \; ,$$

$$A_{2i-1, 2k} = \overline{u_k v_i} \; , \tag{19}$$

$$A_{2i, 2k} = \overline{v_i v_k} + \sigma^2 \delta_{ik} \; , \text{ and}$$

$$\overline{\psi_j u_i} \; , \; \overline{\psi_j v_i}$$

If we set up weights for ease of calculation as we did in (14) and
(15), then

$$\tilde{\psi}_j = \sum_{k=1}^{N} \left\{ \overline{\psi_j u_k} \, \xi_k + \overline{\psi_j v_k} \, \eta_k \right\} \tag{20}$$

where

$$\xi_k = \sum_{i=1}^{N} \left\{ [A^{-1}]_{2i-1, \ 2k-1} \ u_i + [A^{-1}]_{2i-1, \ 2k} \ v_i \right\}$$

(21)

$$\eta_k = \sum_{i=1}^{N} \left\{ [A^{-1}]_{2i, \ 2k-1} \ u_i + [A^{-1}]_{2i, \ 2k} \ v_i \right\}$$

For a statistically isotropic field, the covariance function of ψ can be expressed as

$$f_{ij} = f(r) = \overline{\psi(x + r \cos \alpha, \ y + r \sin \alpha) \ \psi(x, \ y)}$$

(22)

$$r = \left[(x_i - x_j)^2 + (y_i - y_j)^2 \right]^{\frac{1}{2}}, \quad \alpha = \cos^{-1} (x_i - x_j)/r$$

Then the desired velocity cross-covariance can be expressed in terms of $f(r)$ as

$$\overline{u_i u_j} = \overline{\frac{\partial \psi}{\partial y}\bigg|_i \ \frac{\partial \psi}{\partial y}\bigg|_j} = -\left[\cos^2 \alpha \ \frac{d^2 f}{dr^2} + \sin^2 \alpha \ \frac{1}{r} \frac{df}{dr} \right]$$

$$\overline{u_i v_j} = \overline{u_j v_i} = \sin \alpha \ \cos \alpha \ \left[\frac{d^2 f}{dr^2} - \frac{1}{r} \frac{df}{dr} \right]$$

(23)

$$\overline{v_i v_j} = -\left[\sin^2 \alpha \ \frac{d^2 f}{dr^2} + \cos^2 \alpha \ \frac{1}{r} \frac{df}{dr} \right]$$

Thus, a mapping of the stream-function field ψ can be effected. If the vector velocity field is desired directly, we use the same ξ's and η's and find

$$\tilde{u}_j = \sum_{i=1}^{N} \left\{ \overline{u_i u_j} \ \xi_i + \overline{v_i u_j} \ \eta_i \right\}$$

(24)

$$\tilde{v}_j = \sum_{i=1}^{N} \left\{ \overline{u_i v_j} \ \xi_i + \overline{v_i v_j} \ \eta_i \right\}$$

(25)

This procedure was carried out during MODE-I to evaluate velocity fields in the region of the field experiment (see MODE-I, 1975).

SUMMARY

A mapping procedure has been discussed which involves only linear combinations of the data. The same techniques can be extended to any higher order, but not without skyrocketing complexities. In fact, even for a second order interpolation in two-dimensions, the correlations needed to complete the calculations are beyond reach. For a second-order interpolation scheme, we would estimate Θ as

$$\tilde{\theta}_j = \sum_{i=1}^{N} \alpha_{ij} \, \phi_i + \sum_{i=1}^{N} \sum_{k=1}^{N} \beta_{ijk} \, \phi_i \phi_k$$

The usual minimum squared-error calculation yields the equations in α_{ij} and β_{ijk}:

$$\sum_{i=1}^{N} \alpha_{ij} \, \overline{\phi_i \phi_n} + \sum_{i=1}^{N} \sum_{k=1}^{N} \beta_{ikj} \, \overline{\phi_i \phi_k \phi_n} = \overline{\theta_j \phi_n}$$

$$\sum_{i=1}^{N} \alpha_{ij} \, \overline{\phi_i \phi_n \phi_m} + \sum_{i=1}^{N} \sum_{k=1}^{N} \beta_{ikj} \, \overline{\phi_i \phi_k \phi_n \phi_m} = \overline{\theta_j \phi_n \phi_m}$$

The quantity $\overline{\phi_i \phi_k \phi_n \phi_m}$ is a four-point correlation function (a fourth-order tenser) which should be known or estimatable. This is almost never the case. Consequently, even this lowest of the higher order schemes is intractable.

It is clear that this optimal mapping technique be limited to the linear case. But such a restriction is more than compensated for by the fact that prior knowledge, i.e., a known covariance function, can be used as additional information. Even when prior knowledge does not exist, it is the statistics and not the data themselves which determine the coefficient. These features make this approach unique.

REFERENCES

Bretherton, F. P., R. E. Davis, and C. B. Fandry, 1976. A technique for objective analysis and design of oceanographic experiments applied to MODE-73. *Deep-Sea Research* 23: 559-582.

Cavendish, J. C., 1974. Automatic triangulation of arbitrary planar domains for the finite element method. *Int. J. Num. Methods in Eng.* 8: 679-696.

Gandin, L. S., 1965. *Objective Analysis of Meteorological Fields.* Israel Program for Scientific Translation, Jerusalem. 242 pp.

Karweit, M., 1976. Some visual interpretations of MODE (16 mm color computer graphics movie). MODE Contr. No. 10-T. MODE Executive Office, Dept. Meteorology, Massachusetts Institute of Technology.

Kolmogorov, A. N., 1941. Interpolated and extrapolated stationary random sequences. *Izvestiya AN SSSR, seriya matematicheskoya* 5: 3-14.

MODE-I. *Dynamics and Analysis of MODE-I.* Unpublished report. MODE Executive Office, Dept. Meteorology, Massachusetts Institute of Technology.

Thompson, P. D., 1956. Optimum smoothing of two-dimensional fields. *Tellus* 8: 384-393.

Wiener, N., 1949. *Extrapolation, Interpolation, and Smoothing of Stationary Time Series.* Technology Press of the Massachusetts Institute of Technology. 163 pp.

Two-Dimensional Data Sets and Image Processing

Anil K. Jain

INTRODUCTION

In stochastic modeling of images, an image is considered to be a sample function of a two-dimensional random process called a *random field*. This permits a description of a group, class or ensemble of images and is desirable whenever one is interested in developing processing techniques which perform equally well on an entire class rather than on an individual member. The random field representing a class of images is modeled by a joint probability density function. Practical considerations force the probability density model to be Gaussian so that the mean and covariance functions provide sufficient statistics.

COVARIANCE MODELS

An approach used often involves specifying the image mean and covariance functions by parametric formulas which can be identified easily by measuring a small number of statistical parameters. For example, a popular model considers the random field to be *stationary* with

$$Eu_{i,j} = \mu \tag{1}$$

$$r_{k,\ell} \triangleq \text{cov}[u_{i,j}, u_{i+k,j+\ell}] = \sigma^2 \rho_1^{|k|} \rho_2^{|\ell|} \tag{2}$$

where μ is the mean and σ^2 is the variance of the random field denoted by $u_{i,j}$, $i,j \in I$ and I is a set of two-dimensional indices. This model is commonly called the *separable model* and is specified by the four parameters μ, σ^2, ρ_1, ρ_2 where ρ_1 and ρ_2 are the one-step correlations in the i and j direction respectively. For a large variety of grey level images $\rho_1 \simeq \rho_2 \simeq 0.9$ to 0.95. An alternative model, called the *isotropic model* is of the form

$$r_{k,\ell} = \sigma^2 \rho^d, \quad d = \sqrt{k^2 + \ell^2} \tag{3}$$

This also depends only on μ, σ^2 and the one-step correlation ρ. The separability property of (2),

$$r_{k,\ell} = r_1(k) r_2(\ell), \quad r_i(k) = \sigma \rho_i^{|k|}$$

makes it mathematically attractive, since many two-dimensional algorithms become equivalent to two one-dimensional algorithms, one for each direction. Further, it also gives certain finite order difference equation realizations of the random field which yield computationally attractive algorithms. The isotropic model of (3) has been shown by Jain (1977) to be a better estimate for a variety of images. The complexity of the algorithms associated with this model, however, is substantially greater. Neither model is completely satisfactory for representing the covariance statistic of images.

The basic problem of covariance modeling is referred to as Spectral Estimation, which is, "Given a set of image data, find an estimate of its covariance, or equivalently, the spectral density function." This is a hard problem in two or more dimensions because a desirable estimate should be described by several parameters.

PARTIAL DIFFERENCE EQUATION MODELS

Conceptually, under the assumption Gaussian statistics, mean and covariance functions are all that need to be specified. Accurate specification of a two-dimensional covariance function or, equivalently, the spectral density function, is a difficult task and requires one to consider spectral estimation techniques. Moreover, even if a covariance function is given, analytically or otherwise, it is desirable to have a difference equation characterization of the random field. Such a representation is useful in obtaining computationally feasible algorithms as well as in adapting the processing technique (via the model) by updating the model parameters with any statistical changes in the data. Thus, the problem here is to find a stochastic difference equation model that realizes the co-variances of the given ensemble accurately. This is also known as the spectral factorization problem.

In one dimension, given a rational spectral density, its finite order difference equation realization can always be found. In two dimensions this is not always possible. Therefore, we start by con-sidering various types of two-dimensional difference equation re-presentations and studying the type of covariance realizations they achieve. From the theory of partial differential equation (PDE), we know that two-dimensional PDE's may be classified into three cate-gories: 1) hyperbolic; 2) parabolic; and 3) elliptic. In analogy with this classification, we study these types of discrete random fields. Let L denote a PDE operator and let D be its discrete approximation obtained via a suitable numerical differencing scheme. Then,

$$D[u_{i,j}] = \varepsilon_{i,j}$$

represents a discrete random field where $\varepsilon_{i,j}$ is an elementary ran-dom field such as a white noise field or a finite order moving average field.

CAUSAL MODELS

These models arise when we consider hyperbolic equations. For example, the equation

$$u_{i,j} - a_1 u_{i-1,j} - a_2 u_{i,j-1} + a_3 u_{i-1,j-1} = \varepsilon_{i,j} \tag{4}$$

represents a finite difference approximation of a hyperbolic wave equation. If $\{\varepsilon_{i,j}\}$ is a white noise random field and $a_3 = a_1 a_2$, then (4) is an exact realization of the separable covariance model (for $a_1 = \rho_1$, $a_2 = \rho_2$) i.e., $u_{i,j}$ of (4) satisfy (2). Hyperbolic equations can be solved as initial value problems so that the asso-ciated discrete model such as (4) is causal. It can be shown that if we define 'past' at a location i,j, as

$$I^- = \{k,\ell: \ k < i, \ \forall \ell \ or \ k=i, \ \ell < j\} \tag{5}$$

then

$$\hat{u}_{i,j} \overset{\Delta}{=} a_1 u_{i-1,j} + a_2 u_{i,j-1} - a_3 u_{i-1,j-1} \tag{6}$$

is the best mean square predictor of $u_{i,j}$. Thus (4) provides a local predictor and could be used to design a DPCM scheme for data compression of images. The difficulty with causal models is that a very high order model may be required to realize a given set of covariances. Higher order hyperbolic equations may be used to ob-

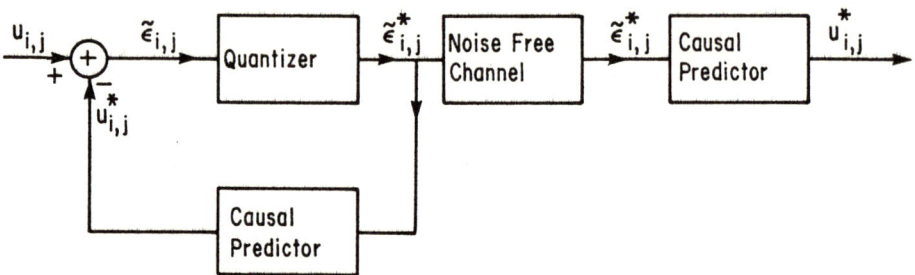

FIGURE 1: DPCM image data compression via causal (hyperbolic) P.D.E. models.

tain higher order, stable, causal models. Note that causality defined in (5) is not a single quadrant causality, but is defined on a nonsymmetric half plane. (Figure 1 shows a DPCM method using causal models.)

SEMICAUSAL MODELS

A semicausal model is causal in one direction and noncausal in the other. For example, the equation

$$u_{i,j} = \alpha(u_{i-1,j} + u_{i+1,j}) + \gamma u_{i,j-1} + \varepsilon_{i,j} \tag{7}$$

is causal in j and noncausal in i for $|\alpha| < \frac{1}{2}$ and $|\gamma + 2\alpha| < 1$. In other words, by simply re-indexing the variables and writing (7) as a seemingly-causal equation, e.g.,

$$u_{i+1,j} = \frac{1}{\alpha} u_{i,j} - \frac{\gamma}{\alpha} u_{i,j-1} - u_{i-1,j} - \frac{1}{\alpha} \varepsilon_{i,j} \tag{8}$$

(8) does not become a causal representation. This is because, if solved as an initial value problem, it is unstable. However, (7) is stable as an initial-boundary value problem and arises by numerical differencing of the well-known diffusion equation (Jain, 1977). Semicausal models of this type lead to vector DPCM or the so-called hybrid algorithms. Here, a transform coding, or filtering, is performed along the noncausal variable and predictive coding (or recursive filtering) is performed in the other direction (Jain and Wang, 1977; Jain and Jain, 1978). Figure 2 shows a typical semicausal model-based hybrid coding. From each image column u_j, first the boundary effects (i.e., the information about u_j contained in the boundary points) are subtracted. The residual u_j^o is transformed by its Karhunen Loeve (KL) transformation and each successive transformed sample is coded predictively by an independent DPCM channel. It can be shown that 1) the KL transform vectors are the eigenvectors of the noncausal part of the operator, and 2) the eigenvalue distribution associated with these eigenvectors determines the

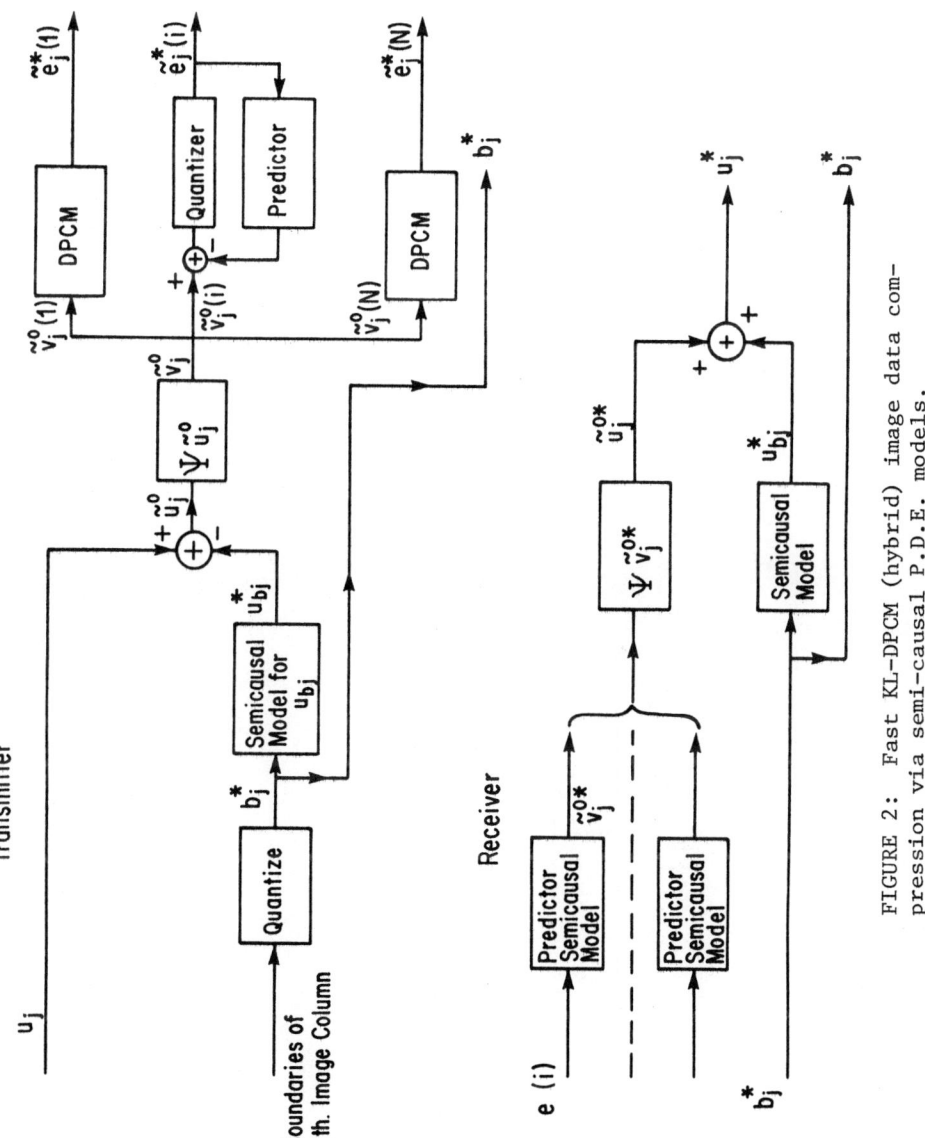

FIGURE 2: Fast KL-DPCM (hybrid) image data compression via semi-causal P.D.E. models.

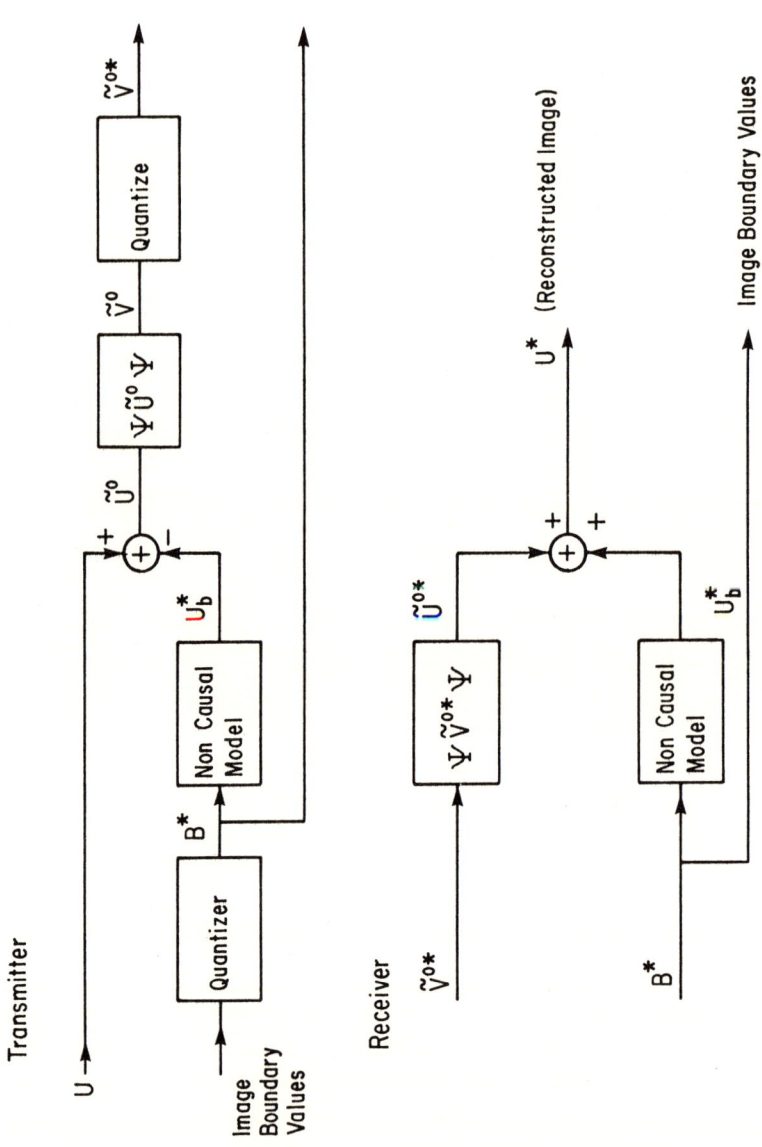

FIGURE 3: (Fast KL) Transform image data compression via non-causal P.D.E. models.

optimal bit allocation for predictive quantization of transformed samples. Extensive studies for compression and filtering of images, using semicausal models have been made. Details may be found in Jain and Wang (1977) and Wang (1979).

NONCAUSAL MODELS

These models are associated with elliptic PDE's. For example,

$$u_{i,j} = \alpha(u_{i-1,j} + u_{i+1,j} + u_{i,j-1} + u_{i,j+1}) + \varepsilon_{i,j} \tag{9}$$

is a discrete approximation of a Poisson equation. When $\{\varepsilon_{i,j}\}$ is a white noise field, this represents a Markov-2 field. If $\{\varepsilon_{i,j}\}$ is a certain moving average process with covariances

$$r_\varepsilon(k,\ell) = \beta^2 \begin{cases} -\alpha, & k=0, \ell=\pm 1 \quad \text{or} \quad \ell=0, k=\pm 1 \\ \\ 1, & k-\ell=0 \end{cases} , \tag{10}$$

then (9) is a Markov-1 field. These models are useful in transform coding (or filtering) and easily realize isotropic covariance functions (Jain, 1977). A fast KL transform coding algorithm (Jain, 1976) introduced earlier was based on such a representation. Figure 3 shows this method for general noncausal models. The KL transform is obtained from the eigenvectors associated with the noncausal operator D and the eigenvalues determine the bit allocations for quantization of transform samples. Details may be found in Wang (1979).

SUMMARY AND CONCLUSIONS

In summary, there are three basically different classes of realizations of random fields. The covariance functions realized by one class need not be realizable by a rational operator from the other two classes. The differences in spatial structures of these algorithms give rise to three different types of algorithms. Causal models give predictive coding such as DPCM algorithms. Semicausal models yield hybrid algorithms (e.g., KL/DPCM coding) and noncausal models yield transform coding algorithms. The coder design, e.g., KL transform, bit allocation etc., is obtained by studying the properties of the partial difference equation operator D. Extension to adaptive coding is possible by updating the model with statistical changes or by using a different model in different areas of the image.

We conclude by saying that the model classes suggested here offer a unifying framework for representation and processing of images. Further work is required to develop techniques for identification of such models.

REFERENCES

Jain, A. K., 1976.   A fast Karhunen Loeve transform for a class of
     random processes.   *IEEE Trans. Communications* COM-24:   1023-1029.
Jain, A. K., 1977.   Partial differential equations and finite dif-
     ferences in image processing.   Part I: Image representation.
     J. Optimization Theory and Appl. 23:   65-91.
Jain, A. K. and J. R. Jain, 1978.   Partial differential equations
     and finite differences in image processing.   Part II: Image
     restoration.   *IEEE Trans. Aut. Control* AC-23:   817-834.
Jain, A. K. and S. H. Wang, 1977.   Stochastic image models and
     hybrid coding.   *Dept. of Electrical Engineering, SUNY at Buf-
     falo, Amherst, New York, Final Rep., Contract N00953-77-C-003-
     MJE.*
Wang, S. H., 1979.   Application of stochastic models in image coding.
     Ph.D. Thesis, Dept. of Electrical Engineering, SUNY at Buffalo.

SATELLITE AND AIRCRAFT-BASED REMOTE SENSING

Synoptic Measurements in Marine Biology

Robert W. Johnson

ABSTRACT: Experiments conducted in the United States Coastal Zone indicate that certain pollution and oceanographic features have distinctive spectral characteristics. Remotely sensed wide area synoptic coverage of these features provides information that is not readily available by other means. Remote sensor data outputs that may be of particular interest to oceanographers include 1) synoptic high resolution surface measurements of particulates, salinity, temperature, chlorophyll a, and possibly plankton diversity from airborne multispectral, passive microwave, and active laser sensors; and 2) repetitive ocean surface data from environmental satellites (e.g., LANDSAT, SEASAT 1, NIMBUS 7, etc.).

High resolution aircraft remotely sensed data may be interpreted by two methods. First, quantitative analyses using multiple regression techniques were applied to develop statistically significant relationships between concurrent sea-truth measurements and remotely

sensed data from multispectral scanner, passive microwave and active laser systems. Calibrated regression equations, which are the
results of these analyses, have been used to
map distributions of parameters such as suspended solids, chlorophyll a, temperature, and
salinity in the coastal zone. Second, qualitative analysis techniques were applied in
which features have been located, identified,
and mapped without concurrent sea-truth. In
these qualitative analyses an in-scene background elimination technique has been developed
that "normalizes" environmental effects. This
approach potentially provides a means of identification of features, such as river plumes
or pollutant sources (e.g., from ocean dumping
of wastes), that is independent of the remote
sensor used or the specific scene monitored.

Repetitive coverage data from environmental
satellites have been used to study features
using suspended solids as tracers. LANDSAT
data have been used to study river plumes,
and to locate ocean dump plumes and study
their long-term drift in and around the dump
zones. The primary oceanographic applications
in the coastal zone for current satellite optical range remote sensors appear to be for
wide area and/or repetitive measurements of
well defined features such as sediment plumes
and spectrally different ocean water masses.

INTRODUCTION

    Experiments conducted in United States Coastal Zones indicate
that certain pollution and oceanographic features have distinctive
spectral characteristics. Remotely sensed wide area synoptic coverage provides information on these features that is not readily
available by other means. Data outputs from remote sensors may
be analyzed and interpreted to provide information of interest to
oceanographers and environmental managers. This information includes 1) synoptic high resolution surface measurements of particulates, salinity, temperature, chlorophyll a, and possibly phytoplankton diversity from airborne multispectral (optical range),
passive microwave and active laser systems and 2) repetitive ocean
surface data from environmental satellites (LANDSAT, SEASAT 1,
NIMBUS 7, etc.). McClain (1977) reviewed the National Oceanic and
Atmospheric Administration (NOAA) operational satellites; however,
low spatial resolution limits application of the data in coastal
zone oceanographic experiments.
    Duntley (1971) indicated that spectral responses at different
wavelengths could be related to changes in chlorophyll a concen-

trations. Various investigators have applied classification or
regression techniques to calibrate satellite (LANDSAT) and aircraft multispectral scanner data and to map distributions of water
quality parameters in inland and coastal zone systems, Williamson
and Grabau (1973), Yarger et al. (1973), Klemas et al. (1974),
Bowker et al. (1975), Rogers et al. (1975), Johnson (1977a), and
Johnson and Bahn (1977). Subsequently, experiments were conducted
in the United States Coastal Zone to determine the applicability
of aircraft and satellite remote sensing systems 1) to locate, identify, and map features without the requirement for concurrent sea-
truth measurements and 2) to evaluate previously developed analysis
techniques for determining quantitative distributions of specific
parameters. As a part of (1) above, multispectral analysis techniques were developed that may be used to identify features at a
variety of locations and to assess the effects of environmental
conditions (e.g., sun brightness and elevation, atmospheric haze
and background) on the analysis and interpretation of remotely
sensed data (Johnson, 1977b).

Remote sensing in the microwave spectral range from airborne
platforms has been reviewed by Swift and Sorrell (1978). Experiments using L and S band (1.43 and 2.65 GHz) radiometers have been
conducted by the National Aeronautics and Space Administration's
(NASA) Langley Research Center to determine distributions of salinity and temperature. These measurements, which are based on "microwave brightness" radiation measurements, may be made over a wide
range of environmental conditions (e.g., cloudy and hazy days, at
night), thus increasing the opportunity for monitoring the coastal
zone.

Evaluations of active systems include the Airborne Lidar
Oceanographic Probing Experiment (ALOPE) (Brown et al., 1977) which
measures the induced fluorescence in four optical range spectral
bands to estimate total chlorophyll a concentrations and chlorophyll
distribution among four phytoplankton color groups (green, golden
brown, red, and blue-green) thereby providing a measure of phytoplankton diversity. Following successful laboratory evaluations,
this helicopter-mounted instrument is in the early stages of a
field evaluation program. As an active system, the ALOPE is not
limited to daylight or clear weather operation, as in the case of
passive optical systems which depend on reflected radiation. Concurrent sea level measurements of water attenuation coefficients
are required for instrument calibration.

Investigations by Bowker and Witte (1977) using satellite
(LANDSAT) data indicate similar analysis techniques are applicable
to digital data from aircraft and satellite systems. In the spacecraft data, features with high radiance contrast such as river discharges and ocean disposal plumes are readily discernable. Repetitive LANDSAT data have been analyzed by Klemas et al. (1977) to
study the drift of plumes from dumping of industrial acid wastes
(which has a persistence of up to about 3 days, under some conditions) in and around the designated dump zone.

It is the purpose of this paper to review results from investigations that utilized remote sensing to locate features and/or
measure parameter gradients in the coastal zone. Data collection,

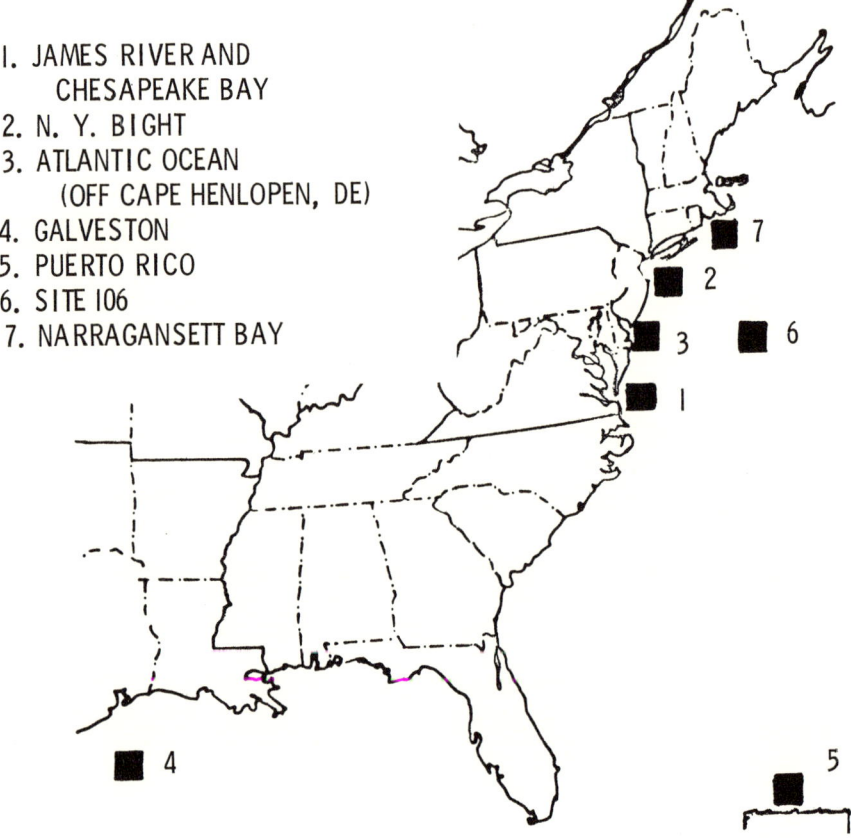

1. JAMES RIVER AND
   CHESAPEAKE BAY
2. N. Y. BIGHT
3. ATLANTIC OCEAN
   (OFF CAPE HENLOPEN, DE)
4. GALVESTON
5. PUERTO RICO
6. SITE 106
7. NARRAGANSETT BAY

FIGURE 1: Locations of experiments in the United
States Coastal Zone.

format, and computerized analysis techniques that use digital data
from generally available instruments such as multispectral scanners
(on aircraft and satellite platforms) will be discussed from a user
viewpoint. Much of this information also will be applicable to instruments under development such as passive microwave and active
laser sensors.

EXPERIMENTS

     Locations of several experiments are shown on Figure 1 and per-
tinent experiment features are listed in Table 1, along with references for more detailed information. Selected experiments will
be described in detail in subsequent sections to illustrate remote
sensing data collections, data handling, analysis techniques, and
products.

TABLE 1: Experiment locations and other features.

Experiment Location (See Fig. 1)	Sensor*	Parameter Measured	References
1	MSS, M2S, camera	SS**, chloro a	Bowker et al. (1975) Johnson (1978) Johnson and Bahn (1977)
	Microwave	Temperature, salinity	Blume et al. (1977)
2	MSS, OCS, M2S, camera	SS, chloro a	Bowker and Witte (1977) Johnson (1977a) Johnson (1977b) Johnson (1978) Johnson and Ohlhorst (in press)
3	MSS, M2S, camera	SS	Bowker and Witte (1977) Klemas et al. (1977) Johnson and Ohlhorst (in press)
4	M2S, camera	SS	Johnson and Ohlhorst (in press)
5	M2S, camera	SS	Johnson and Ohlhorst (in press)
6	M2S, camera	SS	***
7	ALOPE	Chloro a, diversity	Farmer et al. (in press)

*MSS - Satellite (LANDSAT)
 OCS, M2S - Fixed Wing Aircraft
 Microwave - Fixed Wing Aircraft
 ALOPE - Helicopter
**Suspended Solids
***Data Being Analyzed

Synoptic High Resolution Airborne Measurements

Remotely sensed data were collected in conjunction with sea-truth measurements over environmentally different areas of the coastal zone, 1) to study distributions of water quality parameters and 2) to assess the applicability of remote sensors for monitoring ocean dumping. Emphasis will be placed on two experiments that determined quantitative distributions of parameters of interest to the marine biologist (e.g., chlorophyll a and suspended solids) in the James River, Virginia, and the New York Bight.

On May 28, 1974, NASA and the Environmental Protection Agency (EPA) conducted a joint experiment over the James River, Virginia (site 1, Fig. 1). Remotely sensed data were collected by a Modular Multispectral Scanner (M2S) onboard a Bendix Aerospace Systems Division aircraft at a flight altitude of 2.4 kilometers (km) (8,000 ft.). The M2S has 11 bands in the visible, near infrared (IR), and thermal IR spectral ranges. Only those data in the visible and near IR spectral ranges will be discussed. Bandwidths and wavelengths of the M2S are listed in Table 2, along with spatial coverage information at the 2.4 km flight altitude. Picture element (pixel) size and resolution were about 7 m (23 ft). Sea-truth measurements of chlorophyll a and suspended solids concentrations along with corres-

ponding radiance values in the M2S bands were made for 21 stations.
Measurements of these and other parameters investigated have been
reported by Johnson and Bahn (1977).

From April 7-17, 1975, a joint experiment was conducted by
NASA, the National Oceanic and Atmospheric Administration (NOAA),
and EPA in the New York Bight (site 2, Fig. 1) to study marine
processes and pollution effects. On April 13, 1975, remotely
sensed data were collected by a 10 band Ocean Color Scanner (OCS)
onboard a NASA Ames Research Center (ARC) U-2 aircraft which flew
at an altitude of 19.7 km (65,000 ft.). The OCS has 10 bands in
the visible and near IR spectral ranges, with center wavelengths
from 430 nanometers (nm) to 772 m for the nominally 20 nm wide
bands. Bandwidths and wavelengths along with spatial coverage information for the OCS are shown in Table 2. Sea-truth measurements for this experiment were collected by NOAA personnel on-
board a Coast Guard helicopter over stations in the New York
Bight apex. Sea-truth measurements of chlorophyll a and suspended
solids concentrations along with corresponding radiance values in
the OCS bands were made. Only bands 1 through 8 of data were
available for the analysis. Eighteen sets of observations were
usable in the chlorophyll a analysis and 22 for suspended solids.
Measurement of these and other parameters, such as particle size
distributions, have been reported by Johnson (1977a).

Additional experiments have been conducted at sites 3 through
6 (see Fig. 1), to investigate the applicability of remote sensors
for monitoring ocean dumping and for measuring the effects of
these discharges on the marine ecosystem. Experiment objectives
included development of monitoring techniques (e.g., location,
identification, and mapping without concurrent sea-truth) and
evaluation of quantitation techniques (using concurrently collected
sea-truth) for studying spatial and temporal dispersion characteristics of plumes resulting from dumping of sewage sludge and industrial waste materials (Johnson et al., 1979; Johnson and Ohlhorst,
in press).

Data Preprocessing   Representative radiance values corresponding to the sea-truth measurements were determined by locating
the sampling station as nearly as possible in the imagery, then
taking the average within a pixel field centered at that location
to obtain the representative value. For the May 28, 1974, James
River and April 13, 1975, New York Bight experiments an 11 x 11
pixel field was used to determine representative radiance values.
This field size was empirically determined for each remotely sensed
data set as the area to compensate for uncontrollable spectral and
spatial noise or uncertainty. Inflight calibrations were used to
determine average radiance values in each of the M2S or OCS bands.

Quantitative Data Analysis and Mapping   Multiple regression
analysis, specifically stepwise regression, was used to determine
calibrated regression equations for quantitatively relating sea-
truth measurements to remotely sensed data, as in Johnson and Bahn
(1977). In stepwise regression analysis (SWRA) the program selects
the independent variable (radiance in one scanner band) that has

TABLE 2: Spectral and spatial characteristics of
multispectral scanners.

I. M2S (2.4 km altitude)	
Spectral	
Band	Range
1	380 – 440 nm
2	440 – 490 nm
3	495 – 535 nm
4	540 – 580 nm
5	580 – 620 nm
6	620 – 660 nm
7	660 – 700 nm
8	700 – 740 nm
9	760 – 860 nm
10	970 – 1,060 nm
Thermal	8000 – 13,000 nm

Spatial

Field of view	
width, m	6,800
length, m	Continuous
Resolution, m	7

II. OCS (19.7 km altitude)	
Spectral	
Band	Range
1	422 – 444 nm
2	460 – 482 nm
3	486 – 513 nm
4	535 – 560 nm
5	570 – 595 nm
6	607 – 533 nm
7	651 – 673 nm
8	688 – 708 nm
9	722 – 745 nm
10	760 – 783 nm

Spatial

Field of view	
width, m	25,000
length, m	Continuous
Resolution, m	75

the highest correlation with the dependent variable (e.g., chlorophyll a). A number of variables are then selected consecutively
until each of the independent variables that makes a significant
contribution to determining the dependent variable are in the regression equation and the others are "outside" of the regression.
Limiting the regression equation to significant variables reduces
the analysis time and improves the accuracy of the results. The
criterion for inclusion of variables is a 95% confidence level (90%
for the OCS data collected on April 13, 1975, due to apparently
higher "noise") as determined by the statistical "F" test (see
Draper and Smith (1966) for a discussion of the SWRA and "F" test).
     Results of SWRA applied to the May 28, 1974, James River data
set, first with chlorophyll a as the dependent variable and second
with suspended solids as the dependent variable, are as follows:

Water Quality Parameter	M2S Bands in Re- gression Equation	Standard Error of Estimate	Correlation Coefficient	Correlation Coefficient to Suspended Solids	Range of Sea-Truth Measure- ments
Chloro- phyll a	M2SR2, M2SR6, M2SR8	1.75 mg/m^3	0.96	0.88	1.61–19.5 mg/m^3
Suspend- ed So- lids	M2SR2 M2SR6	5.23 mg/l	0.90	–	8.60–47.60 mg/l

where M2SRN is the radiance in M2S band N (i.e., M2SR2 is radiance
in band 2); standard error of estimate is a measure of the scatter about the fitted regression line; correlation coefficient is a
measure of the relative change among variables; correlation coefficient to suspended solids is the linear correlation coefficient
of chlorophyll a to suspended solids; and range of sea-truth measurements is for the water quality parameter being analyzed. The regression equation for chlorophyll a is

$$\text{Chloro } a, \text{ mg/m}^3 = 25.63 - 18.06 \text{ M2SR2} + 7.23 \text{ M2SR6} + 4.46 \text{ M2SR8}$$

     Comparison of the remotely sensed values, calculated from the
linear regression equation, and measured sea-truth values for chlorophyll a and subsequently suspended solids, indicated approximately
random distribution about the fitted regression lines for these parameters in these ranges of measurements.
     Results of SWRA applied to the April 13, 1975, New York Bight
data set first with chlorophyll a as the dependent variable and sec-

ond with suspended solids as the dependent variable, were as follows:

Water Quality Parameter	M2S Bands in Regression Equation	Standard Error of Estimate	Correlation Coefficient	Correlation Coefficient to Suspended Solids	Range of Sea-Truth Measurements
Chloro-phyll a	OCSR3, OCSR6	3.87 mg/m^3	0.83	0.90	2.20-24.30 mg/m^3
Suspend-ed So-lids	OCSR6	1.39 mg/l	0.79	-	0.46-8.38 mg/l

where OCSR6 is radiance in OCS band 6, etc.; and the other symbols
are as defined previously. The regression equation for chlorophyll
a is

$$\text{Chloro } a, \text{ mg/m}^3 = -4.51 - 4.96 \text{ OCSR3} + 11.44 \text{ OCSR6}$$

Comparison of the analysis results presented above indicates
several interesting features. First, the most significant spectral
regions are similar, from about 450 to 750 nm, for the two scanners
which were flown at different altitudes. However, specific equation
coefficients are different due to the number of bands in the regression equations, their specific wavelengths, and the generally
higher radiance levels in the New York Bight experimental area. Second, the M2S indicates higher correlations between remotely sensed
data and sea-truth measurements for both chlorophyll a and suspended solids. A number of factors such as spectral and spatial
resolution and environmental factors make interpretation of this
effect difficult. Finally, correlations among sea-truth and remotely sensed measurements will be discussed in the section on quantitative mapping.

Quantitative mapping of water quality parameter concentration
distributions may be determined from the regression equations. For
each water quality parameter, concentrations are determined for each
pixel (or equal spacings of pixels on lines and columns). These
data are typically smoothed to remove local spectral and spatial
noise features, and then a contour map is developed by computer.
The smoothing used in these analyses is an averaging on a line-by-
line and column-by-column basis in the data field where the middle
value is replaced by the arithmetic mean of it and the two adjacent
pixels. Edge values remain the same. In the James River data set,
the data field used every fourth pixel in each eighth line with five
smoothing passes; in the New York Bight data set, each pixel was
used in every third line with two smoothing passes. The actual
spacing and smoothing were determined empirically for each set of
data.

Quantitative distributions of chlorophyll a in the James River

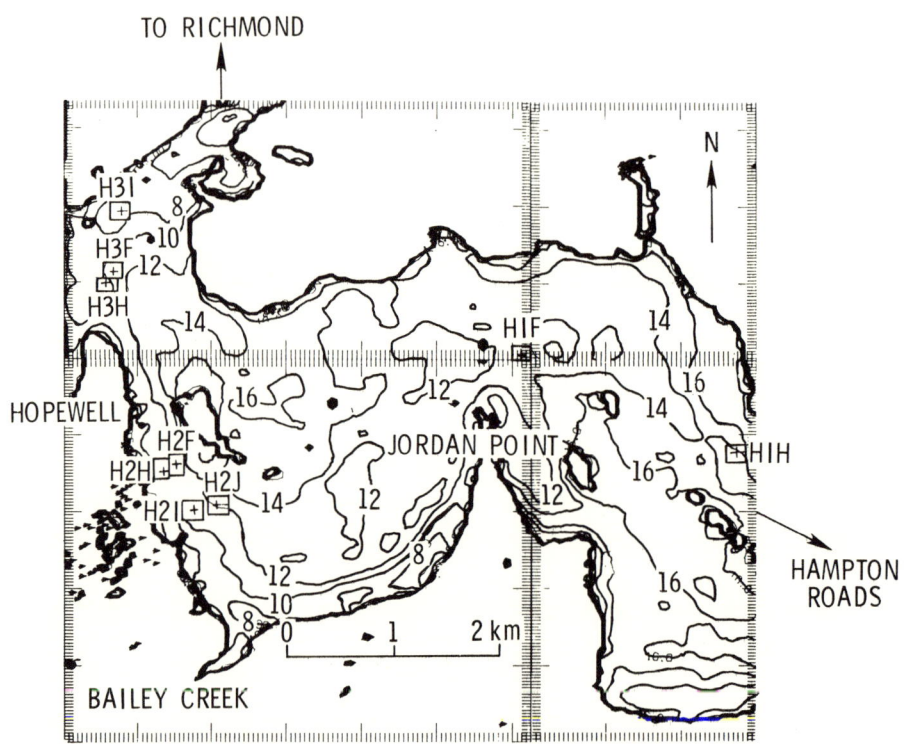

FIGURE 2: Quantitative distributions of chlorophyll a (concentrations in mg/m^3) in the James
River, Virginia, on May 28, 1974. Truth stations
also are located. (Taken from Johnson, 1978).

near Hopewell and for the New York Bight apex are shown in Figures
2 and 3, respectively. Sea-truth measurement locations are also indicated (Johnson, 1978).

A feature of particular interest in the James River experimental area is Bailey Creek, located in the lower left corner of
the map (Fig. 2). Bailey Creek is a source of sewage treatment
plant and industrial effluent and its plume is characterized by
lower chlorophyll a concentrations, probably due to toxic materials.
In the New York Bight experiment (Fig. 3) high chlorophyll a concentrations are indicated in the near-shore areas. The Hudson
River plume as it flows into the New York Bight is also characterized by higher chlorophyll a concentrations. Another feature
of interest in this scene is the acid waste plume that is qualitatively mapped in the lower center of Figure 3 (qualitative mapping will be discussed in a later section).

Analysis results, presented previously, may be used to evaluate whether the distributions shown in Figures 2 and 3 are unique
to chlorophyll a. One approach to evaluation is by comparison of

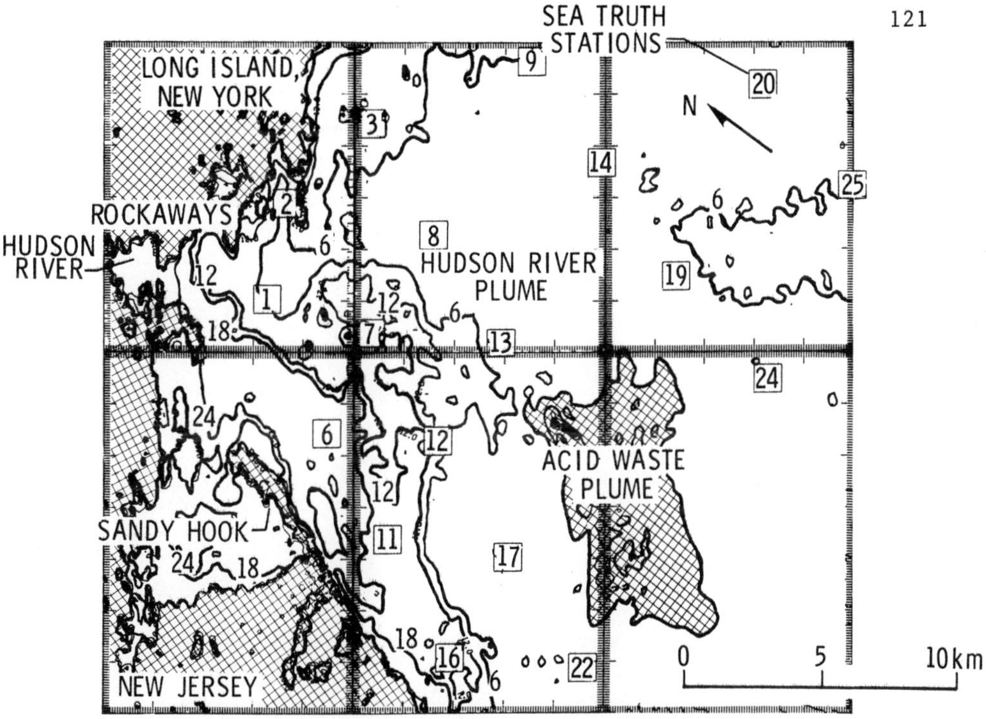

FIGURE 3: Quantitative distributions of chlorophyll a (concentrations in mg/m³) in the New York
Bight on April 13, 1975. Sea truth stations also
are located. (Taken from Johnson, 1978).

correlation coefficients (recall that correlation coefficients
measure the relative changes among variables) as shown below:

	James River Experiment	New York Bight Experiment
Correlation coefficient, chlorophyll a to suspended solids	0.88	0.90
Correlation coefficient, chlorophyll a to remotely sensed data	0.96	0.83
Correlation coefficient, suspended solids to remotely sensed data	0.90	0.79

For both experiments correlation of sea truth to sea-truth measurements (chlorophyll a to suspended solids) is about the same. However correlations between sea-truth measurements and remotely sensed

data are appreciably higher in the James River experiment than those
for the New York Bight. The higher correlation of chlorophyll a to
remotely sensed data indicates a unique chlorophyll a mapping in the
James River experiment. The lower correlation coefficient values of
both chlorophyll a and suspended solids to remotely sensed data in
the New York Bight indicates a combined response in which neither
parameter may be unambiguously separated. This conclusion is further
supported by the relatively high standard error of estimate for chlorophyll a in the New York Bight experiment. The specific reason for
this difference in the analysis results cannot be identified; however, it is probably related to combined environment (e.g., atmospheric and/or pollutant mixtures in the water) and instrument (e.g.,
spectral and spatial resolution) effects.

Qualitative Analysis and Mapping   Qualitative analysis of features such as plumes from ocean dumping and river discharges includes
location, identification, and mapping their extent. In general,
these pollution features have radiance levels different from those of
the background water in one or more spectral regions. In almost all
cases, plumes in the coastal zone have higher radiances than the background water that are due to suspended materials in the plumes (e.g.,
Klemas et al., 1974). These differences have been observed in photographic and digital remotely sensed data.

Identification of features without concurrent sea-truth measurements requires consideration of atmospheric as well as pollutant
spectral responses. One method of "normalizing" environmental effects (e.g., atmospheric, sun angle and background) between scenes
is to use an in-scene background elimination (Johnson, 1977b). This
approach, also used in the results reported here, was to determine
ratios of plume radiances to background ocean water for the same remotely sensed scene in each of the multispectral scanner bands.
These ratio values as a function of wavelength indicate distinctive
characteristics that, when combined with geographical information,
may be used for identification of plumes. Example radiance ratio
curves for ocean dumped sewage sludge and three types of industrial
waste (acid, petrochemical and pharmaceutical) plumes are shown in
Figure 4 (Johnson and Ohlhorst, in press). Note that the pharmaceutical waste plume indicates lower radiance values than the ocean
water (radiance ratio values less than 1.0), probably due to absorption of the liquid or particles in the waste. After the plumes
have been located and identified, they may be mapped using equal-
radiance mapping and/or classification techniques.

Passive Microwave

Experiments have been conducted in the Chesapeake Bay (location
1, Fig. 1) using passive microwave radiometers to determine temperature and salinity distributions. Demonstration tests of a microwave
system, incorporating nadir viewing L and S band (1.43 and 2.65 GHz)
radiometers by Blume et al. (1977), indicated that surface temperature can be measured to within 1°C and salinity with a mean deviation of 1 ppt (for salinity greater than 5 ppt). To obtain these
accuracies, the radiometers were calibrated before each flight and

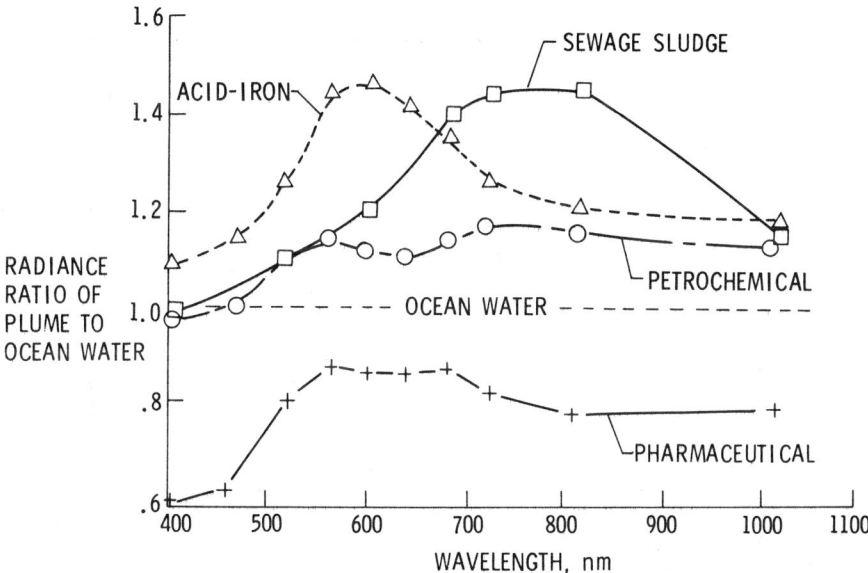

FIGURE 4: Spectral characteristics of ocean
dumped materials.

the radiometric data were corrected for extraterrestrial background
radiation, atmospheric effects, sea surface roughness, and antenna
beam efficiency. By flying a series of parallel east-west flight
lines a data grid was developed and used to plot isothermals (Fig.
5) and isohalines (Fig. 6) of the lower Chesapeake Bay and nearby
ocean water.

Active Laser

An Airborne Lidar Oceanographic Probing Experiment (ALOPE) system has been developed and laboratory tested (Brown et al., 1977) at
the NASA Langley Research Center. The laser in the system is a unique
four-color dye laser pumped by a single linear xenon lamp. Fluorescent dyes, which lase at 454, 539, 598, and 617 nm, are the active media for the separate dye lasers, which are separately activated using a rotating intercavity shutter. Induced fluorescence from chlorophyll a in the water is collected through a telescope, passed through
a narrow band pass filter (centered at 685 nm), digitized and recorded.
Results of laboratory tests using a mixture of four species (one from
each of four color groups; blue-green, golden brown, green, and red)
indicated correlation coefficients (between cell counts and ALOPE
measurements) of 0.994 for total chlorophyll a and from 0.81 to 0.99
for the individual color groups.
More recently, results from local field tests and an NASA/Univer-
sity of Rhode Island/EPA Narragansett Bay Experiment (site 7, Fig. 1)
confirm the applicability of the ALOPE system for monitoring total
chlorophyll a and phytoplankton diversity in estuarine systems (Farmer
et al., in press). The local field experiments were conducted with

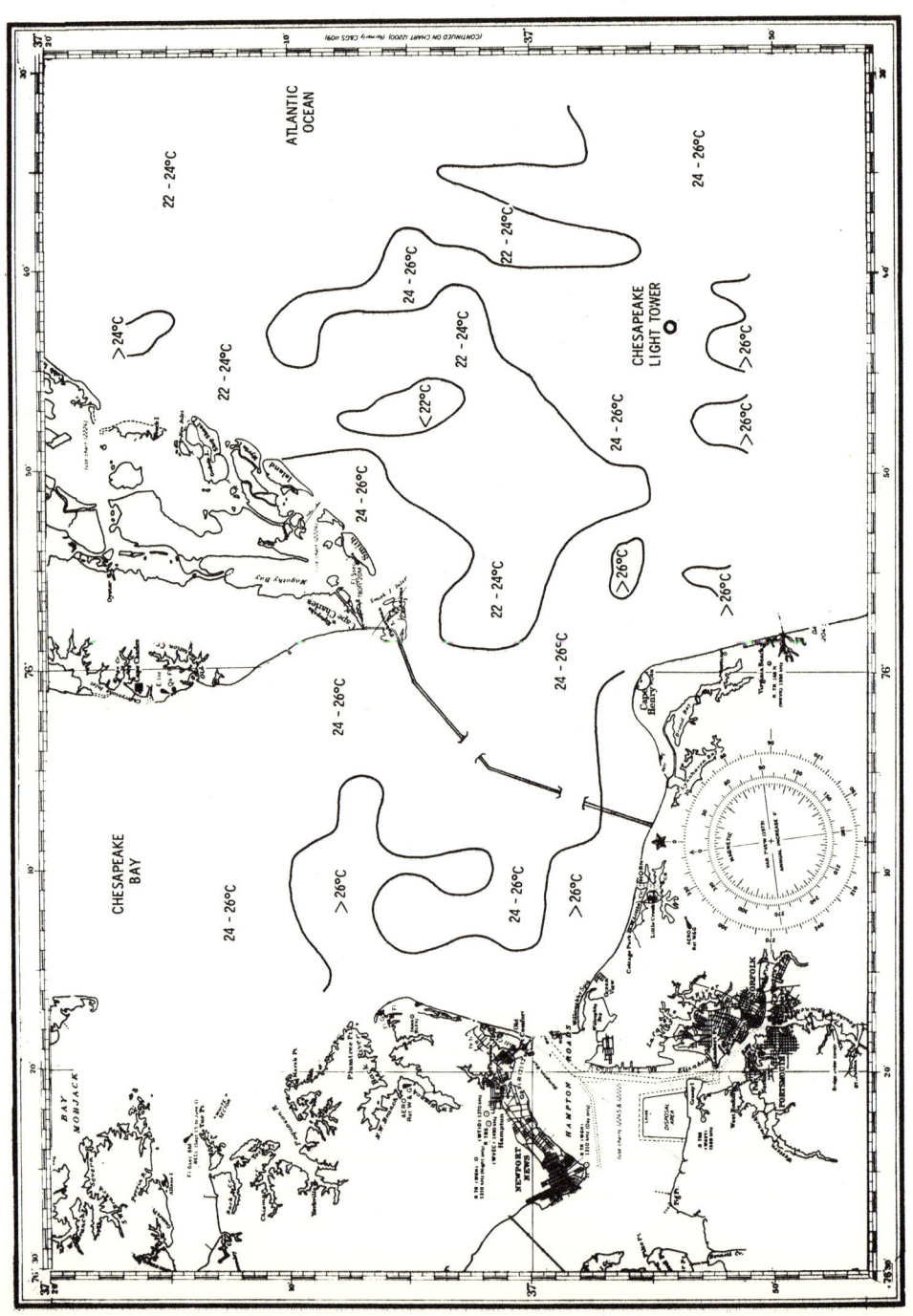

FIGURE 5: Isothermals of the lower Chesapeake Bay on August 24, 1976. (Taken from Blume et al., 1977).

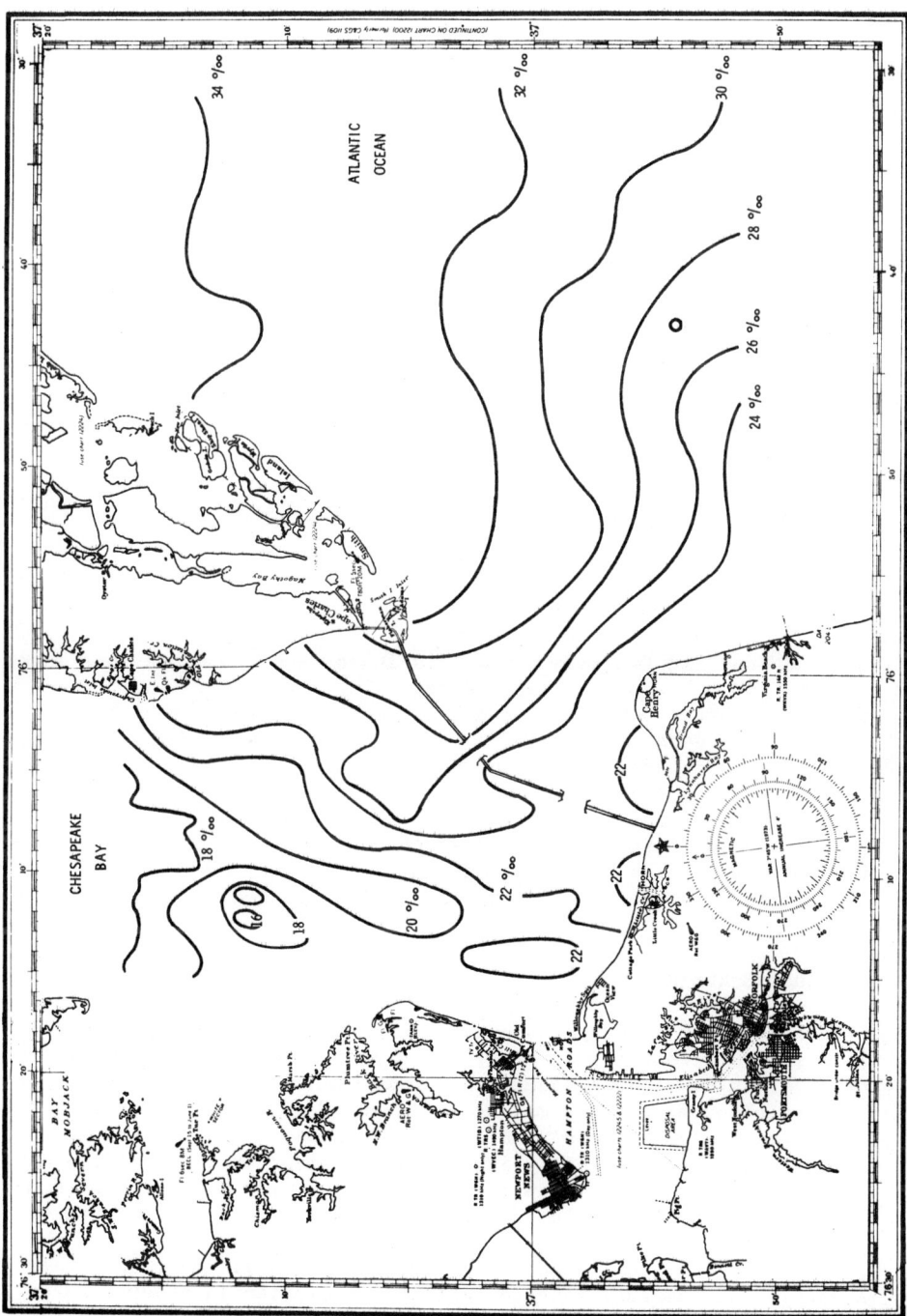

FIGURE 6: Isohalines of the lower Chesapeake Bay on August 24, 1976. (Taken from Blume et al., 1977).

the ALOPE system mounted on docks or other fixed structures. In the
Narragansett Bay experiment, the instrument was mounted on a helicopter platform (nominal flight altitude 30 m). Sea-truth measurements
for calibration of the remotely sensed data include *in situ* water attenuation coefficients.

Environmental Satellites

Multispectral data collected over estuarine and coastal zones
from LANDSAT environmental satellites have been evaluated to study
water quality and pollution features, such as suspended solids, chlorophyll a, and plumes resulting from ocean dumps. Using suspended
solids as a tracer, Klemas et al. (1974) studied current flow and
plumes in the Delaware Bay and nearby coastal areas. Klemas et al.
(1977) also used LANDSAT data to study the drift of plumes resulting
from ocean dumping of industrial acid waste at a site in the Atlantic
Ocean about 40 mi east of Cape Henlopen, Delaware (site 3, Fig. 1).
Bowker et al. (1975) and Bowker and Witte (1977) investigated
the utility of repetitive LANDSAT coverage for monitoring 1) changes
in the lower Chesapeake Bay and 2) ocean dumping. Industrial acid
waste dump plumes which have a persistence of up to 5 days were located in a number of cases; however, other industrial wastes and sewage sludge were only rarely detected, probably due to short persistence.
LANDSAT digital data are available on Computer Compatible Tapes
(CCT's) through the EROS Data Center. The format (typically nine-
track, 800 bytes per inch (bpi))is essentially the same as for the
airborne M2S data and many of the preprocessing and analysis techniques described earlier also apply.
Increased coverage frequency and improved spectral and spatial
resolution in planned satellite sensors will provide more detailed
measurements. Information on present and proposed satellite systems
that may provide coastal zone data of interest to oceanographers is
summarized in Table 3 (Johnson and Harriss, personal communication).

CONCLUDING REMARKS

Results of experiments conducted in the United States Coastal
Zone indicate that remote sensing may provide data and measurements
of interest to oceanographers and environmental managers. High resolution airborne sensors may be utilized to provide information on
oceanographic parameters such as particulates, salinity, temperature,
chlorophyll a, and possibly phytoplankton diversity. Repetitive data
collections from environmental satellites provide synoptic and temporal information on high radiance contrast features such as river
plumes and plumes resulting from ocean dumping.
Airborne optical spectral range sensors (e.g., multispectral
scanners and cameras) measure reflected radiation in several spectral bands. Parameters that have indicated radiance changes with
concentration changes include chlorophyll a and suspended solids.
Quantitative analysis techniques, using multiple regression, have
been developed to relate sea-truth measurements to corresponding

TABLE 3: Satellite platforms, sensors, and oceanographic parameters.

	Satellite and Sensor Characteristics					Oceanographic Parameter, Range (accuracy)				
Satellite	Coverage Frequency	Sensor	Spectral Range	Spatial Resolution (at Nadir), m	Swath Width, km	Suspended Solids, mg/ℓ	Chlorophyll a Conc., mg/m³	Diversity	Temperature Degrees K	Salinity ppt
Landsat (9 Day for 2)	18 Day	MSS	VIS,	70	185	>5 (±5)	"Bloom"(>10)	–	–	–
			TIR (*)	240	185	–	–	–	260-340(±1.5)	–
AEM-A	5 Day	HCMR	VIS, NIR	500	700	Yes (**)	–	–	–	–
			TIR	500	700	–	–	–	Yes (**)	–
Nimbus-7	6 Day	CZCS	VIS, NIR	800	1200	Yes (**)	Yes (**)	–	–	–
			TIR	800	1200	–	–	–	Yes (**)	–
SMS/GOES	30 min	VISSR	VIS	1000	3500	Qual.	–	–	–	–
			TIR	8000	3500	–	–	–	180-315(±2)	–
TIROS-N	0.5 Day	AVHRR	VIS, NIR	1000	2800	Qual.	–	–	–	–
			TIR	1000	2800	–	–	–	180-315(±1)	–

VIS – Visible; 0.3 – 0.7 µm (Typical)
NIR – Near IR: 0.7 – 1.1 µm (Typical)
TIR – Thermal IR; 10.5 – 12.5 µm (Typical)
* Landsat C only; one of two detectors operational
** Recent Launch, data being evaluated
– No identified capability

remotely sensed data. Results of these analyses, calibrated regression equations, have been used to map distributions of parameters such as chlorophyll a and suspended solids.

Qualitative analyses use photoanalysis or other techniques
in which the interpretations do not require concurrent sea-truth
measurements. In general, features of interest, such as a river
plume or a plume from ocean dumping of waste materials have radiance levels different from surrounding water. After these features have been located, spectral analysis techniques assist the
identification. The extent of the feature may be mapped using
equal radiance mapping and/or classification techniques. An environmental application of these techniques is the monitoring of
ocean disposal of waste materials.

Microwave spectral range measurements provide information on
temperature and salinity of surface waters. Synoptic coverage
from parallel aircraft flight lines provides data for mapping distributions of these parameters. Microwave measurements may be
made over a wide range of environmental conditions. Current configurations require calibration for surface water roughness and
other factors.

Helicopter mounted active laser remote sensors such as the
ALOPE may be used to determine total chlorophyll a density and
to estimate phytoplankton diversity. Laboratory results have
indicated statistically significant correlations between the remotely sensed and the measured densities and diversity. Preliminary evaluations of data from a field test in the Narragansett
Bay support results obtained in the laboratory. As an active
system, the ALOPE is not limited to daylight operations, as are
passive optical spectral range sensors. At the present time, *in
situ* attenuation coefficient measurements are required.

Repetitive remotely sensed measurements from environmental
satellites provide wide area synoptic measurements on features
that exhibit high radiance contrast. Proposed environmental and
meteorological satellites will provide improved spectral and spatial resolution as well as increased coverage frequency. Presently, the primary applications for optical range remote sensors
appear to be for wide area and/or repetitive measurements of well
defined features such as sediment plumes and spectrally different
ocean water masses.

REFERENCES

Blume, Hans-Jurgen C., Bruce M. Kendall, and John C. Fedors, 1977.
    Sea-surface temperature and salinity mapping from remote mi-
    crowave radiometric measurements and brightness temperature.
    NASA TP 1077. 26 pp. National Technical Information Service,
    Springfield, Virginia.
Bowker, D. E. and W. G. Witte, 1977. The use of Landsat for moni-
    toring water parameters in the coastal zone. *Proceedings of
    the AIAA Joint Conference on Satellite Applications of Marine
    Applications,* American Institute of Aeronautics and Astro-
    nautics. pp. 193-198.

Bowker, D. E., W. G. Witte, P. Fleischer, T. A. Gosink, W. J. Hanna, and J. Ludwick, 1975. An investigation of the waters in the lower Chesapeake Bay area. *Proceedings of the Tenth International Symposium on Remote Sensing of Environment,* Vol. 1. pp. 411-420. Environmental Research Institute of Michigan, Ann Arbor, Michigan.

Brown, Clarence A., Jr., Franklin H. Farmer, Olin Jarrett, Jr., and Weldon L. Staton, 1977. Laboratory studies of *in vivo* fluorescence of phytoplankton. *Proceedings of the Fourth Joint Conference on Sensing of Environmental Pollutants.* *American Chemical Society* 782-788.

Draper, N. R. and H. Smith, 1966. *Applied Regression Analysis.* John Wiley and Sons, Inc., New York.

Duntley, S. Q., 1977. Optical methods of water pollution. *Proceedings of the Environmental Quality Sensor Workshop* at Western Environmental Research, Environmental Protection Agency, Washington, D. C. II-15-II-27.

Farmer, F. H., C. A. Brown, Jr., O. Jarrett, Jr., J. W. Campbell, and W. L. Station, in press. Remote sensing of phytoplankton density and diversity using an airobrne fluorosensor. *Advanced Concepts in Ocean Measurements for Marine Biology,* F. P. Diemer, F. J. Vernberg and D. Z. Mirkes (eds.). Univ. of South Carolina Press, Columbia.

Johnson, R. W., 1977a. Mapping the Hudson River Plume and an acid waste plume by remote sensing in the New York Bight Apex, April 1975. In: *Results from the National Aeronautics and Space Administration Remote Sensing Experiments in the New York Bight - April 7-17, 1975,* (J. B. Hall, Jr. and A. O. Pearson, compilers), pp. 106-129. NASA TM X-74032. National Aeronautics and Space Administration, Langley Research Center, Hampton, Virginia.

Johnson, R. W., 1977b. Multispectral analysis of ocean dumped materials. In: *Proceedings of the Eleventh International Symposium on Remote Sensing of Environment,* pp. 1619-1625. Environmental Research Institute of Michigan, Ann Arbor, Michigan.

Johnson, R. W., 1978. Mapping of chlorophyll *a* distributions in coastal zones by remote sensing. *Photogrammetric Engineering and Remote Sensing* 44: 617-624.

Johnson, R. W. and G. S. Bahn, 1977. Quantitative analysis of aircraft multispectral-scanner data and mapping of water-quality parameters in the James River in Virginia. *NASA TP 1021.* National Technical Information Service, Springfield, Virginia. 31 pp.

Johnson, R. W. and C. W. Ohlhorst, in press. Application of remote sensing to monitoring and studying dispersion in ocean dumping. In: *Proceedings of the First International Symposium on Ocean Dumping.* University of Rhode Island, Kingston, Rhode Island.

Johnson, R. W., I. W. Duedall, R. M. Glasgow, J. R. Proni, and T. A. Nelsen, 1977. Quantitative mapping of suspended solids in wastewater sludge plumes in the New York Bight Apex. *Journal Water Pollution Control Federation* 49: 2063-2073.

Johnson, R. W., R. M. Glasgow, I. W. Duedall, and J. R. Proni, 1979. Monitoring the temporal dispersion of a sewage sludge plume. *Photogrammetric Engineering and Remote Sensing* 45: 763-768.

Klemas, V., M. Otley, W. Philpott, C. Wethe, and R. Rogers, 1974. Correlation of coastal water turbidity and circulation with ERTS-1 and skylab imagery. In: *Proceedings of the Ninth International Symposium on Remote Sensing of Environment*, pp. 1289-1317. Environmental Research Institute of Michigan, Ann Arbor, Michigan.

Klemas, Vytantas, Gary R. Davis, and Robert D. Henry, 1977. Satellite and current drogue studies of ocean-disposed waste drift. *Journal Water Pollution Control Federation* 49: 757-763.

McClain, E. P., 1977. Recent progress in earth satellite data applications to marine activities, Volume 1: 14A-1-5. *Oceans '77 Conference Record*, Marine Technology Society and Institute of Electrical and Electronics Engineers, Los Angeles, California.

Rogers, R. H., N. J. Shah, J. B. McKeon, C. Wilson, L. Reed, V. E. Smith, and A. Thomas, 1975. Application of Landsat to the surveillance and control of eutrophication in Saginaw Bay. In: *Proceedings on the Tenth International Symposium in Remote Sensing of Environment*, pp. 437-446. Environmental Research Institute of Michigan, Ann Arbor, Michigan

Swift, C. T. and F. Y. Sorrell, 1978. Monitoring of the ocean and other aquatories. In: *High Resolution Passive Microwave Satellites*, (David H. Staelin and Philip W. Rosenkranta, eds.), Research Laboratory of Electronics, MIT: 4-1 to 4-48.

Williamson, A. N. and W. E. Grabau, 1973. Sediment concentration mapping in tidal estuaries. In: *Third Earth Resources Technology Satellite-I Symposium, W shington, D. C.* NASA SP-351, pp. 1347-1386. National Technical Information Service, Springfield, Virginia.

Yarger, Harold L., James R. McClauley, Gerald W. James, and Larry M. Magnuson, 1973. Water turbidity detection using ERTS-1 imagery. In: *Symposium of Significant Results Obtained from the Earth Resources Technology Satellite-I, Goddard Space Flight Center, New Carrollton, Maryland.* NASA SP-327, pp. 651-658. National Technical Information Service, Springfield, Virginia.

Distribution and Characteristics of Fluorescent and Bioluminescent Matter in the World's Oceans

G. Daniel Hickman

ABSTRACT: A review has been made of the current
status of research concerning the distribution
and characteristics of the luminescence (fluores-
cence/bioluminescence) in the world's oceans. One
of the goals of this program was to determine the
feasibility of: a) using fluorescence techniques
to identify and map distributions of chlorophyll,
red tides and Gelbstoff; b) sensing bioluminescence from aircraft and satellites; and c) using
such data as an indirect measure of temperature
and salinity. Requirements exist for *in situ*,
laboratory, and remote sensing experiments of each
subject category listed. Aircraft remote sensing experiments should be designed and coordinated with extensive ground truth measurements made
from ships. Such aircraft experiments should include a variety of optical sensors such as multispectral scanners, low light level TV and lasers.
The results of these aircraft/ground truth programs will provide a sound basis for future
laboratory and field measurement programs in
luminescence.

INTRODUCTION

Optical gradients which exist both within the water and at the water surface are important to remote sensing applications using optical techniques, e.g., those applications in which visible light is used to detect and measure parameters of the hydrosphere. It is desirable then, that these optical gradients can be identified. Presently, hue, brightness and saturation of colors serve to determine optical gradients. It is probable that even more selective signatures can be obtained utilizing luminescence (fluorescence/bioluminescence) spectroscopy. The high degree of selectivity which is obtainable using luminescence results from the fact that the intensities and wavelengths of emission from a particular material or organism are specifically coupled to the intensity and wavelength of the illuminating source. This fact enhances the probability of detecting and identifying specific substances or organisms in the water. A basic knowledge of the distribution and characteristics of fluorescent material and bioluminescence in the waters of the world is, therefore, of prime interest.

Which parameters of the hydrosphere can be derived from measurement of fluorescent and bioluminescent signatures, and can these signatures be excited and detected by aircraft and spacecraft platforms? First, remote sensing of fluorescent matter either floating on the water or within the water column may yield information on existing currents and eddies at various water depth. Second, since temperature and salinity affect the intensities and spectral character of the fluorescence and various other signals, (Raman, Rayleigh and Brillouin) it is possible that an analysis of these signals can directly yield temperature and salinity. The possibility exists that a remote sensing measurement of luminescent spectra might be used to indirectly yield surface/subsurface water temperature. For instance, various species of dinoflagellates are found in different locations throughout the world's oceans. One explanation for this occurring could be difference in temperature and salinity. Therefore, a measurement of a specific spectrum may yield information as to the water temperature. An example of an indirect measurement of temperature could occur with red tides. A red tide could come into existence when the temperature and salinity changes at a given location are just right. The temperature of the water in the bloom areas is typically 2 to 5 degrees warmer than the surrounding water mass. Detection of a red tide (either by the color of the water or by fluorescence) may be an indication of water conditions. Conversely, if a red tide is already in existence the detection of bloom degradation may yield valuable information as to the water temperature and salinity. Third, it is possible that any underwater disturbances which result in stirring and altering the distribution of luminescent matter could be detected by observing changes in the distribution of luminescence. This would be possible only if the background luminescence is known for that location.

A possible use of fluorescence as an indicator of subsurface disturbances is given by the following example: the majority of the upper ocean is thermally stratified, sterile and blue in color. Nutrients and cold water are transported to the sunlit layers when this stratification is overturned as by natural currents, human activity, or meteorological events. Through the process of photosynthesis a biological chain reaction occurs, which results in the accumulation of chlorophyll and other biochromes. In highly productive areas, the freshly upwelled waters are cold and clear. However, this water gradually warms up and, due to an increase of surface algae, turns green. Thus, a measurement of the chlorophyll concentration, either by observing reflected sunlight (passive measurement) or by sensing the light-induced fluorescence (active measurement), might be used to determine the nature of the subsurface disturbance.

The fluorescence of sea water is from both inorganic and organic material. The inorganic portion is composed of water and various inorganics such as dissolved rare earths and uranyl salts. The inorganic portion may also include suspended undissolved fluorescent matter such as calcium tungstate or zinc sulfide. The fluorescence from organics in the sea may arise from a number of sources. Among the most important and damaging sources to the marine environment, especially in shallow waters, are the hydrocarbons (both natural and manmade sources). The main organic sources of luminescence in the ocean's waters which are addressed in this paper are a) chlorophyll, b) red tides, c) Gelbstoff and d) bioluminescence.

A comprehensive bibliographic investigation covering the years 1950-1978 on the fluorescence and bioluminescence of the world's oceans has been completed by Hickman. Annotated bibliographies on these subject areas will be available in the near future. A technical assessment of this bibliographic search is still in progress with the main goal of determining the feasibility of: a) identifying chlorophyll, red tides, Gelbstoff and bioluminescence from data obtained from aircraft and satellites, and b) using such data as an indirect measure of temperature and salinity.

Brief summaries of each major subject area that contributes to ocean luminescence: chlorophyll, red tides, Gelbstoff and bioluminescence constitute the remainder of this paper. Known remote sensing studies and measurements of luminescence have been identified. Recommendations are given for research programs that will strengthen our knowledge of remote sensing of ocean luminescence.

FLUORESCENCE AND OCEAN COLOR

Laboratory and *In Situ* Measurements

Chlorophyll   Chlorophyll is an important constituent of terrestrial and marine plants. Some forms of chlorophyll are also found to exist in bacteria and algae. Chlorophyll, with other related compounds, forms an important link in the photosynthetic process in plants. Since the first half of the last century, a great deal of research has been conducted on the various chemical structures and properties of chlorophyll; nevertheless, our knowledge on this subject is far from complete.

Chlorophyll is easily altered or modified by the medium in which it is found. In fact the very method that extracts it from

plants may create new variations of this compound which are not
know to exist *in vivo* in the plant from which it was extracted. It
is reasonable to argue, on the other hand, that some of the chlorophyll forms that do exist in plants may be destroyed by the process of extraction.

Arguments that may be forwarded in support of using chlorophyll
as a remote indicator of water temperature and salinity are 1) its
presence indicates plant life, and 2) it is abundant in the oceans.
The literature presents evidence showing the dependence of the
fluorescence and absorption spectra of chlorophyll on temperature.

Several types of chlorophyll have been identified. No attempt
will be made here to detail the various chemical and physical properties of chlorophyll. *The Chlorophylls* (Vernon and Sealy, 1966)
is an excellent reference on this subject. The following section
describes what is known about the spectroscopic characteristics
of chlorophyll.

An experiment was conducted in which the water absorption in
the blue region of the spectrum was measured as a function of phytoplankton pigment. Maximum transmittance was found to shift to
the green region. Red band absorption of chlorophyll did not influence the water color (Yentsch, 1960).

Chlorophyll *a* is the most abundant of the chlorophylls. Blue
green algae, yellow green algae and some species of red algae contain small quantities of chlorophyll *a*. Chlorophyll *a* is also always found with a little chlorophyll *c* in diatoms, a symbiotic algae
of a sea anemone, dinoflagellates and all brown algae. Chlorophyll *a*
is accompanied by chlorophyll *b* in the green algae *Euglena*.

Chlorophyll *a* exhibits two absorption bands; the maximum of
one band occurs at approximately 430nm while the other is at 680nm.
The major fluorescence that occurs in algae *in vivo* is due to chlorophyll *a*. The fluorescence spectrum of chlorophyll *a* shows a relatively sharp peak at 685nm, with a broad less intense band occurring in the 730nm region. The specific wavelength of maximum fluorescence, as in the case of the absorption spectra, depends on the
form of the chlorophyll *a*. Studies have shown that both the pH and
temperature of chlorophyll *a* solutions affect the intensity of
fluorescence (Papageorgiou and Govindjee, 1971).

Chlorophyll *b* occurs as a minor constituent of *Euglena* and in
higher plants. Chlorophyll *b* has absorption bands at 453nm and
642nm.   These absorption measurements were made for chlorophyll
extracts in ether solutions where the ratio of the peak of the
Soret band to that of the red absorption band (642nm) is approximately 2.84 (Strain, 1963).

Chlorophyll *c* is found in numerous marine algae and diatoms.
It occurs with chlorophyll *a* in diatoms, dinoflagellates, brown algae, and a certain symbiotic algae of sea anemones. Organisms carrying chlorophyll *c* perform most of the photosynthesis in the sea,
although the chemistry of this chlorophyll is not well known.
Chlorophyll *c* has a prominent absorption peak at about 446nm and
minor peaks in the red region. The ratio of the peak of the Soret
band to that of the most prominent peak in the red is approximately
10.

Bacteriochlorophyll *a* is a major chlorophyll of various photosynthetic bacteria. This chlorophyll is also found in green bacteria as a minor constituent comprising 5% to 10% of the total chlorophyll content of this bacteria. Bacteriochlorophyll a extracts have absorption bands in the UV at 358nm and 391nm, another peak in the orange at 577nm, and one in red at 772nm.

Red Tides The term "red tide" or "red water" is a term used loosely to describe a situation in which a single dinoflagellate species is present in extremely high concentrations (Steeman Nielsen, 1962). This organism, which normally is present throughout the year undergoes a period of rapid growth (blooming) and concentrates in very dense patches that in certain cases may reach concentrations of over a million cells per liter. The colors of these patches range from yellow to red to brown, and in some instances may even bioluminesce (Serafin, 1967); Tsujita, 1968). Direct contact with red tides causes severe irritation to skin, discolors beaches and coastal areas, and is fatal to fish and other forms of marine fauna (Hart, 1966; Hirayama and Namaguchi, 1973; Ilzuka, 1973; and Wardle et al., 1975). The interest in red tide is twofold, economical and health related. Depletion of certain species of fish have an economic impact on fish industries, while the poisoning of other species poses health hazards to the population at large (Grindley and Taylor, 1970 and Stuart et al., 1967).

Understanding red tides is essential in attempting to control the side effects that result from their outbreak. A great deal of time and money has been invested in research on the mechanisms by which red tides form (Rounsefell and Nelson, 1966). Wind, nutrients, temperature and salinity are factors which have been identified as being critical to the formation and spreading of red tides (Irie, 1973; Mulligan, 1975; Murphy et al.; and Steidinger, 1975). A brief description of the effect of the above environmental parameters on the formation of red tides is given below. A mathematical model for the formation of a red tide is given by Wyatt and Horwood (1973).

Wind Red patches generally are formed in areas where the wind velocities are relatively low. Maclean (1975) reports the *Pyrodinium bahamense* Plate, a toxic dinoflagellate that blooms in Papua, New Guinea, is formed only in sheltered portions of the coastline. Pomeroy et al., (1956) observed that at Sapelo Island, Georgia, the red tide occurred in the headwater of the Dublin River, an almost stagnant mass of water. This red tide also appeared in a down wind section of a boat basin at the island. Studies on the mechanisms of red tide patches have been made on the bioluminescent red tide in Oyster Bay, Jamaica. These patches dispersed when the wind drove them into a well mixed part of the Bay. In False Bay, South Africa (Grindley and Taylor, 1970), onshore winds concentrated the red tide patches while offshore winds dispersed them. Pingree et al. (1975) reported that a widespread red tide which occurred in the English Channel coincided with a period of calm weather. A long period of calm wind was given by Hickel et al. (1971) as the reason behind extensive *Gymnodinium* blooms occurring in the Helgoland Bight.

One conclusion is that wind velocity is responsible for dis-

persing the bloom patches if it is strong enough, while it can aid
in forming blooms if the wind is blowing in a direction towards an
enclosed area. High wind velocities are often reported before the
outbreak of a red tide in areas where wind actions disturb the bottom sediments of the coastal areas. These sediments are the source
of nitrates and silicates required for sustaining a bloom.

   Nutrients   The occurrence of upwelling water that is rich in
nitrates and phosphates preceeding red tide blooms has been reported
by Grindley and Taylor (1970). Tidal mixing yielding similar type
blooms was reported by Fung and Trott (1973), Blasco (1977) and
Pingree et al. (1977). River discharges and terrestrial effluent
form another source of needed nutrients (Pomeroy et al., 1956;
Collier, 1958; and Maclean, 1975). A detailed study of the role of
nitrates in a bloom was thoroughly discussed by Harrison (1976) in
relation to a bloom of *Gonyaulax polyedra*. Other important studies
of the effect of nutrients on the outbreak of red tides have been
reported by Hammer, 1968; Fukozo, 1970; Zernova, 1970; and Eppley
and Harrison, 1975.

   Salinity and temperature   Dinoflagellate blooms often occur
after periods of major changes in temperature and salinity. These
blooms occur after a period of unfavorable conditions for the growth
of the other organisms normally present in the water (Grindley and
Taylor, 1970). Sieburth (1967) performed a study on the temperature effects on bacterial selection. Hulburt and Rodman (1963)
studied the distribution of phytoplankton as a function of salinity. Major changes in salinity and temperature are found to have
different effects on various phytoplankton species. Under specific
values of temperature and salinity, it is possible to have red tide
organisms grow at a faster rate than other organisms, the exact
conditions for a red tide. Several authors have studied the temperature and salinity ranges needed to support specific red tide
blooms (Fonds and Eisma, 1967; Holmes et al., 1967). Grindley and
Taylor (1970) described a plankton bloom in False Bay, South Africa
in which the dinoflagellate *Gonyaulax polygramma* became abundant at
temperatures above 17°C while the final bloom occurred at temperatures greater than 20°C. *Noctiluca scintillans* was found to bloom
at lower temperatures (15°-18°C) than *Gonyaulax polygramma*, while
Gephyracapsa oceanica blooms occurred in waters that were still
cooler. In the northern lakes of the St. Lucia estuary system in
Natal, an extremely large increase in salinity (89°/oo) followed by
a decrease to 34-42°/oo, occurred just prior to a red tide bloom of
Noctiluca scintillans. Hickel et al., (1967) reported modest increases in salinity (26°/oo to 28°/oo) and temperature (14.2° to
17.5°C) for several days prior to a bloom. Pingree et al.(1977)
found *Gyrodinium aureolum* blooms in areas where the water temperature was 15°-16°C. Pomeroy et (1956) reported a bloom of a
mixture of *Gymnodinium splendens* and *Amphidinium fusiforme* in a
temperature zone of 25°-29°C. Wilton and Barham (1968) showed a
complex pattern for temperature profiles (in both depth and time)
and a corresponding distribution profile for *Gymnodinium flavum*.
Maclean (1975) reported the bloom temperature range of *Pyrodinium
bahamense* as 26.2°-30.7°C and the salinity range 28.6°/oo to

31.5⁰/oo. This dinoflagellate can tolerate lower and higher salinity values than the range given above. Blasco (1977) found a
Gonyaulax polyedra bloom that was bounded by a 33.8⁰/oo salinity
isopleth. A systematic study of the effects of temperature, salinity, pH, and nutrients on the growth rate of eight species of neritic red tide flagellates was conducted by Iwasaki (1973). In
general, blooms occurred during conditions of high temperature and
low salinity.

The organisms responsible for the red tide outbreaks are not
the same everywhere. For instance, *Gymnodinium flavum*, a dinoflagellate specie that causes a yellow water bloom was first reported
off the coast of California. *Gonyaulax polyedra*, another dinoflagellate responsible for California red tide is the most abundant of
the dinoflagellates. Red tide *Noctiluca* blooms have been reported
in European waters and off the coasts of Africa and India. In the
Gulf states, red tides of *Gymnodinium breve* have caused large fish
mortality. *Gymnodinium splendens* red tide blooms occurred in 1928
in Delaware Bay, New Jersey and off the Ivory Coast in 1969
(Dandonneau, 1970), while a similar bloom has been reported in
Georgia estuaries (Pomeroy et al., 1956). *Gonyaulax sp.* was reported in 1942 as the organism responsible for the mortality of
fish in Offats Bayou near Galveston, Texas, while *Noctiluca* caused
the bloom on the Catalan Coasts (Lopez and Arte, 1971). A discussion of red tides occurring in Southern California during the past
25 years is given by Sweeney (1975). A brief summary of the red
tide research follows:

† Red tide blooms are caused by various dinoflagellates.
† Laboratory investigations dealing with red tides are
  limited.
† Different red tide organisms bloom at different temperatures
  and salinities.
† The growth rates of blooms as a function of nutrients are
  largely unknown.
† Few (if any) measurements of fluorescence have been made.
† No remote sensing (aircraft/satellites) measurements have
  been made.

Gelbstoff   Gelbstoff is a term used to describe a number of
acidic organic complex compounds found in inland waters or along
coastal shores. Several authors agree on the factors affecting the
increase in the presence of such compounds, but disagree about
their properties and origin. The factors affecting the increase of
Gelbstoff are of particular importance in predicting outbreaks of
red tide marine dinoflagellates. As stated above, such outbreaks
may cause appreciable economic impact on fisheries and may precipi-
tate health hazards to the population at large. The formation of
Gelbstoff is influenced by the temperature and salinity of the medium in which it is found. Also, it is well known that the yellow
organic compounds fluoresce whenever they are irradiated with UV
sources.

These substances were orginally thought to be terrestrial in
origin and were given the name Gelbstoff in 1937 by Kalle. Later,
Jerlov and Kalle reached the conclusion that these substances were
of marine origin. Sieburth and Jensen (1969) concluded that their
research indicates similar mechanisms for the formation of Gelbstoff
or humic material, whether it is of marine or terrestrial origin.
The literature shows some confusion to exist in the differences
between the organic acidic compounds found in various water masses.
Shapiro (1957) discussed the similarity between the yellow organic
acids found in lake water with those of soil origin. He differentiated between the yellow organic acids and the so called humic
acids. Ghassemi and Christman (1968) discussed the yellow organic
acids of natural water without studying their origin. Humic substances give coastal waters a yellowish brown color but Prakash and
Raschid (1968) found it to be different from Gelbstoff. Egan and
Cassin (1973) and Egan (1974) associated Gelbstoff in terrestrial
waters with humic acids and cell decomposition, while it is formed
from carbohydrates and amino acids in the open ocean. Otsuki and
Wetzel (1973) also stated that the yellow organic acids are a major
component of the soluble humic acids. Kirk (1976) identified the
yellow organic substances, in fresh water, as Gelbstoff.

Stuermer and Payne (1976) compared the spectra of marine and
terrestrial fulvic acids. Sieburth and Jensen (1968, 1969) measured absorption spectra for various terrestrial and marine waters
in the range of 200-500nm for various pH values. Differences in
the spectra were found, depending on the pH value and on the sample
origin. Kirk (1976) obtained the absorption spectra in the photosynthetic band (350-700nm) of water samples taken from various water bodies. The spectra were plotted relative to distilled water.
Bladh (1972) showed that the Gelbstoff concentration usually decreased with increasing salinity, and was generally not found in
the open oceans. However, large absorbance coefficients of Gelbstoff
have been measured for the Baltic Seas, the peak absorption occurring at 380nm.

Yellow materials extracted from lake waters had a fluorescence
color of both yellow-green and brown (Shapiro, 1957). Long wavelength UV radiation was used as the source of excitation. Fulvic
acid, which is extracted from soil, fluoresces when submitted to
UV excitation (Ghassemi and Christman, 1968). Egan's (1974) paper
presented evidence that Gelbstoff can be detected using short wavelength UV radiation as the excitation source and a broadband detector with good sensitivity in the visible region. Hornig and Eastwood
(1973) present fluorescence and emission spectra of chlorophyll and
Gelbstoff in natural ocean waters from the Gulf, Atlantic and Pacific
coasts. The peak fluorescence intensity of Gelbstoff was given by
Sieburth and Jensen (1968) as approximately 263nm. Karabashev and
Zangalis (1971) present a plot of absorption and fluorescence spectra of samples of Gelbstoff obtained from the Baltic Sea. The fluorescence spectra of these samples varied from 400nm to 600nm, and
were dependent on the excitation wavelength.

Various studies reviewed by Kalle (1966) have shown that water
which contains a large concentration of Gelbstoff also contains a
fluorescence component. The ratio of the fluorescence to Gelbstoff

was found to be a function of salinity; the ratio increases with
increasing salinity, although both the intensity of fluorescence
and the concentration of Gelbstoff decreased with increasing salinity. One set of experiments showed this ratio to increase by a
factor of eight as the salinity increased from 0°/oo to 10°/oo.
The possibility exists that a measurement of this ratio can be used
as an indirect determination of salinity.

Sieburth and Jensen (1969) reached the conclusion that water
temperature and salinity appeared to have selective effects on the
exudation of *Fucus vesiculosus*, a marine plant which is commonly
found in Gelbstoff. Malmberg (1964) measured Gelbstoff by beam
transmittance as a function of the time variations of temperature
and salinity at hydrographic stations located in the Kattegat and
the inner Skagerrak between Sweden and Denmark. The results showed
an inverse relationship between salinity and yellow substance.
Otto (1967) measured salinity, temperature and fluorescence of water in the North Sea as a function of the attenuation of light
through Gelbstoff.

Raising the pH from 3 to 11 increased the intensity of the
color by 50% (Shapiro, 1957). Ghassemi and Christman (1968) stated
that pH effected the color as well as the fluorescence of yellow
organic acids. At low color values (fixed pH), the intensity of
the fluorescence varied linearly with color concentration. Sieburth
and Jensen (1969) found that increasing the pH of exuded solutions
of the brown seaweed *Fucus vesiculosus* resulted in increased
Gelbstoff concentration.

Distribution of Fluorescent Matter

Fluorescence was detected in water at high levels of plankton
concentration (Traganza, 1969). This study also showed that a
large concentration of dissolved organics in water is not found at
times of medium plankton concentration. Other fluorescent distributions have been observed by Ivanoff et. al., (1961) in the
Mediterranean and by Karabashev (1970) in the Black Sea. In these
cases the surface fluorescence was generally less than the subsurface fluorescence.

The fluorescent distributions are in general much different in
coastal waters, where a high percentage of the fluorescence is derived from substances that are released into fresh waters. An important aspect of this effect is that these substances do not
decompose rapidly in the sea. Measurements of fluorescent substances in the coastal zone have been reported by Duursma (1974) and
Højerslev (1974). Research programs such as Tyler (1964) have been
involved with *in situ* measurements of ocean color using photometers
in areas known to contain large concentrations of Gelbstoff.

Besides the research on the fluorescence of organics in the
sea, there has been a growing interest in the distribution and origin of organic matter in the oceans. As previously noted, many of
the organics are directly fluorescent or can be made to fluoresce
in the right environment. Therefore, much of the information which
is available on organics in the sea could be of interest to those
concerned with the distribution of fluorescent matter.

Remote Sensing Measurements

The majority of research investigations on the fluorescence of constituents in the water has been performed either in the labora- tory or *in situ* using spectrofluorometers. During the past few years rapid advancements have been made in the general area of re- mote sensors. Many of these sensors, both passive and active, have been used successfully to detect water quality. Sensors have been flown on airplanes, helicopters and spacecraft.

Following the initial experiments by Hickman and Moore (1970) who used a laser to excite the fluorescence in algae, several in- vestigators have utilized various types of active laser systems for detection of phytoplankton (Hickman et al., 1973; Kim, 1973; Mumola et al., 1973; and O'Neil et al., 1975). An airborne laser fluorosensor system was developed (Hickman and Kim, 1973) for de- tection of an oil spill. Other investigators have studied oil fluorescence by various active and passive sensors (Hornig and Eastwood, 1973 and Millard and Arvesen, 1972). High power lasers used to excite fluorescence in specific molecules can also excite Raman emissions. Experiments have shown the potential of Raman signals for measuring water temperature and salinity. It may also be possible to use Raman signals to calibrate the intensity of fluorescence emissions. Although Raman scattering arises from a physical effect different from that of fluorescence, Raman-excited emissions are important since they can cause interference with de- tection of fluorescent signals from phytoplankton, oil and other organics in sea water.

The correlation between measurements of water color obtained remotely on the Great Lakes and *in situ* parameters of water quality has been examined by McNeil and Thompson (1974). Austin et al., (1975) developed a theoretical model to investigate the effect of chlorophyll concentration on ocean color. The study of ocean color changes due to chlorophyll concentration (Hovis et al., 1974) was made from aircraft both at high and low altitudes. Stoertz et al. (1969) developed and used the Fraunhofer Line Discriminator (FLD) for detection of chlorophyll. The FLD is a passive sensor that utilizes sunlight to excite the fluorescence in the chlorophyll. One of the disadvantages of this instrument is that it is strictly a daylight sensor, while, in general active sensors can be used both at night and day. Clarke et al. (1970) used the backscatter from chlorophyll as an indicator of water color, which he in turn was able to correlate with phytoplankton concentrations. This method has been used to detect low (0.1 mg/m) chlorophyll concen- trations. Knowledge of the chlorophyll concentration allows one to make a prediction of the intensity of the chlorophyll fluorescence for a given intensity source illumination. Several investigators have attempted to use multi-color imagery of LANDSAT to relate ocean color to plankton activity.

Plass et al. (1978) performed extensive calculations on the influence of the presence of hydrosol, chlorophyll and Gelbstoff on upwelling light. Such calculations are useful in predicting the feasibility of deploying airborne remote sensors for detection of Gelbstoff, either by color or fluorescence. Several studies have been performed involving aircraft and satellite imagery for measurement of ocean color. One such program was performed by Maul and Gordon (1975), where observations of the Gulf Stream in the Gulf Stream in the Gulf of Mexico were obtained in synchronization with LANDSAT-1. Although not designed explicitly to detect Gelbstoff, the results of such programs are very useful in defining feasibility type airborne investigations of Gelbstoff.

BIOLUMINESCENCE

Bioluminescence in the sea has been observed for centuries dating back to the time of the Greek philosopher Aristotle. Descriptions of bioluminescent events are found in the western literature as far back as the fifteenth century. The scientific interest in bioluminescence goes back to the last century where several scientists tried with varying degrees of success to understand the nature of bioluminescence. Fishermen have located schools of fish by the luminescence they induce in other marine organisms. Luminescent wakes created by the passage of surface ships and submarines has been used by the U. S. Navy for tracking the passage of ships.

Marine luminescent organisms include such diverse species as fish and dinoflagellates, however, not all species are luminescent. For example, some red tide dinoflagellates exhibit biolumescent activities while others do not. Dinoflagellates and bacteria, together, due to their large scale abundance in the oceans of the world, contribute the largest share of marine bioluminescence (Gitel'zon, 1970). Reports of incidents of bioluminescence in freshwater masses are few and in general, not reliable. Apparently, some degree of salinity is needed to trigger the biological mechanisms of bioluminescence of aquatic organisms.

Bioluminescence is sometimes spontaneous and triggered by internal mechanisms of the organism. Most bioluminescent organisms exhibit a diurnal rhythm of luminescence activities at night that are more frequent and intense than those occurring during the daytime (Sweeney and Hastings, 1957).

Luminescence, on the other hand, can be triggered by a large number of external factors, such as reduction in light intensity (on cloudy days) or by mechanical stimulation resulting from a passing ship or a large marine animal. Luminescence can be triggered in the laboratory environment by light, electrical pulses, mechanical disturbance, and addition of salt or chemicals. The literature is full of evidence on the effects that temperature and salinity have on the intensity and pulse characteristics of the luminescence emitted by such organisms.

Harvey's (1952) excellent reference, *Bioluminescence*, provides a valuable critique of research efforts in all aspects of this field. Other more recent valuable review articles have appeared. Tett and Kelley, 1973; Cormier et al., 1975; and Johnson, 1975. A brief review is given below of the properties and characteristics of luminous organisms.

Laboratory and *In Situ* Measurements

Light Inhibition   Harvey (1952) stated that certain dinoflagellates lose their ability to luminesce in daylight or sunlight
and that bacteria are less sensitive to changes in light intensity.
Swift and Taylor (1967) produced quantitative data on the effect of
light on the bioluminescence of the dinoflagellate *Pyrocystis
lunula*. Dinoflagellates that were moved from light to dark areas
increased in bioluminescence, while transferring them back to light-
ed areas resulted in a decrease. Few quantitative measurements on
light effects on dinoflagellates appear in the literature; however,
the existing data indicate that light inhibits the bioluminescence
of both bacteria and dinoflagellates (Hardy, 1964).

Luminescence Spectra   Luminescence spectra of some species
of bacteria are available; for good review articles on bacterial
bioluminescence, see: McElroy and Strehler, 1954; McElroy, 1961;
and Hasting and Nealson, 1977. Spruit-van der Burg (1950) gives
the emission spectra of three species of photobacterium; *Photobacterium phosphoreum*, *Photobacterium splendidum* and *Photobacterium
fischeri*. The peaks of the emission spectra of these organism occur at 460nm, 480nm and 490 nm, respectively. The bioluminescent
spectrum of bacterium *Achromobacter fischeri* was found to peak at
490nm (Lee and Seliger, 1955), while that of bacterium *Issatchenko*
was at 480nm (Voitov et al., 1960) . Spectra for various dinoflagellates have also been studied. For instance, *Gonyaulax* cells
emission spectrum peaks were found to occur at 475nm (Bode and
Hastings, 1963), while those of *Pyrocystis lunula* were at 477.5nm
(Swift and Taylor, 1967).

Salinity   Freshwater bacteria luminesce when the water is
made slightly saline. Marine bacteria needs water with high salt
concentration to luminesce. Marine bacteria, when added to distilled water, lose their luminescence and fail to survive. In one
experiment, the luminescence of the bacteria *A. fischeri* reached
its peak when the salt concentration of the water was raised to
150% that of seawater. Concentrations of 10% and 400% reduced the
luminescence to virtually zero.

Dinoflagellates, on the other hand are more sensitive than
bacteria to a decrease in salinity. *Noctiluca* emits a faint glow
(distress sign) at salinities at 40% that of seawater. The organisms revive when the salinity is increased. At salinities of 30%
of seawater organisms emit distress signals and die. *Gonyaulax
polyedra* bioluminescence was studied by DeSa and Hastings (1968).
Different salts were tested on *G. polyedra* with regard to inhibition of scintillon luminescence. Copper, silver and zinc chlorides
were found to inhibit bioluminescence by 85%, 90% and 75%, respectively at concentrations of 1×10^{-3} moles/litre. Chloride salts of
Cs, Mg, Li, K, Na, Ba, Al, Cr, Fe, Co and Ni were not found to
effect luminescence.

Temperature   Temperature appears to have an influence on the
development of luminous dinoflagellates and bacteria. Areas of

uniform high temperatures, like those of the Eastern Caribbean, report luminence all year round. In the Gulf of Mexico, which is influenced by the colder northern winds, bioluminescence develops mostly in spring and summer. The effect of temperature on biolumi- nescence shows up in the waters surrounding Norway where the tem- perature of the Gulf stream has been reduced. In south Norway the bioluminescence is restricted to two months during the summer sea- son. The Baltic Sea is an example where low salinity results in few reports of bioluminescence. The Mediterranean and Black Seas are given as examples of regional variations of bioluminescence with temperature. The bioluminescence in the western basin of the Mediterranean occurs between October and June while the biolumines- cence in the Black Sea is restricted to the fall season.

If the temperature is relatively low, luminous bacteria grow in almost any medium which is not too acid or alkaline. Early au- thors measured the temperature range of bioluminescence for differ- ent species of bacteria (see Harvey, 1952). In general, the tem- perature limits are found to range between $-25°$ and $+45°C$. Bacte- ria are warmed, they start to luminesce without displaying any apparent injury. Zirpole in 1932, found that the Bacterium *Baccil- lus pierantonii* survived liquid helium temperature $(-271°C)$ for a period of several days. The bacteria began to luminesce as the temperature was increased. These bacteria have an upper tempera- ture limit of survival which is well below the boiling point of water. The luminescence of the bacterium *Acromobacter fischeri* has received some attention. Johnson (1954) described the luminescence of *A. fischeri* as a function of temperature. The maximum lumines- cence occurred at a temperature of $28°C$. The luminescence intensi- ty fell to 3% of its peak at $39°$ and $8°C$. The photobacterium *Phosphoreum* was found to have its maximum luminescence at $22°C$.

The luminescence of dinoflagellates as a function of tempera- ture has been studied by numerous investigators. Stimulated flash intensities were recorded by Eckert (1965) at $30°$, $20°$, and $10°C$. The peak luminescence intensities at $20°$ and $10°C$ were 60% and 30%, respectively, of their intensity at $30°C$. DeSa and Hastings (1968) compared the flash produced by the scintillons of this dinoflagel- late at $20°$ and $2°C$. The peak luminescence intensity was found to drop by 68% as the temperature was lowered from $20°$ to $2°C$ at a pH of 8.2.

Remote Sensing Measurements

One of the only remote sensing experiments involving biolumi- nescence that was located in the literature was the work of Cram (1973). In this experiment, aircraft were used to locate *Sardinops ocellata* schools near the coast of Southwest Africa. Therefore, there is a real need for "good" remote sensing - ground truth mea- surements (aircraft/satellites) in the general area of biolumines- cence.

SUMMARY

Although several research programs have investigated the effects of salinity, temperature and nutrients on the growth of chlorophyll, there is a lack of similar data on red tides, Gelbstoff and bioluminescence. In addition, there is a lack of fluorescence signatures (spectral, intensity and temporal) of both the living organisms and other matter in the water, especially those induced by light sources such as lasers.

There is a need for both *in situ* and laboratory experiments on chlorophyll, red tides, Gelbstoff, and bioluminescence. The purpose of the *in situ* measurements is to generate a data bank of appropriate fluorescence signatures under various environmental conditions. The laboratory experiments should consist of measuring the fluorescence spectra and intensities for various values of temperature, salinity, light levels and nutrients. In general, the samples for these measurements should come from the natural environment. Special care must be used to preserve the samples so that they retain their properties found in the natural environment. In some cases, such as Gelbstoff, it may be both instructive and interesting to perform similar laboratory experiments on artificially made material. Typical experimental temperature and salinity ranges of interest are 5° to 45°C and 15°/oo to 35°/oo, respectively. The results of both the *in situ* and laboratory measurements are important in estimating the feasibility and performance of various remote sensors for detecting and identifying changes in the fluorescence characteristics of the water, due to changes in environmental conditions.

Numerous authors have performed aircraft and satellite measurements of ocean color and have attempted to relate these results to chlorophyll concentrations. With few exceptions, these programs have been void of ground truth data, thereby making it impossible to make definitive conclusions. McCluney (1975) describes remote sensing techniques that could be used as part of an advanced warning system for major red tides.

There is a need for airborne remote sensing experiments which have been designed and coordinated with ground truth. It is recommended that these remote sensing experiments be performed in areas that are known for outbreaks of red tides or Gelbstoff, high concentrations of chlorophyll or for their highly visible bioluminescence. It would be advantageous to fly aircraft containing a variety of optical remote sensors, such as multispectral scanners, low light level TV, and lasers over these test sites. If suitable, these areas would also have good satellite coverage. Ground truth data taken by *in situ* instrumentation and water samples would be taken simultaneously from ships. These measurements would be used to determine the fluorescence levels and concentration of organisms. The data would be correlated with the results obtained by the remote sensors, and used to guide the design of future equipment and experiments.

ACKNOWLEDGMENTS

This work was supported by the Office of Naval Research under
Contract N00014-76-0711.

REFERENCES

Austin, R. W., S. Q. Duntley, C. F. Edgerton, S. E. Moran and W. H.
    Wilson, 1975. Ocean color analysis. *Scientific and Technical
    Reports* 12: 2945-2946.
Bladh, J. O., 1972. Measurements of yellow substance in the Baltic
    and neighboring seas 1970-1972. Lysenkil, Sweden. Havsfiske-
    laboratoriet. *Meddelande*, 138: 15.
Blasco, D., 1977. Red tide in the upwelling region of Baja
    California. *Limnol. Oceanogr.* 22: 255-263.
Bode, V. C. and J. W. Hastings, 1963. The purification and proper-
    ties of the bioluminescence system in *Gonyaulax polyedra*. *Arch.
    Biochem. Biophys.* 103: 488-499.
Clarke, G. L., G. C. Ewing, and C. J. Lorenzen, 1970. Spectra of
    backscattered light from the sea obtained from aircraft as a
    measure of chlorophyll concentration. *Science* 167: 1119-1121.
Collier, A., 1958. Some biochemical aspects of red tides and re-
    lated oceanographic problems. *Limnol. Oceanogr.* 3: 35-39.
Cormier, M. J., J. Lee, and J. E. Wampler, 1975. Bioluminescence:
    recent advances. *Ann. Rev. Biochem.* 44: 255-272.
Cram, D., 1973. Pilchard stocks surveyed by remote sensors. *Fish.
    News Internat.* 12: 35-36.
Dandonneau, Y., 1970. Un phenomene d'eaux rouges au large de la
    Cote D'Lvoire cuase par *Gymnodinium splendens*. Doc. Scient.
    Centre Rech. Oceanogr. *Abidjan* 1: 11-19.
DeSa, R. and J. W. Hastings, 1968. The characterization of scin-
    tillons; bioluminescent particles from the marine dinoflagellate,
    Gonyaulax polyedra. *J. Gen. Physiol.* 51: 105-122.
Duursma, E. K., 1974. The fluorescence of dissolved organic matter
    in the sea. *Optical Aspects of Oceanography* (N. G. Jerlov and
    E. Steemann-Nielsen, eds.), pp. 237-256, Academic Press.
Eckert, R., 1965. Biolectric control of bioluminescence in the
    dinoflagellate *Noctiluca I.* Specific nature of triggering mecha-
    nisms. *Science* 147: 1140-1142.
Egan, W. G., 1974. Measurement of the fluorescence of Gulf Stream
    water with submerged *in situ* sensors. *Mar. Tech. Soc.* 8: 40-47.
Egan, W. G. and J. M. Cassin, 1973. Correlation of *in situ* fluo-
    rescence and bioluminescence with biota in the New York Bight.
    Biol. Bull. 144: 262-275.
Eppley, R. W. and W. G. Harrison, 1975. Physiological ecology of
    Gonyaulax polyedra, a red water dinoflagellate of Southern
    California. *Proceedings of the First International Conference
    on Toxic Dinoflagellate Blooms* (V. R. LoCicero, ed.) pp. 11-22,
    Massachusetts Science and Technology Foundation, Wakefield,
    Massachusetts.
Fonds, M. and D. Eisma, 1967. Upwelling water as a possible cause

ɔf red-plankton bloom along the Dutch coast. *Neth. J. Sea Res.*
3: 458-463.

Fukozo, U., 1970. Relation between environmental characteristics
of waters and changes of phytoplankton in an estuary where red
tide often appears. *Bulletin of the Plankton Society of Japan*
16: 89-98.

Fung, Y. C. and L. B. Trott, 1973. The occurrence of a *Noctiluca
scintillana* (Macartney) induced red tide in Hong Kong. *Limnol.
Oceanogr.* 18: 472-476.

Ghassemi, M. and R. F. Christman, 1968. Properties of the yellow
organic acids of lake waters. *Limnol. Oceanogr.* 13: 583-597.

Gitel'zon, I. I., 1970. *Bioluminescence of the Sea.* Joint Publications Research Service. Washington, D. C. (51226). 24 August,
1970.

Grindley, J. R. and F. J. R. Taylor, 1970. Factors affecting
plankton blooms in False Bay. *Transactions of the Royal Society
of South Africa* 39: 201-210.

Hammer, L., 1968. Investigations on red tides in the Caribbean Sea.
Kieler Meeresforschungen 24: 33-39.

Hardy, A. C., 1964. Experimental studies of plankton luminescence.
J. Mar. Biol. Assoc. 44: 435-484.

Harrison, W. G., 1976. Nitrate metabolism of the red tide dinoflagellate *Gonyaulax polyedra* Stein. *J. Exp. Mar. Biol. Ecol.* 21:
199-209.

Hart, T. J., 1966. Some observations on the relative abundance of
marine phytoplankton populations in nature. *Some Contemporary
Studies in Marine Science* (H. Barnes, ed.), pp. 375-393.

Harvey, E. N., 1952. *Bioluminescence.* Academic Press, Inc., New
York.

Hastings, J. W. and K. N. Nealson, 1977. Bacterial bioluminescence.
Ann. Rev. Microbiol. 31: 549-595.

Hickel, W., E. Hagmeier and G. Drebes, 1971. *Gymnodinium* blooms in
the Helgoland Bight (North Sea) during August 1968. *Helgolander
wiss. Meeresunters* 22: 401-416.

Hickman, G. D., J. E. Hogg, E. J. Friedman, A. H. Ghovanlou, 1973.
Application of a pulsed laser for measurements of bathymetry
and algal fluorescence. *8th International Symposium on Remote
Sensing of the Environment, Ann Arbor, Michigan, October 2-6,
1972, Proceedings* 1, Environmental Research Institute of
Michigan. pp. 617-637.

Hickman, G. D. and H. H. Kim, 1973. *An Airborne Laser Fluorosensor
for the Detection of Oil on Water,* ISA-JSP6720. p. 369.

Hickman, G. D. and R. B. Moore, 1970. *Laser Induced Fluorescence
in Rhodamine B and Algae.* Presented at the 13th Conference on
Great Lakes Research, Buffalo, New York, March 31-April 3.

Højerslev, N. K., 1974. Inherent and apparent optical properties
of the Baltic. *Institute of Physical Oceanography, University
of Copenhagen, Report 23.*

Holmes, R. W., P. M. Williams and R. W. Eppley, 1967. Red water in
La Jolla Bay, 1964-1966. *Limnol. Oceanog.* 12: 503-512.

Hornig, A. W. and D. Eastwood, 1973. *A Study of Marine Luminescence
Signatures* 1. NASA-CR-114578.

Hovis, W. A., M. L. Forman, L. R. Blaine, 1974. Detection of ocean color changes from high altitudes. *Scientific and Technical Aerospace Reports* 12: 661.

Hulburt, E. M. and J. Rodman, 1963. Distribution of phytoplankton species with respect to salinity between the coast of Southern New England and Bermuda. *Limnol. Oceanogr.* 8: 263-269.

Ilzuka, S., 1973. *Gymnodinium* type - '65 red tide in occurring a anoxic environment of Omura Bay. *Bulletin of the Plankton Society of Japan* 19: 22-23.

Irie, H., 1973. The present state and problems on the red tide research in Japan. *Bulletin of the Plankton Society of Japan* 19: 115-124.

Ivanoff, A., N. G. Jerlov, and T. H. Waterman, 1961. A comparative study of irradiance in beam transmittance and scattering in the sea near Bermuda. *Limnol. Oceanogr.* 6: 129-148.

Iwasaki, H., 1973. The physiological characteristics of neritic red tide flagellates. *Bulletin of the Plankton Society of Japan* 19: 104-114.

Johnson, F. H., 1975. *Research on Bioluminescence*. Princeton University, New Jersey. Department of Biology: for Office of Naval Research, Arlington, Virginia.

Kalle, K., 1966. The problem of the Gelbstoff in the sea. *Oceanogr. Mar. Biol. Ann. Rev.* 4: 91-104.

Karabashev, G. S., 1970. Methods of photoluminescence studies of the sea water. *Okeanologiya* 10: 883-888.

Karabashev, G. S. and K. P. Zangalis, 1971. Some results of research on photoluminescence spectra of sea water. *Izv, Atmospheric and Oceanic Physics* 7: 671-672.

Kim, H. H., 1973. New algae mapping technique by the use of an airborne laser fluorosensor. *Applied Optics* 12: 1454-1459.

Kirk, J. T. O., 1976. Yellow substance and its contribution to the attenuation of photosynthetically active radiation in some inland and coastal South-Eastern Australian waters. *Aust. J. Mar. Freshwat. Res.* 27: 61-71.

Lee, J. and H. H. Seliger, 1965. Absolute spectral sensitivity of phototubes and the application to the measurement of the absolute quantum yields of chemiluminescence and bioluminescence. *Photochem. Photobiol.* 4: 1015-1048.

Lopez, J. and P. Arte, 1971. Aquas rojas en las Costas Catalanas (red waters on the Catalan Coasts). *Inv. Pesq.* 35: 699-708.

Lynch, R. V., III, 1978. The occurrence and distribution of surface bioluminescence in the ocean during 1966 through 1977. *Naval Research Laboratory Report* NRL 8210: 45. Washington, D. C.

Maclean, J. L., 1975. Red tide in the Morobe District of Papua New Guinea. *Pacific Science* 29: 7-13.

Malmberg, S. A., 1964. Transparency measurements in the Skagerack. *Kungl. Vetenskaps-OCH Vitterhets-Samhalles Handlingar*, F6, Ser. B, 9.

Maul, G. A. and H. R. Gordon, 1975. On the use of the earth resources technology satellite, Landsat-1, in optical oceanography. *Remote Sensing of Environment* 4: 95-128.

McCluney, W. R., 1975. NASA red tide remote sensing research. Proceedings of the Florida red tide conference. *Fla. Mar. Res. Publ.* 8: 10-11.

McElroy, W. D., 1961. Bacterial luminescence. In: *The Bacteria 2: Metabolism* (I. C. Gunsalus and R. Y. Stanier, eds.), pp. 479-508, Academic Press, New York and London.

McElroy, W. D. and B. L. Strehler, 1954. Bioluminescence. *Bacteriol. Rev.* 18: 177-194.

McNeil, W. R. and K. P. B. Thomson, 1974. Remote measurement of water color and its application to water quality surveillance. Proceedings of the Third Conference of Earth Resources Observations and Information Analysis System, Tullahoma, Tennessee, *Remote Sensing of Earth Resources* 3.

Mulligan, H. F., 1975. Oceanographic factors associated with New England red tide blooms. *Proceedings of the First International Conference on Toxic Dinoflagellate Blooms* (V. R. LoCicero, ed.), Massachusetts Science and Technology Foundation, Wakefield, Massachusetts.

Mumola, P. B., O. Jarrett, Jr. and C. A. Brown, Jr., 1973. Multi-wavelength laser induced fluorescence of algae *in vivo* - a new remote sensing technique. Joint Conference on Sensing of Environmental Pollutants, 2nd. Washington, D. C., December 10-12, 1973. *Proceedings of the Instrumental Society of America*: 53-63.

Murphy, E. B., K. A. Steidinger, B. S. Roberts, J. Williams and J. W. Jolley, Jr., 1975. An explanation of the Florida East Coast *Gymnodinium breve* red tide of November 1972. *Limnol. Oceanogr.* 20: 481-486.

O'Neil, R. A., A. R. Davis, H. G. Gross and J. Kruss, 1975. A remote sensing laser fluorometer - for detecting oil, ligninsulfonates, and chlorophyll in water. In: *NASA Wallops Station - The Use of Lasers for Hydrographic Studies*: 173-196.

Otsuki, A. and R. G. Wetzel, 1973. Interaction of yellow organic acids with calcium carbonate in freshwater. *Limnol. Oceanogr.* 18: 490-493.

Otto, L., 1967. Investigations on optical properties and water masses of the Southern North Sea. *Neth. J. Sea Res.* 3: 532-552.

Papageorgiou, G. and Govindjee, 1971. pH control of the chlorophyll *a* fluorescence in algae. *Biochem. Biophys. Acta* 234: 428-432.

Pingree, R. D., P. R. Pugh, P. M. Holligan and G. R. Forster, 1975. Summer phytoplankton blooms and red tides along tidal fronts in the approaches to the English Channel. *Nature* 258: 672-677.

Pingree, R. D., P. M. Holligan and R. N. Head, 1977. Survival of dinoflagellate blooms in the Western English Channel. *Nature* 265: 266-269.

Plass, G. N., T. J. Humphreys and G. W. Kattawar, 1978. Color of the ocean. *Applied Optics* 17: 1432-1446.

Prakash, A. and Raschid, 1968. Influence of humic substances on the growth of marine phytoplankton: dinoflagellates. *Limnol. Oceanogr.* 13: 507-514.

Rounsefell, G. A. and W. R. Nelson, 1966. Red tide research summarized to 1964 including an annotated bibliography. *U. S. Fish Wildlife Service, Special Report* 535.

Serafin, M., 1967. Observations preliminares sobre la cosmosicion Y_n asper to del "turbio" o marea roja en las costas orientales de Venezuela. *Mem. Soc. Nat. La Salle, Venezuela* 27(76): 37-45.

Shapiro, J., 1957. Chemical and biological studies on the yellow organic acids of lake waters. *Limnol. Oceanogr.* 2: 161-179.

Sieburth, J. McN. and A. Jensen, 1969. Studies on algal substances in the sea, II. The formation of Gelbstoff (humic material) by exudates of phaeophyta. *J. Exp. Biol. Ecol.* 3: 275-289.

Sieburth, J. McN. and A. Jensen, 1968. Studies on algal substance in the sea, I. Gelbstoff (humic material) in terrestrial and marine waters. *J. Exp. Biol. Ecol.* 2: 174-189.

Sieburth, J. McN., 1967. Seasonal selection of estuarine bacteria by water temperature. *J. Exp. Mar. Biol. Ecol.* 1: 98-121.

Spruit van der Burg, A., 1950. Emission spectra of luminous bacteria. *Biochim. Biophys. Acta* 5: 175-178.

Staples, R. F., 1966. The distribution and characteristics of surface bioluminescence in the ocean. *Naval Oceanographic Office, Technical Report* TR-184. Washington, D. C.

Steemann-Nielsen, E., 1962. The relationship between phytoplankton and zooplankton in the sea. *Rapp. et process-verbaux reunions. Conseil Perman. Internat. Explorat. Mer* 153: 178-182.

Steidinger, K. A., 1975. Basic factors influencing red tides. *Proceedings of the First International Conference on Toxic Dinoflagellate Blooms* (V. R. LoCicero, ed.), pp. 153-162, Massachusetts Science and Technology Foundations, Wakefield, Massachusetts.

Stoertz, G. E., W. R. Hemphill and D. A. Markle, 1969. Airborne fluorometer applicable to marine and estuarine studies. *J. Mar. Tech. Soc.* 3: 11-26.

Strain, H. H., M. R. Thomas, and J. J. Katz, 1963. Spectral absorption properties of ordinary and fully deuteriated chlorophyll *a* and *b*. *Biochem. Biophys. Acta* 75: 306-311.

Stuart, P., P. T. Chandler, E. B. Kahan, A. R. Loeblish, III, G. Fuller and A. A. Benson, 1967. Food value of red tide. *Science* 158(3802): 789-790.

Stuermer, D. H. and J. R. Payne, 1976. Investigation of seawater and terrestrial humic substances with carbon-13 and proton nuclear magnetic resonance. *Geochem. Cosmochim. Acta* 40: 1109-1114.

Sweeney, B. M., 1975. Red tides I have known. *Proceedings of the First International Conference on Toxic Dinoflagellate Blooms* (V. R. LoCicero, ed.), pp. 225-234, Massachusetts Science and Technology Foundation, Wakefield, Massachusetts.

Sweeney, B. M. and J. W. Hastings, 1957. Characteristics of the diurnal rhythm of luminescence in *Gonyaulax polyedra*. *J. Cell. Comp. Physiol.* 49: 115-128.

Swift, E. and W. R. Taylor, 1967. Bioluminescence chloroplast movement in the dinoflagellate *Pyrocystis lunula*. *J. Phycol.* 3: 77-81.

Tett, P. B. and M. G. Kelley, 1973. Marine bioluminescence. *Oceanogr. Mar. Biol. Ann. Rev.* 11: 89-173.

Traganza, E. D., 1969. Fluorescence excitation and emission spectra of dissolved organic matter in sea water. *Bull. Mar. Sci.* 19: 897-904.

Tsujita, Tokimi, 1968. On red water oceanography (review). *Bulletin of the Plankton Society. Japan.* 15(2): 1-10.

Tyler, J. E., 1964. Color of the ocean. *Nature* 202: 1262-1264.

Vernon, L. P. and G. R. Seely, eds., 1966. *The Chlorophylls.* Academic Press, New York

Voitov, V. I., A. A. Egorova and N. I. Tarasov, 1960. Luminescence of cultures of the free-living Black Sea bacteria *Bacterium issatchenkoi*. Egorova. Doklady Akad. Nauk USSR (English translation: *Doklady Biological Science Section* 132: 452-453.

Wardle, W. J., S. M. Ray and A. S. Aldrich, 1975. Mortality of marine organisms associated with offshore summer blooms of the toxic dinoflagellate *Gonyaulax monilata* Howell at Galveston Texas. *Proceedings of the First International Conference on Toxic Dinoflagellate Blooms* (V. R. LoCicero, ed.), pp. 257-263, Massachusetts Science and Technology Foundation, Wakefield, Massachusetts

Wilton, J. W. and E. G. Barham, 1968. A yellow water bloom of *Gymnodinium flavum* Kofoid and Swezy. *J. Exp. Mar. Biol. Ecol.* 2: 167-173.

Wyatt, T. and J. Horwood, 1973. Model which generates red tides. *Nature* 244: 238-240.

Yentsch, C. S., 1960. The influence of phytoplankton pigments on the colour of sea water. *Deep-Sea Res.* 7: 1-9.

Zernova, V. V., 1970. On water discoloring in the Gulf of Mexico caused by development of plankton algae. *Okeanol. Issled. Rezult. Issled. Mezhd. Geofiz. Proekt.* 20: 105-109.

Remote Sensing of Phytoplankton Density and Diversity Using an Airborne Fluorosensor

Franklin H. Farmer
Clarence A. Brown, Jr.
Olin Jarrett, Jr.
Janet W. Campbell
Weldon L. Staton

ABSTRACT: A prototype remote fluorosensor was
flown in Narragansett Bay during the 1978 winter-
spring diatom bloom. Concurrent water column data were obtained using standard techniques. The
fluorosensor, mounted in a helicopter, obtained
remote fluorescence at 685 nm at hover over the
boat stations, and in line-of-flight between
stations. Remote data obtained at the hover points
showed excellent correlation with *in situ* measured
in vivo fluorescence, chlorophyll *a*, and total
cell count, while data obtained in flight correlated well with *in situ* fluorescence. Chlorophyll *a* concentrations were computed from the remote fluorescence using laboratory determined
fluorescence cross sections. These "remote" chlorophyll *a* concentrations showed excellent correlation with the chlorophyll *a* measured at the
stations. Remote data showed much more patchiness
of phytoplankton in the Providence River than in
the West Passage or lower bay. Data on the gross
taxonomic diversity of phytoplankton obtained
from analysis of the relative amounts of fluorescence produced by the multiwavelength excitation

indicated a predominantly golden-brown population
throughout the western half of the bay, with small
but significant green species content in the non-
bloom areas. This same trend was noted in cell
counts and identifications from samples obtained
in situ. The remotely measured *in vivo* fluorescence data and the chlorophyll *a* concentrations
calculated from it were found to be equivalent to
their respective *in situ* values as indicators of
phytoplankton density.

INTRODUCTION

During the last quarter-century, there has been a trend of
increasing interest by biological oceanographers in the spatial
heterogeneity of phytoplankton, a phenomenon also known as "patchiness", and the interactions of biological, chemical, and physical
processes which cause these distributions (Steele, 1978). Two indicators of this increasing interest have been the development of
theoretical models (Kierstead and Slobodkin, 1953; Parsons et al,
1967; Wroblewski and O'Brien, 1976), and specifically designed instrumentation and measurement techniques (Lorenzen, 1966; Clark et al.,
1970). The latter have been designed to obtain quasi-synoptic mapping of the sea-surface chlorophyll distribution which can be used
as inputs to these models. In response to this need for synoptic
or near-synoptic measurements of both phytoplankton density and diversity, a remote sensor is being developed by the National Aeronautics and Space Administration's (NASA) Langley Research Center
(LaRC). This aircraft-borne system utilizes narrow-band light from
multiple dye lasers to excite selected algal photopigments and then
measures the resultant fluorescence emitted from chlorophyll *a* in
the region of 685 nm. The wavelengths of the exciting light were
selected on the basis of recognized differences in the fluorescence
excitation spectra of species belonging to the four major algae
color groups. The theoretical basis of this remote fluorosensor
and the details of the system design and construction are provided
by Mumola et al., 1973 and Brown et al., 1977. These authors also
present some of the results of laboratory investigations performed
during the early phases of system development.
After about two years of laboratory tests of the system using
pure and mixed cultures of marine algae in 100 l tanks (Brown et al.,
1977) and a series of static field tests from a pier and a bridge,
a major field experiment was planned and executed in conjunction
with personnel at the University of Rhode Island Graduate School of
Oceanography and the Environmental Protection Agency's National
Environmental Research Center, Kingston, Rhode Island. The objectives of that experiment follow: (1) to obtain a synoptic view of
the horizontal variation or "patchiness" of the phytoplankton community in Narragansett Bay, Rhode Island, by remotely measuring the
variation of chlorophyll *a* concentration (density); (2) to measure
at the same time the variation of the concentration of the major
components of the phytoplankton community (diversity); (3) to as-

certain if sensor motion (flight) would have any detrimental effects on the accuracy and sensitivity of the fluorosensor system;
(4) to determine if the sensitivity of the system would be greater
in the less turbid areas of this bay than in Chesapeake Bay; and
(5) to investigate the variation of the remote fluorescence to *in
situ* chlorophyll *a* ratio over a large estuarine area. This paper
presents and discusses some of the results of that experiment.

MATERIALS AND METHODS

     The data presented in this paper were obtained on the morning
of March 16, 1978, during the winter-spring diatom bloom in
Narragansett Bay, Rhode Island. The selection of this time and
place for the first flight test of the remote fluorosensor was
based on requirements for wide ranges of chlorophyll *a* concentration, optical attenuation, and nutrient levels.

     Sampling Procedure

     Both remote and water column data were obtained at and between
11 stations ranging from the mouth of Narragansett Bay (in Rhode
Island Sound), up the West Passage of the bay, and up the Providence
River to the head of navigation (Fig. 1). Water column samples were
obtained by University of Rhode Island personnel on three boats
which started at stations 1, 7, and 15, respectively. At these
three stations, "simultaneous" samples were obtained by the boats
and fluorescence measurements by the helicopter-borne fluorosensor.
At the other stations, boat data were obtained at free floating
markers dropped by the Army National Guard helicopter at the point
where remote data had been obtained (lead times were as much as an
hour). The helicopter started at station 1 and flew up the bay,
getting remote fluorescence data at each station while hovering at
an altitude of about 25 m, then flying the intervening distance
straight line and obtaining a set of data every 41 m. An exception
to this regimen was made between stations 7 and 9, where a criss-
cross pattern was used.

     Water Column Measurements

     At each station, a water sample was taken with a Niskin bottle
at 0.5 m below the surface. The principal measurements made on the
sample on station were temperature and *in vivo* fluorescence (Lorenzen,
1966) subsequently called *in situ* fluorescence. The Lorenzen method was modified by the additional use of narrow-band excitation
filters at each laser wavelength. Twenty ml of sample were fixed
with Lugol's Iodine for subsequent phytoplankton cell counts and
identifications. Four 1 of sample were transported to the laboratory where measurements were made for total chlorophyll *a* (Holm-
Hansen et al., 1965) subsequently called "direct chlorophyll *a*",
optical attenuation at 454, 539, and 685 nm (Brown et al., 1977),
salinity (using a hand-held A/O Spencer refractometer), and pH
(using a Corning pH meter with glass electrode). Cell counts and

154

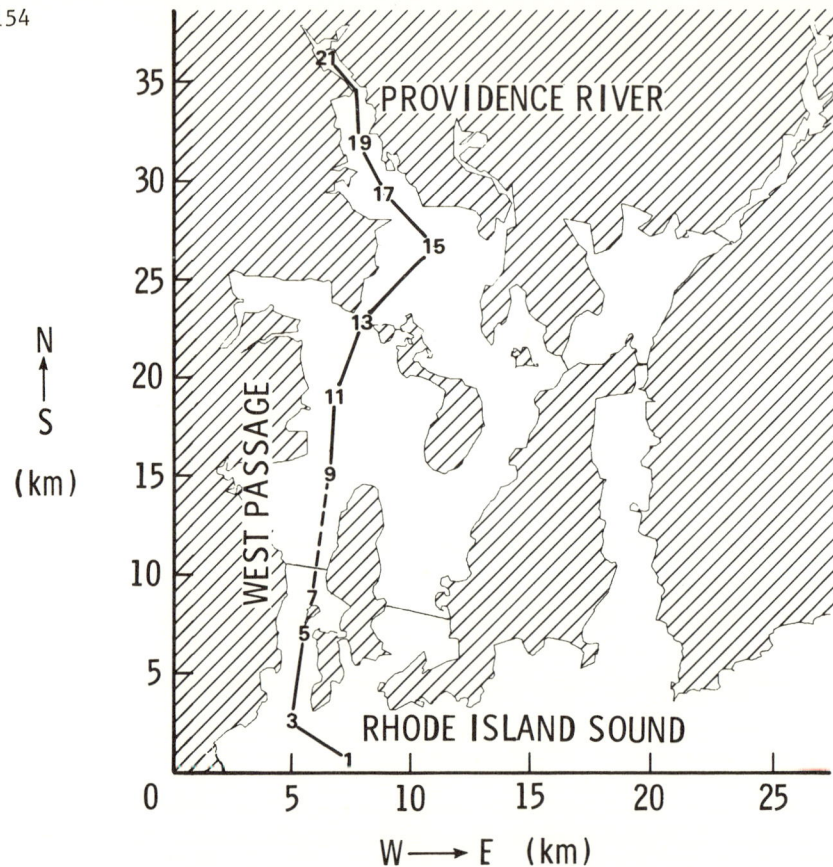

Figure 1.   Location of helicopter flight path and
            sea truth stations for the March 16,
            1978 Narragansett Bay experiment.

identifications for stations 1, 5, 9, 13, 17, and 21 were done on
1 ml samples in a Sedgewick-Rafter Chamber at 100 and 200x by
Dr. Gabriel Vargo, GSO University of Rhode Island, and for stations
3, 7, 11, 15, and 19, by Dr. Robert Jordan, Virginia Institute of
Marine Science, by settling chamber methods. The latter counts
were made using an inverted microscope at 100x, 200x and 400x.

    Between stations pulsed samples were taken from the water sur-
face at 3 min intervals. Distance between samples varied with boat
speed. *In vivo* fluorescence was the only measurement made on
these samples.

    Remote Measurements

    Remote fluorescence was measured using the NASA LaRC Remote
Airborne Fluorosensor (RAF). This system is designed to excite
phytoplankton sequentially with light of four different wavelengths
(454, 539, 598, and 617 nm) and to measure the returning fluorescence at 685 nm resulting from each excitation. The excitation bandwidth is 5 nm and the fluorescence bandwidth is 9 nm. A

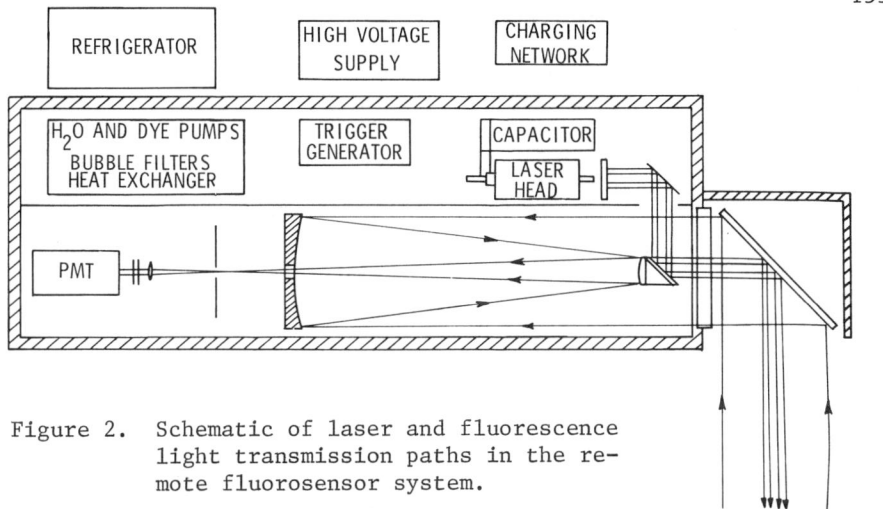

Figure 2.   Schematic of laser and fluorescence
light transmission paths in the remote fluorosensor system.

schematic plan of the fluorosensor is shown in Figure 2.   The excitation light is produced by four pulsed dye lasers which are excited by a common linear flashlamp.   Each time the flashlamp fires,
all four dyes are excited, but a rotating intracavity shutter
only allows lasing in one dye cuvette at a time.   For the
Narragansett Bay experiment, however, the system was modified
to operate at only two wavelengths (454 and 539 nm) because
phytoplankton of only two color groups (green and golden-brown)
were expected.   This allowed the same dye to be used in two lasers, which doubled the power of the exciting light at both
wavelengths and increased the sensitivity of the system.   This
required a modification to the shutters to permit simultaneous
laser firing.

The fluorescence from algae is diffuse and only a small portion is captured by the fluorescence telescope.   This portion is
inversely proportional to the square of the distance from the telescope to the water surface.   The fluorescence is concentrated by
the telescope, passed through a narrow-band optical filter which
transmits light in the region of 685 nm only, and detected by a
photomultiplier tube.   In order to compensate for the background
radiance at 685 nm, two separate signal integrators are used with
the detector.   One measures only the background light (i.e., the
return signal when the laser is not firing).   The other measures
the background plus induced fluorescence (i.e., the returned signal
immediately after laser firing).   The difference between these two
signals is the induced fluorescence of chlorophyll a, subsequently
called "remote fluorescence".   Although phaeophytin a does contribute to the remote fluorescence at 685 nm, this contribution is believed to be minor for the following reasons: (1) Absorption of
phaeophytin a at 454 (±2.5) nm is slight and at 539 (±2.5) is negligible in $vivo$; (2) Fluorescence of phaeophytin a is off the 685 nm
peak of chlorophyll a fluorescence enough to reduce the amount recovered in the 680.5 to 689.5 nm region substantially; (3) The fluorescence efficiency of phaeophytin a is less than half that of chlorophyll a; and (4) Phaeophytin a content of Narragansett Bay water

samples was usually less than five percent and never exceeded 14 percent of the chlorophyll a concentration.

Remote fluorescence was recorded as counts on magnetic tape along with the laser output power and the altitude. Data reduction consisted of converting the counts to joules, and then correcting the remote fluorescence for the effects of mirrors and optical filters within the system and for variations in laser output power. Other factors taken into consideration were the attenuation of the laser light and the returned fluorescence in the water column by light scattering and absorption, and the altitude of the sensor above the water surface.

Calculations

Computations of the "remote" chlorophyll a for each color group is achieved by the simultaneous solution of two equations, each with two unknowns, using two sets of remote fluorescence data (F_{454}, F_{539}). These equations are discussed in detail in Brown et al., (1977). The parameter which relates fluorescence at 685 nm to the chlorophyll a concentration, the fluorescence cross-section, has been intensively investigated in our laboratories and mean spectra for the four major color groups have been computed (Fig. 3). The fluorescence cross-sections used to compute the "remote" chlorophyll a concentrations in this experiment were derived from laboratory data from experiments in which the remote fluorosensor was the fluorescence detector and 100 l tanks of pure cultures of species representing the four color groups served as the excited medium (Brown et al., 1977). Total "remote" chlorophyll a is the sum of the "remote" chlorophyll a calculated for all color groups present.

RESULTS AND DISCUSSION

The conditions on the morning of March 16, 1978, were very close to optimum for the operation of the remote fluorosensor and provided the range of chlorophyll a, optical attenuation and nutrient desired for the system test. The spring diatom bloom had reached its maximum and had begun to deteriorate slightly. Maximum chlorophyll a concentration was 43.8 µg/l at station 13 in the upper end of the West Passage, decreasing gradually in both directions and reaching lows of 2.7 µg/l at station 1 and 9.4 µg/l at station 19. Table 1 presents most of the pertinent *in situ* data. Note that while the optical attentuation of the water (at 632.8 nm) increases gradually from station 1 to station 19, it almost doubles between stations 19 and 21. It appears that phytoplankton are the principal light scatterers in the West Passage, but in the Providence River other materials begin to predominate, particularly at station 21. Although the total phytoplankton counts followed the same general trends as the direct chlorophyll a concentrations, some disparity was observed between the two methods of counting, particularly with regard to the green species; the latter difference was related to both magnification and individual interpreta-

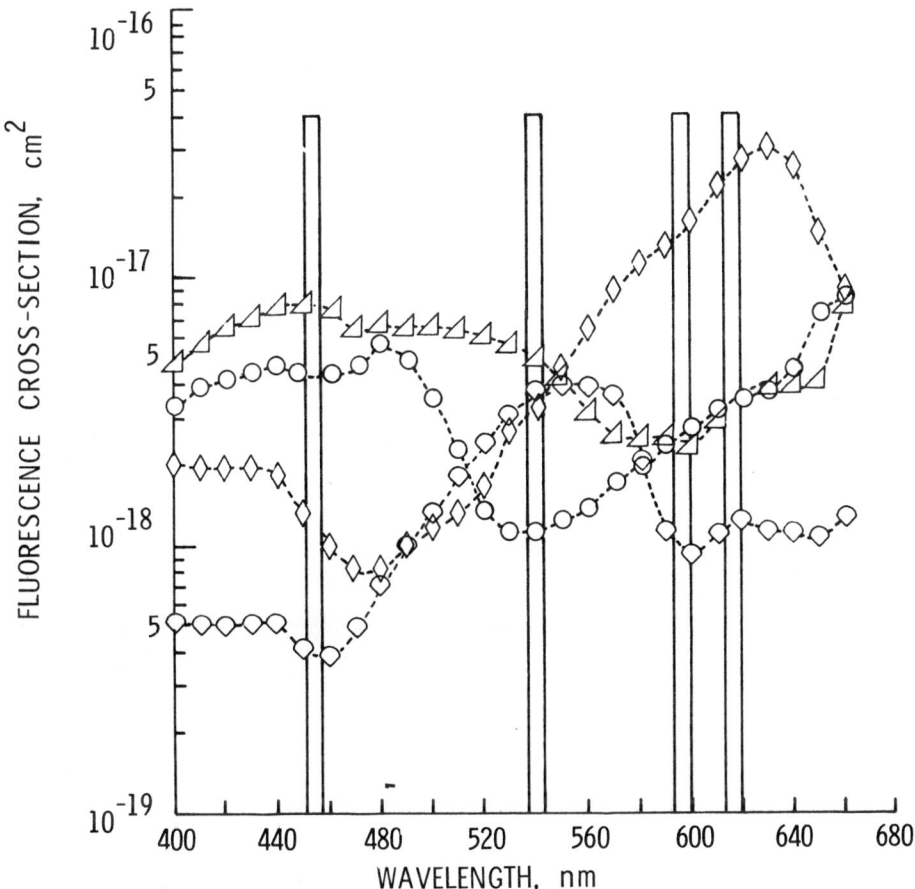

Figure 3.   Mean fluorescence cross-sections of
            the four major algae color groups,
            golden-brown ($\triangle$), green (O), blue-
            green ($\Diamond$), and red ($\bigcirc$). Vertical
            bars indicate excitation wavelengths
            of the remote fluorosensor.

tion. It is interesting to note that the number of cells of green
species noted by Jordan in the bloom area (stations 7 to 17) is
less than the number either in the lower bay or upper Providence
River. This trend is even more striking when the content of greens
is viewed as a percent of the total count.
     The fluorescence of chlorophyll a at 685 nm resulting from ex-
citation at 454 nm and measured remotely by the fluorosensor along

TABLE 1. Narragansett Bay data for March 16, 1978.

Station #	Direct Chlorophyll a (μg/l)*	Salinity 0/00	pH	Water Column Optical Attenuation** at 632.8 nm	Total Phytoplankton (cells/ml)	Green Species (%)
1	2.7	31.0	8.14	0.97	682	–
3	7.1	30.0	8.17	1.32	3581	26.0
5	8.4	30.0	8.23	1.53	2143	–
7	15.5	30.0	8.36	1.77	4921	2.1
9	22.2	30.0	8.42	2.29	5287	–
11	29.8	28.5	8.45	3.09	8845	5.6
13	43.8	26.0	8.40	4.34	6737	–
15	37.0	23.0	8.23	4.09	9980	2.3
17	31.5	21.0	8.14	4.78	5392	–
19	9.4	–	–	4.87	3147	23.8
21	24.6	9.0	7.58	9.15	3248	–

*Mean of two total chlorophyll a values.

**Per meter.

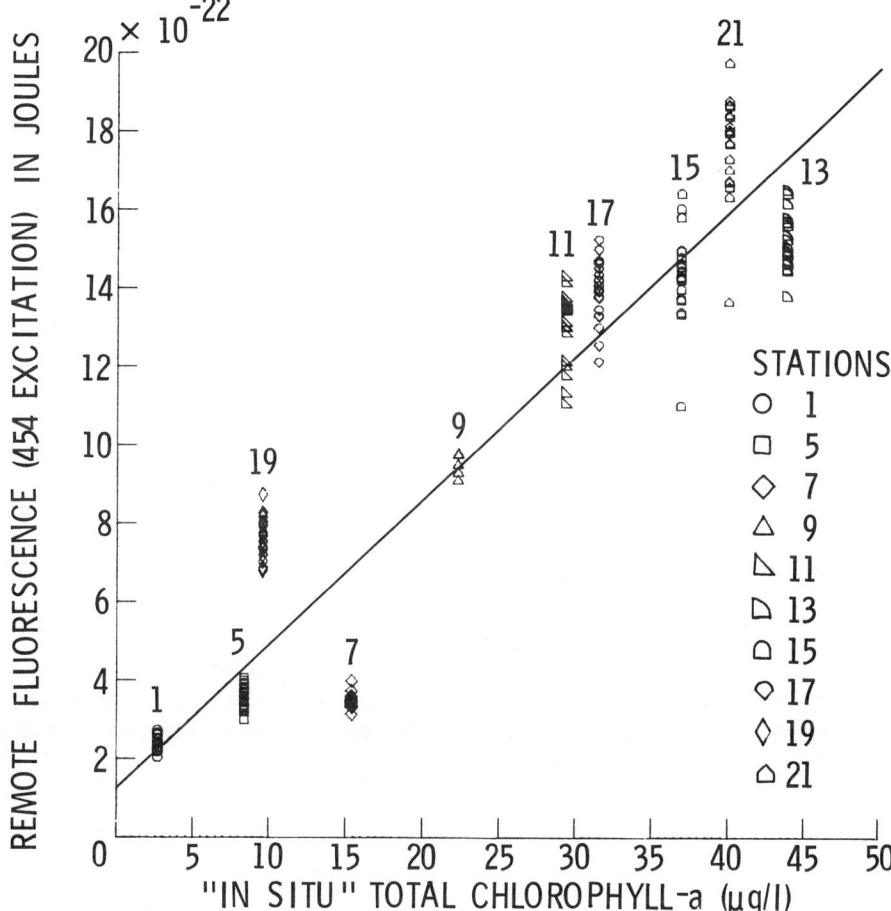

Figure 4.   Remote fluorescence resulting from excitation at 454 nm (flight and hover
data) along flight path from station 1
to station 21.

the flightpath from stations 1 to 21 is presented in Figure 4.   The
average remote fluorescence for the hover points is also shown.
The general trend of the remotely measured fluorescence is the same
as that noted with all the indicators of phytoplankton density (direct chlorophyll a, cell count, and in $situ$ fluorescence), a slow
steady increase from stations 1 to 13 (28 km) followed by a general
decrease going up the Providence River.   The latter trend, however,
shows considerably more variability which may indicate alternating
regions of growth stimulation and inhibition in that region.   A
linear regression analysis of the relationship between the remote
hover data and in $situ$ fluorescence indicated a correlation coeffi-
of 0.925 between the two measurements.   While this is considerably
less than the degree of agreement observed in laboratory studies
with this system (0.990), it still indicates a high degree of cor-

relation between these two measurements under near optimum operating conditions.

A good agreement between the remote and *in situ* fluorescence of chlorophyll a does not automatically indicate a similar agreement between remote fluorescence and direct chlorophyll a. The ratio between *in situ* fluorescence and chlorophyll a concentration has been reported to be extremely variable in deep ocean (Strickland, 1968), estuarine and freshwater environments. Both Kiefer (1973), in Pacific waters and Lake Tahoe, and Loftus and Seliger (1975), in Chesapeake Bay, reported up to a tenfold variation in this ratio. Blasco (1973) has shown this variation to be due, at least in part, to the effects of nutrient limitation, ambient light levels, and species variation on the fluorescence quantum yield of chlorophyll a. A comparison of the remote fluorescence excited by laser light at 454 nm and the direct chlorophyll a concentration at each of the 10 stations is made in Figure 5. These data, obtained under widely varying light level and nutrient conditions, show very little variation in the remote fluorescence to direct chlorophyll a ratio throughout the region except at stations 7 and 19. The correlation coefficient between the two sets of data was 0.932. The maximum difference in the ratio (at similar direct chlorophyll a concentrations) was about 3.0 (highest remote fluorescence value at station 19 divided by the lowest at station 5). An examination of the possible causes of the greater variation from the mean ratio at stations 7 and 19 revealed an interesting point. The ratio of *in situ* fluorescence to direct chlorophyll a concentration also exhibits maximum deviation at these stations, being 0.13 and 0.30, respectively, in relation to a mean ratio of 0.21 and a standard deviation of 0.05. This indicates that the source of variation in the above ratio at these stations is not the remotely sensed data, but rather something happening in the water column, or perhaps an error in direct measurement of chlorophyll a.

The "remote" total chlorophyll a calculated from remote fluorescence (from both 454 nm and 539 nm excitation) obtained on the flight legs between stations 1 and 21 is presented in Figure 6. Included are over 600 independent estimates of the chlorophyll a concentration. The average direct chlorophyll a measurement at each station is shown for reference. The "remote" chlorophyll a concentration follows the same trends as the remote fluorescence. A linear regression analysis of the average direct chlorophyll a against the first and last "remote" chlorophyll a estimate at each station, (i.e., the last value on the previous flight leg and the first on the subsequent leg) gives a correlation coefficient between the two sets of data of 0.926. Note that the direct measurement of chlorophyll a concentration at station 19 (41.75 km) is very low when compared to the other direct values in the Providence River, but is closely matched by several points of "remote" chlorophyll a concentration. In addition, the direct chlorophyll a concentration at a traditional station (15a) on the Providence River Channel made independently by the University of Rhode Island crew (17.3 µg/l) showed good agreement with the nearest "remote" chlorophyll a concentration (20.5 µg/l) at 35.664 km.

The minimum detectable chlorophyll a concentration for the RAF

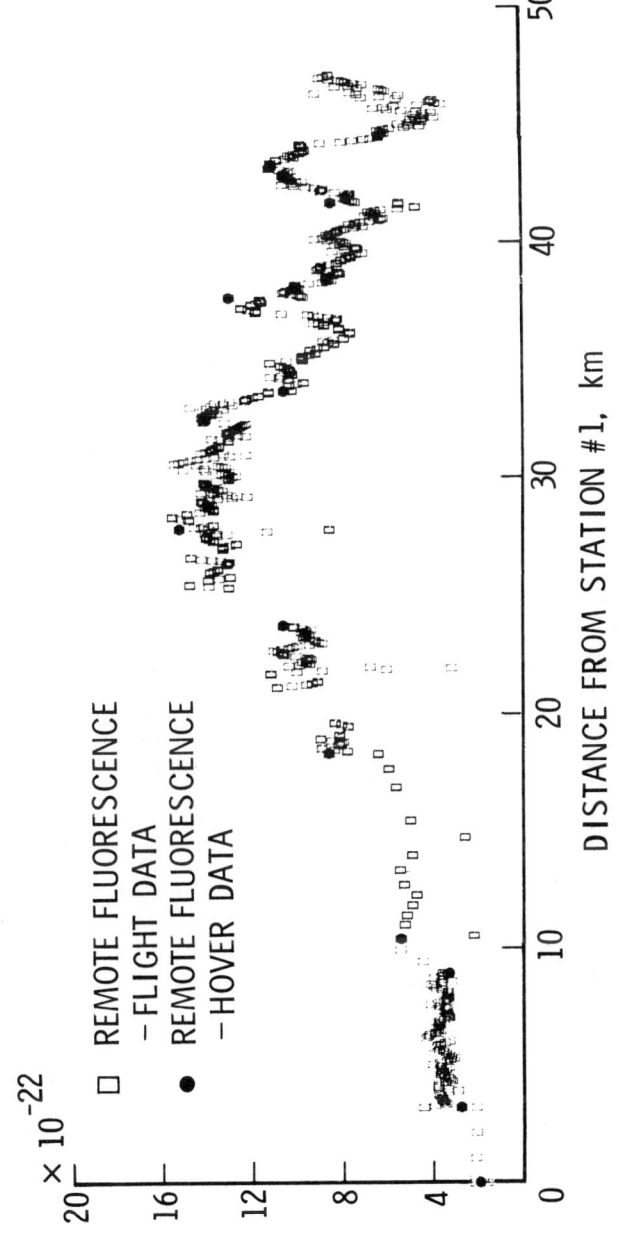

Figure 5. Remote fluorescence resulting from excitation at 454 nm (hover data) versus
in situ chlorophyll *a* for 10 stations
in the West Passage and Providence
River.

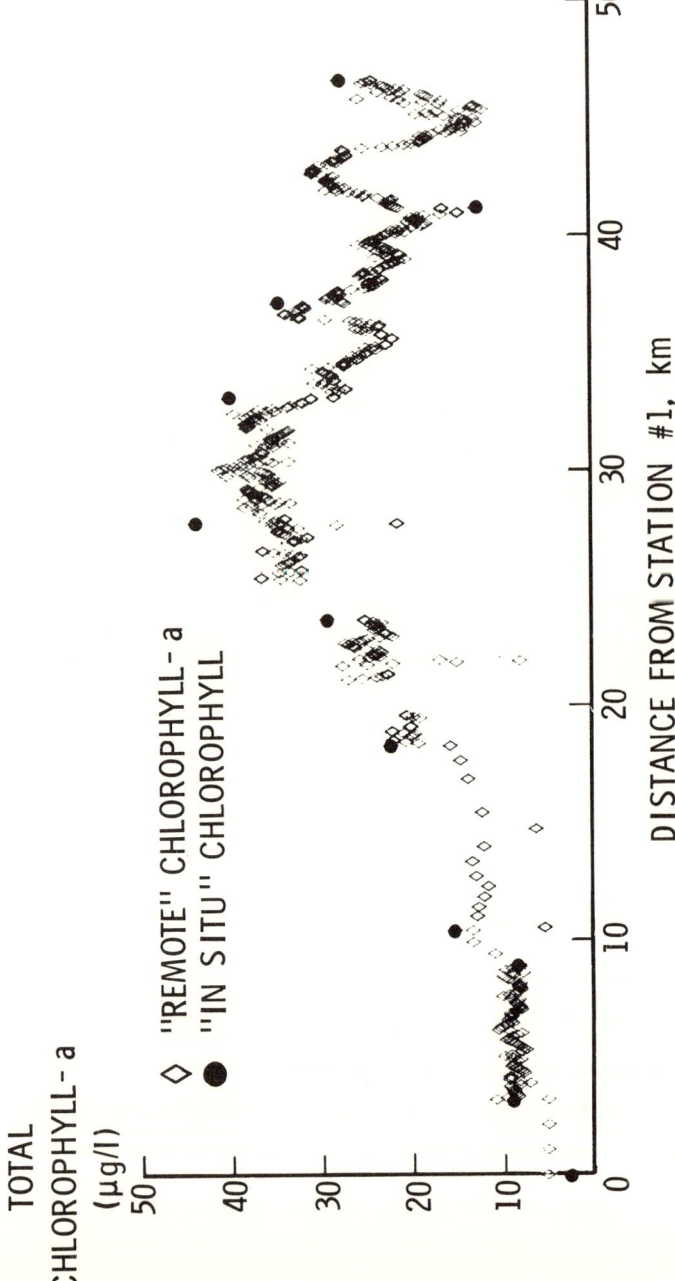

Figure 6. "Remote" chlorophyll *a* along flight path and *in situ* chlorophyll *a* at the 10 sea truth stations.

at present output power levels is one of the parameters of interest to those researchers who may have applications for its remote data. A minimum detectable chlorophyll a concentration was calculated for the conditions prevalent at station 1 on March 16, particularly background light level and water turbidity. This computation was based on the assumption that the minimum detectable signal would be three times the standard deviation of the background (17.5 counts), or 52.5 counts. Using a sensitivity of 108 counts/μg of chlorophyll a (293 counts/2.7 μg/1 chlorophyll a), the minimum detectable signal was computed to be equivalent to 0.48 μg/1 (52.5/108). However, the effect of increased ambient light and water turbidity on the minimum detectable chlorophyll a, as indicated by values of 2.8 and 3.8 μg/1. calculated for stations 5 and 13, respectively, must always be taken into consideration.

Although this experiment was designed with the foreknowledge that the diversity of the phytoplankton population at this particular time of the year would be minimal with the population being dominated by the diatoms, the diversity information obtained from the remote fluorescence measurements was of considerable interest. The basis of the diversity computation is illustrated in Figure 7, where the ratio of the remote fluorescence measured after excitation at 539 nm to that measured after excitation at 454 nm is presented for all stations. These data were obtained in the hover mode. Each point represents a sequential pair of data points, up to 20 for each station, and the mean ratio at the station is indicated by the solid symbol. Also shown are two lines representing the fluorescence (539 nm)/fluorescence (454 nm) ratios previously obtained in the laboratory for the golden-brown and green color groups (i.e., the ratios of the fluorescence cross-sections for these color groups). The location of a data point relative to the lines is indicative of the distribution of the component color groups in that sample. For example, the mean ratios for stations 7 and 9 are close to the golden-brown, indicating a very high content of species from this color group, while the mean ratios for stations 1, 5, and 19 are much closer to the green line, indicating a much higher component of green species than at stations 7 and 9. Note that all mean ratios for stations in the diatom bloom area (7-17) are near the golden-brown line, while those for stations located outside the bloom area (1, 5, and 21) are located closer to the green line. This distribution between the color groups at the these stations is fully in agreement with *in situ* cell count/identification data presented in Table 1 for stations 3, 7, 11, 15, and 19.

The flight data obtained between stations are further indicative of the above trend. In Figure 8, where the total "remote" chlorophyll a of the green species is presented as a function of distance from station 1, the only substantial green species concentrations were in the lower bay and upper Providence River transects. In the region in between, the golden-brown bloom species predominated and the chlorophyll a contribution from the green species was at or near zero.

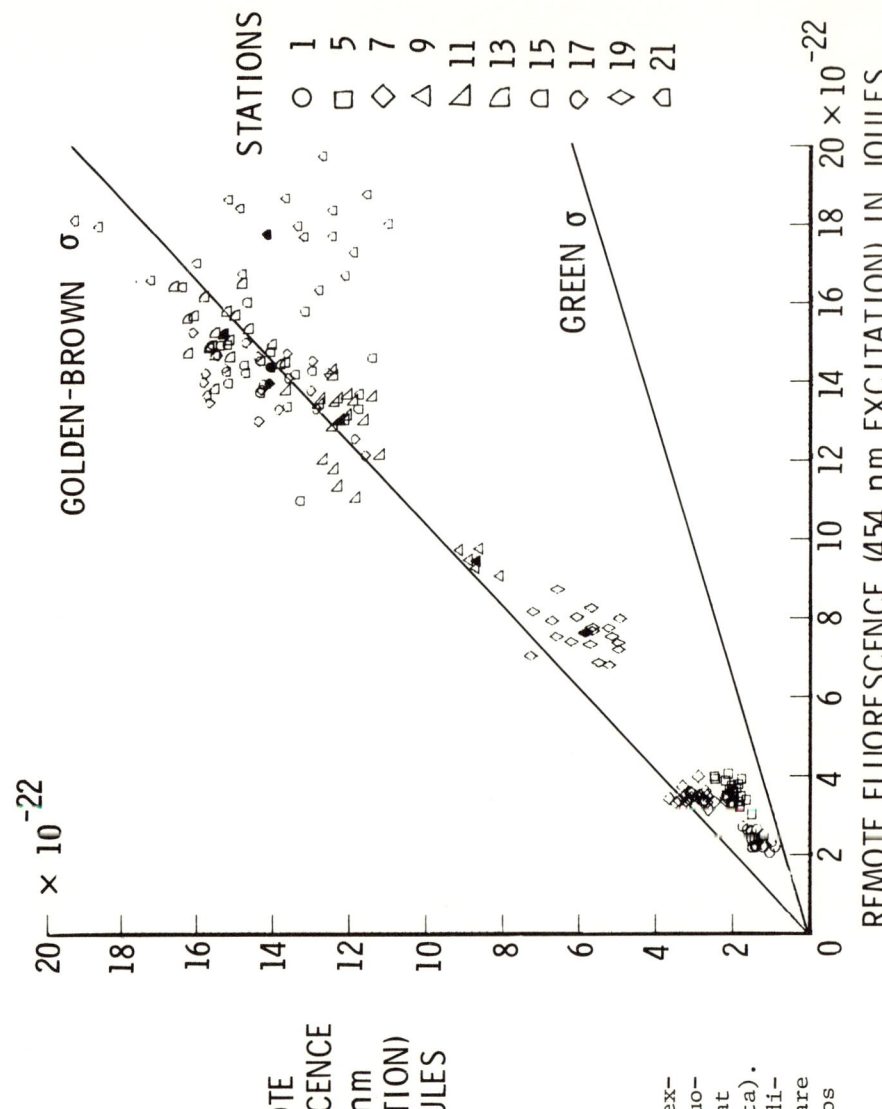

Figure 7.

Remote fluorescence resulting from excitation at 454 nm versus remote fluorescence resulting from excitation at
539 nm at the 10 stations (hover data).
Mean values for each station are indicated by solid symbols. The lines are
the fluorescence cross-section ratios
for these wavelengths obtained from
previous laboratory studies.

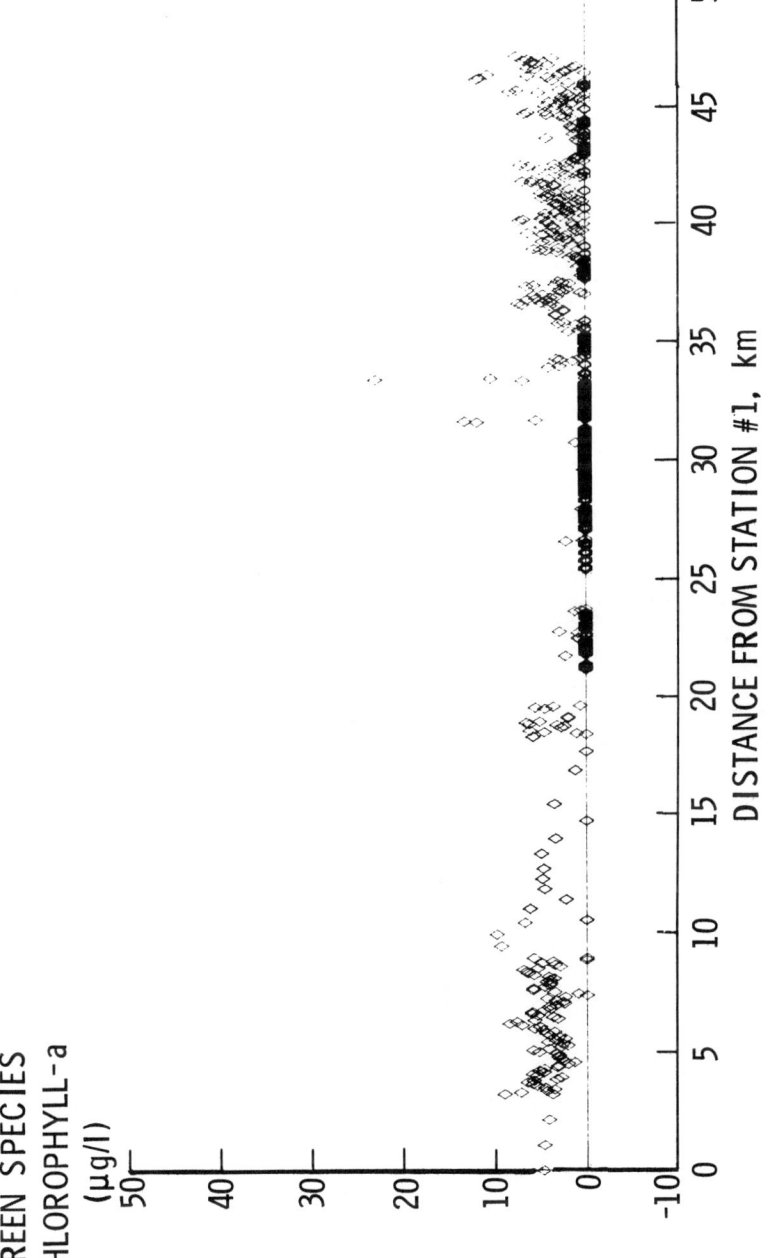

Figure 8. "Remote" chlorophyll a for green species only (flight data).

CONCLUSIONS

Several observations about the phytoplankton patchiness in
Narragansett Bay and the applicability of the remote fluorosensor
(RAF) to studies of the variation of both the density and diversity
of phytoplankton can be made on the basis of this single, but extensive, experiment. First, the remotely measured fluorescence of
chlorophyll a at 685 nm excited by laser light at 454 and 539 nm
mirrors the trends indicated by *in situ* fluorescence, direct chlorophyll a concentration, and total cell counts, but at the same
time gives more detailed information on the major scale variations
in phytoplankton density. This was particularly true in the Providence River Channel where regions of growth stimulation and/or inhibition covering several kilometers were missed entirely by the *in
situ* measurements, but were clearly defined by dozens of remote
measurements. In some areas, features as small as 500 m in length
were indicated by the remote data. Examination of both the variation of remote fluorescence and "remote" chlorophyll a indicated
much more patchiness in the river than in the West Passage and
lower bay. Second, the ratio of remote fluorescence / direct chlorophyll a concentration was less variable than expected, even though
the factors reported to effect this ratio varied widely over the
almost 50 km track covered by the helicopter-borne fluorosensor.
Third, the distribution of the total "remote" chlorophyll a between
the two major phytoplankton color groups which was determined from
the remote fluorescence data, showed three distinct areas within
the bay. The lower bay and the Providence River showed small, contiguous, green populations, while the middle and upper West Passage
showed almost 100% golden-brown species. These data agreed qualitatively with direct counts and identifications. Fourth, the transition from a hover or stationary mode to a flight mode evidently
does not degrade the accuracy of the fluorosensor. Although further studies should be conducted at other speeds, the 50 knot speed
has no detectable effect. Fifth, the minimum concentration of chlorophyll a detectable with the remote fluorosensor varies with the
environmental conditions, particularly ambient light and water turbidity, at the point where the remote measurement is made. Based
solely on data obtained in Narragansett Bay on March 16, 1978, it
is estimated that the minimum detectable concentration of chlorophyll a with the present system can range from about 4 µg/l under
conditions of high ambient light and turbidity to considerably
less than 1 µg/l in the early morning or late afternoon in clear
water.

REFERENCES

Blasco, D., 1973. Estudio de las variaciones de la relacion fluo-
    rescencia *in vivo*/chlorofila *a*, y su aplicacion en oceanografia.
    Influencia de la limitacion de diferentes nutrientes, efecto
    del dia y noche y dependecia de la especie estuiada. *Invest.*
    Pesa. 37: 533-556.
Brown, C. A., Jr., F. H. Farmer, O. Jarrett, Jr., and W. L. Staton,
    1977. Laboratory studies of *in vivo* fluorescence of phytoplank-
    ton. *Proc. Fourth Joint Conf. Sens. Environ. Pollutants*: 782-
    788.
Clarke, G. L., G. C. Ewing, and C. J. Lorenzen, 1970. Spectra of
    backscattered light from the sea obtained from aircraft as a
    measure of chlorophyll concentration. *Science* 167: 1119-1121.
Holm-Hansen, O., C. J. Lorenzen, R. W. Holmes, and J. D. H.
    Strickland, 1965. Fluorometric determination of chlorophyll.
    J. Cons. Perm. Int. Explor. Mer. 30: 3-15.
Kiefer, D. A., 1973. Fluorescence properties of natural phytoplank-
    ton populations. *Mar. Biol.* 22: 263-269.
Kierstead, H., and L. B. Slobodkin, 1953. The size of water masses
    containing plankton blooms. *J. Mar. Res.* 12: 141-147.
Loftus, M. E., and H. H. Seliger, 1975. Some limitations of the *in*
    vivo fluorescence technique. *Chesapeake Sci.* 16: 79-92.
Lorenzen, C. J., 1966. A method for the continuous measurement of
    in vivo chlorophyll concentrations. *Deep Sea Res.* 13: 223-227.
Mumola, P. B., O. Jarrett, Jr., and C. A. Brown, Jr., 1973. Multi-
    wavelength laser induced fluorescence of algae *in vivo*: a new
    remote sensing technique. *Proc. Second Joint Conf. Sens.*
    Environ. Pollutants: 53-62.
Parsons, T. R., R. J. LeBrasseur, J. D. Fulton, 1967. Some obser-
    vations on the dependence of zooplankton grazing on the cell
    size and concentration of phytoplankton blooms. *J. Oceanogr.*
    Soc. Japan 23: 10-17.
Steele, J. H., ed., 1978. Spatial pattern in plankton communities.
    NATO Conference Series IV 3: 470, Plenum Press, New York.
Strickland, J. D. H., 1968. Continuous measurement of *in vivo*
    chlorophyll; a precautionary note. *Deep Sea Res.* 15: 225-227.
Wroblewski, J. S., J. J. O'Brien, 1976. A spatial model of phyto-
    plankton patchiness. *Mar. Biol.* 35: 161-175.

Interactive Processing, Display, and Interpretation of Remotely Sensed Data

Roland E. Nagle

ABSTRACT: Remote sensing of the environment from
spacecraft has rapidly advanced from a subjective
art to a quantitative science. The massive amount
of data produced by satellites has previously relegated consideration of quantitative processing of
these data only to large computer systems at centralized facilities. The paradox in this situation
is that the highest quality data are not available
to the ultimate operational or research users.
However, the application of minicomputer, microprocessor technology in interactive graphics display devices offers a means to alleviate this
dilemma.

This paper described a hardware/software, mini-
computer-based interactive graphics display device comprising a system that can be segmented
functionally into three basic areas: 1) data
acquisition, 2) data processing, and 3) information display. Data acquisition aspects of
these systems are discussed and an example of a
geostationary satellite direct readout station
is shown. The data processing and information

display are accomplished on the Naval Environmental Prediction Research Facility's (NEPRF)
Satellite Data (SATDAT) Processing and Display
System (SPADS). The SPADS is described to illustrate the types of hardware and software
that are required in such systems. Examples
are shown to illustrate the power and flexibility of hands-on manipulation of remotely sensed
data. The potentials of this type of syttem for
research and operational use are discussed.

BACKGROUND

The beginnings of remote sensing of the environment from space
can be traced back to 1 April 1960 when TIROS-I (Television and Infrared Observational Satellite) was first placed into orbit. The
success of this spacecraft subsequently led to two generations of
meteorological satellites (the TIROS and ESSA series) which employed
vidicon detection systems. These instruments, although they could
not be calibrated, produced images that revealed in detail the organized structure of the earth's cloud cover, and assured the con-
tinuance of the meteorological satellite program.

A significant advance in the technology occurred with the
launches of the third generation of polar orbiting satellites, the
ITOS/NOAA and the Defense Meteorological Satellite Program (DMSP)
Block-5 series, in which the vidicon sensors of previous spacecraft
were replaced with scanning, high resolution radiometers. The most
important advantage of these latter instruments was that they could
be calibrated; using the instrument response curves determined prior
to launch and the onboard calibration data provided in the telemetry
down-link, the data could be handled digitally for conversion to absolute measures of the energy received at the sensor in each observing channel. Thus, remote sensing entered the quantitative era
where consideration could be given to deriving precise measurements
of discrete elements of the phenomena being observed.

The threshold of the fourth generation of operational environmental satellites was recently crossed with the launch of the DMSP
Block-5D (Nichols, 1975) and the TIROS-N (Kidwell, 1979) satellites.
In these newest systems, changes from the previous generation of
spacecraft are evolutionary rather than revolutionary. The DMSP
Block-5D spacecraft incorporates a new visible and infrared imager
that compensates for changes in resolution along the scan lines.
The 15 μ vertical sounding instrument carried by the previous Block-
5C satellites has been augmented with channels in moisture absorption
bands to allow atmospheric moisture profile extraction. Additional
improvements have been programmed for the Block 5D-2 vehicles; the
digitalization rate of the imaging IR channel will be increased from
6 to 8 bits and the spectral window will be narrowed to 10.5-12.5 μ
thereby greatly improving the instruments' sensitivity to sea surface temperatures. A significant change is provision for full on-
board digitalization of the data and the inclusion of onboard processing capability. This capability effectively precludes contami-

nation of the data during transmission from the spacecraft via analog communications channels.

Corresponding upgrades have been incorporated in the TIROS-N vehicles. The Advanced Very High Resolution Radiometer (AVHRR) is a
4-5 channel instrument whose channels have been selected specifically
for high resolution quantitative mapping of sea surface temperature.
The TIROS-N Operational Vertical Sounder (TOVS) should be a significant improvement over previous sounding instruments. Channels in
both the 4.3 and 15.0 μ carbon dioxide absorption bands as well as
microwave channels are included in the TOVS, thereby offering the
potential of improved vertical and absolute resolution in the retrieved atmospheric temperature and moisture profiles.

The spacecraft and instruments discussed to this point were
designed primarily for atmospheric sensing; with few exceptions,
most of the data from these spacecraft have been processed for
meteorological purposes. However, several instruments designed
for oceanographic sensing have been flown on recent research satellites. The passive electronically scanning microwave radiometers
flown on both NIMBUS-5 and -6 were sensitive to ocean roughness.
A six channel Coastal Zone Color Scanner (CZCS) was flown on
NIMBUS-7 specifically for the measurement of ocean color (see
Madrid, 1978). The multispectral scanner carried onboard the
LANDSAT series, while effectively utilized for land resource
studies, has distinct capabilities in marine biological surveillance. SEASAT-A (Anon, 1976), launched in June of 1978 was a
"proof-of'concept" vehicle designed specifically for oceanographic remote sensing. The instrument complements of SEASAT-A were
not designed for direct marine biology measurements; however, a
multichannel Coastal Zone Scanner is being considered for inclusion on the follow-on satellite to SEASAT-A, the National Oceanic
Satellite System (NOSS, 1979), which is currently in the planning
stage.

Thus far, only low altitude polar orbiting satellites have
been discussed. High altitude satellites are another class of
platform used for environmental observation. These spacecraft
are placed in orbit in the earth's equatorial plane at an altitude of some 35,000 km above the earth. At this altitude, the orbital period of the satellite matches the rotations of the earth
and the satellites appear to remain stationary over one location
on the earth's surface. The United States currently operates
three Geostationary Observational Environmental Satellites (GOES)
(Corbell et al., undated). The GOES satellites obtain data at
1.0 km and 5.0 km resolution in the visible and infrared spectrum,
respectively every 30 min. Whereas, complete global coverage cannot be obtained from one geostationary satellite, the high frequency repetitive observations over the same area have certain
distinct advantages over observations from polar orbiting satellites.

Significant advances have taken place in the past 20 years
in environmental spacecraft and remote sensing instrumentation,
and accelerated advances can be anticipated in the near future
when the space shuttle commences operation. Paralleling these
advances, the ground processing of remotely sensed data has also

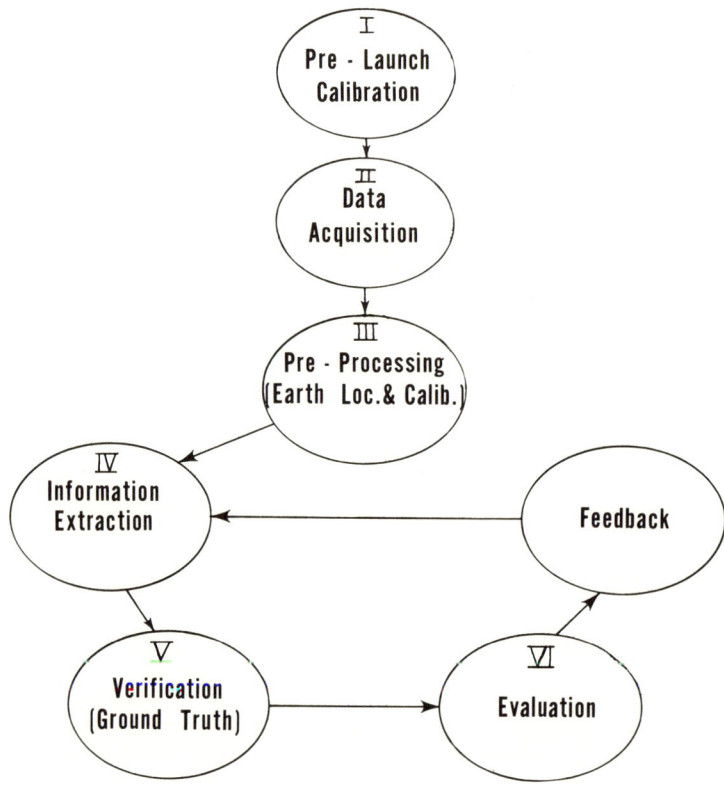

FIGURE 1: Systems aspects of quantitative application of remotely sensed data.

undergone notable advances. In the first days after the launch
of TIROS-I, the handling of the data from this spacecraft was
entirely manual. Pictures were earth-located and gridded using
perspective grids and visually identifying landmarks within the
images. From these crude and very labor-intensive beginnings,
the technology, aided by advances in computer capacity, has
rapidly progressed to a fully automated quantitative science. However, because of the vast amounts of data produced by environmental
satellites, consideration of full digital processing has previously
been relegated to only the largest of centralized computer facilities. Even with some of the largest computers currently in existence, it is not possible to process all the data produced by even
one satellite system such as the DMSP on a global scale and in quasi-
real time. The advent of interactive graphics display devices incorporating minicomputers, microprocessors and special function processors now offers a mechanism for the fuller exploration of the current and expanding remote sensing capabilities on both the research
and operational levels. The purpose of this paper is to describe one

example of this type of system, the SATDAT Processing and Display System (SPADS) under development at the Naval Environmental Prediction Research Facility (NEPRF), and to show some illustrations of the SPADS capabilities.

SYSTEMS ASPECTS OF REMOTE SENSING

Remote sensing observations are unconventional in that they are not direct measurements of traditional geophysical parameters. For example, conventional meteorological measurements are of such parameters as winds, pressure, temperature, and moisture, whereas scanning radiometers detect reflected and/or emitted energy from the land, sea, and cloudy and clear atmosphere. The most familiar medium for displaying remotely sensed data is grey-shade or color coded, hard-copy images; therefore, the extraction of information requires comparisons with "verifying" data and/or interpretations by trained personnel. Even though the data are displayed in an image format, the original data are digital and can be treated as a precise scientific measurement.

The functional areas that must be controlled in order to insure uniqueness in the process of extracting quantitative information from remotely sensed data are illustrated in Figure 1. Functions I through IV are usually performed by the agency operating the satellite and performing the ground data processing. The researcher or operational user usually has access only to a processed image and has little or no control over the antecedent manipulation of the data. Not only may the user have difficulty in acquiring the data in the format desired, but he has little knowledge of how the data have been handled in producing the image. There may be infinitely more information relevant to a particular phenomenon contained in the raw data than is portrayed in an image, information that is unrecoverable from the image because of the data processing procedures. Therefore, control by the user of functions III and IV of Figure 1 is essential to the optimal extraction of information from the data.

INTERACTIVE GRAPHICS DEVICES

Hardware Components

Over the past ten years, there has evolved a class of devices based on minicomputers, microprocessors and Cathode Ray Tube (CRT) displays which greatly facilitate the quantitative use of remotely sensed data. The Man Machine Interactive Display System (McIDAS) (Smith, 1975) developed by the University of Wisconsin was one of the first applications of this technology to remotely sensed environmental data. Other systems are under development at Stanford Research Institute (Serebreny et al., 1970), Goddard Space Flight Center (Billingsley and Hasler, 1975), and Naval Oceanographic Research and Development Agency (Pressman and Holyer, 1978).

The basic components of these devices are shown in Figure 2.

174

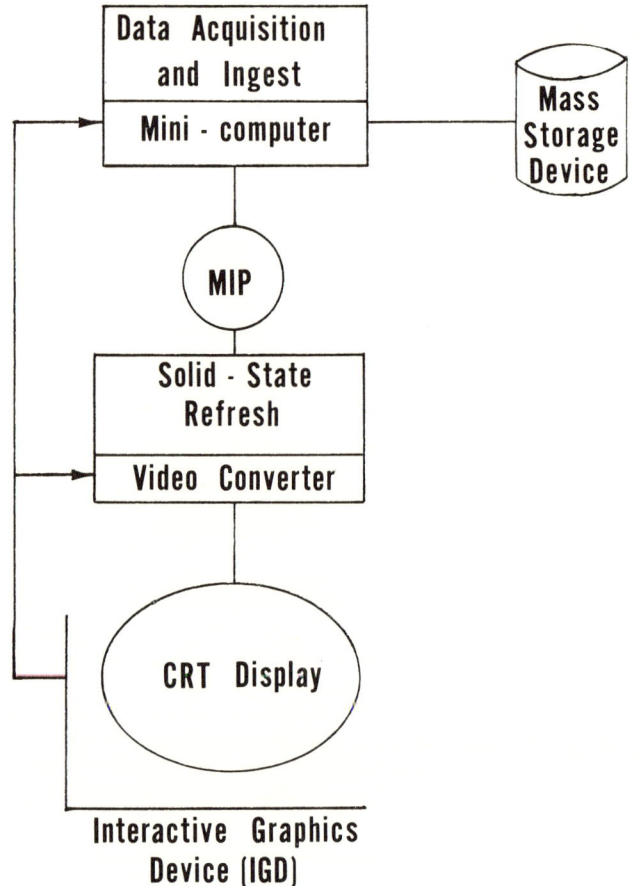

FIGURE 2: Basic components of interactive graphics
devices.

Data acquisition and ingestion can consist of a full satellite acquisition system and/or tape units to access digital data acquired
by other agencies. The minicomputers are usually 16 bit/word machines, although a 24 bit/word machine is used in the McIDAS. The
manufacturers have rapidly increased the speed and capacity of
these machines to the point that they can now challenge the computational capability of third generation, large main-frame computers. The massive amounts of data produced by satellites require
large-capacity storage devices; these can range from "floppy disks"
to 300 megabyte systems. The function of these devices is to store
the digital satellite imagery, systems software, and applications
programs.

    The key component of state-of-the-art interactive graphic de-
vices is the solid state refresh (SSR) memory, which consists of
computer memory boards that store the digital imagery. The capa-

city of the SSR can range from one 3-bit image (8 grey shades or colors) to multiple 8-bit images (256 grey shades). The micro-information processor (MIP) is an essential interface component between the minicomputer and the SSR. This is a microprocessor with a very high speed execution cycle ($\approx$150 nanoseconds) and access to a small amount of high speed memory. The primary function of the MIP is to perform the address computations required to load the digital imagery into the SSR as it is passed from the minicomputer. The MIP also can perform certain elementary operations on the data (e.g., enhancements).

The video converter accesses the data stored in the SSR and converts the digital data raster-by-raster into an analog video signal that drives the CRT display. The more advanced video converters contain a microprocessor that is capable of performing manipulation of the data at TV refresh rates. This is a significant advantage because such operations as data enhancements can be performed without reaccessing the data through the minicomputer from the mass storage device.

The final component is the interactive graphics device, whose function is to allow the human operator to interact with the computer system. Several devices, including track-balls, light-pens, or graph tablets, can be used to perform this function.

The NEPRF Hardware System

A diagram of the system under development at NEPRF is shown in Figure 3. Currently, there is a capability to acquire data directly from two satellite systems, the GOES and TIROS-N satellites. A manually operated antenna/receiving system is available to acquire Automated Picture Transmission (APT) data from the TIROS-N satellites; these data are transmitted directly from the satellite in an analog mode on a VHF channel. The APT data are digitized directly by a NEPRF-developed A/D converter and the data are recorded on a 7-track tape unit for later ingestion into the SPADS.

The GOES Acquisition and Data Handling System (GADHS) is a full readout station for acquiring "stretched-mode" digital data from either of the GOES geosynchronous satellites. The GADHS consists of an hour angle over declination 8.5 m parabolic dish, receiver, demodulator, bit and frame synchronizers, a sectorizer, and a minicomputer controller with CRT keyboard. A high capacity digital recorder is currently being added to the system. The sectorizer extracts a 640-line by 956-element array of data from either the visible or IR channel of the full-disc GOES image. A sector is selectable either under thumb-switch or automated control from the GADHS controller. Resolution options include 1.0, 2.0, 4.0, 8.0 or 16.0 km for the visible data; or 4.0, 8.0, or 16.0 km for the infrared data. The areal coverage varies inversely with the resolution.

The SPADS display system, diagrammed in the lower half of Figure 3, is a dual minicomputer system with one side driving a black and white CRT and the other driving a color CRT. The computers are linked in two manners: a hard-wired computer-to-computer link that permits intercommunication between the two processors and a second

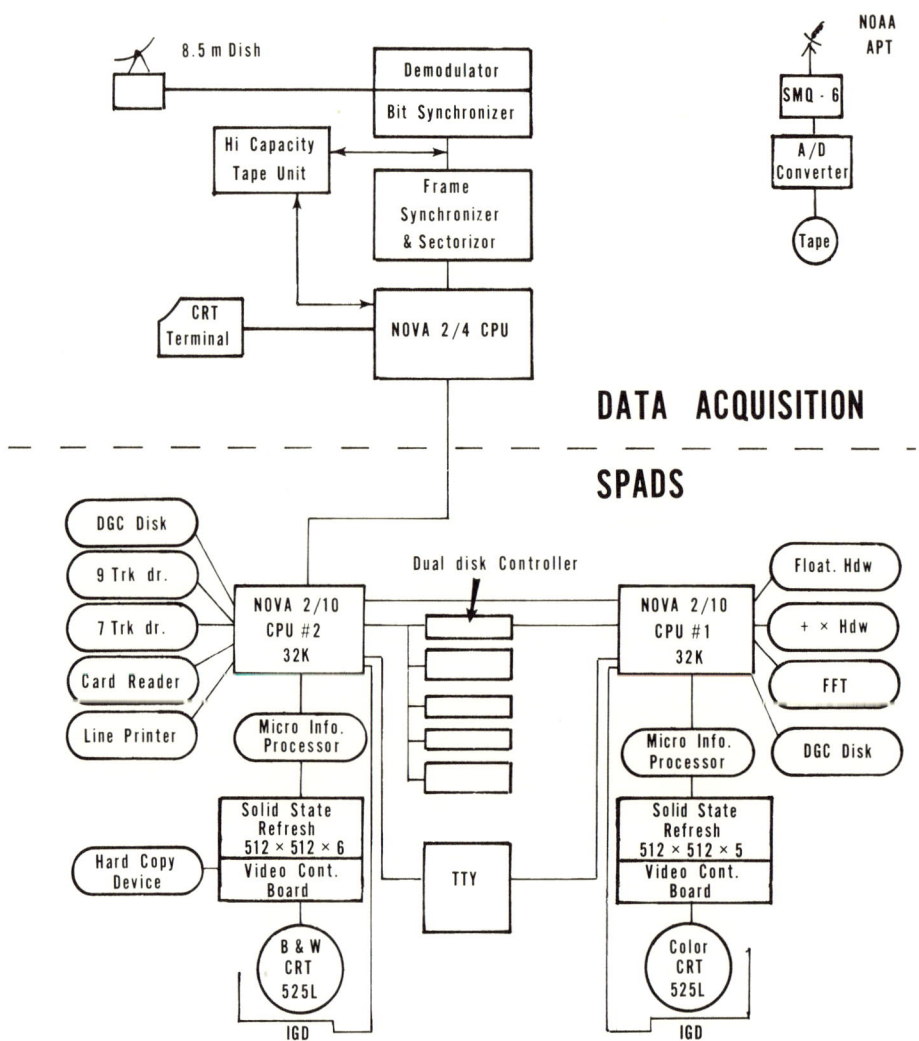

FIGURE 3: Functional diagram of the NEPRF Satellite
Data Acquisition and Display System (SPADS).

method through the disk system. The dual disk controller has one
port attached to each computer; this allows both computers to access
(although not simultaneously) data stored on the disk system. The
SSR memories are configured in a slightly different manner on the
two sides of the system. The black and white SSR consists of six
512 x 512 x 1 bit planes; four of these planes are reserved for
imagery, while the remaining two are used for overlays. The SSR
on the color side contains five 512 x 512 x 1 bit planes, three for
imagery and two for overlays.
    The system has a full complement of peripheral devices including
7- and 9-channel tape units, a card reader, an electrostatic plotter,

a. GOES antenna

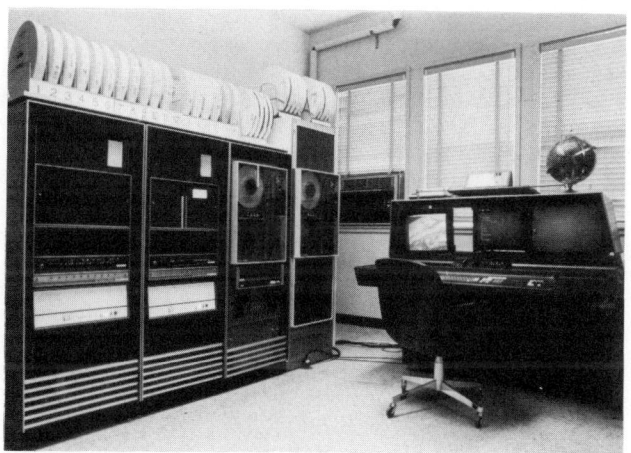

b. SPADS

FIGURE 4: GOES antenna and SPADS.

and a relatively low quality hard-copy device. Of particular importance is the inclusion of three hard-wired special processing units;
one is for floating point operations, the second is for multiple/di-
vide operations, and the third is an array processor. This latter
device, among other functions, performs fast Fourier Transforms in
hardware. These devices permit orders-of-magnitude speed-up in the
computations.

The GOES receiving antenna is shown in Figure 4a and the SPADS
display system is shown in Figure 4b. The antenna, shown in its
stow position, is capable of being trained on either the eastern or
western GOES satellites; antenna repositioning requires about three
minutes.

Computer Systems Software Requirements

Three categories of computer software are required to make these devices function: computer systems, satellite preprocessing, and applications.

Computer Systems Software

Computer systems software is defined as software that makes the hardware function; it is usually transparent to users. With minicomputers, the availability of manufacturer-provided, systems-level software is essential. Most major manufacturers have extensive libraries of systems software for their minicomputers, including such modules as disk or tape operating systems, FORTRAN compilers, editors, debuggers, etc. These types of software may not be available for "nonstandard brand" minicomputers and the development of such software is an extremely expensive undertaking. The microprocessors similarly require machine language coding, and it is only recently that integrated microcomputer and minicomputer systems with exchangeable software have become available.

Two operating systems resident on the SPADS are a Data General Corp. Realtime Disc Operating System (RDOS), and a NEPRF-written system called SEAOPS. The RDOS must operate from a standard Data General Corp. disk; a small auxiliary disk is available on the SPADS to support this function. RDOS permits the generation, editing, and execution of FORTRAN code on CPU 1 of the SPADS.

The SEAOPS, which is resident on both CPU's of the SPADS, is a stripped-down operating system that is more efficient than RDOS. The RDOS is utilized for software development, while SEAOPS is used for execution of fully checked-out programs.

The microprocessor programs available on the SPADS are internally generated and written in machine language. This software is very difficult to check out because no output devices are attached to the MIP.

Satellite Preprocessing Software Requirements

Figure 5 shows the structure of the SPADS satellite pre-processing and applications software. The basic software concept is designed to allow the utilization of the applications modules on the raster formatted data regardless of the satellite of origin.

Data ingest modules for each satellite and each instrument (or derived product) are resident on removable disk packs. These modules are written to accommodate the particular formats of the data from the different satellite systems. As appropriate, the data are calibrated using onboard calibration information and/or pertinent conversion tables. Calibration procedures can absorb considerable amounts of computer time; as a consequence, it is more efficient for high-data-rate sensors such as the NOAA Very High Resolution Radiometer (VHRR) to produce an intermediate calibrated data tape. This process eliminates repeated calibration of the same data at great savings of computer time.

After calibration, the data are staged onto the disk system in a universal format. This is a raster formatted file of 512 lines

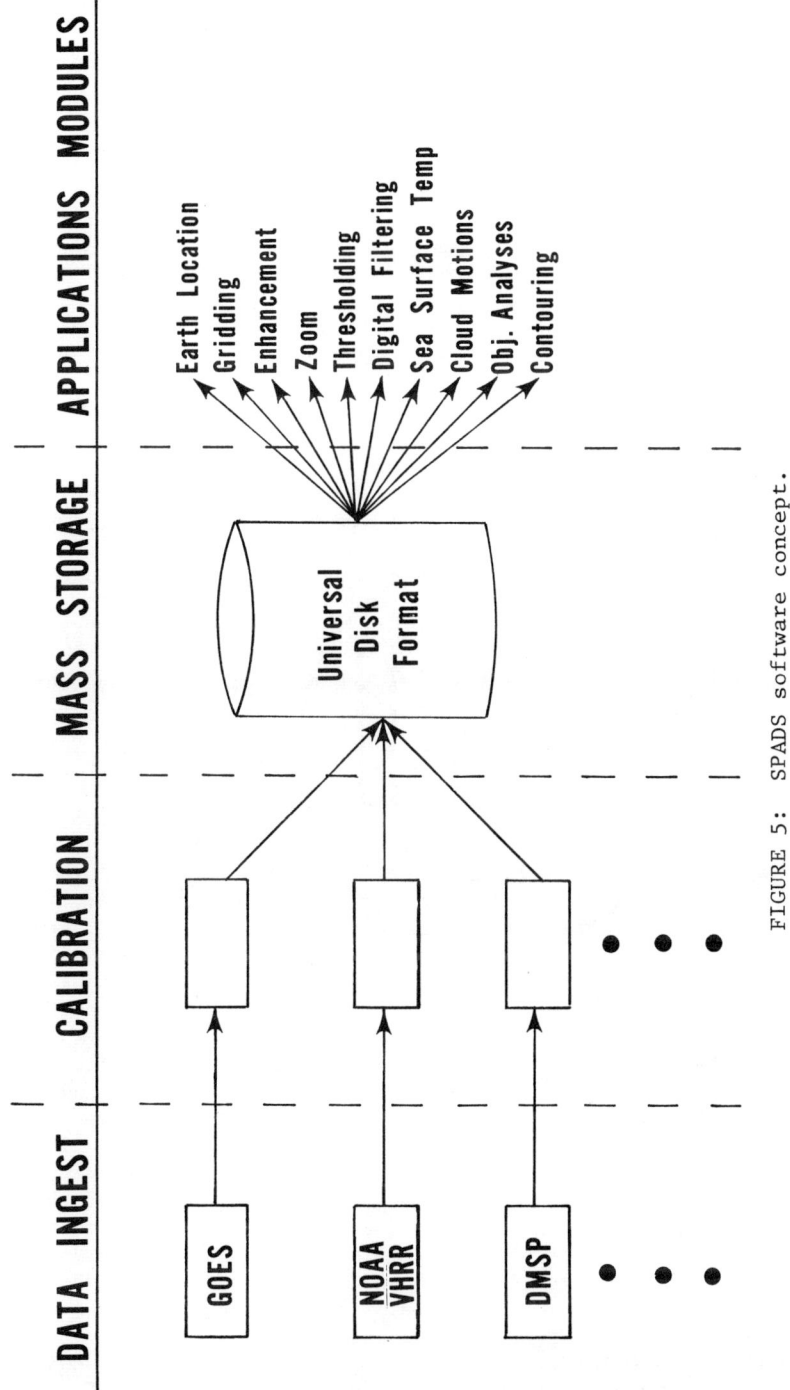

FIGURE 5: SPADS software concept.

of data consisting of 512 6- or 8-bit elements. All of the applications modules assume the data are structured in this format.

Applications Software Requirements

Earth Location   A critical prerequisite to the quantitative
use of remotely sensed data is accurate location of data in the
earth coordinate system. In order to accomplish this function, a
precision, general-purpose ephemeris program is resident on the
SPADS. Given the Keplerian elements for a particular satellite
for a particular epoch in time, the latitude and longitude of the
subpoint locations of the satellite can be predicted. The program
computes the subpoint locations at one-min intervals for eight consecutive orbits. These points are transformed into a polar stereographic coordinate system and can be displayed on the SPADS with
super-imposed geography and latitude/longitude lines. Figure 6a
illustrates this product for the NOAA-4 satellite.

Under computer control, a particular orbit and area can be
selected for further processing. Figure 6b shows a ¼ hemisphere
sector surrounding 24°N, 124°W of orbit number 102; Figure 6c shows
a 1/16 hemisphere sector for the same location. The latitude/longi-
tude of selected scan spots are derived and a new display is generated for the selected area. An example of this product is shown in
Figure 6d.

The functions of these programs are to permit the association
of raw scanline data to a particular earth area, to allow definition of the areal coverage of the data, and to provide the basic
information from which the data may be earth-located and/or gridded.

Gridding   Grid information is provided with each GOES frame
and it is a simple matter to superimpose a grid onto a GOES image.
An example of a gridded GOES visible image is shown in Figure 7a.
The grid does not always correspond precisely to landmarks; however, as the grid is resident in an overlay plane in the SSR and
is not melded into the image, it can be adjusted under manual control to give a precise fit to the landmarks. This is accomplished
by using the interactive graph tablet to meld the grid into the
data.

Gridding of polar orbiting image data is a more complicated
and time-consuming process. The current procedure for gridding
polar orbiting satellite data is to superimpose a "relative" grid
onto the data. The "relative" grid is precomputed for a 5° lati-
tude/longitude mesh using a nominal satellite altitude and inclination. With knowledge of the time-after-ascending node of the
beginning scan line of an image, and one earth-located element
along the subpoint track, the section of the "relative" grid
corresponding to the satellite data can be extracted and superimposed onto the image. An example of this product is shown in
Figure 7b. It must be emphasized that this grid represents 5°
latitude/longitude increments from the one known position; that
is, the lines are not integer multiples of exact 5° latitude/
longitude lines.

Work is currently in progress to generate programs that will

provide superimposition of coastline, geography, and latitude/
longitude overlays on both polar-orbiting and geostationary satellite imagery. This software will also permit the association of a
particular pixel element with a given latitude/longitude or vice-
versa.

   Enhancements   Digital remote sensing imagery can be enhanced to bring out features that are of particular interest to
the user. The procedure is to select the range of the measurement pertinent to the phenomena it is desired to display; then
the selected range of values are displayed across the full dynamic range of the CRT. A GOES IR sector is shown in Figure 8a in
unenhanced mode; the 256-count range of the original data has
been portrayed in 64 grey shades in this figure. An enhanced
version of the same image is shown in Figure 8b. In this latter
case, the range of counts between 50 and 95 (about 10° to about 25°C)
has been displayed over the full 64 grey shades of the hard-copy display; features less than a count of 50 are displayed in solid white,
while features greater than 95 are displayed as solid black. Note
that there is information contained in the original data which cannot
be portrayed in the unenhanced mode. Enhancements can be performed
on the SPADS in less than ten seconds. The enhancement illustrated
is linear; other enhancements, e.g., logarithmic, require only a simple change in a look-up table.

   Thresholding   In many remote sensing applications, it is desirable to manipulate two channels of data simultaneously. For example, in order to extract sea surface temperature information from IR
data, cloud-contaminated IR samples must be eliminated from data.
This can be accomplished by using the brightness in the visible channel as a screening criteria for the presence of clouds. This function is readily accomplished on the SPADS by the process illustrated
in Figure 9.
   Figure 9a shows an enhanced 2 n mi IR sector from the western
GOES satellite covering the eastern Pacific Ocean. The lighter grey
shades are indicative of both clouds and cold sea surface temperatures.
Clouds with cloud-top temperatures significantly colder than the range
of sea surface temperatures can be readily identified. However, the
tops of low clouds (e.g., fog or stratus) can have temperatures within
the acceptable range of sea surface temperatures and it is not possible to uniquely identify such clouds using only the IR information.
   The visible spectral range information corresponding to Figure 9a
is shown in Figure 9b. It readily can be seen that the entire California coast is free of clouds and that the lighter grey shades in this
area are associated with cold surface temperatures. The visible and IR
channels are used in combination in the following manner: the numerical
value of each visible pixel is tested to ascertain if it exceeds a specified threshold; if the test is true, a bit is set in an overlay plane
corresponding to the location of the pixel in the image. This process
is then performed on the entire 512 x 512 array or any subregion thereof, and the cloud areas are masked out of the IR image using the electronic logical "OR" capability of the video control board. The decision
tree for this logic is shown in Figure 9c. The effect of this process

is to mask out areas whose brightness exceeds the specified threshold. This is illustrated in Figure 9d, where all pixels with a brightness count greater than 15 have been masked out of the IR data.

  Point Temperature Extraction   The SPADS capabilities greatly facilitate the manipulation of imagery. However, in order to utilize remote sensing data fully in the scientific process, it is essential to extract the information in a quantitative manner. The interactive graphics device (IGD) on the SPADS provides an eminently efficient method for extracting numerical values from the digital imagery.

  In this process, a cursor is displayed on the CRT and moved to a desired location with the IGD. When the stylus of the IGD is depressed, the I, J location of the stylus is transmitted to the minicomputer. The sample corresponding to this location is retrieved from the disk file and converted to an absolute value, and this value is displayed at the corresponding location on the CRT. The cursor can be moved rapidly to other locations to repeat the process. An example of this capability as applied to the NOAA-4 VHRR IR data is shown in Figure 10a; the enhanced IR image is shown in Figure 10b for comparison. It must be emphasized that these temperatures are calibrated IR temperatures that have not been corrected for atmospheric moisture attenuation.

  Horizontal Profiles Information   In certain applications, the horizontal gradients of a measurements are more important than point absolute values. The capability to derive horizontal profiles between two points on an image has been implemented on the SPADS. The IGD is used to specify the beginning and end points of a line across which a profile is desired. The computer system retrieves each element along the specified line, scales the data, and displays the result in a graph format. An example of this capability as applied to GOES IR data is shown in Figure 11. This chart can also be put out as a line chart on the electrostatic plotter. Note than even though the quantization of these data are about $0.5^{\circ}C$, the spot-to-spot precision (gradients) are on the order of $0.25^{\circ}C$ or less.

  Zoom Capability   High resolution remote sensing instruments can produce images whose line densities greatly exceed the capability of the SPADS display. Therefore, it is not possible to display a full image on the SPADS at one time. For certain applications, it is desirable to display a large area of an image at reduced resolution and then to "zoom in" on a particular area and display the data for that area at higher resolution. Figure 12 shows 8 km visible and enhanced IR sectors from the eastern GOES satellite. As can be seen in Figure 12a, the western Atlantic Ocean north of $37^{\circ}N$ is relatively cloud-free and the enhanced IR shows some interesting structure in the sea surface temperatures in this area. The data enclosed within the box shown in Figure 12b are extracted and displayed at higher resolution.

  Figure 13 shows the data for this sector at 4 km resolution[*].

---

[*]These zooms are actually produced by duplicating the samples along the line and then repeating each line. The installation of the high capacity tape recorder on the SPADS will give the capability to actually extract the data at higher resolution.

A particularly interesting feature in Figure 13b is the appearance
of a Gulf Stream ring in the left center of the image. In order to
illustrate the flexibility of SPADS software, the cloud thresholding
option has been applied to these zoomed data and IR temperatures have
been extracted from the cloud-free areas. This product is shown in
Figure 13c. Similarly, the Profiling function has been applied to
these data and the temperature cross-section across the Gulf Stream
ring is shown superimposed upon the enhanced IR data in Figure 13d.

Digital Filtering   The image data presented so far have been
relatively noise-free. Noise can be introduced into the data from
a variety of sources, and the SPADS computer system can be used to
eliminate certain of these contaminants. The NOAA APT data are
particularly prone to noise because they are transmitted from the
satellite via an analog VHF channel. Figure 14a shows an enhanced
NOAA APT IR image evidencing two types of noise. The first is a
low frequency line-by-line change in the base amplitude of the data,
and the second is a high frequency pattern superimposed over the entire image. Digital filtering has been applied to a sector of this
image and the result is shown in Figure 14b. The low frequency noise
has been removed by using the FFT capability of the SPADS and applying a low frequency band-pass filter in reconstructing the image.
Experimentation has shown that the high frequency noise is random
and cannot be removed by FFT techniques. The high frequency noise
has been reduced by spatial averaging.

Superimposing of Contours   As discussed in the section on
Systems Aspects and Remote Sensing, an essential component of working with remote sensing data is the requirement for verification
using conventional observations or analyses. One method of accomplishing this function is to compare remotely sensed data directly
with contour values from analyses of conventional data. A limited
capability of superimposing contours onto imagery is currently
available on the SPADS. Figure 15a shows a mosaic of the IR data
from fourteen NOAA-4 orbits transformed into a polar stereographic
projection. This display is a 512 x 512 array (about 25 km resolution) for the Northern Hemisphere that was extracted from the original product (accessed from tape) in a 2048 x 2048 array (about
12 km resolution). The latitude/longitude and geography grid was
superimposed on the SPADS. Utilizing the IGD, a ¼ or 1/16 hemisphere sector can be randomly selected by the operator; an example
of a sector selection is shown in Figure 15b. The original data
is retrieved from a disk file and displayed on the CRT. The geography and latitude/longitude grid corresponding to this sector is
also generated and the combined image is displayed as shown in Figure 15c. All options previously demonstrated can be performed on
these data. In addition, contours of analyzed fields for the selected sector can be superimposed using the second overlay plane
of the SSR. In Figure 15d, the Fleet Numerical Weather Central
(FNWC) sea level pressure chart has been displayed over the satellite data. Any of FNWC's other charts may be similarly displayed.

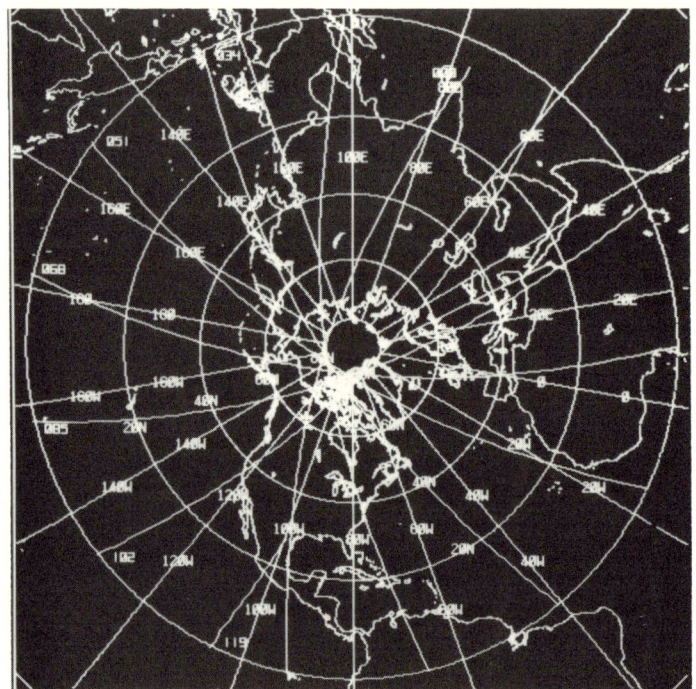

a. Sub-point tracks

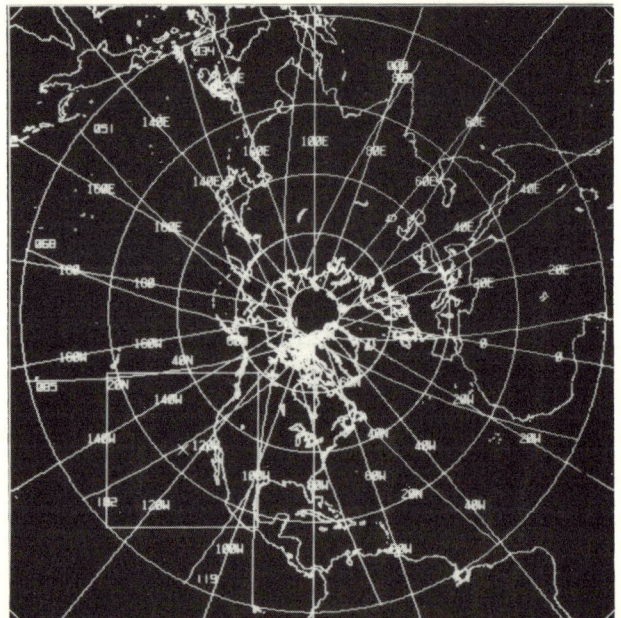

b. Sub-region definition (2x)

FIGURE 6: Subpoint track and scan line element
display.

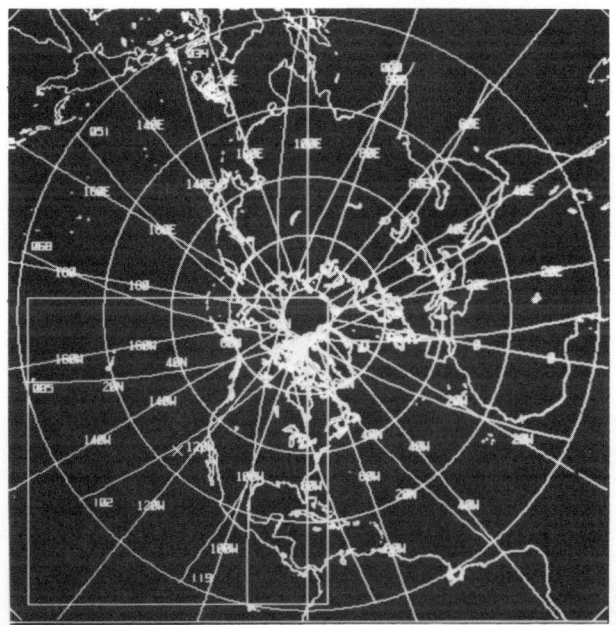

c.   Sub-region definition (4x)

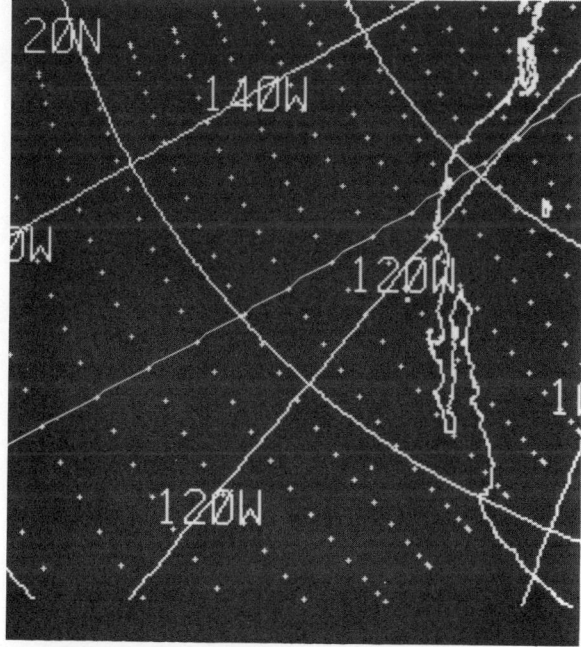

d.   Swath coverage

a. Gridded GOES image

FIGURE 7: Example of gridding capability on SPADS.

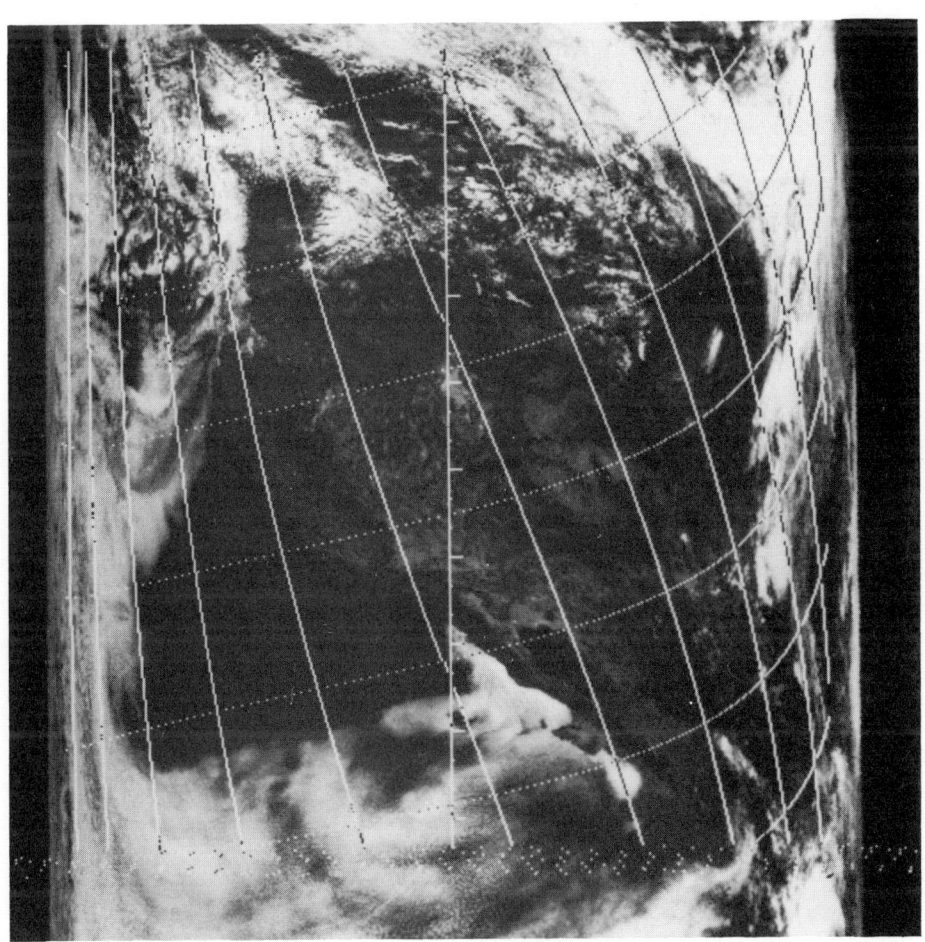

b. Gridded NOAA APT

a.   Unenhanced IR

FIGURE 8: Example of linear enhancement.

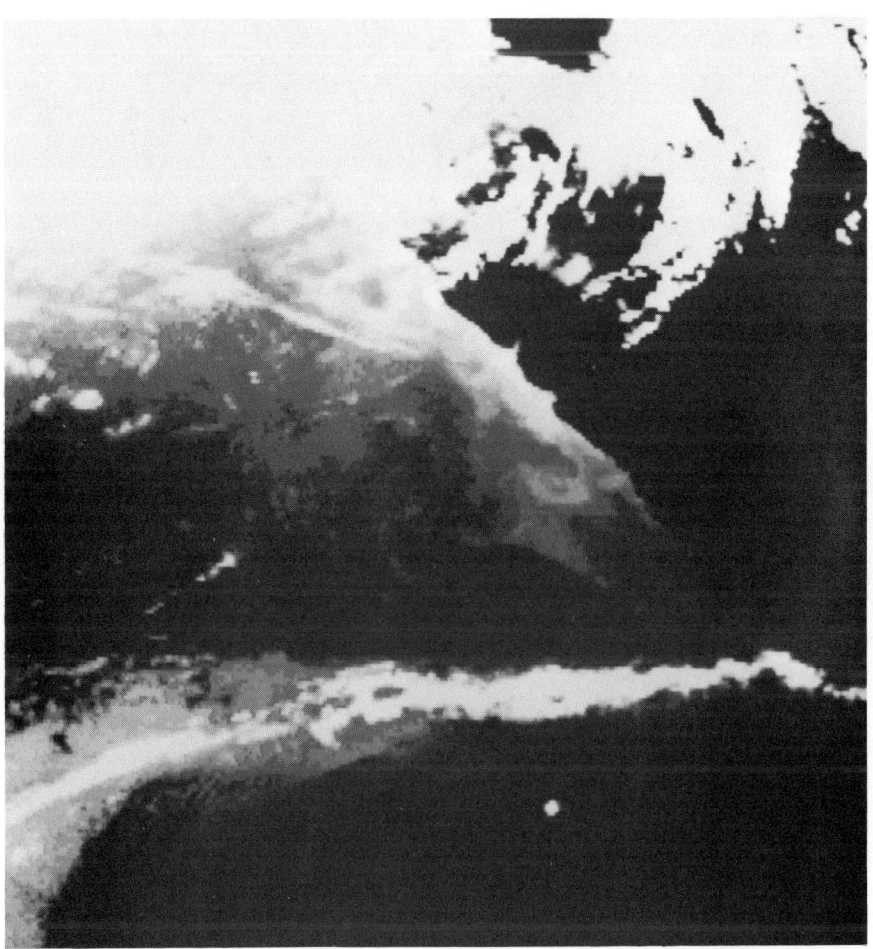

b.   Enhanced IR

a. Enhanced IR

b. Visible

FIGURE 9: Dual-channel thresholding.

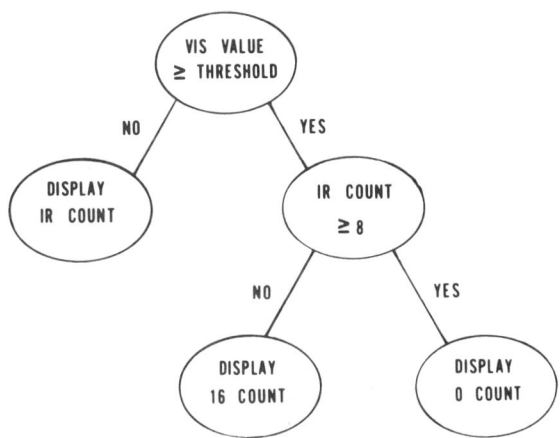

d.   Threshold IR

a. Point temperature values

FIGURE 10: Point numerical value extraction.

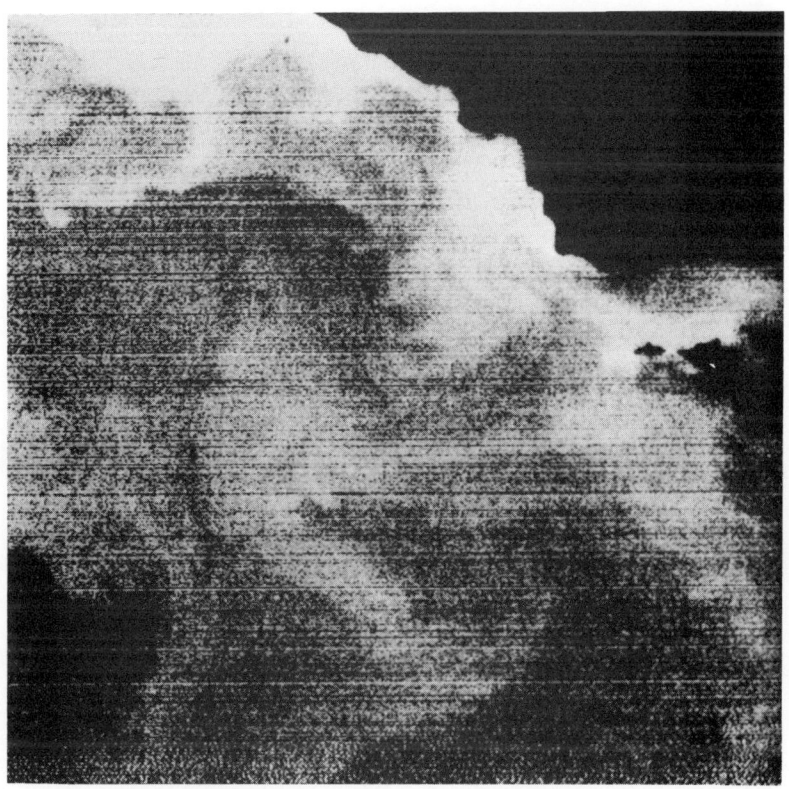

b. Enhanced IR

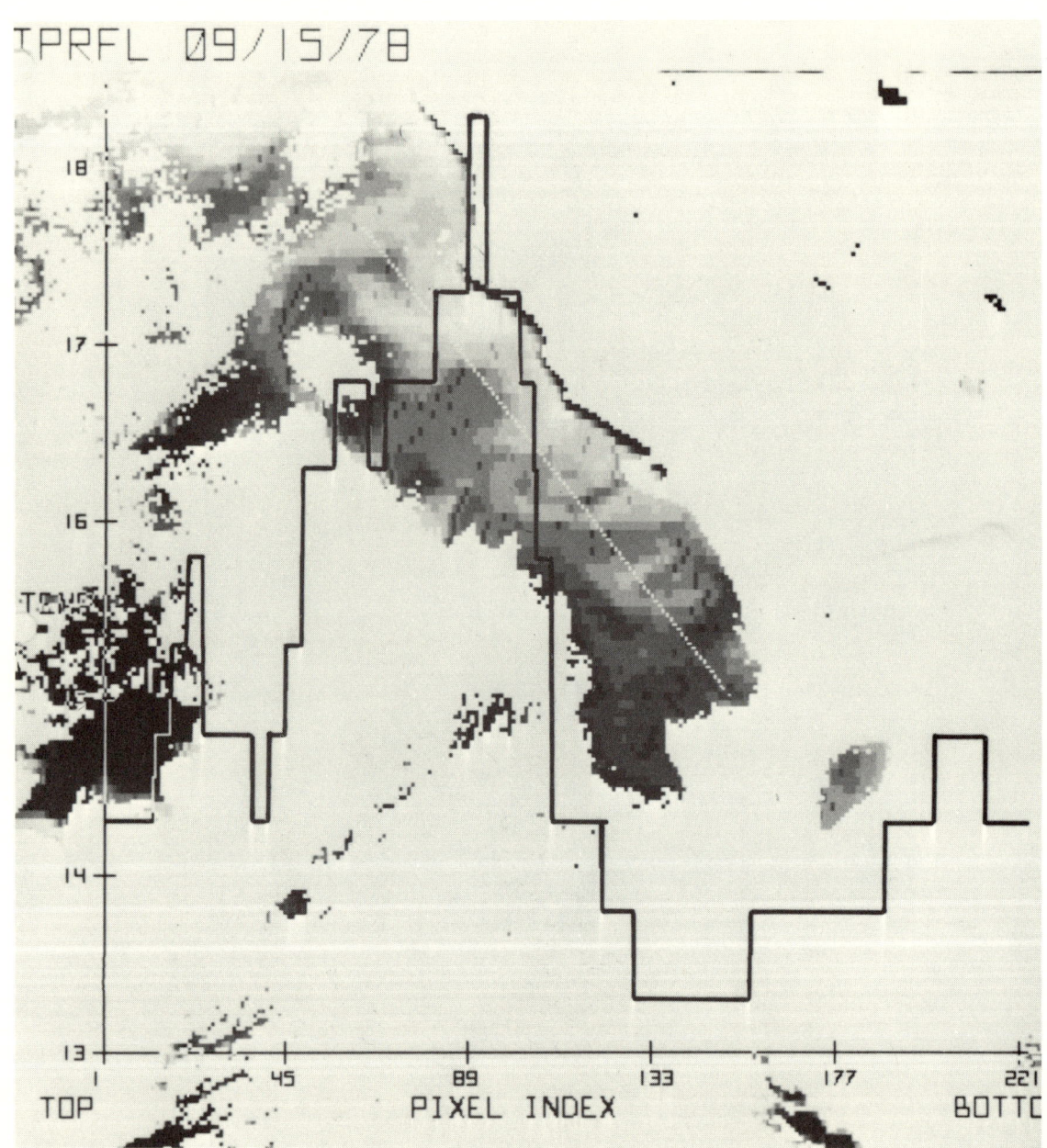

FIGURE 11:   Temperature profile determination.

FIGURE 12:   Sector definition for ZOOM.   (opposite page)

a. Visible

b. Enhanced IR

196

a. Zoomed VIS

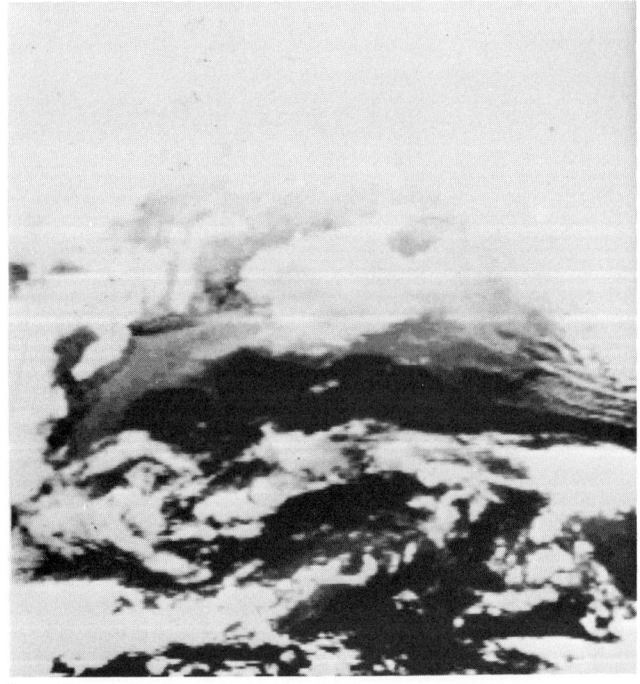

b. Zoomed IR

FIGURE 13: Multiple option capabilities.

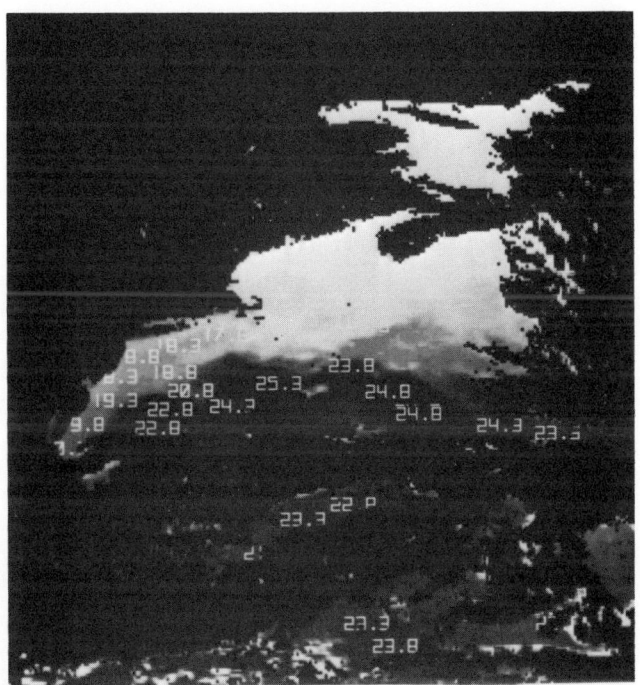

c. Point temperatures

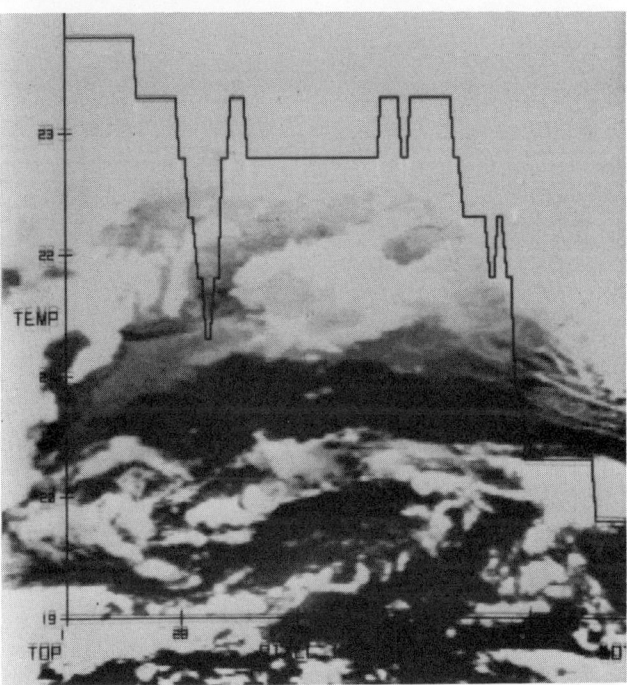

d. Temperature profiles

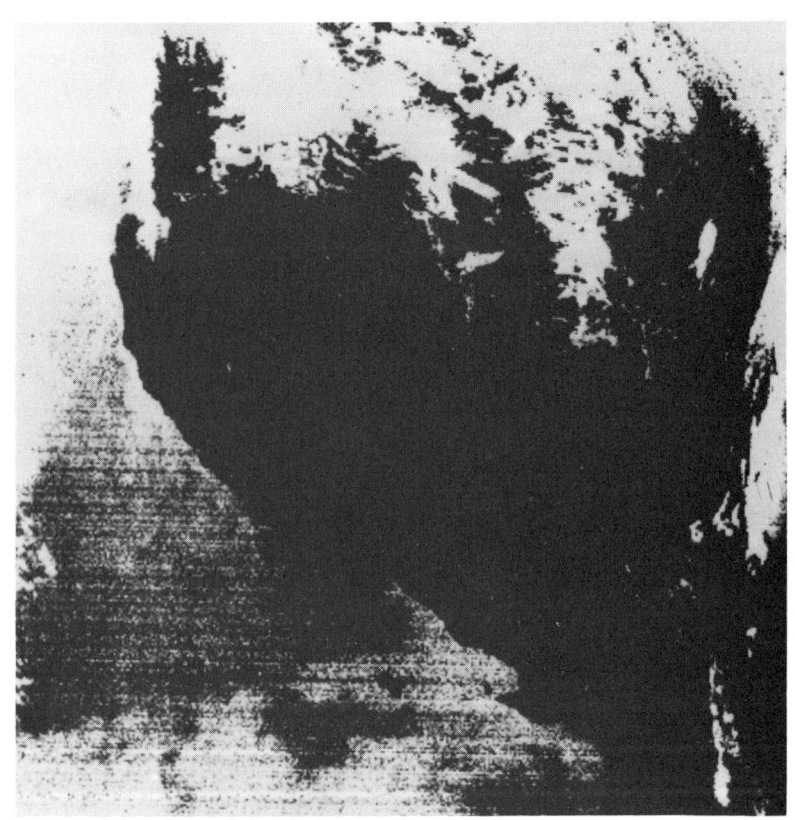

a.   Enhanced NOAA IR

FIGURE 14:   Digital filter with an array processor.

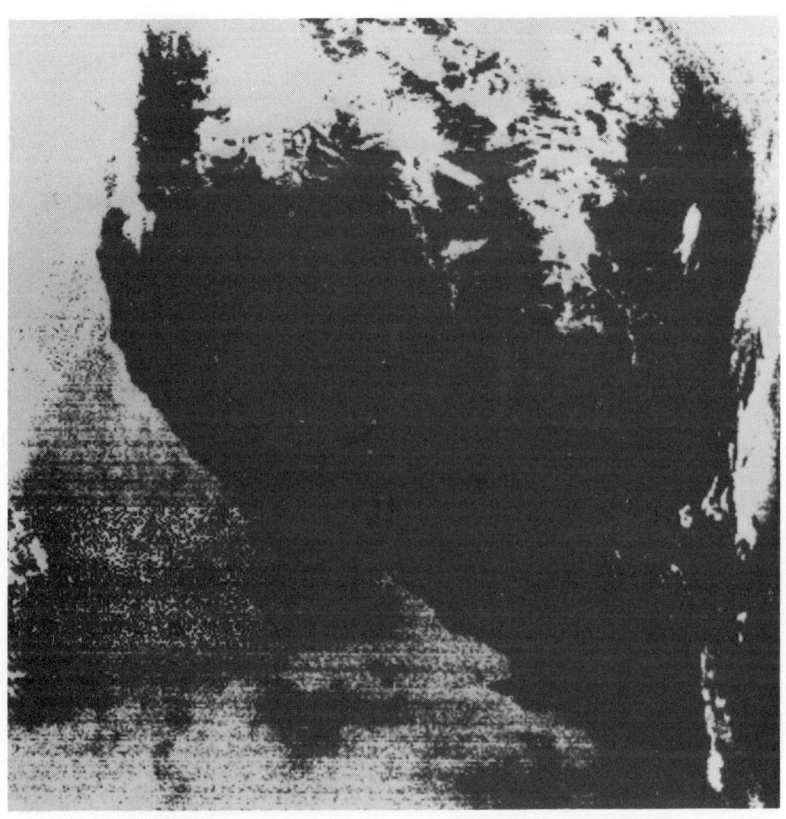

b. Digitally filtered

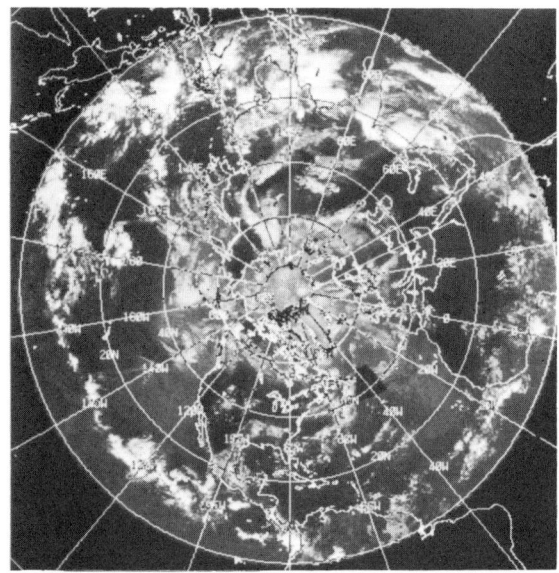

a. Gridded data

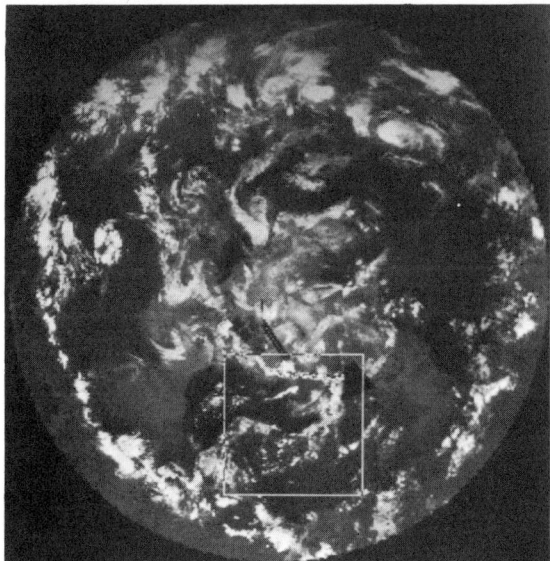

b. Sub-region definition

FIGURE 15: Superimposing contour analyses on satellite imagery.

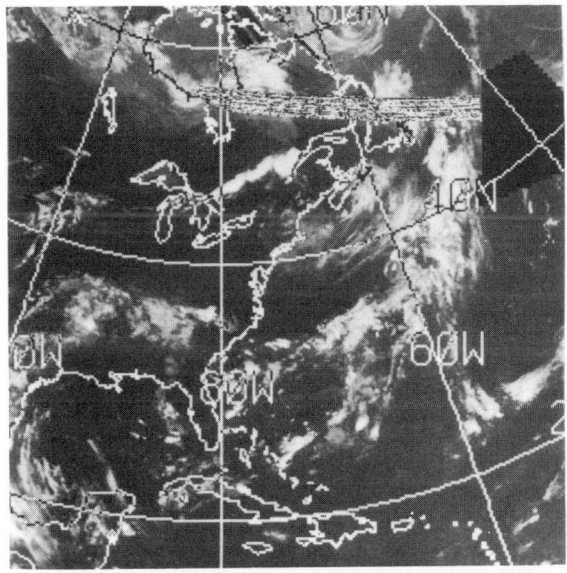

c. Sub-region expansion

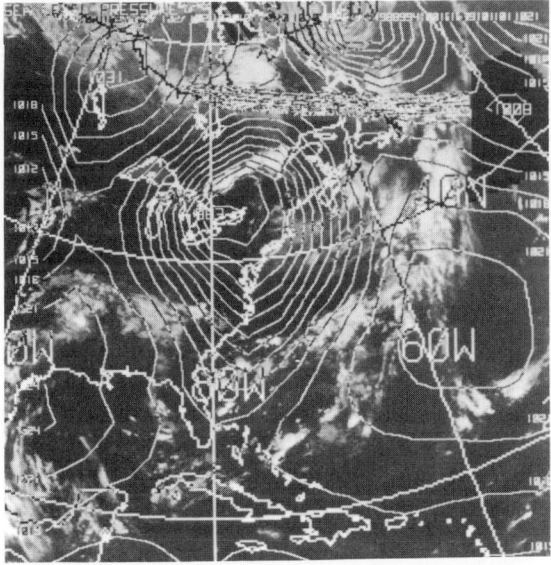

d. Contour overlay

SUMMARY

The architectural structure of a minicomputer, interactive graphics display device has been discussed. Examples of some applications of these types of devices to the quantitative extraction of information from digital satellite imagery have been shown. It is unfortunate that constraints of time and cost do not permit illustration of these examples in the color capabilities of the SPADS, because these displays are even more definitive in color.

The application of this new technology offers a unique tool to the researcher interested in remote sensing. Given the systems capabilities described, the researcher no longer needs to depend on the processing methods of other agencies, because he can now control the processing of the data to his own standards. This is considered absolutely essential in the quantitative use of these data. Further, the capability to manipulate the data rapidly places the control of *what* is being extracted from the data under the control of those personnel best qualified to assess the significance of the information.

The current generation of minicomputers, microprocessors, and special function processors, when combined with interactive CRT displays, are fully capable of ingesting, calibrating, earth-locating, enhancing, convoluting, and extracting inherent information from remotely sensed data. Major advanced in this technology are anticipated in future years. An order-of-magnitude increase in the speed of microprocessors is anticipated within the next ten years. The size and cost of microprocessors have been drastically reduced over the past five years and this trend is expected to continue. This offers the potential for fully self-contained and portable satellite processing and display units for both research and operational applications.

REFERENCES

Anon, 1976. *Project SEASAT-A, Support Instrumentation Requirements Document*. National Aeronautics and Space Administration, Washington, D. C.

Billingsley, J. B. and A. F. Hasler, 1975. *Interactive Image Processing for Meteorological Applications at NASA/Goddard Space Flight Center*. Report X-933-75-97, Goddard Space Flight Center.

Corbell, R. P., C. J. Callahan, and W. J. Kotsch, eds., undated. *The GOES/SMS Users Guide*. National Oceanic and Atmospheric Administration, NESS, and National Aeronautics and Space Administration, Washington, D. C.

Kidwell, K. B., 1979. *NOAA Polar Orbiter Data (TIROS-N) Users Guide - Preliminary Version*. National Climatic Center, Satellite Data Services Division, Washington, D. C.

Madrid, C. R., ed., 1978. *The NIMBUS-7 Users Guide*. Goddard Space Flight Center, National Aeronautics and Space Administration, Washington, D. C.

Nichols, D. A., 1975. *Block 5D Compilation*. Defense Meteorological
  Satellite Program, AFSC, SAMSO, Los Angeles, California.
NOSS: National Oceanic Satellite System. A Joint Effort by the
  National Aeronautics and Space Administration, Department of
  Commerce and Department of Defense, Washington, D. C., March
  1979.
Pressman, A. E. and R. J. Holyer, 1978. *Interactive Digital Satel-
  lite Image Processing System for Oceanographic Applications*.
  NORDA Tech. Note 23, Naval Ocean Research and Development Acti-
  vity, NSTL Station, MS.
Serebreny, S. M., E. J. Weigmad, R. G. Hadfield, and W. E. Evans,
  1970. Electronic system for utilization of satellite cloud
  pictures. *Bull. Amer. Meteor. Soc.* 51.
Smith, E. A., 1975. The McIDAS System. *IEEE Transactions of Geoscience Electronics* GE-13(3).

Remote Sensing of Marine Fisheries Resources

V. Klemas
D. S. Bartlett
W. D. Philpot

ABSTRACT: Remote sensors on aircraft and satellites are being used effectively to map important
physical and biological surface properties synoptically over large coastal and estuarine areas.
With color and color infrared photography one can
delineate tidal wetland boundaries and plant
types. Multispectral analysis techniques have
been developed for estimating the emergent biomass and species composition. Multispectral scanners and laser fluorosensors are being tested for
monitoring primary productivity of coastal surface waters and the outflow of plankton and detritus from marshes and estuaries.

Satellites can locate some high probability areas
for fish availability, such as the nutrient rich
upwelling areas off the coasts of Peru, western
United States and Africa, primarily due to the
strong thermal and spectral gradients caused by
the colder upwelling water and its spectrally
different nutrient/chlorophyll content. There
also appears to be some correlation between coastal water properties such as water color, turbid-

ity, chlorophyll concentration and the presence
of fish.

A review of the state of the art indicates that
remote sensing can be used effectively to map the
plant species, productivity and quality of wetland habitat. However, techniques being developed for determining surface water productivity,
detritus/nutrient flow and fish availability require additional field testing to establish their
reliability.

INTRODUCTION

    Coastal wetlands and estuaries are among the most productive
ecosystems in the world. They provide food and shelter to finfish,
shellfish, birds, and mammals. Wetlands produce vitamins and growth
regulators, as well as regenerate, recycle, and store nutrients. In
addition, wetlands act as buffers to coastal erosion and, to some
extent, control water quality.
    To manage the production of living marine resources, one must
monitor and evaluate the fish-habitat relationships and be aware
of the trends in extent and quality of available wetland habitat.
Obtaining field measurements of such factors has proven to be costly
in terms of time, manpower, and funds. Also, the results are sample
site specific and may not be representative of surrounding areas.
    Remote sensing can provide considerable information for minimal
cost and time when large coastal areas are to be surveyed. However,
remote sensing, like any other technique, also has some serious limitations. The objective of this article is to give a fair appraisal
of the potentials and limitations of remote sensing as a technique
for synoptic inventories of fishery-related resources.

WETLANDS MAPPING

    Both photographic and satellite data sources have advantages
and limitations with respect to providing all data elements in an
accurate cost-effective manner. LANDSAT satellite data processing
is a least-cost method of producing coastal land cover maps and
tabular data for large areas. Planning studies, however, often
require more detailed coastal vegetation/habitat information at an
accuracy level that is difficult to provide consistently over a
range of categories through the LANDSAT data-extractive process.
Manual interpretation of aerial photography is a more expensive
and time-consuming process than digital multispectral processing,
but it can yield the more detailed categorization of wetlands/habitats
that many planning activities require.
    Operational wetland mapping programs designed to meet rigorous
cartographic standards typically employ photo-interpretation of low
altitude color and color-infrared photographs supplemented by ground
surveys. Most east coast states in the USA have mapped their wet-

lands to define a wetlands boundary and to inventory major marsh plant species at scales ranging from 1:2,400 to 1:24,000 (Anderson and Wobber, 1973; Bartlett et al., 1976; McEwen et al., 1976). The United States Fish and Wildlife Service is also conducting an inventory of the nation's wetlands (U. S. Fish and Wildlife Service, 1975). Aircraft have also been used in various ecological studies, including species composition of wetland vegetation, wetland productivity, wildlife habitat, diversity, impact of man-made structures on wetlands and mosquito breeding habitats (Kolipinski et al., 1969; Egan and Hair, 1973; Reimold et al., 1973).

More recently, the potential of Skylab imagery and LANDSAT multispectral scanner (MSS) data for mapping and inventorying tidal wetlands has been demonstrated (Erb, 1974; Anderson et al., 1975).

The LANDSAT imagery is produced by the four channel multispectral scanner having the following bands: band $4 = 0.5$-0.6 μm; band $5 = 0.6$-0.7 μm; band $6 = 0.7$-0.8 μm; and band $7 = 0.8$-1.1 μm. The satellite passes over each ground point every 18 days and, imaging from an altitude of 920 km, the MSS covers a swathwidth of 185 km. Visual analysis can be performed on MSS imagery, or the data stored on magnetic tape can be analyzed directly, in digital form, using computers. Digital and visual analyses of LANDSAT data have been used in several studies of coastal wetland habitat (Carter and Schubert, 1974; Anderson et al., 1975; Klemas et al., 1975). Results are generally good but are restricted by the limited spatial resolution of LANDSAT data (one picture element ≈ 0.49 hectares = 1.1 acres) and by overlapping spectral signatures of some vegetation types at some times of the year. Klemas et al. (1975) found that classification accuracies derived by comparison of interpreted LANDSAT data with existing maps and photographs were above 80% for most categories.

In comparison, Skylab's S190A Multispectral Photographic Facility had a resolution of 0.09 hectare and its S190B Earth Terrain Camera about 0.02 hectare (Klemas et al., 1976), providing valuable information for detailed wetland mapping. In a comparative study, Skylab photographs were used to identify five wetland plant species and drainage patterns were mapped in more detail than using LANDSAT (Anderson et al., 1975). Spatial resolution was still insufficient to satisfy most state inventory statues which require geometric accuracy approaching National Map Accuracy Standards.

Data collected by orbital platforms seems more suited to periodic updating of inventories and to monitoring of large scale habitat dynamics. A significant advantage of the LANDSAT system in this regard is the repetitive coverage which permits observations of physical and morphological characteristics of plants over their entire growing cycle. Although weather conditions and limitations in foreign data acquisition reduce the actual frequency of LANDSAT coverage to more than the nominal 18-day cycle, its repetitive nature is extremely valuable for inventory updates and when seasonal factors influence the spectral characteristics of the resource being monitored. An example of seasonal influences on spectral reflectance signatures has recently been noted during research on reflectance characteristics of wetland plant canopies in Delaware. Spectral discrimination of the two major wetland communities present,

(a merged category containing *Distichlis spicata* and *Spartina patens*)
and cord grass (*S. alterniflora*), has been found to depend on major
differences in their canopy structures (Carter and Schubert, 1974;
Bartlett et al., 1977). However, seasonal changes in canopy height
of cord grass and the percentage of live and senescent vegetation in
both communities can produce relative signature convergence or divergence during different portions of the year resulting in reduced
or enhanced potential for discrimination in the LANDSAT spectral wave
bands. Unpublished data developed recently suggests that optimal
spectral discrimination of the salt hay from the cord grass communities in Delaware will occur in November and December when a differential in rates of senescing vegetation in the two communities produces
high reflectance contrast between them in all wave bands. Lowest
contrast between these two communities occurred in May and June as a
result of differing responses of canopy reflectance to increasing
amounts of green vegetation during the spring. Traditional reliance
on data collected during the growing season for analysis of wetlands
vegetation at temperature latitudes is thus subject to reevaluation.

A further consequence of spectral response to canopy variables,
particularly in cord grass (*S. alterniflora*), gives promise to estimation of emergent biomass by remote sensing. Reimold et al. (1973)
found that general biomass classes could be established for this species using tonal variations in color-infrared aerial photography.
Spectral measurements in the LANDSAT wave bands using a field radiometer have confirmed that various components of emergent biomass
systematically affect canopy reflectance (Klemas et al., 1978). Figure 1 shows the linear regression relationship between emergent
green biomass, as measured by the harvest method, and the ratio of
canopy reflectance in MSS bands 7 (0.8-1.1 µm) and 5 (0.6-0.7 µm).
Bands 5 (red) reflectance was shown to be inversely correlated with
the percentage of green biomass in the canopy due to chlorophyll absorption of red radiation by green plant material. Band 7 (infrared)
reflectance was positively correlated with canopy height. As plant
leaves are generally highly transmissive and reflective in the near
infrared, reflectance in this spectral region has been shown to be
sensitive to the number of stacked leaf layers within a canopy
(Gausman et al., 1976) and thus, in this case, to the thickness of
the natural canopy. As the product of canopy thickness and percentage of green vegetation is highly correlated with green biomass, so
is the product of infrared reflectance and the inverse of red reflectance as shown in Figure 1. Thus, "remote sensing" would appear
to hold promise for useful estimation of emergent green biomass of
this wetland species, whether from satellites or in the field using
a hand-held radiometer. In either case the speed of sampling inherent in the technique will allow more cost-effective and extensive
sampling than is now possible by harvesting measured quadrants.
Greater uncertainty in estimating the biomass of individual samples
will be offset somewhat by more effective characterization of large
areas than can be achieved using small numbers of harvested samples.
In addition, as red reflectance is correlated with the percentage of
green/total biomass (r^2= 0.72), estimates of total emergent biomass
may also be possible using multispectral data.

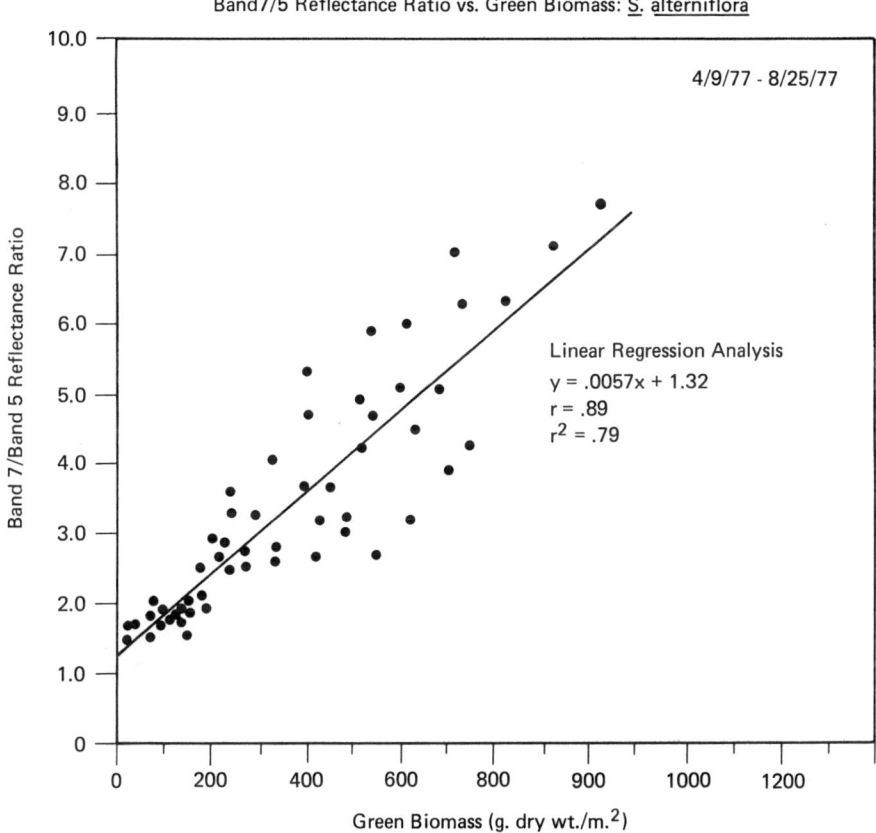

FIGURE 1: Plot of measured LANDSAT MSS band 7/ band 5 reflectance ratio as function of green biomass in grams of dry weight per square meter.

SURFACE WATER PRODUCTIVITY

Aircraft and satellite imagery have been used with notable success in observing surface water features which are relevent to the dynamics of phytoplankton distribution and the productivity of surface waters. The remotely sensed data is limited to surface waters because of the limited penetration of light into the water. Even with this limitation, there is a large amount of useful information available.

Single-Band Imagery

One of the simplest and most immediately useful forms of remote sensing data is single-band aircraft or satellite imagery. This category includes anything from black and white aerial photography to thermal, digital satellite imagery. In coastal and estuarine waters

where the sediment load in the surface waters is relatively high,
the sediment can be used as a tracer for surface currents. The patterns of sediment distribution are indicative of fronts, shoals, currents, eddies, and upwelling regions. Klemas et al. (1976) used
LANDSAT imagery of Delaware Bay to observe variations in the current
patterns over a tidal cycle. Klemas and Polis (1977) have also used
the LANDSAT imagery as well as aerial photography in a study of the
location and dynamics of fronts in Delaware Bay. In the above work,
remote sensing was an essential part of the study because the geographic regions were too large and/or the time scales too small to
allow adequate coverage by any other method.

Color imagery adds another dimension to the observation of surface waters. The apparent water color is related to the types of
materials suspended or dissolved in the water column. Unfortunately,
apparent water color is also affected by the sun angle, atmospheric
conditions, and the sea state. Nonetheless, properly treated ocean
color analysis is potentially one of the most effective means of remote observation of water properties.

Classification of Coastal Water Characteristics Using Spectral
Reflectance

The use of spectral reflectance characteristics to identify substances in the water has attracted considerable attention in the past
several years (McCluney, 1974; Gordon et al., 1975; Mueller, 1976;
Hovis, 1977; Gordon, 1978; Plass et al., 1978). Our approach is most
similar to that suggested by Mueller (1976)--eigenvector (principal
component) analysis. Eigenvector analysis has been described by a
number of investigators including Mueller (1976) and Simmonds (1963).
Perhaps the best and most complete presentation can be found in
Morrison (1976) and the reader is referred to this text for a detailed
discussion of the technique.

Much of what is unique about our approach is an indirect result
of the use of LANDSAT/MSS digital data. LANDSAT was not really intended for use in water observations. The gain is low making the dynamic range of the sensor very limited. The four spectral channels
were selected for land use applications and are hardly ideal for water
observations. Yet, there is a surprising amount of information in the
LANDSAT imagery. LANDSAT data has been used to map sediment distribution patterns (Klemas et al., 1976), to observe the occurrence of
estuarine fronts (Klemas and Polis, 1977) and to observe the occurrence of internal waves (Apel, 1974), to cite only a few of the many
papers in which LANDSAT imagery has been used in sensing of water.
The history of utilization, coupled with the high-probability of the
continuation of the LANDSAT program for many years to come, is sufficient incentive for trying to extract as much as possible from the
data.

One major reason for using eigenvector analysis is that it allows
the reduction of significant variates with minimal loss of information.
With LANDSAT/MSS data there are only four spectral bands and therefore
only four variables to begin with and the analysis will rarely reduce
this number by more than one. However, the eigenvectors can also provide an efficient representation of variations in water color which can

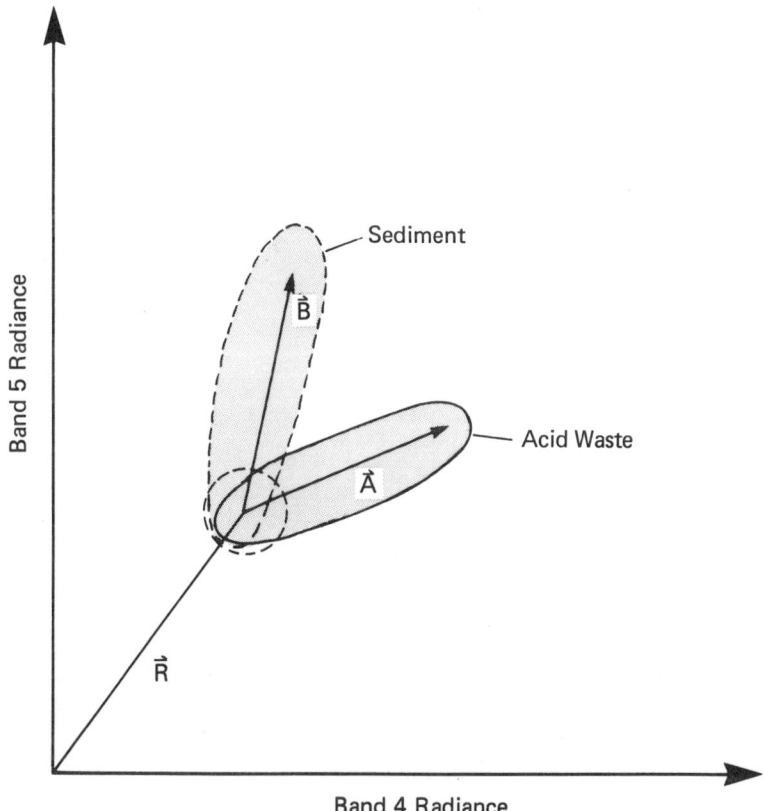

FIGURE 2: Diagram of the geometry of the eigenvector analysis. The origin has been placed at
the position of the "clear" water standard.

be readily adapted to an automatic classification process. It is in
this facile adaptation to automated classification that the eigenvector
method carries the most promise for application to LANDSAT data.

Eigenvector Analysis of LANDSAT Data

     To illustrate the technique, one can imagine a body of water,
part of which is clear, part of which is heavily sediment-laden, and
part of which contains pollution of some sort. A LANDSAT image of
the area would show the clear water as relatively dark while both
the sediment and pollutant would appear relatively bright. The sediment might show up brighter than the pollutant in Band 5 (0.6 μm-0.7 μm)
and the reverse might be true in Band 4 (0.5 μm-0.6 μm). If one were to
plot the radiances observed in both bands for each picture element
(pixel) the result would appear as in Figure 2. Here 0 represents the
clear water pixels and the two lobes represent sediment pixels ($\bar{B}$) and
pollutant pixels ($\bar{A}$). Clear water has a particular spectral signature

TABLE 1: Angular separation (in degrees) between primary eigenvectors.

			Iron-Acid Waste							Sediment				Clouds				Ice
			24 Feb 76	19 Jan 76	21 Oct 75	19 Aug 75	17 Nov 75	15 Mar 74	Average	24 Feb 76 (North)	24 Feb 76 (South)	19 Jan 76	Average	19 Jan 76	19 Aug 76	15 Mar 74	Average	19 Jan 76
			1	2	3	4	5	6	7	8	9	10	11	12	13	14	15	16
Acid	24 Feb 76	1	--															
	19 Jan 76	2	14.4	--														
	21 Oct 75	3	03.9	13.7	--													
	19 Aug 75	4	12.7	06.2	10.5	--												
	17 Nov 75	5	04.4	13.6	07.8	13.8	--											
	15 Mar 74	6	06.3	10.7	08.6	11.6	03.3	--										
	Average	7	08.5	11.1	08.6	10.3	08.8	07.9	--									
Sediment	24 Feb 76	8	36.1	22.9	33.9	23.7	36.4	33.5	31.5	--								
	24 Feb 76	9	37.9	24.8	35.7	25.4	38.3	35.4	33.3	02.0	--							
	19 Jan 76	10	34.2	22.6	31.4	21.8	35.2	32.7	30.1	06.1	06.5	--						
	Average	11	36.1	23.5	33.7	23.6	36.6	33.8	31.8	03.6	03.9	05.1	--					
Clouds	19 Jan 76	12	47.2	33.8	45.9	35.7	46.1	42.9	42.2	19.8	18.9	24.1	21.0	--				
	19 Aug 76	13	31.6	18.5	30.8	21.3	30.2	26.9	27.0	16.9	17.8	20.2	18.3	16.2	--			
	15 Mar 74	14	39.4	25.7	38.1	27.8	38.4	35.2	34.5	13.5	13.4	18.0	15.1	08.3	09.2	--		
	Average	15	39.8	26.7	38.7	28.8	38.8	35.5	35.0	16.9	16.9	20.9	18.3	10.5	10.8	7.2	--	
Ice	19 Jan 76	16	42.6	28.6	41.1	30.6	41.9	38.7	37.6	11.2	10.5	16.5	13.0	09.4	14.3	5.9	10.4	--
	Average																	

and addition of any material to the water will cause that signature
to deviate from the clear water signature--the more material added, the
greater the deviation. If the deviation is in different directions for
two materials then they will be distinguishable to some extent. Vectors
$\bar{A}$ and $\bar{B}$ represent the first eigenvectors associated with each material.
The simplest measure of the spectral separability of the two materials
is the angular separation of these two vectors. The eigenvector analysis also provides measures of the dispersion of the data about the axis
of the first eigenvectors. The whole procedure is covered in considerable detail by Klemas et al. (1978).

For our present purposes we will limit ourselves to the angular
separation of the eigenvectors as a measure of spectral separability.
Table 1 shows the results of analyzing six different coastal Delaware
LANDSAT scenes for sediment, ice, clouds, and an iron-acid industrial
waste. There is some dispersion among vectors identifying each material
resulting solely from the use of data acquired on different days. This
dispersion amounts to ∿6° for sediment, ∿10° for the acid waste and
∿10° for clouds. The angular separation between different substances
is always significantly higher, however, between acid and sediment it
is ∿35°, between acid and clouds it is ∿40°.

To demonstrate what this means in terms of classification, two of
these scenes were chosen in which there was some uncertainty as to what
was cloud and what was acid. The eigenvector analysis was used to
classify each pixel in both scenes as either acid, sediment, clouds,
or clear water. The results are shown in Figures 3 and 4. In Figure
3a and 4a the clouds and acid are both plotted as dark points. The
light areas correspond to clear water. (Less than 20 points out of
tens of thousands were classified as sediment in both cases. These
points were treated as clear water.) In Figures 3b and 4b only those
pixels actually classified as acid were plotted. The results are particularly striking in Figure 3b. The pattern that is seen is the
course followed by the acid barge while dumping. There is some noise
in the background and there are some gaps in the pattern caused by
clouds directly over the dump track, but generally the distinction
is quite good. The results illustrated in Figure 4b are still good
although much more noisy. The waste had been in the water for better
than six hours as compared to a fresh dump in the earlier example.
Because of dispersion and settling, the signal is not as clear, as is
apparent in the relatively higher noise level. Still, it is clear
that the acid was dumped in a straight line track on this day and
that the other linear feature in the scene was a cloud.

It is likely that this approach can be extended with the LANDSAT
data to include several other substances, and that considerably better results could be achieved using spectral channels more appropriate
for analysis of water, such as those on the Coastal Zone Color Scanner
(Hovis, 1977).

Laser Fluorosensing

Laser fluorosensing systems have been used experimentally for
several years for the remote detection and characterization of oil
slicks (Fantasia et al., 1971; Rayner and Szabo, 1978) and phyto-

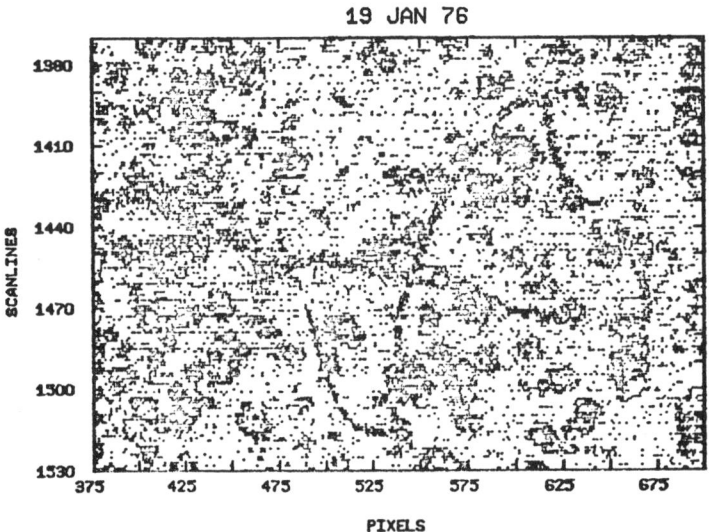

FIGURE 3a: Digital printout of cloud-obscured
iron-acid waste plume imaged by LANDSAT on 19
January 1976.

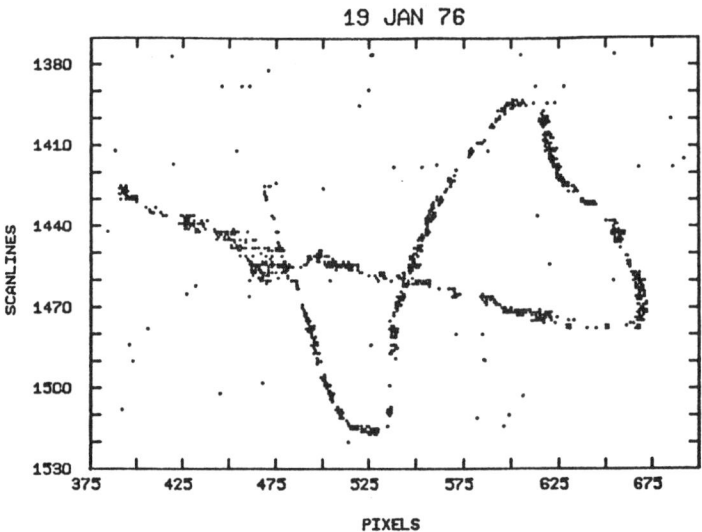

FIGURE 3b: Enhancement of the iron-acid waste
plume of 19 January 1976, against a cloud background.

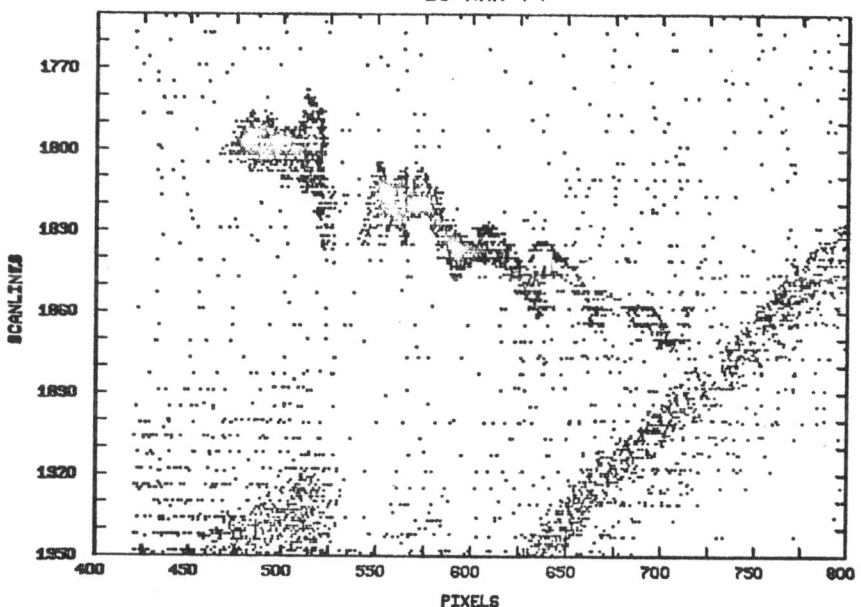

FIGURE 4a. Digital printout of cloud-obscured
iron-acid waste plume imaged by LANDSAT on 15
March 1974.

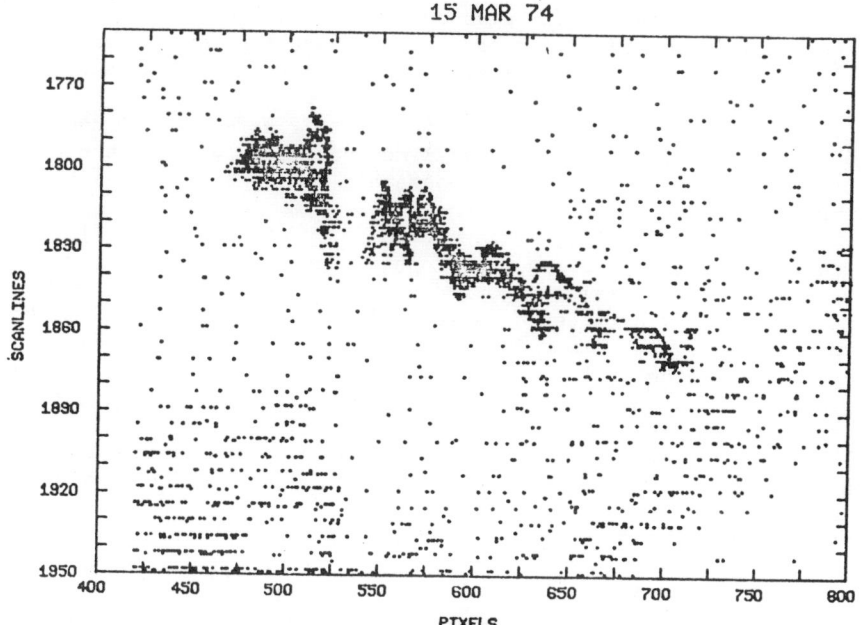

FIGURE 4b: Enhancement of the iron-acid waste
plume of 15 March 1974, against a cloud background.

plankton (Mumola et al., 1973). We will be concerned here only with
remote detection of phytoplankton.

Any fluorosensing material, when excited by light at one wavelength, will almost immediately reemit the energy as light at longer
wavelength than the exciting wavelength. Generally, the emission is
strongest over a rather narrow range of wavelengths and for a particular excitation wavelength. The success of a laser fluorosensing
system depends on the fact that the excitation spectra and emission
spectra are distinctly different for different materials. Thus, in
phytoplankton, each pigment will have different fluorescence characteristics. Since each species of phytoplankton has its own characteristic balance of various pigments, it may be possible not only to
detect the presence of phytoplankton, but to identify the color group
if not the specific species.

Brown et al. (1977) have developed a multiwavelength laser system in order to make use of this species-dependent wavelength variability in the fluorescence. By using four excitation wavelengths and
observing the fluorescence at the four corresponding emission peaks,
they have been able to detect the presence of chlorophyll and have
also been able to make reasonable estimates of the chlorophyll concentration in the water.

EFFECTS OF SURFACE CURRENTS AND FRONTS ON FISH EGG AND LARVAL MOVEMENT

Many finfish and shellfish which spawn offshore depend on surface
currents to transport their eggs and larvae into estuarine nursing
grounds. This period of egg and larval drift represents the most critical survival period for certain marine nekton. When surface currents do not provide favorable transport, or fronts prevent their
drift, the respective fishery may be severely affected. A recent investigation by Nelson et al. (1977) demonstrated that surface transport was the most important oceanographic factor affecting menhaden
recruitment along the Atlantic Coast.

At the present time, surface transport for fishery applications
is calculated by estimating geostrophic wind field from surface atmospheric pressure fields prepared by groups such as the U. S. Fleet
Numerical Weather Service. Sea surface stress is normally inserted
into the appropriate Ekman formulation, including the Coriolis parameter, to obtain an estimate of the surface transport. However, this
approach is limited by about 300 km in space and by about one month
in time (Brucks and Leming, 1977).

Satellite remote sensors can be used to monitor certain oceanographic features synoptically over wide areas and with frequent temporal coverage. Specifically it appears that a SEASAT type scatterometer could be used to map sea surface stress with the following advantages: a) synoptic coverage over a large area (2 x 500 km swathwidth); b) reasonably high resolution (50 km); c) a 36-hour repetition
frequency; and d) a direct measurement of sea surface stress which can
be converted into surface transport for correlation with actual fish
larvae movement (drift). SEASAT-A scatterometer algorithms for estimating wind-induced surface layer transport are being modified for
coastal conditions and their accuracy evaluated. At the present time

we are proposing to track the actual movement of water masses containing fish eggs and larvae along the east coast of the United States
in order to calibrate and verify in real time the circulation and
transport model being developed by the National Marine Fisheries Service (Brucks and Leming, 1977).

Another important aspect to be considered is the influence of
coastal and estuarine fronts on the movement of fish eggs and larvae.
Various scientists, including our own group, have found that coastal
and estuarine fronts seriously influence the drift and dispersion of
oil slicks, phytoplankton, and other suspended matter. In recent experiments in Delaware Bay, oil slicks and phytoplankton were found to
line up along convergent fronts rather than follow the drift pattern
predicated by a model using wind and current information (Klemas and
Polis, 1977b). Convergent fronts have been observed regularly along
the east coast on the shelf and in the Delaware and Chesapeake Bays.
Fronts and their movements can be monitored by NOAA-5, VHRR, LANDSAT,
MSS and aircraft cameras. The aircraft/satellite/ship techniques
proposed will help determine the influence of such fronts on fish
egg and larvae drift as was done with oil slicks and phytoplankton
in previous studies (Klemas et al., 1977a).

LOCATION OF COASTAL FISH SCHOOLS

Spotter pilots flying light aircraft are regularly used to guide
twin purse seine boats and other fishing vessels to large schools of
fish such as menhaden (*Brevoortia patronus*). The pilots direct the
boats by radio to a particular fish school and notify the captains
when to encircle a school with a purse seine. An actual school of fish
cannot be readily observed from satellite altitudes owing to resolution, atmospheric transmission, and surface reflection problems. However, satellites and aircraft can detect secondary indicators of highly
productive coastal waters, such as chlorophyll-a, sea-surface temperature, and turbidity (Maul, 1974). The assumption that fish distribution is governed by certain oceanographic parameters detectable from
satellites has been substantiated by Kemmerer (1976) and Savastano and
Leming (1975). Oceanographic conditions reflected in ground truth
measurements of surface temperature, chlorophyll-a, salinity, water
color (Forel-Ule), and turbidity (Secchi disc transparency) from two
study areas were compared to determine which ones correlated with fishery data collected from fishing vessels at sites of menhaden capture.
The assumption was that if menhaden were caught in the same kind of
water with respect to one or more of the parameters, then the parameters showing consistency probably were affecting fish distribution.

Forel-Ule water color, turbidity, and chlorophyll-a concentrations were similar at locations of menhaden capture in both study
areas; salinity and temperature were not. As the first three parameters can be identified in LANDSAT MSS imagery, a spectral pattern
recognition technique was used to determine if water containing menhaden could be recognized from space. Locations of menhaden schools
were translated into the LANDSAT coordinate reference system so that
areas with and without menhaden could be identified. Radiance values
from each of the four spectral channels were extracted from the data

for these areas so that a computer algorithm could be developed. Digital LANDSAT MSS data were then classified into high and low probability fishing areas with the algorithm.

Menhaden school locations used to develop the classification algorithm were limited to those identified within ±2 hr of satellite coverage. Twenty-five of the 29 school locations satisfying this temporal limitation fell within or immediately adjacent to the high probability fishing areas, and 16 out of 19 other schools located outside the allocated time period fell within or next to these areas. A correlation analysis applied to menhaden and MSS spectral data provided correlation coefficients of 0.65, 0.75, 0.67, and 0.61 for bands 4, 5, 6, and 7, respectively, all significant at the 99% confidence level (Savastano and Leming, 1975; Kemmerer, 1976).

In one mission, computer classification of LANDSAT MSS data into high probability fishing areas was completed and disseminated to the fishing fleet 21 hr after satellite reception. The fleet reported that menhaden were concentrated in these areas and the test was successful. This report was verified by plotting locations of menhaden captures and observations on the prediction chart. Most locations fell into or adjacent to the high probability areas (Kemmerer, 1976).

MONITORING COASTAL UPWELLING

Upwelling is a process of vertical water motion in the sea whereby subsurface water moves toward the surface. Upwelled water can introduce large quantities of nutrients (phosphates, nitrates, etc.) to the euphotic or light zone; thus, upwelling is conducive to high organic production. Knowledge of the location and prevailing conditions of upwelling areas is important for fishing fleets. For example, especially extensive fishing areas and kelp beds are found in upwelling areas off the African and North and South American continents. In addition, considerable bird populations, whose guano is of economic importance, occur off Peru. Near the Antarctic Convergence, particularly in the Atlantic, the abundant nutrients support an unusually large standing crop of diatoms and flagellates which ultimately support krill, the main food of whales, seals, and other species (Fairbridge, 1966).

Upwelling may take place anywhere, but it is more common along the western coasts of continents. Upwelling may be caused by wind displacing surface water away from the coast or by currents impinging on each other or on land masses. The most pronounced coastal upwellings are found off the western United States, Peru, Morocco, the Somali Republic, and the west coast of Africa (Fig. 5).

The ability to identify upwelling areas from aircraft and satellites depends on the fact that deep water, which is brought up to, or near, the sea surface, has different properties from surface water. The most distinguishing feature of the upwelled water is that it is colder and denser than the adjacent surface water and may contain chlorophyll and other nutrients at concentrations exceeding background levels.

When nutrients and cold water are brought into the sunlit layers,

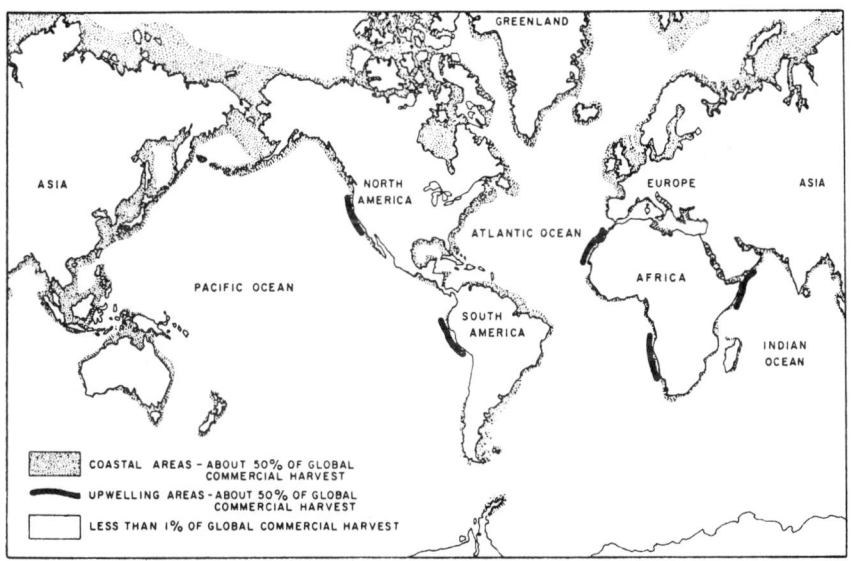

FIGURE 5: Distribution of the world's fisheries
and upwelling areas.

photosynthesis initiates a biological chain reaction that gives rise
to accumulation of chlorophyll and other biochromes. In highly productive areas, the freshly upwelled water is initially cold and clear
but gradually warms up and turns greenish with increased surface age.
The fade-out of thermal contrast and the build-up of color contrast
are supplementary processes (Clarke et al., 1969).

Remote sensing systems that simultaneously measure sea surface
temperature and chlorophyll coloration have provided valuable new information about the distribution in space and time of the biological
activity in upwelling areas (Huebner, 1971). Particularly successful
have been surface temperature observations from the NOAA series satellites (NOAA-2 to NOAA-5) which have been mapping the temperatures of
vast coastal regions with accuracies within several degrees Kelvin
(Fig. 6). Under relatively cloud-free conditions, the Very High Resolution Radiometer (VHRR) aboard NOAA-5, which has a spatial resolution of 1 km at nadir, gathers data in both the visible and infrared
channels. The visible channel (0.6-0.7 μm) measures the reflected so-
lar radiation from the earth, while the infrared channel operates both
day and night to measure the radiation emitted from the earth's surface in the 10.5-12.5 μm wavelength region. The orbiting motion of
the satellite (near-polar and sun-synchronous at an altitude of 1460
km), together with the day-night operation of the VHRR scanner, provides thorough coverage of North America and the adjacent ocean areas
out to 1000 mi or more from shore twice daily at approximately 0900
and 2100 hours local time. The direct readout capability of this instrument, when the satellite is within range of the NOAA command and
data acquisition ground stations at Wallops Island, Virginia; Fair-

FIGURE 6: Portion of a NOAA-5 satellite. Very High Resolution Radiometer thermal image of Gulf Stream, eddies, oceanographic fronts, and colder continental shelf water along the United States east coast. (13 April 1977)

banks, Alaska, and San Francisco, California; allows immediate use of the data. The information content of the gray-scale images is extracted to produce charts that display several significant ocean surface thermal features. Major water masses, fronts, upwelling areas, currents, and eddies can be identified and located (LaViolette, 1974; Maul, 1974). LANDSAT-C, which is expected to be launched in September 1977, will be particularly valuable for discriminating smaller coastal surface water features, since the multispectral scanner on this satellite will have a thermal band (10.4 μm to 12.6 μm) capable of 100 m resolution on the ground.

Detection and monitoring of photosynthetic productivity in upwelling areas has been tried with aircraft, LANDSAT, and Skylab, with some success in locating chlorophyll-rich upwelling areas (Clarke et al., 1970; Szekielda et al., 1975). Attempts to quantify chlorophyll concentrations in aquatic suspension have had mixed results. Inaccuracy is partly due to the fact that chlorophyll and inorganic sediment are indistinguishable. However, measurements of marine photosynthetic organisms show a "hingepoint" at approximately 0.52 μm, below which chlorophyll in suspension reflects strongly and above which absorption is dominant. Sediment, on the other hand, acts as a broadband backscatter. Thus, the use of two bands, separated at approximately 0.52 μm, may allow discrimination of chlorophyll from inorganic sediment. In summary, one can say that aircraft and satellite remote sensors can rapidly locate nutrient-rich upwelling areas for fishing fleets, but cannot reliably quantify the chlorophyll concentration or photosynthetic productivity of the water.

SUMMARY AND CONCLUSIONS

The applicability of remote sensing techniques to monitoring and managing marine productivity and fisheries resources is summarized in Tables 2 and 3.

Remote sensing can provide considerable information for a minimal cost and time when large coastal areas are to be surveyed to determine available habitat, nutrient flow, and fish biomass. Color and color infrared photography have been used to map tidal wetland boundaries, vegetation species, and net primary production. Multispectral digital satellite techniques have been employed to map plant types density and height of the standing crop; and other properties related to the quantity and quality of marsh biomass. Multispectral scanners and laser fluorosensors are being tested for monitoring of primary productivity of coastal surface waters and the outflow of plankton and detritus from marshes and estuaries.

Satellites have been employed to locate areas of high probability for fish availability, such as the highly productive, nutrient rich, upwelling areas off the coasts of Peru, western United States, and Africa. Upwelling areas can be mapped by multispectral scanners primarily due to their strong thermal and spectral gradients caused by the colder upwelling water and its spectrally different nutrient/chlorophyll content. There also appears to be a correlation between certain coastal water properties such as water color, turbidity, chlorophyll concentration, and the presence of fish.

From available results it is reasonable to conclude that remote sensing techniques can be used effectively and reliably to map the location, size, and quality of wetlands, including marine habitat. However, remote sensing techniques being developed for determining surface water productivity, detritus/nutrient flow and fish availability require additional field testing to establish their reliability.

TABLE 2: Summary of remote sensing techniques used to determine wetland habitat, water productivity and fish availability.

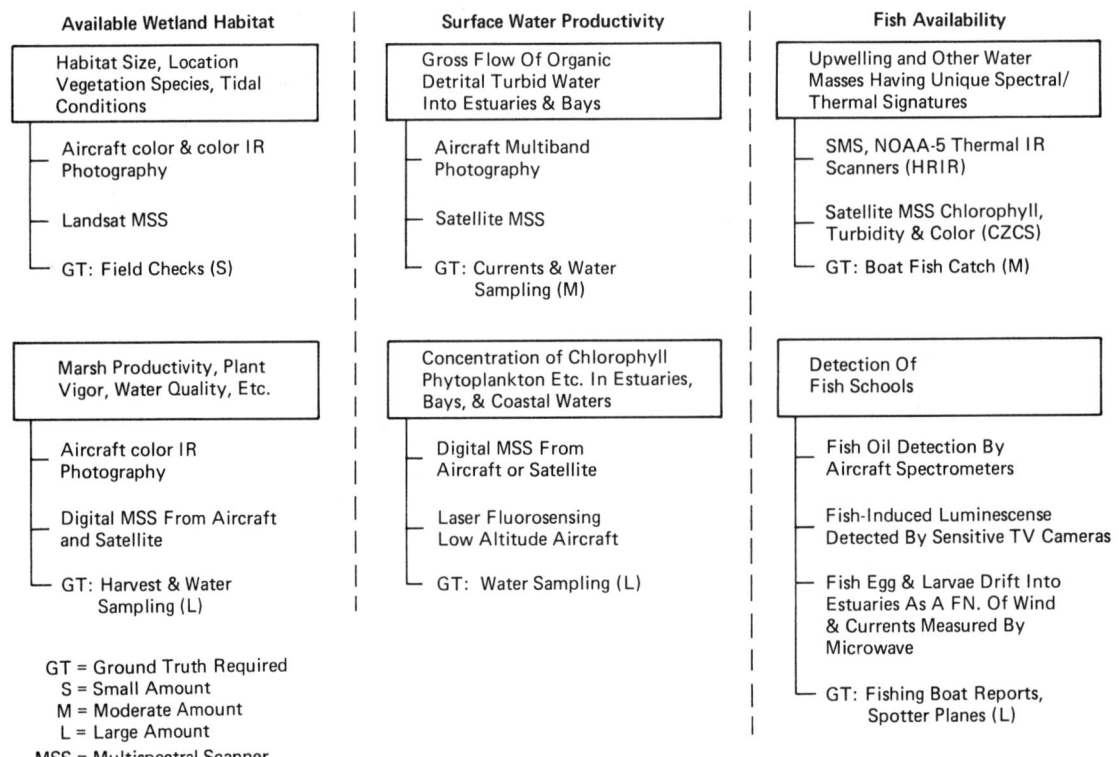

Available Wetland Habitat

Habitat Size, Location Vegetation Species, Tidal Conditions
- Aircraft color & color IR Photography
- Landsat MSS
- GT: Field Checks (S)

Marsh Productivity, Plant Vigor, Water Quality, Etc.
- Aircraft color IR Photography
- Digital MSS From Aircraft and Satellite
- GT: Harvest & Water Sampling (L)

GT = Ground Truth Required
S = Small Amount
M = Moderate Amount
L = Large Amount
MSS = Multispectral Scanner

Surface Water Productivity

Gross Flow Of Organic Detrital Turbid Water Into Estuaries & Bays
- Aircraft Multiband Photography
- Satellite MSS
- GT: Currents & Water Sampling (M)

Concentration of Chlorophyll Phytoplankton Etc. In Estuaries, Bays, & Coastal Waters
- Digital MSS From Aircraft or Satellite
- Laser Fluorosensing Low Altitude Aircraft
- GT: Water Sampling (L)

Fish Availability

Upwelling and Other Water Masses Having Unique Spectral/ Thermal Signatures
- SMS, NOAA-5 Thermal IR Scanners (HRIR)
- Satellite MSS Chlorophyll, Turbidity & Color (CZCS)
- GT: Boat Fish Catch (M)

Detection Of Fish Schools
- Fish Oil Detection By Aircraft Spectrometers
- Fish-Induced Luminescense Detected By Sensitive TV Cameras
- Fish Egg & Larvae Drift Into Estuaries As A FN. Of Wind & Currents Measured By Microwave
- GT: Fishing Boat Reports, Spotter Planes (L)

TABLE 3: Remote Sensing Systems for Fisheries Applications

	Temperature	Salinity	Chlorophyll	Color	Suspended Sediment	Sea State	Fronts	Patchiness	Oil
LANDSAT MSS	0	0	1	2	3	1	2	2	2
NIMBUS – G (CZCS)	0	0	2	3	3	1	2	1	1
NOAA – 5 HRIR	3	0	0	0	0	1	3	1	1
DMSP	3	0	0	0	0	1	3	1	1
HCMM	3	0	0	0	0	0	2	1	1
SEASAT	3	1	0	0	0	3	2	2	2
U-2 MSS (OCS)	0	0	2	3	3	1	2	3	3
U-2 Photographic	0	0	1	1	2	1	2	3	2
NP3A Photographic	0	0	1	2	2	2	2	3	2
NP3A MSS (M2S)	3	0	2+	3	3	2	3	3	3
NP3A Infrared (Thermal)	3	0	0	0	0	1	3	2	2
NP3A Microwave	2	2	0	0	0	3	2	1	2
NP3A Radar	0	1	0	0	0	3	1	2	2
Helicopter Fluorosensor	0	0	2+	1	1	1	2	2	3
Small Aircraft Photographic	0	0	1	1	2	2	2	3	3

0 = Not applicable
1 = Limited value (future potential)
2 = Needs additional field testing
3 = Reliable (operational)

U-2 = High altitude aircraft
NP3A = Medium altitude aircraft (or C-130)
HCMM = Heat capacity mapping mission satellite
DMSP = Defense meteorological satellite
MSS = Multispectral scanner
OCS = Ocean color scanner (10 bands)
M2S = Modular multispectral scanner (11 bands, including thermal infrared)
Photographic = Zeiss and Mitchell = Venten film cameras

REFERENCES

Anderson, R. R., L. Alsid, and V. Carter, 1975. Comparative utility
    of LANDSAT-1 and Skylab data for coastal wetland mapping and
    ecological studies. In: *Proc. of NASA Earth Resources Survey
    Symp.*, Vol. III, Summary Reports, pp. 469-478. NASA, Houston,
    Texas.
Anderson, R. R. and F. J. Wobber, 1973. Wetlands mapping in New
    Jersey. *Photogrammetric Engineering* 39: 353-358.
Apel, J. R., R. V. Charnell, and R. J. Blackwell, 1974. Ocean in-
    ternal waves off the North American and African coasts from
    ERTS-1 In: *Proceedings of the Ninth International Symposium on
    Remote Sensing of Environment*, pp. 1345-1352. Ann Arbor, Michigan.
Bartlett, D. and V. Klemas, 1977. Variability of wetland reflectance
    and its effect on automatic categorization of satellite imagery.
    Proc. Am. Soc. Photogram. 43rd Annual Meeting, pp. 70-89.
    Washington, D. C.
Bartlett, D. S., V. Klemas, O. W. Crichton, and G. R. Davis, 1976.
    Low-cost aerial photographic inventory of tidal wetlands. Uni-
    versity of Delaware Marine Studies Report CRS-2-76. pp. 1-29.
Brown, C. A., F. H. Farmer, O. Jarrett, and W. L. Station, 1977.
    Laboratory studies of in vivo fluorescence of phytoplankton.
    *Proc. Fourth Joint Conference on Sensing of Environmental
    Pollutants, New Orleans.*
Brucks, J. T. and T. D. Leming, 1977. SEASAT-A wind stress measure-
    ments as an aid to fisheries assessment and management. MARMAP
    Contrib. No. 149. pp. 1-4.
Carter, Virginia and Jane Schubert, 1974. Coastal wetlands analysis
    from ERTS MSS digital and field spectral measurements. In: *Pro-
    ceedings of the Ninth International Symposium on Remote Sensing
    of Environment*, Volume II. University of Michigan, Ann Arbor,
    Michigan. 1241 pp.
Clarke, G. L., G. C. Ewing, and C. J. Lorenzen, 1969. Remote measure-
    ment of ocean color as an index of biological productivity. In:
    *Proceedings of the Sixth International Symposium on Remote Sensing
    of the Environment*, pp. 991-1002. University of Michigan, Ann
    Arbor, Michigan.
Clarke, G. L., G. C. Ewing, and C. J. Lorenzen, 1970. Spectra of
    backscattered light from the sea obtained from aircraft as a
    measure of chlorophyll concentration. *Science* 167: 1119-1121.
Egan, W. G. and M. E. Hair, 1973. Automated delineation of wetlands
    in photographical remote sensing. In: *Proceedings of the Seventh
    International Symposium on Remote Sensing of the Environment*, pp.
    2231-2251. University of Michigan, Ann Arbor, Michigan.
Erb, R. B., 1974. The ERTS-1 investigation (ER-600): ERTS-1 coastal/
    estuarine analysis. NASA Report TMX-58118. Houston, Texas. 258
    pp.
Fairbridge, R. W., 1966. *The Encyclopedia of Oceanography*. Van Nos-
    trand Reinhold Co., New York. 1021 pp.
Fantasia, J. F., T. M. Mard, and H. G. Ingrao, 1971. An investigation
    of oil fluorescence as a technique for the remote sensing of oil
    spills. DOT/Transportation Systems Center, USCG Report TSC-USCG-71-
    7.

Gausman, H. W., R. R. Rodriquez, and A. J. Richardson, 1976. Infinite reflectance of dead compared with live vegetation. *Agron, J.* 68: 295-296.

Gordon, H. R., 1978. Remote sensing of optical properties in continuously stratified waters. *Applied Optics* 17: 1893-1897.

Gordon, H. R., O. B. Brown, and A. Jacobs, 1975. Computer relationships between the inherent and apparent optical properties of a flat honogeneous ocean. *Applied Optics* 14: 417-427.

Hovis, W. A., Jr., 1977. Remote sensing of water pollution. In: *Proceedings of the Eleventh International Symposium on Remote Sensing of the Environment,* pp. 361-362. University of Michigan, Ann Arbor, Michigan.

Kemmerer, A. J., 1976. The application of color remote sensing to fisheries. In: *Summary Report of the Ocean Color Workshop.* NOAA-AMOL. Miami, Florida.

Klemas, V., D. Bartlett, and R. Rogers, 1975. Coastal zone classification from satellite imagery. *Photogrammetric Engineering and Remote Sensing* 41: 499-512.

Klemas, V., D. Bartlett, and R. Rogers, 1976. A Skylab and ERTS-1 investigations of coastal land-use and water properties in Delaware Bay. In: *Progress in Astronautics and Aeronautics,* M. I. Kent, E. Stuhlinger, and S-T. Wu, eds.), pp. 351-371. Scientific investigations on the Skylab satellite.

Klemas, V., D. S. Bartlett, and W. Philpott, 1978. Remote sensing of coastal environment and resources. *Proc. Coastal Mapping Symp.,* Am. Soc. Photogrammetry.

Klemas, V., G. Davis, J. Lackie, W. Whalen, and G. Tornatore, 1977a. Satellite, aircraft and drogue studies of coastal currents and pollutants. *IEEE Transactions on Geoscience Electronics* 2: 119-126.

Klemas, V., W. Philpot, and G. R. Davis, 1978. Determination of the spectral signatures of substances in natural waters. Final Rept., NASA Grant NSG 1149, College of Marine Studies, University of Delaware, Newark, Delaware. 99 pp.

Klemas, V. and D. F. Polis, 1977b. Remote sensing of estuarine fronts and their effects on pollutants. *Photogrammetric Engineering and Remote Sensing* 43: 599-612.

LaViolette, P. E., 1974. A satellite-aircraft thermal study of the upwelled waters off Spanish Sahara. *J. Physical Oceanography* 4: 676-684.

McEwen, R. B., W. J. Kosco, and V. Carter, 1976. Coastal wetland mapping. *Photogrammetric Engineering and Remote Sensing* 42: 221-232.

McCluney, W. R., 1974. Ocean color spectrum calculations. *Applied Optics* 13: 2422-2429.

Maul, G. A., 1974. Applications of ERTS data to oceanography and marine environment. In: *Proc. of COSPAR Symp. on Earth Survey Problems,* pp. 335-347. Akademie-Verlag, Berlin.

Morrison, D. F., 1976. *Multivariate Statistical Methods,* Second Edition. McGraw-Hill, New York.

Mueller, J. L., 1976. Ocean color spectra measured off the Oregon coast: characteristic vectors. *Applied Optics* 15: 395-402.

Mumola, P. B., O. Jarrett, Jr., and C. A. Brown, Jr., 1973. Multiwavelength laser induced fluorescence of algae *in vivo*: a new remote sensing technique. In: *Proc. of the Second Joint Conference on the Sensing of Environmental Pollutants*, Washington, D. C.

Nelson, W. R., M. C. Ingham, and W. E. Schaaf, 1977. Larval transport and year-class strength of Atlantic menhaden: *Brevoortia tyrannus*. *Fish. Bull.* 75: 23-41.

Plass, G. N., P. J. Humphreys, and G. W. Katawar, 1978. Color of the ocean. *Applied Optics* 17: 1432-1446.

Proc. of the Symp. on Remote Sensing in Marine Biology and Fishery Resources, 1971. (G. L. Huebner, ed.), Texas A&M University Publ. TAMU-SG-71-106. College Station, Texas. 299 pp.

Rayner, D. M. and A. G. Szabo, 1978. Time-resolved laser fluorosensors: a laboratory study of their potential in the remote characterization of oil. *Applied Optics* 17: 1624-1630.

Reimold, R. J., J. L. Gallagher, and D. E. Thomson, 1973. Remote sensing of tidal marsh. *Photogrammetric Engineering* 39: 477-489.

Savastano, K. J. and T. D. Leming, 1975. The feasibility of utilizing remotely sensed data to assess and monitor oceanic gamefish. In: *Proceedings of the NASA Earth Resources Survey Symposium*, Volume I-D. Houston, Texas.

Simmonds, J. L., 1963. Application of characteristic vector analysis to photographical and optical response data. *J. Opt. Soc. Amer.* 55: 968-974.

Szekielda, K. H., D. J. Suszkowski, and P. S. Tabor, 1975. Skylab investigation of the upwelling off the northwest coast of Africa. *Proc. of the NASA Earth Resources Survey Symp.*, pp. 2005-2022, NASA, Houston, Texas.

U. S. Fish and Wildlife Service, 1975. *National Wetland Classification and Inventory Workshop*. College Park, Maryland.

A Radio Whale Tag

William A. Watkins
Douglas Wartzok
Hugh B. Martin III
Romaine R. Maiefski

ABSTRACT: A radio tag for whales has long been
needed to permit assessment of individual behavior,
group movement, and population distribution. Such
a tag was developed for tracking large whales at
sea, using both shipboard and airborne tracking
systems. A series of tests have shown that the
tags were retained by both finback and humpback
whales for periods of two to three weeks and
provided new information on the behavior of these
whales. The radio tag had a 200-mwatt, 27 to 30-
MHz transmitter housed in a tubular (1.9 x 24 cm)
stainless steel pressure case with a 45-cm whip
antenna. The whale tag was launched from a modified 12-gauge shotgun and implanted into the
blubber of large whales at distances up to 30
meters. The tag was implanted leaving only the
antenna protruding. A seawater switch turned
the transmitter off underwater to give a
potential life of up to 12 months from the power
supply of three organic lithium batteries. Tagged
whales could be tracked by ship at distances of 5
to 25 km depending on antenna orientation and
exposure. Telemetry and satellite tracking could

This is Contribution Number 4259 from the Woods Hole
Oceanographic Institution.

extend the usefulness of a longer term radio whale
tag.

INTRODUCTION

A means of following the movement of whales is important to
an understanding of their biology as well as being necessary for
providing a sound basis for their conservation and regulation
(Norris et al., 1974). The emphasis in the development of a
radio tag for whales has been to find a system that could be
successfully used on finback whales (*Balaenoptera physalus*) at
sea. This is a stringent goal, since these whales generally
are very difficult to approach. Finback whales were chosen as
the target species because they are among the largest of the
mysticete whales still being hunted, and identity of their
populations as well as numbers of animals in the populations
is critically needed for management and conservation. A radio
tag which is successful on this species should be applicable
to several other genera (e.g. *Eubalaena* and *Megaptera*).

Ideally, a radio whale tag (Fig. 1) should provide an
identifying signal whenever the whale is at the surface so
that the tagged animal can be relocated (telemetered data
from the periods between surfacings is an obvious refinement).
The tag should be attachable from at least a distance of tens
of meters since whales are not easily handled at sea, and many
cannot be approached closely enough to attach the tag manually.
The tag should disturb the whale as little as possible, and the
life of the tag should be long enough to provide information on
relatively long segments of the whale's life pattern - several
months, at least.

EARLY TAGS

Efforts to develop such a tag began in 1961 with the
building of a small transmitter and consideration of methods
of attaching it to right whales, *Eubalaena glacialis* (Schevill
and Watkins, 1966). During 1962, 1964, and 1965, a succession
of radio tags was designed for use on right whales, and though
we were unsuccessful in tracking the animals, it was a good
introduction to radio tagging of whales. The 1965 tag was in
a 1.5 cm x 15.5 cm cylindrical case with a wire antenna at one
end and a barbed point at the other (Fig. 2). The tag was
attached by dropping it on a weighted pole from a helicopter
so that the tag penetrated to the base of the antenna and the
pole was then released and pulled back. The transmitter circuitry
(140 MHz at 1 mwatt) operated only when the antenna was clear of
the water surface.

Although the tags were successfully implanted in the whales,
tracking was frustrated by damaged tags, competing radio-
frequency noise, and movement of the whales away from our area.
The main difficulty, however, was the lack of adequate directional

FIGURE 1: A radio tag implanted on a humpback
whale provides individual identification and a
means of following the movement of that whale.
The tag is implanted in the blubber of the whale
leaving only the antenna external to the whale.
A colored plastic streamer attached to the radio
tag provides a visual identification of the tagged
whale.

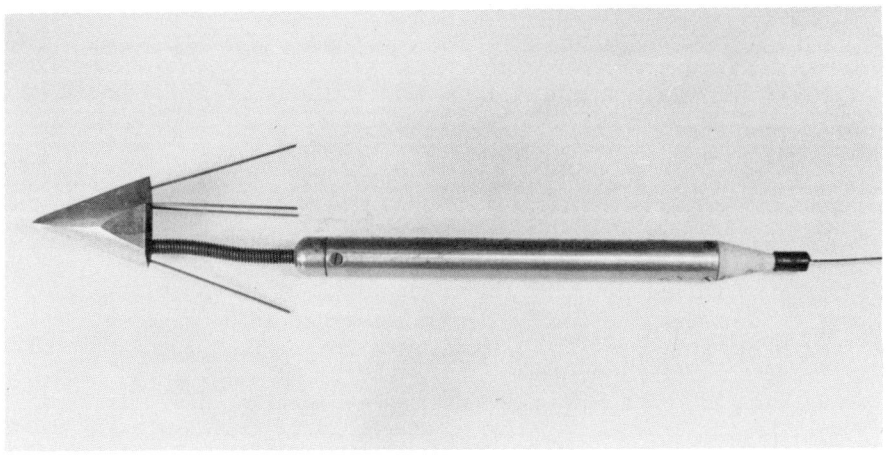

FIGURE 2: The 1965 radio whale tag (1.5 x 15.5
cm), for right whales, was attached by dropping
it on a weighted pole from a helicopter.

receiving gear. A rapid indication of direction was needed for
the very short (2 sec or less) signals that were transmitted when
the tag appeared at the surface.

DEVELOPMENT OF THE TAG

During the next few years small radio beacons were developed
for use in the recovery of instruments at sea (Martin and Kenny,
1971) and they were adapted to the tracking of porpoises (Evans,
1971). Portable automatic radio direction finding (ADF) receivers
also were developed for Evans by Ocean Applied Research Corporation,
San Diego (Martin et al., 1971). These systems were used in tracking several species of the smaller cetaceans. The methods of attachment of the radios required that the animal be captured and the
equipment fastened in place; therefore only animals that could be
caught and handled could be tagged. For example, radios were
attached to calves of larger whale species that were restrained
(Norris and Gentry, 1974; Norris et al., 1974). A radio tag
attached to a small captive gray whale provided temperature and
depth information as well as a track of its movement after release
(Evans, 1974). More recently, similar radio tracking has been done
on killer whales in the Seattle area (Erickson, 1978).

Building on this background (Watkins and Schevill, 1977;
Watkins, 1978), we began in 1973 to develop a radio whale tag
(Fig. 3) that could be used on large whales at sea. The development of the radio whale tag has been a cooperative venture with
three groups providing most of the input of ideas and resources -
the manufacturer, Ocean Applied Research Corporation (Hugh Martin
and Romaine Maiefski), The Johns Hopkins University (G. Carleton
Ray and Douglas Wartzok), and Woods Hole Oceanographic Institution
(William E. Schevill and William A. Watkins).

We considered a variety of methods for remote attachment of
the tag to the whale (Schevill and Watkins, 1966), and finally
decided to develop a tag which when fired from a gun would implant
with only the antenna outside the whale. Electronics, power
supply, and antenna had to be designed and supported in ways that
would tolerate the rapid accelerations and decelerations of firing.
Propellants were devised to permit relatively slow starts, and high
enough velocities to deliver the tag at distances of 25 m or more.
The ballistics development was monitored with high-speed (1200
pictures/sec) motion picture photography.

The radio frequency of the tag was chosen to maximize the
distance that the signal of a tagged whale could be received from
a ship. The design of the launching system limited the antenna
length of the tag to about 50 cm, an efficient quarter-wave antenna
length for 150 MHz. But at such high frequencies, radio transmission
is essentially line-of-sight so that the curvature of the earth
limits reception to about 12 km over calm seas, from a 15 m antenna
height. Seas are rarely calm and previous experience (Schevill and
Watkins, 1966) showed that even small wavelets blocked the transmissions to reduce reception distances further. Near-surface
atmospheric inversions also limited transmission ranges at these
frequencies. We tried much lower frequencies (2-3 MHz), which
permitted more reliable transmission over longer distances, but
they required unacceptably large or inefficient antennas on the
radio transmitters. In order to avoid these restrictions, we
chose an intermediate 27 to 30 MHz frequency whose wavelength was

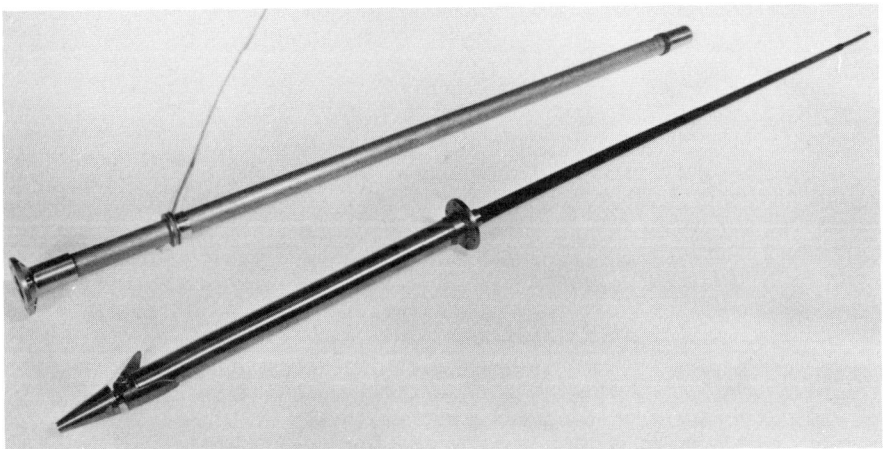

FIGURE 3. The 1978 radio whale tag (lower) is
shown with the pushrod (upper) used over the
antenna as the tag is launched. The case of
the tag is 29 cm long and the antenna is 45 cm.
A line attached to a ring that slides along the
pushrod allows the retrieval of a missed tag and
the removal of the pushrod after a tag is implanted.

long with respect to the usual wave height, and whose transmission
characteristics permitted reception somewhat beyond the horizon.
Relatively efficient short antennas could be utilized at 27 to 30
MHz, and automatic radio direction-finding receivers were available
for these frequencies. The specific frequencies for individual tags
were chosen to permit operation at frequencies which had been found
experimentally to be relatively free of radio noise, and the tag
transmitters were operated under Federal Communications Commission
experimental licenses.

Tests of the radio tag were conducted on whale carcasses for
two seasons in Iceland, in 1976 and 1977 (Watkins and Schevill,
1977). Live whales were successfully tagged in the Gulf of St.
Lawrence in 1976 (Ray et al., 1978), during two series in Southeast
Alaska in 1976 and 1977 (Reports of the Marine Mammal Division,
National Marine Fisheries Service - Seattle, 1976 and 1977) and in
Prince William Sound, Alaska, in 1978 (Watkins et al., 1978).

The 1978 tests of the radio whale tag were on both finback
and humpback whales and demonstrated the tag's ability to provide
detailed behavioral information as tagged whales were tracked for
two to three weeks. This proved to be the effective duration of
the tags on the whales. After implantation, the radio tags
gradually worked their way out of the blubber, but we are hopeful
that new modifications to the tag will improve retention, and
permit extended tracking periods. The details of the radio tags,
the attachment procedure, and the tracking arrangements given

below are of the system that was used during the experiments in
June 1978.

THE TRANSMITTER

     The transmitter circuit is shown in Figure 4. It consists of
a radio frequency section with crystal-controlled oscillator and
power amplifier, a timing section, a power switch, and an emersion
sensor to shut off the transmitter when the transmitter is underwater. In the design, attention was paid to power efficiency in
both the on and off conditions. Transistors Q4 and Q5 formed the
crystal-controlled oscillator and power amplifier. The only
unusual feature of the oscillator was that the output and the
feed-back tap were derived from the same tuned circuit, with
sufficient isolation and different impedances provided between
the two by means of both a tapped coil and a capacitive divider.
The pulse timing for the transmitter was from the low power
operational amplifier (CA 3078), and the transmission "on" and
"off" times were set by the circuit time constants (R4, C1 and
R5, C1). The bias control on pin 5 of the operational amplifier
was used to shut the system down when there was a seawater
connection between the antenna tip and ground. Emersion was
sensed by the differential transistor pair Q1 and Q2 operating
at very low current (about 100 μa) levels. Transistor Q3 served
as the switch for the transmitter driven by the output of the
operational amplifier.
     The antenna element was a coil wound on a fiberglass
strength member and tuned to an electrical ¼ wave length. The
coil was kept as high as possible on the antenna structure to
keep the voltage field up away from the animal and the water.
A second larger diameter winding was incorporated at the base
of the antenna and connected to ground. This coil was tuned
also to provide an improved counterpoise for the main antenna
element. With a good ground reference, this antenna maintained
its efficiency even at orientation angles as low as 20° from
the horizontal, and its efficiency deteriorated relatively
slowly as the whole transmitter moved away from the ground
plane.
     The power supply for the radio tag used three hermetically
sealed organic lithium batteries (Mallory LO-32S), 9 v at a
nominal 1 amp hrs. The transmitter pulse length was set at
about 50 msec for consistent and reliable directional response
from the ADF receivers. The pulse rate was made 2/sec to try
to provide two to four pulses for each antenna exposure. And
the transmitter power to the antenna was set at 200 mwatts to
provide sufficient signal for tracking as well as a small
enough power drain to assure relatively long battery life.
Previous experiments and observations had indicated that we
could expect a well-placed tag to transmit for about 2% of
the time on finback whales. During bench tests, the tags have
operated continuously at apparently full output for up to 180
hrs, so the potential life on a whale could be as long as
12 months.

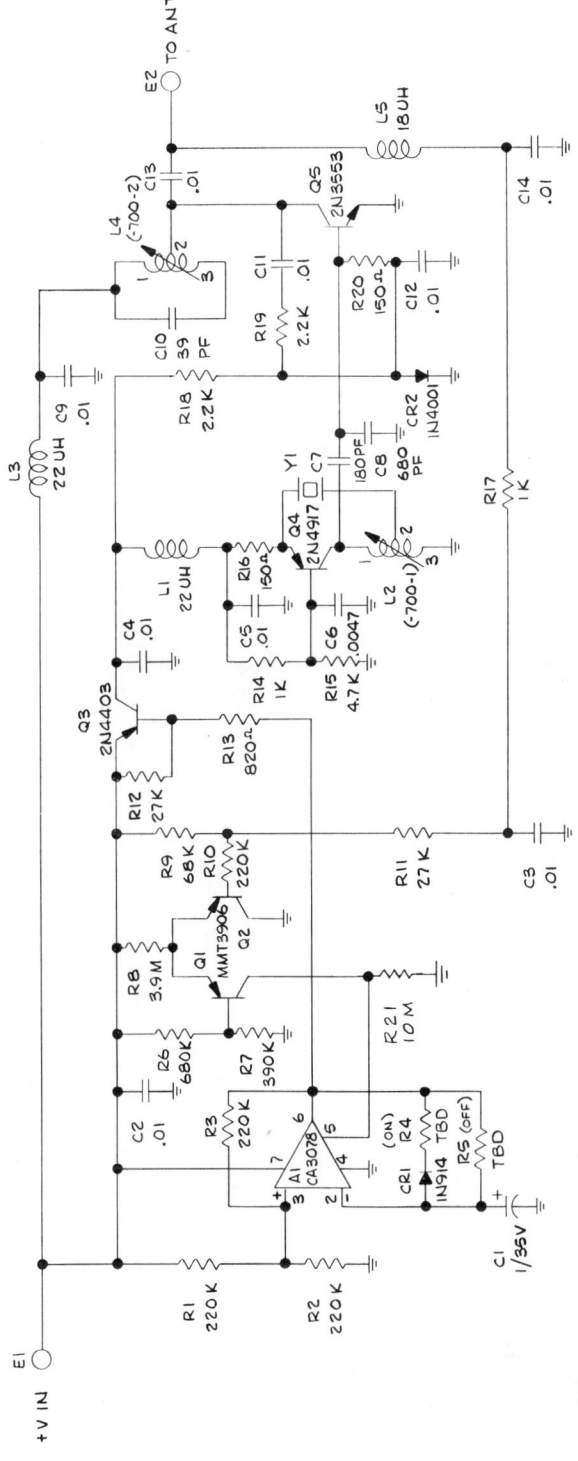

FIGURE 4: The heart of the radio whale tag is a 200 mwatt transmitter at 27 to 30 MHz, producing 50 msec pulses at a rate of 2/sec. A seawater switch clamps the oscillator off while the antenna is submerged. Schematic prepared by Ocean Applied Research Corporation. (BT-243-500).

THE PRESSURE CASE

     The cylindrical configuration of the tag (Fig. 3) was
chosen to meet requirements of pressure since whales sometimes
dive deeply, and the minimum cross section was maintained for
simplicity in ballistics and blubber penetration. The pressure
case provided separate compartments for the transmitter and for
the power supply. The diameter of the tag (1.9 cm) was dictated
by the size of the batteries. End-caps with "O" ring seals
provide access at both ends of the pressure case. The antenna
was moulded in polyurethane with a core of linear fiberglass,
inserted into the upper end-cap and was protected by an external
water-sealing sleeve. A flange 3.8 cm in diameter at the base
of the antenna prevented implantation beyond that point. The
body of the tag was fabricated from stainless steel (#304, 19 mm
outside diameter and a wall thickness of 1.27 mm). The lower
(forward) end-cap was fitted with two folding 3 cm toggles to
keep the tag in place after implantation, and a threaded projection
to allow attachment of the penetration point. A variety of point
shapes were tested to determine the one that would penetrate
blubber best at low angles of incidence (Watkins, in press). Thus,
the complete tag measured 29 cm from point to flange, and was
designed to be imbedded in the blubber with only the 45-cm
antenna protruding. The radio whale tag was pressure tested at
the Woods Hole Oceanographic Institution's hydrostatic pressure
facility (8 June 1977) to a depth of at least 3,750 m, recycled
four times to a pressure of 350 kg/cm^2.

LAUNCHING SYSTEM

     The launching system for the radio tag used a detachable
hollow pushrod (Fig. 3) that fit over the antenna and into the
barrel of a modified 12-gauge shotgun (Fig. 5). The gun (Harring-
ton and Richardson, with a 7-cm chamber) was fitted with an adjust-
able-height rear sight and given 1.5 kg additional weight to reduce
the recoil of firing the 550-gm tag and pushrod projectile. A
chamber adapter was used to make sure that the tags were fired with
specially loaded cartridges and not fired with standard shotgun
loads which would produce dangerously high barrel pressures.
     The special cartridge used to fire the tag was loaded to
provide a relatively soft start and continuous acceleration throughout the length of the barrel. Modified magnum cases (Winchester 300)
were loaded with magnum rifle primers, two grains of Bullseye pistol
powder initiator, and 14.5 grains of rifle powder (Hodgen H-414 Ball).
Cotton packing was used to assure powder contact with the primer. A
plastic 12-gauge shot-cup/gas-seal was used over the end of the pushrod .and seated against a shoulder within the chamber adapter close to
the end of the cartridge. The gas seal fit snugly into the chamber
adapter and kept the end of the pushrod centered in the barrel.
     The radio whale tag was propelled at a muzzle velocity of about
68 m/sec, slowing to only 60 m/sec at 25 m, as determined from high-
speed photographic measurements. Because the drop in trajectory with

FIGURE 5. A weighted 12-gauge shotgun fires a
specially loaded cartridge to launch the radio
whale tag. The pushrod fits into the barrel and
is attached to the tag by nylon screws. Upon implantation, the screws (Fig. 6) part to release
the tag, and the pushrod is retrieved by a line
that pays out from the cannister below the barrel.

distance was 1 m at 25 m (Ray and Wartzok, 1975), estimation of
distance to the target was critical for vertical accuracy, but horizontally the trajectory of the projectile was consistent and accurate.
Therefore, with practice and good distance estimates, the tag could be
placed within 5 cm of a stationary target from a hand-held gun (Watkins
and Schevill, 1977). Because of the vertical kick of the muzzle of the
gun and the relatively slow start, the tail of the pushrod was lifted
up as it left the barrel. This produced a vertical wobble in the
trajectory of the projectile that appeared to be repeatable. The
trajectory wobble was a potential problem for penetration at low
angles of incidence particularly if the tag was turning upward as
it struck. However, the use of a point that penetrated blubber well
apparently obviated these effects (Watkins, in press). The wobble
could be reduced by adding up to 2.5 kg to the gun to dampen the
vertical excursion of the muzzle, but the heavier gun was more
difficult to use from the bow of a bouncing boat.
    The pushrod was made of aluminum (#2024-T3) tubing with a
nominal outside diameter of 16 mm, a wall thickness of 1.65 mm,
and heat treated to T-6 condition. It was attached to the tag

FIGURE 6. The tag is fastened to the pushrod by
three nylon screws that break away after the tag
is implanted. Springs allow this connection to
flex without breaking the screws during firing yet
keeps the projectile together in flight. The
flange serves as a penetration stop.

by means of three nylon screws (#4-40) with 1.5-cm coil springs
beneath their heads (Fig. 6). This spring-loaded connection
permitted sufficient movement to keep the screws from breaking
with the vertical kick of the barrel during firing, and yet it
kept the tag and pushrod together in flight. The nylon screws
had a combined breaking strength of about 51 kg, the maximum
direct pull needed to separate the pushrod from the tag. In
practice, relatively little pull was required to release the
pushrod after a tag was implanted since one or more screws
usually sheared during impact.
    A line ring that slid freely along the pushrod was used
for attachment of the retrieval line. The line ring fit snugly
into the end of the barrel and served to center the pushrod at
that point. As the rod moved out of the barrel during firing,
the line ring was caught and carried along by a flange near the
end of the rod. A crush washer between the ring and this flange
absorbed some of the impact of the sudden acceleration, and
minimized deformation of the flange and line ring. The end of
the retrieval line was wrapped in a channel around the ring and
secured with a bowline knot. The rest of the 35 m of nylon
braided cord (#60 - breaking strength 250 kg) was wrapped on a
tapered spindle (removed before firing) and payed out from the
center of the spool held in a cannister below the barrel. The

retrieval line served as drag to help the projectile fly straight, as well as a means of retrieving a tag that missed, or pulling the pushrod free from an implanted tag.

TRACKING SYSTEM

For tracking the radio-tagged whales, automatic radio direction-finding receivers were necessary since the whales' surfacing times were often too short to permit the transmission of more than one or two pulses. Our most successful tracking system utilized an ADF receiver equipped with a phase-lock loop and phase sensitive detector for signal processing (Ocean Applied Research Corporation, ADF-922). Directional information was clearly displayed on a cathode ray tube.

An important feature of the tracking systems was the beat-frequency oscillator (BFO) which provided an audible tone during the reception of each radio pulse. The audio signal from the BFO served to focus attention on the ADF, so that the operator could be alerted to note the direction of the next signal pulse. Without a good BFO, the operator had to remain alert and watch for the signals on the ADF display. This made tracking and particularly searching both physically and psychologically tiring. A good audio BFO tone at a comfortable listening level made tracking very much easier.

The best of our shipboard receiving antenna systems used pairs of Adcock antennas at right angles. The 30-MHz Adcock antennas occupied about 5 cubic meters and were mounted at heights of 8 to 10 m. Since other antennas and parts of the super-structure of the ship influenced the reception pattern, we found it useful to make a correction chart of the reading on the ADF compared to the true direction to a transmitter.

For aircraft, we used crossed ferrite core antennas in a 31 x 47 x 4 cm package attached to the exterior of the aircraft. We have used this antenna on the belly of a Cessna 170 aft of the landing gear as well as on the top of the forward portion of the cabin roof on a pontoon-equipped Cessna 185. At these frequencies, the pontoons created too much interference for the antenna to function well when positioned anywhere near them. Regular landing gear interfered less but also produced some distortion in the reception pattern. The reception ranges usually were greater when the antenna could be mounted under the aircraft, probably because of elimination of effects of any shielding by the body or wings of the plane.

Radio reception distances from tagged whales have depended on so many variables that it has been difficult to predict use-ful ranges. With the transmitter floating vertically in the water with only the antenna above the surface, the signal was strong to at least 65 km from aircraft at 300 m and to 25 km or more from ships with 10-m antenna heights. Sometimes these ranges could have been realized from transmitters on whales, but usually the distances have been very much shorter. Orientation of the transmitter antenna to the ground plane or to the body of

the whale, relative electrical resistance in the flesh of the
animal, and proximity to wavelets or to portions of the whale,
all had an effect. Ideally, perhaps, the tag should be placed
vertically on the midline of the forward portion of the whale's
back so that it will be lifted well clear of the water at each
surfacing. Practically, this has been difficult to achieve.
The whales were seldom that cooperative during tagging so that
the antenna orientation usually has been at relatively low angles
with the tag implanted to the side of the animal. Thus, the
actual reception ranges usually have been much shorter than the
potential distances. During the 1978 tests with tag antennas
mostly at low angles, shipboard tracking often was reliable only
within about 5 km and aircraft tracking only to perhaps 15 km.
These short signal reception distances, prolonged submergences by
the whales, and unpredictable changes in behavior often made
continuous tracking very difficult.

The effective radiation of the tag antenna on a whale
suffered because of absorption by the body of the whale, a
variable unknown. It also suffered because of a variable but
probably significant ground resistance provided by the flesh of
the whale, and there was no way to provide a separate counterpoise for the antenna. In addition, the orientation of the
antenna was seldom advantageous since it was dictated by the
angle at which the transmitter was implanted as well as its
position on the whale as it surfaced. Thus, in spite of careful
design of both tracking and transmitting systems, the weakest
link was probably the radiation from the whale tags themselves.

LOOKING AHEAD

At the present stage of radio tag development, it is
necessary to maintain close contact with the animals to monitor
changes in the orientation of the antenna, to document the
effects of tag placement, and to observe the retention characteristics of the transmitter in the whale. But the ultimate goal, of
course, would be remote tracking, perhaps by satellite, to supplement ship and aircraft operations. Ship tracking is expensive and
time-consuming, it is possible only over limited ranges, and only a
few animals can be tracked at one time. Currently, however, there
are several problems in considering satellite tracking of whales.
1) Satellite RAMS systems that probably would be used in animal
tracking are at higher frequencies than are useful for surface-to-
surface tracking. Therefore, a different whale tag with multiple
frequencies would be required. 2) In order to achieve successful
transmission to a satellite, more consistent signal strengths
would seem necessary so that transmitting antenna orientation and
placement of the tag would have to be optimum. Precise implantation
of the radio tags depends on the cooperation of the whales. 3) The
two to three week retention time of the transmitter in the animal is
still too short to justify satellite tracking. 4) In addition,
different locating systems would appear to be required for tracking
whales from satellites because of the whales' irregular and short
surfacing times and their long unpredictable submergence times.

Some modifications (such as reduction in the size of the transmitter) will improve the future long-term data return from the tag
but are not necessary to its development. The present major problem
is the relatively short time duration of the tag in a whale. The tag
seems to be rejected at least partly by tissue necrosis accentuated
by the constant pressure on the toggles from hydrodynamic drag. The
pressure on the tissue in contact with the toggles can be reduced in
two ways: 1) increase the surface area of the toggles and add
another set of toggles to hold the tag in the blubber; and 2) reduce
the hydrodynamic drag of the parts of the tag which are external
to the surface of the animal by removing the flange at the base
of the antenna. The flange now must experience a considerable
drag after the tag has backed out a bit. The function of this
flange is to prevent implantation beyond the base of the antenna;
instead the base of the pushrod could be used for this purpose.
With only the antenna left outside the whale, the drag should be
significantly reduced.

The present radio whale tag is not yet the year-long-migration-
path package that was our design goal. So we may want to reassess
the trade-offs between pulse lengths, repetition rates, and power
output. It may be that increased output power would provide a
significant increase in reception range so that even poorly positioned tags would provide usable tracks, and increased pulse
repetition rate might provide better tracks. More short-term
information perhaps could be collected even though the potential
life of the tag were reduced. A higher transmitting frequency
would provide greater antenna efficiency in manageable sizes so
it might be advantageous to switch primarily to aerial tracking
and forego continuous surface-to-surface tracking. In addition,
the use of hybrid construction techniques would reduce the size
of the transmitter substantially in order to provide in the same
tag a 30 to 50% increase in battery capacity, or to produce a
smaller tag.

There are obvious additions we would like to make to the tag
as soon as the initial problems of duration and range have been
solved. We would want to telemeter information on a variety of
parameters, such as temperature, heart rate and depth of dive.
More sophisticated electronics would allow information on the
depth-time integral of the preceding dive and give range from
the transmitter to the receiving antenna to permit precise
plotting of the distance to the whale at each breath.

WHAT CAN A RADIO TAG TELL US?

Attached to a whale, a radio transmitter itself without
additional telemetry can provide a great deal of presently
unavailable information regarding whale populations and the
behavior of individual animals, such as daily routines, changes
in social behavior, breathing patterns, local movements, and
rhythms of activity. An important question which can be
addressed immediately is that of group structure. Does a given
group of animals stay together for a significant period of time?
What is the relationship between different small groups observed

in an area? Expanding these questions over longer periods and
in more detail, we can approach the study of possible gene flow
between different subpopulations of a whale species. For example,
is there interbreeding between the population that summers in
Alaska and the whales seen in Hawaiian waters in the winter?
Preliminary observations based on coloration of the dorsal surface
of the flippers suggest that these subpopulations are separate
(Herman and Antinoja, 1977) but radio-tagged whales could provide
direct evidence.

     If the range of the tag transmitter were great enough to
allow reliable reacquisition of the signal and enough whales
were tagged, population estimation could be attempted through
the use of the radio signal in a mark-recapture context. Aerial
surveys could determine both the number of radio signals (re-
captures) and the number of visually sighted animals.

     The whale radio transmitter in its present state of develop-
ment can provide heretofore unobtainable data on the behavior and
movements of individual marked whales over two to three week
periods of time and over 5 to 15-km transmission ranges. The
modifications that are planned should increase the retention time
of the transmitters and improve reception distances. This will
give access to additional information on group behavior, population
structure, and perhaps even migration patterns.

ACKNOWLEDGMENTS

     The radio whale tag has progressed through several cycles of
development and testing (over the last 17 years, at Woods Hole
Oceanographic Institution) and we have had the benefit of encourage-
ment and help from many of our colleagues. Particularly we want to
mention William E. Schevill and Karen E. Moore of Woods Hole
Oceanographic Institution, G. Carleton Ray of The Johns Hopkins
University, James H. Johnson and Michael F. Tillman of the National
Marine Fisheries Service - Marine Mammal Division, Edward D.
Mitchell of Environment Canada, and William E. Evans of Hubbs Sea
World Research Institute. Support for the tag development has come
through several sources, but consistent help has been from the
Oceanic Biology Program of the Office of Naval Research (ONR),
Contract N00014-74-C0262 NR 083-004 to Woods Hole Oceanographic
Institution, as well as support to The Johns Hopkins University,
ONR Contract N00014-75-C0701, and from NASA Contract NAS 2-9300
to JHU. Karen E. Moore has taken the photographs for the figures
and prepared the manuscript.

REFERENCES

Erickson, A. W., 1978. Population studies of killer whales (*Orcinus
     orca*) in the Pacific Northwest: A radio-marking and tracking
     study of killer whales. *Final Report to the U. S. Marine
     Mammal Commission*, Report No. MMC-75/10. Reproduced by
     National Technical Information Service, No. PB 285 615. 34 pp.

Evans, W. E., 1971. Orientation behavior of delphinids: radio telemetric studies. *Ann. N. Y. Acad. Sci.* 188: 142-159.

Evans, W. E., 1974. Radio-telemetric studies of two species of small odontocete cetaceans. In: *The Whale Problem: A Status Report*, (W. E. Schevill, ed.), pp. 385-394. Harvard University Press, Cambridge, MA.

Herman, L. M. and R. C. Antinoja, 1977. Humpback whales in the Hawaiian breeding waters: Population and pod characteristics. *Sci. Repts. Whales Res. Inst.* 29: 59-85.

Marine Mammal Division, NMFS, 1976 and 1977. Radio-tagging of humpback whales. *Report to the National Marine Fisheries Service (NMFS)* in compliance with Permit No. 136. Northwest and Alaska Fisheries Center, NMFS, Seattle. 4 pp.

Martin, Hugh, W. E. Evans, and C. A. Bowers, 1971. Methods for radio tracking marine mammals in the open sea. IEEE '71. *Engineering in the Ocean Environment Conference*, IEEE Publishers, NY. pp. 44-49.

Martin, H. B. and J. E. Kenny, 1971. Recovery of untethered instruments. *Ocean. Internat.* 6: 29-31.

Norris, Kenneth S., William E. Evans, and G. Carleton Ray, 1974. New tagging and tracking methods for the study of marine mammal biology and migration. In: *The Whale Problem: A Status Report*, (W. E. Schevill, ed.), pp. 395-408. Harvard University Press, Cambridge, MA.

Norris, Kenneth S. and Roger L. Gentry, 1974. Capture and harnessing of young California gray whales, *Eschrichtius robustus*. *Mar. Fish. Rev.* 36: 53-64.

Ray, G. C., E. D. Mitchell, D. Wartzok, V. M. Kozicki, and R. Maiefski, 1978. Radio tracking of a fin whale (*Balaenoptera physalus*). *Science 202: 521-524*.

Ray, G. Carleton and Douglas Wartzok, 1975. Tests of an implantable beacon transmitter for use on whales. *Report to the National Marine Fisheries Service* in compliance with Endangered Species Permit No. E4 and Marine Mammal Permit No. 99. The Johns Hopkins University. 8 pp.

Schevill, William E. and William A. Watkins, 1966. Radio-tagging of whales. WHOI Reference No. 66-17. Woods Hole Oceanographic Institution, Woods Hole, MA. 15 pp.

Watkins, William A., 1978. A radio tag for big whales. *Oceanus* 21: 48-54.

Watkins, William A., in press. A point for penetrating whale blubber. *Deep-Sea Research*.

Watkins, William A., James H. Johnson, and Douglas Wartzok, 1978. Radio tagging report of finback and humpback whales. WHOI Reference No. 78-51. Woods Hole Oceanographic Institution, Woods Hole, MA. 13 pp.

Watkins, William A. and William E. Schevill, 1977. The development and testing of a radio whale tag. WHOI Reference No. 77-58. Woods Hole Oceanographic Institution, Woods Hole, MA. 38 pp.

MARINE BIOLOGICAL MEASUREMENT AND INSTRUMENTATION

Problems and Opportunities Connected with Marine Biological Phenomena

Richard V. Lynch

ABSTRACT: The word "phenomenon" means anything
apparent to the senses that can be scientifically
measured. In a laboratory, phenomena can be isolated from one another and individually studied.
In the ocean, individual phenomena undergo complex synergistic and antagonistic interactions,
making isolation difficult. In fact, understanding of biological dynamics requires understanding of the interactions. The gathering of
the large amounts of data needed to study these
interactions requires new methods of measurement.
Remote sensing can provide fast and repeated coverage of large areas and offers the possibility
of sensing several variables simultaneously. Automated instruments can make long-term measurements at selected sites with minimum attention.
Data processing using computers offers a way of
analyzing and storing all these data that is
less cumbersome than individual analysis. The
variety and quantity of knowledge required to
study large-scale processes suggests the use of
a multi-disciplinary team approach. The advantages of this approach and of using the new means

of measurement are outlined in three examples of
current biological interest - pollution studies,
hot spots, and bioluminescence.

INTRODUCTION

The word "phenomenon" means anything apparent to the senses
that can be scientifically measured. Marine biology has been historically concerned with phenomena related to individual organisms
or groups of organisms. Modern concerns of marine biologists involve trying to understand the nature of the interactions between
organisms and their environment (which includes other organisms of
their own and different species as well as the chemical and physical
surroundings) and the dynamics of entire ecosystems. These studies
are of more importance today than ever before, because of the rapid
rate at which man is changing the environment. Such work must be
done largely in the field and requires study of the physics, chemistry, and geology of the ocean as well as its biology. For studies
of the ocean surface, the atmosphere must be considered as well.
Studies performed near shorelines or within the flow of river estuaries may require terrestrial research to achieve full understanding.

From the above statements it can be seen that phenomena connected with marine biology need not be biological in nature. An
approach leading to complete understanding of marine biological
phenomena therefore should be multidisciplinary. The vast amounts
of knowledge required and data collecting that must be done suggest
a team approach to field studies.

Even with a team of people working on a common problem, the
study of interacting phenomena will be slow and tedious unless methods are used to speed up traditional means of data collecting and
processing. One means of speeding up data collecting is by remote
sensing. Remote sensing allows measurements of variables to be made
from a distance and thereby increases the area that a single worker
can cover. Also it allows the use of fast platforms, such as aircraft and satellites, instead of relatively slow ships or footwork.
Multichannel frequency analysis also allows simultaneous measurement
of several variables, which is useful in studying correlated changes
among interacting systems.

At the moment remote sensing is still sufficiently new that on-
the-spot data are required to corroborate its measurements. Furthermore, remote sensing cannot yet be used to measure all the variables
desired. These limitations mean that ground truth data are still
necessary. The gathering of such data can be simplified by using
automated instruments which can be set in place and retrieved at
some later time. Such instruments may even radio back to a central
location their measurements. Use of automated instruments allows
long-term measurements to be made and compliments the quick measurements of remote sensing.

The problems of dealing with interacting systems are a long way
from being solved. However, new techniques and instruments such as
those mentioned above provide opportunities for significant advances

toward solutions. Let us see how they can be applied in three study
areas.

POLLUTION PHENOMENA

     In recent years, as man's population has expanded and land areas
have become more heavily utilized, man has increasingly turned to the
sea to meet his growing needs. In using the sea, man has changed the
natural environment by the introduction of foreign substances or forms
of energy, the stirring up and mixing of sediments and normally distinct bodies of water, the harvesting of certain species in large numbers, and other means. Some of these changes, those involving the
chemical or physical properties of the sea water, can collectively be
called pollution.
     Pollution can occur anywhere in the ocean. Sewage from cities
and radioactive wastes are deliberately dumped in the deep ocean and
pollute the floor. Balls of tar, plastic beads, and other artificial
objects are frequently found floating in areas far from land. Nevertheless the impact of pollution is greatest in shallow water areas of
the continental shelf, since these are the most accessible portions
of the oceans and hence the most easily exploitable. In particular,
estuaries and coastal wetlands have been affected. These areas are
of considerable importance today, because they are among the most
biologically productive areas of our planet. Yet they are endangered.
Wetlands are filled in by developers for housing and industry. Dams
and development reduce water flow from rivers and through the ground,
causing marshes either to dry up or to become more salty. Rivers
carry down heavy loads of silt, sewage, and industrial pollutants.
Even when towns and industries have cleaned up their effluents, run-
off from heavily fertilized fields brings large amounts of chemicals
into coastal waters. Dredging of channels, besides destroying bottoms, releases chemicals trapped in the sediments and rechannels the
flow of water with unknown effects. Air pollution and insecticide
spraying can directly harm plants and animals or can indirectly harm
them by affecting some link in their food web. These are all phenomena with a direct effect on shallow-water marine biology.
     The term, pollution, implies deleterious changes in the environ-
ment. The question is therefore raised, deleterious to what. It is
not necessarily true that all man-made perturbations are harmful.
Some years ago there was a controversy at a nuclear power plant in
New England, which was discharging its coolant water into the sea.
Conservationists argued that the plant should either be shut down
or forced to cool the water before discharge, since the heat was
changing the ecology of the area near the discharge and was harming
some species of fish and other forms of life. Indeed it was. Organisms that could not tolerate the higher temperatures, particularly
in winter, could not live near the discharge area. However, the
warmer water enabled some species that would normally have migrated
south in the winter to remain, and other, more heat-tolerant organisms expanded or moved in to replace those that left. Oyster fishermen were especially happy, because near the discharge area oysters
matured more quickly and became larger and better tasting. From this

anecdote it can be seen that a small change in a physical parameter
caused a serendipitous benefit to oysters. The change could just as
easily have been damaging and in fact was damaging to other organisms.
A means of assessing the effects of man-made environmental changes is
required. Such a means must be based on measurement.

In assessing the effect of, say, a sewage plant on some limited
environment such as a bay, there is a considerable problem in deciding
what to measure. It will not suffice to take a single pollutant from
among the many discharged, and correlate in a laboratory its concentration with the death rate of a single organism. Instead, the complexity of shallow-water ecosystems leads to a bewildering number of
possible interactions with biological effects. As an example of the
problems of measurement and some partial solutions, let us look at
the impact of heavy metal ions on a coastal environment.

The first thing necessary to learn is the distribution of the
metal ions. Time, equipment, manpower, and money are limited, so
that continuous measurements, either in space or time, are impossible. Therefore a sampling program must be set up. In order to
make the most efficient use of our resources, the question of how
many samples must be analyzed in order to get a good distributional
map needs to be answered. In part, the answer to that question
will depend on how rapidly concentration changes occur. Ideally,
one would like to take the concentration of heavy metal ions at a
single place and time and use it to calculate the concentration
elsewhere at some other time. To make this calculation would require, at a minimum, a knowledge of how a substance diffuses
throughout the area with time. This knowledge demands a study of
winds, tides, and currents. One means of gaining this information
would be to release fluorescent dye such as rhodamine-b at some
point and to study its diffusion through time with a fluorometer
mounted on a small boat. Strategically placed flowmeters may also
be useful, provided that there is enough current to measure. In
this manner the distribution of heavy metal ions in a limited
environment can be established to some degree.

It is of little value to know the distribution of the heavy
metal ions without knowing what biological effect particular concentrations have. For this problem it is useful to choose some
indicator organism from among those present. One should try to
choose an organism which is affected in some simple and measurable
way by the pollutant and which is present throughout the environment to be studied. Phytoplankton have frequently been found useful in this regard. Changes in gross phytoplankton concentrations
can be studied rapidly by measuring the concentration of chlorophyll-
a using fluorescence (Ryther and Yentsch, 1957; Udenfriend, 1962;
Yentsch and Menzel, 1963; Slovacek and Hannan, 1977). Heavy metal
effects on individual species may be studied by laboratory toxicity
tests. These can indicate whether or not the metal ions present are
damaging the environment.

It would appear that the study is now complete and that the
environmental harm caused by the discharge of heavy metal ions can
be measured in the field. However, this is not so. These studies
alone are incomplete because they isolate the metal ion-phytoplankton
interaction from all the other interactions that can occur with each.

For example, a particular concentration of ion may have little effect on the phytoplankton, but may injure other more sensitive organisms directly. Also it is possible that higher organisms may suffer from feeding on the phytoplankton and concentrating the metal ion in their bodies to a damaging or lethal level. Furthermore, synergistic effects with other chemicals present or with physical parameters may occur. Therefore, distributions for many other factors must also be mapped and correlated with one another. Furthermore, all variables ideally should be measured simultaneously. The amount of data involved is immense.

Is there any way to simplify the collection and reduction of these data? This question has two parts. The first is, can the data reduction process be simplified? The answer to this question involves inspired use of computers. The second is, can the data-gathering process be simplified and speeded up? One possible means of doing this is through the use of remote sensing techniques. Remote sensors mounted in airplanes can cover large areas with rapidity. Satellite sensing can cover an area rapidly and repeatedly every few hours. In both cases many instruments can be operated simultaneously. Satellite survey techniques over land areas have been worked out with great success, using multichannel data acquisition methods. The LANDSAT satellite is the prime example. An attempt has been made to extend LANDSAT technology to the study of the sea in the form of a SEASAT satellite, but unfortunately the first one failed.

The distribution of heavy metal ions in an area over time is a chemical phenomenon. Winds, tides, and currents which distribute these ions are physical phenomena. Yet all of these may be considered biological phenomena as well, because changes in them can be correlated with changes in the biological community. Remote sensing techniques allow rapid measurements of some of these factors, and hence are useful in marine biological studies. However, the use of remote sensing introduces a whole new set of phenomena unrelated directly to biology into biological studies.

Remote sensing techniques include microwaves, infrared, light, sound, and radio waves. This radiation must pass between the target and the sensor. To understand the validity of the information sensed, one must ask how the radiation is influenced by the medium or media through which it passes. For example, infrared is of no use below the sea surface because it is absorbed so quickly by water, and the speed of sound in water varies with the water temperature. Therefore, to apply remote sensing techniques to marine biology, one must also study the physical phenomena involved in the effects of atmosphere and seawater on various forms of radiation. Furthermore, the standard assumption made in using remote sensing is that homogenous or continuously varying conditions prevail throughout the path of the radiation. This assumption is generally unwarranted. Radiation reaching a satellite sensor, for example, will pass through the ozone layer, whose chemical composition differs from the rest of the atmosphere. It may also pass through layers of water vapor in the form of clouds or other aerosols such as dust or particulate pollution from smokestacks. Sea spray creates an aerosol layer of salt and water vapor near the ocean surface. Localized temperature differences in the atmosphere create a slight amount of radiation scattering. These considerations should not deter us from using remote sensing in field work, but instead should be seen as illustrations of the need to understand phenomena remote from marine biology in order to study biological phenomena.

This discussion has shown that the study of the impact of heavy metal ions on organisms present in a coastal environment is not a simple problem. A complete study involves many disciplines. Chemistry is involved in understanding possible synergistic or antagonistic effects of other chemicals present on the metal ions. Biochemistry is involved in studying how the ions affect various organisms and whether or not they are concentrated somewhere in the food web. Physical oceanography is needed to deal with the problems of tides, winds, and currents. Chemical oceanography will be used in identifying the various chemicals present and their distributions. Physics becomes involved whenever forms of energy are used in the study, as in remote sensing. Atmospheric studies are also important to the correct interpretation of remotely sensed data. Statistics are required to compensate for individual differences in the response of organisms to the stresses imposed. Mathematical topography is needed to properly construct distributional maps. Data processing by computers saves time and effort in analyzing the data. This list of disciplines involved is not exhaustive. Clearly a multi-disciplinary team would be the most effective approach. Also shown has been the potential of remote sensing as a means of fast and frequent data-gathering, which is needed in this area.

HOT SPOTS

Earlier in this paper the effects of a change of temperature on oyster growth was mentioned. As was shown, temperature can have a major effect on marine organisms. Two other physical factors of great importance in the sea are light and pressure. Generally speaking, warm water is associated with regions of low pressure and relatively high light levels. However, near the Galapagos Islands certain unique regions, called "hot spots", have developed on the ocean floor. Hot spots have formed when shifting tectonic plates have created hydrothermal rifts, which in turn have caused pockets of warm water to form at great depths (Corless and Ballard, 1977). This geothermal activity also released into the sea large amounts of sulfur-containing compounds, which ultimately are often converted into hydrogen sulfide. This hydrogen sulfide then can be utilized as an energy source by certain bacteria (Doetsch and Cook, 1973; Vishniac, 1974). Therefore, in these hot spots a source of energy other than photosynthesis or the decomposition of detritus is available. Hence, in the microenvironment near the rifts, rich biological communities of bacteria and higher organisms have developed, in contrast to the sparsely populated area of most of the deep sea and ocean floor.

Hot spots offer a great opportunity to study a community of marine organisms in a limited, clearly defined region. Warm water, low light, high sulfide, and high pressure conditions exist simultaneously. The organisms found are generally those found only in

shallower waters. Many questions immediately come to mind, such
as how did the organisms adapt to the pressure; what are the biological effects of the lack of light, of the high levels of sulfur;
and how did the colonization initially occur. This rare combination of phenomena provides a natural laboratory for many disciplines.
Chemistry would be involved in understanding the production of hydrogen sulfide and biochemistry in the utilization of the hydrogen sulfide by bacteria. Perhaps the high pressure present causes unusual
pathways of conversion. Bacteriology could work on that question.
Geology is involved in explaining the formation of the hot spots.
Biology and biophysics could deal with the adaptation of normally
shallow-water organisms to high pressures and darkness. Again, this
is not a complete list of the disciplines that could be applied to
the study of hot spots. The interaction of multi-disciplinary teams
would again be valuable.

One problem in studying hot spots is access. Because they are
in remote areas and deep waters, they are hard to reach. Remote
sensing devices that can give useful information about them have not
been developed yet. Submersibles, cameras lowered from ships, and
other ship-operated instruments can provide, as always, valuable information; but ships can spend limited time on station, and even
then cannot give continuous observations. This situation calls for
the use of automated instruments to provide continuous sampling.
The advantages of instruments that could be dropped into a hot spot
and left unattended for several months collecting data are obvious.
If this information could be sent back to a shore laboratory while
being collected, the advantage to the scientist would be even
greater.

BIOLUMINESCENCE

The study of bioluminescence at the moment is very broad and
loosely organized. Laboratory study techniques on the mechanism of
the luminescent system in various organisms are quite advanced, although many organisms have not yet been studied. However, field
studies lag behind. For the majority of luminous organisms, *why*
they luminesce is not even known. The answer to this question can
sometimes shed light on other phenomena. For example, the occurrence
of a deep sonar scattering layer is well established and its movements have been studied. One of the important organisms forming this
layer is the luminous myctophid fish. A study of the vertical migration of luminous myctophids at sunrise and sunset has indicated
that they try to maintain themselves at a particular level of light
intensity, which they can match with light from their photophores
(Clarke, 1963). Because they can also rotate their photophores so
that the light coming from them is always aimed straight down,
regardless of the inclination of their bodies (Denton et al., 1972),
and the peak wavelength of their emitted light matches that of the
background light coming from above, it is believed that this migra-
tory behavior acts as a defense against predation by making them
essentially invisible to predation from below (Clarke, 1963;
McAllister, 1967). The biological advantages of this behavior to

the fish are evident. The desire to maintain their position near certain isolumes can account for at least some of the movements in the deep-scattering layer. Furthermore, this information can help extend our knowledge of predator-prey relationships, feeding patterns, and defense mechanisms of organisms.

The data that established this information came from several fields - acoustics, optical oceanography, and biology. A multidisciplinary approach was necessary for fruition. Furthermore, advanced sonar techniques were needed to study the fine movements of the deep-scattering layer. To speak of one layer is actually misleading. Several layers are involved, which can be distinguished by using sonar at several different wavelengths. Sonar can provide a gross picture of the layers and can even give the approximate sizes of the organisms in the layers, but cannot establish what organisms are present. For this purpose net tows must be made. Depth sensors coupled with nets that can be opened and closed on command allow vertical distribution to be studied. Modern nets can be arranged with different sized meshes one after another, so that the size distribution in space may be studied as well (Pearcy et al., 1977). Application of sonar and net tows together provides a powerful tool for studies of distribution. Biological studies such as the above example can then provide reasons for such distributions.

The above example showed how the study of bioluminescence interacts with other disciplines to aid in understanding a phenomenon of general interest. Next let us look at a little understood phenomenon of bioluminescence, luminous displays. Luminous displays can take many forms. They can appear as balls of light rising from the depths and exploding at the surface. They can appear as streaks or rotating wheels of light. Particularly impressive and unusual is the display known as the phosphorescent wheel. Occurring almost entirely within or near the Arabian Sea, Andaman Sea, and Gulf of Thailand (Kalle, 1960; Staples, 1966), it consists of stripes of light revolving around a common center in a manner similar to a pinwheel. Various explanations have been suggested as a cause, among them seismic disturbances (Kalle, 1960), electromagnetic anomalies (Hilder, 1962), and optical illusions due to parallax resulting from the position of the viewer (Verploegh, 1968). However, nothing has been proven.

No scientist has ever observed a wheel. No movies or tape recordings have ever been made. The reports that have appeared in the literature have been made by seamen, ship's passengers, or other untrained observers. The scientific papers concerning wheels that have appeared have been compilations of these reports and theories as to their causes. The cause (or causes) of luminescent wheels would be an interesting fact to establish. First, the causative organism (or organisms) should be established. This would require sampling during a wheel display. The depth at which the sample should be taken is unknown. Perhaps sonar could be used to help establish the depth. It would also be possible to use a bathyphotometer to measure light intensity and correlate that with depth, then tow at the depth showing highest intensities. The greatest problem to overcome, though, is response time. Luminous displays, including wheels, are generally localized, highly transitory phenomena. Unless

one should fortuitously be present during a display, there is little
likelihood that it could be reached in time for study. Since there
is lacking a reliable means of notification of a wheel occurrence and
fast response, the next best thing to do would be to analyze the
distribution and occurrence of reported displays in order to decide
where and when to send out a ship to maximize its chances of being on
the spot at an occurrence. Perhaps a low light level photometer
could be mounted in a satellite to increase the number of reported
displays and thereby increase the reliability of the statistics used.

   Knowledge of the organism involved still does not reveal how the
display is triggered. Several theories, already mentioned, have been
advanced. To test the theory of seismic triggering (Kalle, 1960)
would require interaction with geologists measuring seismic disturbances world-wide and calculating the paths and propagation of the
shock waves. To test the theories of electromagnetic triggering
(Hilder, 1962) and optical illusions (Verploegh, 1968) would require
information from the fields of physics, optics, and oceanography. It
is possible that none of these theories is correct and that the trigger is something in the chemical environment in the ocean or in the
biology of the organism itself. Hence, a team of people from many
disciplines would be most useful in performing this study.

   The study of luminescent wheels is connected with another problem
in bioluminescence. That problem is the phenomenon of its distribution. It would be possible to measure the distribution of light in
the sea by studying the distribution of luminescent organisms and
the conditions under which they luminesce. However, to carry out
such a study would require an enormous amount of time and extensive
knowledge of taxonomy and physiology since bioluminescence occurs in
representatives of practically every major marine taxonomic division
up through fishes (Tett and Kelly, 1973). Furthermore, it would be
incomplete in that it would fail to include species not known to be
luminescent. A more direct approach would be to actually record
bioluminescence using such instruments as bathyphotometers and
low light level TV cameras. Such instruments, like sonar, cannot
tell what organism is present, but can yield measurements of some
property of interest, in this case light production. Furthermore,
the possibility exists of using the pattern of the light flash to
identify the producing species, and so learn their distributions.
Just as firefly species can be identified by the pattern of their
flashes, so may be some species of marine organisms. Spectral
analysis of the flashes and studies of their rise and decay characteristics may also help to identify species. Research in these
areas is not sufficiently advanced to make any certain statements.

   The mapping of the distribution of oceanic bioluminescence is
still a formidable task. Sightings and measurements of bioluminescence have been reported from all over the world at all times
of year (Turner, 1965, 1966; Staples, 1966; Lynch, 1978). In addition, it has been detected to depths of up to 4000 meters, as deep
as bathyphotometers have been lowered (Clarke and Backus, 1956;
Clarke and Hubbard, 1959; Clarke and Bakcus, 1964). Because of the
large number of bioluminescent organisms and their widespread occurrence over area, depth, and time, the use of stations is too slow.
Low light level TV cameras can be flown in aircraft or satellites,

but would only be capable of detecting light emitted spontaneously
or from some form of natural stimulation. Organisms capable of
luminescing but not actually doing so would go undetected unless
some means of remote stimulation can be developed. Furthermore,
airborne instruments will only detect light that reaches the surface,
and present instruments are incapable of determining the depth at
which that light was generated. The development of automated photometers that could be towed behind moving ships or left unattended
and floating at predetermined depths would also be useful. The work
of solving these measurement difficulties requires a knowledge of
physics, oceanography, optics, and electronics, at least, as well as
biology. Many different disciplines must work together to achieve a
solution.

SUMMARY

As long as it can be measured, any process occurring in the
ocean and dealing with marine biology can be considered a phenomenon.
Because of the interactions involved, non-biological phenomena may
also be considered biological phenomena. Therefore, the study of
marine biological phenomena requires a wide range of disciplines, and
a multidisciplinary team approach is best for the study of large-
scale and complex problems. This alone, though, is not enough.
Means of making rapid measurements over large areas and continuous
long-term measurements at selected sites are needed to achieve understanding of the dynamics and interactions of organisms with one another and their environment. Finally, improved methods of data analysis
and storage must be developed to handle the large quantities of data
that will be generated by these approaches.

REFERENCES

Clarke, G. L. and R. H. Backus, 1956. Measurements of light pene-
    tration in relation to vertical migration and records of lumi-
    nescence of deep-sea animals. *Deep-Sea Res*. 4: 1-14.
Clarke, G. L. and R. H. Backus, 1964. Interrelations between the
    vertical migration of deep scattering layers, bioluminescence,
    and changes in daylight in the sea. *Bull. Inst. Oceanog. Monaco*
    64: No. 1318.
Clarke, G. L. and C. J. Hubbard, 1959. Quantitative records of the
    luminescent flashing of oceanic animals at great depths. *Limnol.
    Oceanogr*. 4: 163-180.
Clarke, W. D., 1963. Function of bioluminescence in mesopelagic
    organisms. *Nature* (London) 198: 1244-1246.
Corliss, J. B. and R. D. Ballard, 1977. Oases of life in the cold
    abyss. *Nat. Geog. Mag*. 152: 440-453.
Denton, E. J., J. B. Gilpin-Brown, and P. G. Wright, 1972. The
    angular distribution of the light produced by some mesopelagic
    fish in relation to their camouflage. *Proc. Roy. Soc. Lond*.
    B182: 145-158.
Doetsch, R. N. and T. M. Cook, 1973. Some aspects of the biochemistry

of chemolithotrophy. In: *Introduction to Bacteria and Their Ecobiology*, pp. 256–272. University Park Press, Baltimore, Maryland.

Hilder, B., 1962. Marine phosphorescence and magnetism. *Navigation (J. Aust. Inst. Navigation)* 1: 43–60.

Kalle, K., 1960. Die rätselhafte und „unheimliche" naturerscheinung des „explodierenden" und des „rotierenden" meeresleuchtens – eine folge lokaler seebeben? *Deutsche Hydrog. Z.* 13: 49–77.

Lynch, R. V., 1978. The occurrence and distribution of surface bioluminescence in the oceans during 1966 through 1977. *NRL Report 8210.* 45 pp.

McAllister, D. E., 1967. The significance of ventral bioluminescence in fishes. *J. Fish. Res. Bd. Can.* 24: 537–554.

Pearcy, W. G., E. E. Krygier, R. Mesecar, and F. Ramsey, 1977. Vertical distribution and migration of oceanic micronekton off Oregon. *Deep-Sea Res.* 24: 223–245.

Ryther, J. H. and C. S. Yentsch, 1957. The estimation of phytoplankton production in the ocean from chlorophyll and light data. *Limnol. Oceanogr.* 2: 281–286.

Slovacek, R. E. and P. J. Hannan, 1977. In vivo fluorescence determinations of phytoplankton chlorophyll a. *Limnol. Oceanogr.* 22: 919–925.

Staples, R. F., 1966. The distribution and characteristics of surface bioluminescence in the oceans. *NOO Rept. TR-184.* 48 pp.

Tett, P. B. and M. G. Kelly, 1973. Marine bioluminescence. *Oceanogr. Mar. Biol. Ann. Rev.* 11: 89–173.

Turner, R. J., 1965. Notes on the nature and occurrence of marine bioluminescent phenomena. *NIO Rept. B4.* 46 pp.

Turner, R. J., 1966. Marine bioluminescence. *Mar. Obs.* 36: 20–29.

Udenfriend, S., 1962. *Fluorescence Assay in Biology and Medicine.* Academic Press, New York. 517 pp.

Verploegh, G., 1968. The phosphorescent wheel. *Deutsche Hydrog. Z.* 21: 153–162.

Vishniac, W. V., 1974. Organisms metabolizing sulfur and sulfur compounds. Genus 1. *Thiobacillus.* In: *Bergey's Manual of Determinative Bacteriology*, Eighth Edition, (R. E. Buchanan and N. E. Gibbons, eds.), pp. 456–461. Williams and Wilkins Co., Baltimore, Maryland.

Yentsch, C. S. and D. W. Menzel, 1963. A method for the determination of phytoplankton chlorophyll and phaeophytin by fluorescence. *Deep-Sea Res.* 10: 221–231.

Description of Spatial and Temporal Patterns of Abundance in Open-Ocean Zooplankton: Where Are We Now and Where Do We Go from Here?

Eric Shulenberger

ABSTRACT: One of the most basic problems in a
science is to describe pattern in one's system
so that one may ask what generates and maintains
that pattern. Biological oceanographers are unable to describe effectively patterns of zooplankton distributions (in time, space, and abundance) on scales smaller than a few kilometers.
No tools exist to provide reasonable descriptions
of pattern in three dimensions. The traditional
taxonomic species concept has fueled the almost
exclusive use of nets to sample zooplankton. That
concept and the net appear incompatible with our
desire and need to describe small scale pattern.
For the near-term future, we must develop new
"taxonomies" that will permit new, fast data
collection methodologies which in turn will allow
description of pattern in terms of the new taxa
(="categories"). Ultimately these new devices'
information, taken within new taxonomies, must
mesh with the old taxonomy's data.

INTRODUCTION

Some of the most basic questions in any science are (a) what are the patterns (i.e., what exists in nature, and how the phenomena are distributed in time and space), and (b) why are the patterns as we see them instead of some other way (i.e., what generates and/or controls those patterns?). These amount to elucidation of a system's structure (=description) and functioning (which implies understanding).

I will limit this discussion to one particular, basic problem in one area of oceanic ecology: the description of patterns of spatial and temporal distributions of pelagic organisms, especially epipelagic zooplankton. This discussion is set in the most general of terms, and I will make broad, sweeping statements: to most of these statements there will be particular exceptions. Rather than attempt to cover the literature on problems of either sampling or describing the biology of the ocean, I hope to point out, particularly to the technically-oriented non-biologist, some fundamental problems which biological oceanographers are encountering in their attempts to describe pattern in nature. I choose not to include among these problems the obvious physical factors of wind, ship motion, positioning, inability to track "a parcel" of water, and so forth, nor will I discuss animals' avoidance of sampling systems. Instead I will discuss how biological oceanographers' world view has been set by their equipment and how we must change that view so that we may progress.

This discussion will include several aspects:

a. What are the questions?
b. What have we accomplished so far?
c. What can we do at present?
d. What must we do in the future?
e. What are some possible approaches?

THE QUESTIONS

The questions are simple: how are organisms distributed in time and space, and how do the distribution patterns change with time, location, or other environmental parameters? These questions are posed in the context of marine ecology, particularly open-ocean ecology. Ecology is defined as the study of the interaction of organisms with one another and with their environment. The most important questions about nature, in ecology as in other fields, are those beginning with "why does.......?". The answers to such questions imply that we understand the mechanisms producing the observed patterns. To ask "why?", one must first have *observed* the patterns one seeks to explain: the first question (i.e., what?) must be answered before the second (why?) can be posed (Darwin, 1859; Elton, 1927; Fager, 1963; Elton, 1966; Whittaker, 1975).

Scientists categorize the things they observe. Any system of categorization is a "taxonomy"; the various categories are "taxa" (singular = taxon). The data which constitute the basis of ecological science are 1) what taxa occur in an area, 2) how many individuals there are of each taxon, 3) how those organisms are distributed, and 4) data on environmental parameters which are believed to be important to those organisms. Once these data exist, one may ask the questions beginning with "Why does...?" which are the ecologist's ultimate goal.

Without such information, it is difficult, if not impossible, to even begin consideration of the most basic sorts of ecological questions (Darwin, 1859; Elton, 1927, 1966; Whittaker, 1975):

1. How can we best classify communities?
2. How are species' populations distributed in relation to one another and to communities along an environmental gradient?
3. How are the kinds of communities in an area related to patterns of more than one environmental gradient?
4. How are we to interpret world-wide relationships of community structure to environment?

Throughout much of the open ocean, the description of spatial and temporal patterns of distribution of organisms (the "physiognomy" of the communities therein: Whittaker, 1975) either has not been done or is in an extremely primitive state. We do not know how species' populations are distributed in space either absolutely or relative to one another, or relative to environmental gradients. In short, much of the basic data required for ecological thought do not exist.

Classifying undescribed communities is impossible. We do not know if there are "kinds" of epipelagic communities in an area, and cannot ask if these kinds might be related to patterns of environmental gradients. Only in the relationship of community structure (as represented simply by species lists) to world-scale environmental patterns do we know much about pattern in the open oceans (large scale zoogeography: Ekman, 1953; Brinton, 1962; Fager and McGowan, 1963; Venrick, 1971; McGowan, 1974; Backus et al., 1977).

In addition, there has recently been developed a large literature on ecological theory (e.g., MacArthur and Wilson, 1967; Levins, 1968; May, 1973; Cody, 1974 and contained references). Someday, somehow, this theorizing must be checked against reality in the world's largest habitat, the open ocean. The basic descriptive data under discussion are absolutely vital to such a check, but do not exist.

WHERE ARE WE NOW?

Any science's world view is a function of the tools with which it investigates that world. Since its inception, biological oceanography's window upon the world has been primarily the net, a simple cheap, robust, and remarkably efficient investigative tool. However, as does any simple tool, the net has placed some very subtle constraints upon how we can view "our" world; it has, in fact, limited the questions which we can reasonably pose and answer about nature.

Nets are basically devices intended to ask questions of a
"traditional taxonomic" bent. By this I mean that they are oriented
towards questions involving data couched in a traditional taxonomy
which requires identification of organisms to the species level;
this in turn requires that a physical sample of the organisms be in
hand. Nets will provide physical samples of some kinds of animals,
which allow certain types of questions to be asked; indeed, they
provide the only known method of getting some interesting and important forms of data. (Examples: what species live in a given geographical location? What are the animals' gut contents? Are the
females carrying eggs? Are the individuals healthy? Are they para-
sitized?) In addition, nets have served admirably to detect certain
types of pattern, notably large scale zoogeographical patterns (*op.
cit*). Nets are, however, singularly inappropriate for small scale
(≤ 10 km) descriptions of pattern. With a few modern exceptions, nets
were not designed, nor have they been operated, to either detect
spatial patterns or to provide information on the interactions of
those patterns with the environment on these smaller scales.

Nets suffer primarily from problems involving size scales, and
these problems limit one's ultimate ability to describe pattern
using nets. There is a strong interaction between several factors:
a) population density of oceanic organisms (usually low); b) the
numbers of organisms (usually high) needed per sample to enable one
to state with confidence anything except a species list; c) the resources of money, personnel, and time needed to obtain and adequately
analyze those samples; and d) the distributions and behaviors of the
organisms themselves.

We know that organisms are distributed in patches (i.e., neither
randomly nor evenly distributed) on every time and space scale yet
investigated, from "global/geological" down to "individual/seconds"
(Haury et al., 1978). Some aspects of this patchiness are caused by
physical environmental influences on various scales (advection, concentration in vortices, aggregation on density surfaces, internal
waves, and so forth).

One example will show fairly clearly how these factors interact
to frustrate attempts at pattern description using nets. Because of
the multitude of contributing factors, one optimally would like to
try (initially) to detect and describe pattern in the "simplest"
environment possible. By judicious site selection, one might reduce
to a minimum the probably contribution of the physical environment
to patchiness. The "quietest" epipelagic environment is apt to be
in the central gyres of the oceans, which have minimal horizontal
and vertical advection, great lateral homogeneity of physical oceanographic parameters, relatively calm weather, and no problems with
coastal, human, or bottom topographical influences (Sverdrup et al.,
1942). Unfortunately, such areas have other problems: they are
highly oligotrophic (i.e., nutrient poor) and therefore quite low in
zooplankton biomass or "standing stock". This means that population
densities of organisms are usually low, often orders of magnitude
lower than in more coastal waters. It could also mean that the individual organisms are smaller than in coastal waters: both factors
seem to be important.

In oceanic biology, many parameters having to do with numbers of organisms are highly variable. As a result, large numbers of organisms are needed for relatively simple descriptive tasks. A very basic question is "which taxa are most abundant, and what is the rank order of numerical abundance of the top few taxa?" In a system containing many hundreds of species, one may easily require several thousands of individuals just to obtain accurate rankings one through eight (Wiebe, 1971). Because of low biomass and population densities, obtaining such a sample requires filtering a large volume of water. This requires either a large diameter net or a long towing time and distance. Large diameter nets destroy vertical distributional patterns and are difficult or impossible to handle aboard ship. Long net tows integrate horizontal patchiness.

Nets are basically unable to resolve pattern on a horizontal scale smaller than twice tow length (Steiglitz, 1974). They cannot resolve vertical pattern on scales smaller than twice one's ability to control the net's sampling depth, even assuming the use of opening/closing nets and no internal wave fields, etc. They cannot resolve time scales smaller than twice tow duration (in reality, twice "tow-plus-turnabout" time: see Steiglitz, 1974). This lack of resolving power is independent of the problem of the time involved in microscopically processing each sample to turn it into distributional data.

In addition, nets suffer from severe aliasing problems: they may take a very simple vertical distributional pattern (e.g., thin layers) and present it falsely as apparent horizontal pattern. An even distribution of organisms relative to an environmental parameter which undergoes spatial distortion (e.g., density interfaces distorted by internal wave fields), when sampled by nets, may present a confused picture of apparent spatial pattern. At the very least, a 1000 m long net tow certainly includes as "occurring together" animals from as much as 1000 m apart, which are unlikely to ever have encountered one another outside of the net.

Most net sampling schemes have not involved replicate samples despite the knowledge that such sampling is needed. One of the most highly replicated net sampling schemes yet reported is that of McGowan and co-workers (Scripps Institution of Oceanography, 1974) in the North Pacific central gyre, a location expected to have minimal "physical" contributions to animal distributions. Shulenberger (1978) analyzed 79 of those replicate opening/closing net samples from six depth intervals. Each net tow covered about 1000 m horizontally, about 25 m vertically, and took about 20 min of towing time (plus 1 hr or more of turnabout time). Samples contained from one to 678 individuals (lumped counts for the 83 hyperiid amphipod species counted) and from one to 38 species. Each sample also contained tens of thousands of individuals distributed among hundreds of other species. For eight of the ten most abundant hyperiid species, the author was unable to determine "preferred" depth intervals, unable to discern day/night differences in mean population densities at most depths, and unable to show day/night differences in distributional patterns. Figure 1 shows the range of population densities observed, under as constant and as well-replicated a set of sampling conditions as is ever likely to be obtained (on small

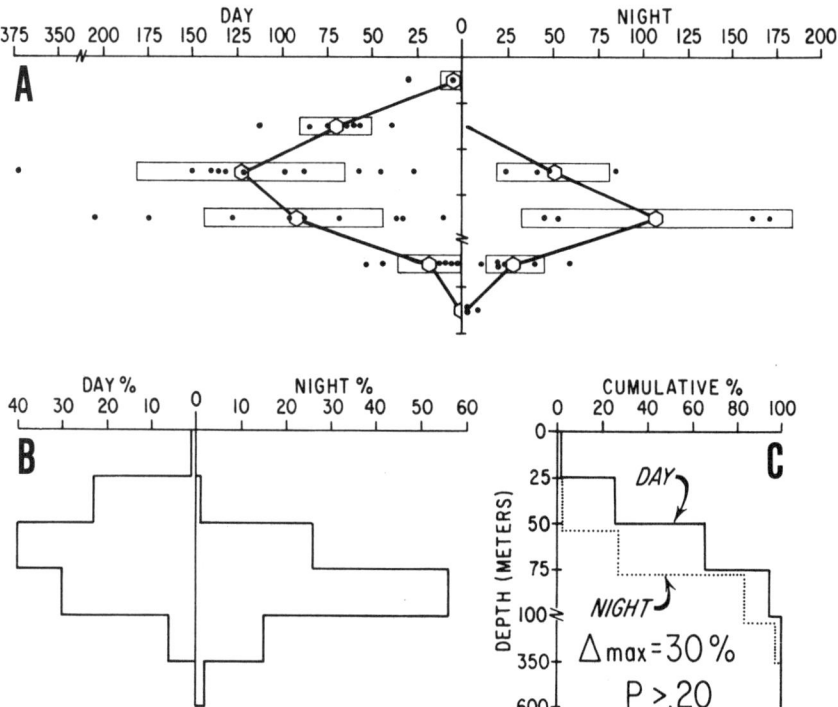

FIGURE 1: Population densities of *Primno
latreillei*, the most abundant hyperiid amphipod in the North Pacific central gyre.
"ROA" = rank order of numerical abundance.
(A). Values are numbers of individuals per
400 m^3 net tow, using opening, closing
"BONGO" nets with 505 μm mesh. All samples
were taken between a pair of parachute drogues
set at 28°N 155°W. Number of dots may
be smaller than number of samples at a time-
depth due to crowding on the graph. Note
scale breaks in both axes. Hexagons = mean
value at a time-depth. Dots = actual counts.
Boxes = ±95% confidence limits for the mean.
(B). Percent of total individuals (i.e.,
summed across all depths within a time of
day) captured in each depth interval, standardized to number of replicates at each time-
depth. (C). Comparison of day/night curves
from B using the Kolmogorov-Smirnov test for
similarity of curves. Null hypothesis: both
curves could be randomly selected representatives of the same underlying universe; accepted, P >.20. See Shulenberger, 1978 for
additional explanation and discussion.

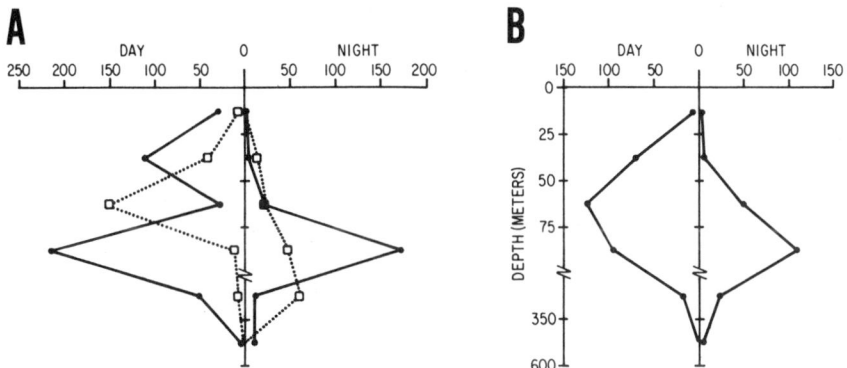

FIGURE 2: (A). Two day and two night vertical
distributions for *Primno latreillei* generated
by randomly selecting one sample per depth interval from data shown in Figure 1: represents
the variable results one might obtain with un-
replicated sampling. (B). Mean abundance versus
depth curve: compare with (A).

scales) using nets. The between-sample variation is enormous,
even in this physically constant or unstructured environment.

Certainly the animals are not uniformly distributed on these
scales (Fig. 1), but just how they *are* distributed is essentially
an open question. For instance, if one randomly took one sample
per depth interval from the values in Figure 1, he could get vertical
distributions as different as those shown in Figure 2. These could
lead to wildly differing conclusions about day/night distributional
patterns and so forth (Fig. 3).

These 79 samples required the services of a large research
vessel for one month ($200,000), many persons at sea for that month
to handle nets ($50,000?), and took many man-years (at $? per) to
analyze just this one small taxonomic group. They answered very
few questions and yielded no effective or convincing description
of the pattern of distribution of the organisms. The rest of the
world ocean is, in general, orders of magnitude "messier".

By definition, ecology includes the interaction of the organisms with their environment. The environmental half of pelagic
ecology has been badly handled. The major problem is lack of
simultaneous collection, *at the same time and place as the sample,*
of environmental information *believed to be relevant to the organisms.* In their environment, various properties of possible importance to zooplankton are not uniformly distributed vertically, horizontally, or temporally. These include light, salinity, temperature,
density, dissolved materials such as oxygen, food supplies (often
represented by chlorophyll measurements, which are a sloppy measure
of plant biomass), and potential or actual predators. A fundamental
but oft-ignored tenet of ecology is that animals are adapted to
where they live. Therefore, their distributions and behavior

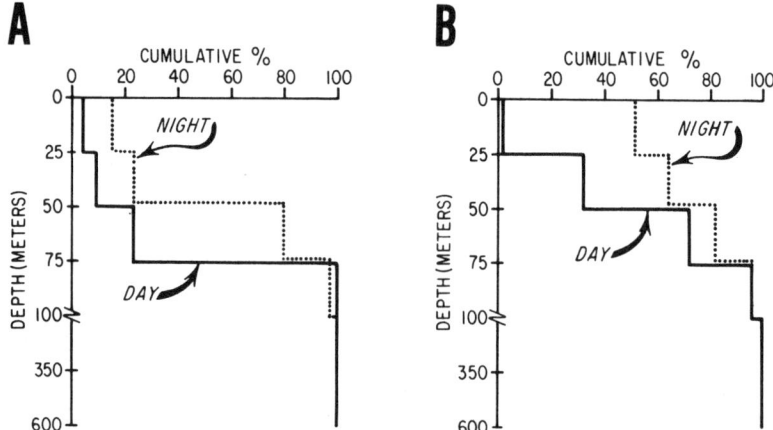

FIGURE 3: Two comparisons of day distribution
versus night distribution, using the non-repli-
cated sampling curves generated in Figure 2.
Note radical difference in apparent vertical
distributions and apparent diurnal migration
patterns, depending on the "luck of the draw".

must be expected *a priori* to reflect those parameters in some way.
One parameter unlikely to be of actual interest *per se* to a zooplankter is depth. Depth is probably of interest to either scientist
of zooplankter only as it covaries with parameters more pertinent to
zooplankters: those covariances are quite weak and therefore even
accurately measured depth information is unlikely to be very useful.
Further, depth is frequently measured only indirectly (e.g., by wire
angle), and is often reported in exact terms when either actual depths
were not known or when depth control during sampling was undocumented.
Yet of all the parameters which *could* be measured, often depth is the
only one which *is* measured. In addition, experience with modern nets
that telemeter their actual depth back to the deck (where a winch
operator can "fly" the net in real time) has proven it to be nearly
impossible to maintain a depth of tow more accurately than about ±5
to 8 m in the upper 200 m.
      Collection of nutrient, chlorophyll, and physical oceanographic
information in the general vicinity of one's net tows at some other
time of day (or week, month, year?) is inadequate in the face of
known variability in all these parameters. Neither does the use of
long-term mean data suffice. For most variables, an average value
is meaningless to an organism: the important parameters are
variance and range of values. This is clearly shown by two examples. 1) Laboratory attempts to rear zooplankton while feeding them
the mean chlorophyll content of their natural environment inevitably
result in the animals starving to death (Mullin and Brooks, 1976;
Dagg, 1977). Obviously they are not encountering in nature what we
measure as a "mean value". 2) Crudely put, while the mean annual
temperature of the Kansas plains may be 45°F, one would not survive
exposure to the annual *variation* in temperature. The picture one
gets of the distribution of animals if one has several replicate
samples and uses mean values (per Fig. 4) is very misrepresentative

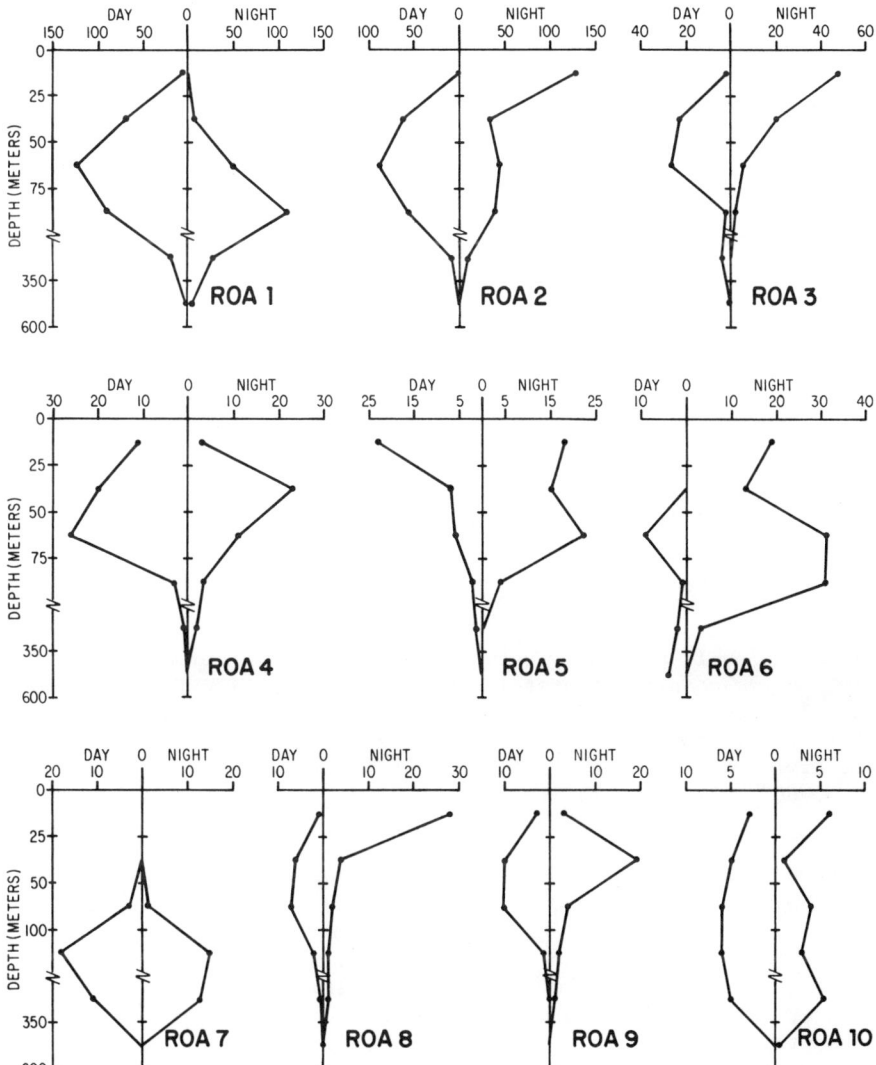

FIGURE 4: Mean value versus depth curves for
the ten most abundant hyperiid amphipod species
in the North Pacific central gyre. "ROA" = rank
order of numerical abundance. Abscissae =
counts per 400 m³ sample: note various scales.
Sampling variability for most species is equal
to (or exceeds) that for ROA 1 (Fig. 1). Despite apparent differences in depth distribution
between species and between day and night within
a species, almost no such comparisons are statistically significant at the 0.05 level. See
Shulenberger, 1978 for species names and measures
of variability for all ten species.

of what the animals are encountering: they probably encounter at
least the variability shown in Figure 1.

Another major problem is partitioning the observed variability
into "signal" versus "noise". Noise and variability are not the
same. I take "noise" to be that portion of overall variability
which remains unexplained and appears to be random (paralleling
"error" in statistics). It may be artifactual, i.e., due to inaccuracies in our measurement techniques, or real. We do not yet
know if the apparently high variability in distributions is "noise"
(unexplainable random variation) or signal that has as yet proven
too complex for us to unravel. If we seek to explain rather than
merely observe this variability it becomes critical that we partition
the variation; this we cannot yet do.

A second major contributor to our problem is the extremely low
rate of data acquisition possible when one datum consists of the
traditional taxonomic analysis (usually by microscope) of the
materials collected by a net in one tow. The present number of
fully analyzed, reasonably well documented zooplankton samples from
the entire world ocean is probably 10^5, plus or, more likely, minus
an order of magnitude. Low rates of data acquisition in an inherently
noisy or highly variable system are perhaps worse than no acquisition
at all, since descriptions derived therefrom are practically guaranteed to be wrong. Low sample rates are acceptable only because *some*
information is generally better than *no* information at all.

A few modern net systems have been designed with some of these
problems in mind. Various opening/closing nets now exist, although
they are in anything but universal service. A few of these provide
multiple-sampling capability, taking samples either in parallel
("BONGO" nets, Scripps Institution of Oceanography (1966) or in   of
series (e.g., the "MOCNESS" net of Wiebe, 1976; "Longhurst-Hardy
Plankton Recorder", Longhurst et al., 1966). These multiple nets
take samples in series at various adjustable tow lengths ranging
from about 50^+ m for the MOCNESS to $15-30^+$ m for the Plankton Recorder. Of the multiple sampling nets, only one or two take any
data other than depth and temperature, and only one or two send
any real-time information to the net operators.

Finally, all these devices use *nets*, and ultimately encounter
the problem of low data acquisition rate, the main bottleneck being
the microscopic analysis of the resulting physical samples. What
one really needs in a noisy/variable system is a tool that will
collect enormous numbers of data points, perhaps hundreds of
millions. Even if those data are crude, with a very large sample
size one may extract trends, correlations, and statistics of
interest and of use. Physical oceanographers' views of the world
are changing radically with the advent of winch-operated conductivity-temperature-depth instruments (the "CTD"), the towed-body CTD,
the expendable bathythermograph ("XBT") and other tools meeting
this criterion. Physical oceanographers have, fortunately, been
able to combine precision with high data rate: if possible, biological sensing systems should strive for the same combination.
The only system on the horizon which may approach such data acquisition rates for biological questions are sonar systems. Certainly
net-and-microscope sampling will never do so.

WHERE DO WE GO FROM HERE?

What then are we to do if we wish to describe pattern in the
zooplankton so that we may ultimately explain or understand it?
Our scientific paradigm (Kuhn, 1968) has at its very core the
traditional species concept. It is this which drives our need to
bring back actual samples of organisms for detailed taxonomic
analysis. I suggest that the description of pattern among zooplankton is not tractable in the traditional taxonomy, and that
nets, vital though they are for other questions, are fundamentally
inappropriate for attacking questions of small scale distributions
of zooplankton. By changing our paradigm to include other sorts
of taxonomies, could we not ask (and perhaps answer) the basic
question much more effectively? A new taxonomy, and its tools,
seem to be needed.

Breaks with traditional taxonomy have been necessary in other
fields. Phytoplankton and bacterioplankton ecologists measure properties which are categorized into non-species taxa: electron transport activity, chlorophyll, primary productivity, ATP, dissolved organic carbon, particulate organic carbon, C:P ratios, and the like.
Even zooplankton ecologists use "displacement volume" on occasion,
as well as "functional groupings" (e.g., "herbivores", "filter
feeders", "small slow particle catchers" etc.). These are all non-
traditional taxonomies, and their users seem to derive useful insights from them.

If we may assume a non-traditional taxonomy, what questions
should we ask? Begin with very simple ones. Call the categories
X_1, X_2, X_3...X_n. Let us ask:

1. How is category X_i distributed? (Evenly? Randomly?)
2. What constitutes a "patch"? Does it have sharp or fuzzy
   edges?
3. What is the size-frequency distribution of patches of X_i?
4. What is the internal distribution of X_i in a patch? Is
   population density within a patch Gaussian? Poisson?
   Other?
5. Are there hierarchies of patchiness (like clusters of
   galaxies)?
6. How are patches of X_i distributed in time and space?
7. What are the diurnal, seasonal, annual, and longer
   scale temporal behaviors of patches of X_i?
8. How are the distributions of X_i related to those of X_j?
   Mutually exclusive? Coincident? Randomly overlapping?
   Lots of small patches of X_i within a large patch of X_j?
9. How are patches of X_i related to environmental parameters?
   How do all the patch parameters correspond to physical
   oceanographic variables? Do the physics of the environment seem to be controlling the biodistributions, and if
   so on what scales and dependent upon what physical parameters?........and so forth almost *ad infinitum.*

Selection of a taxonomy is a nontrivial problem, since it sets
one's world view and defines one's suite of acceptable questions,

and these in turn force one's instrument development.

Examples of "bad" (in the sense of "not very useful") taxonomies are easy to generate: one such might be color (although color
can certainly be very useful at times). "Good" taxonomies should
be clearly tied to some functional aspect of the system being investigated, should be based on as much ecological information as is
available, and should be likely to provide information both unambiguous and interpretable. Ideally they should ultimately mesh well
with traditional taxonomies so that an integration of old and new
may occur.

A very worthwhile approach is to try to set one's self in the
position of the organism one is investigating, and to try to determine
which aspects of the organism's environment are of most critical
importance. A zooplankter's problems include finding food, finding
a mate, avoiding being eaten, and so forth. These are some of the
ecological problems which help generate distributional patterns, and
which should underlie any new taxonomy.

One such ecological concern is feeding. Most oceanic organisms
seem to feed primarily upon organisms smaller than themselves. In
addition, most zooplankton and small fishes appear to be feeding
"generalists", i.e., they will eat anything that comes along, ir-
respective of taxonomy. This behavior is appropriate for an organism living in a food-poor environment. To a zooplankter, therefore,
a most important question is the size of its nearest neighbor. If
that neighbor is larger than the zooplankter, the neighbor is probably a predator upon the zooplankter; if smaller, the neighbor is
the zooplankter's prey. Only after it has fed is the zooplankter
apt to be concerned with the species identity of a neighbor, perhaps for mating purposes. Size, therefore, seems to be a convenient,
defensibly "good" taxonomy. A thorough description of the spatial
and temporal distributions of organisms by size alone could be very
useful and ecologically informative.

Having set up a new taxonomy, practical considerations then
arise: how can one measure accurately, with a high data acquisition
rate, the distributions of various categories of organisms, especially while obtaining relevant environmental measurements in precisely the location of the measured organisms? Is there a single
system which we can devise that will handle all known or expected
characteristics of the organisms we are investigating? The answer is
clearly "no". The taxonomy of size, for instance, is being pursued,
in particular by acousticians (see Holliday, this volume). However,
their measurements are plagued with inverse problems (e.g., did the
echo from volume XYZ come from three large or 18 smaller organisms?),
problems of simultaneous acquisition of environmental information,
and problems with the large numbers of organisms that are nearly
identical to seawater in compressibility.

Other types of systems and other taxonomies are obviously needed
as well, and they must all complement and overlap one another. For
instance, another large, probably important group of zooplankton includes the true jellyfishes, ctenophores, salps, etc. These animals
are gelatinous, transparent, and have compressibilities and indices
of refraction almost identical to those of water; they are in general
so physically delicate that they get strained through the meshes of

our nets. At some times and in some places they can be the dominant
zooplankton. Their abundance in most of the ocean is unknown and can
certainly be an extremely important part of the overall functioning
of a pelagic ecosystem, particularly in terms of energy flow (another
"taxonomy"). We cannot effectively investigate their distributions
or abundances with any presently available techniques.

Stepping away from zooplankton for a moment, we have no way whatever to investigate effectively larger, faster organisms, particularly
those which our nets do not catch. Larger nets do a better job than
present small nets, but suffer increasingly from the same problems as
smaller nets. For large animals, we need to answer questions as
simple as "What is out there?". There are still large animals to
be discovered, an example being the recent accidental discovery of
the 3.5 m-long shark "Megamouth" (a new family, not merely a new
species, of shark: Taylor, 1977). We need methods of exploring
for these large animals.

Of more pressing importance is our inability to assess population densities and distributions of animals such as squid. Squid
are common and information on them is critical to our scientific
understanding of marine ecosystems as well as to our intelligent
economic exploitation or management of those stocks. We have no
techniques for obtaining believable numbers for these sorts of
animals.

No taxonomies or systems presently looming on the horizon, save
perhaps acoustics, will provide a truly three-dimensional look at
distributional pattern; all either take sections (two-dimensional)
or sample long "holes" in the ocean (nets, acoustics, video systems).
The third and fourth dimensions must be built up via multiple passes
and are therefore confounded with one another and with outside factors like our ability to position a ship or follow a "piece of water".

Given a taxonomy, at what spatial scales should one start?
Small scales are best, because one can lump small scale data to
investigate larger scales, whereas doing the inverse is extremely
difficult or impossible. What is "small"? Perhaps the proper
scale for some questions is the daily ambit (i.e., that piece of
the world through which the animal wanders) of a typical organism
selected from those one is investigating. Such an ambit may be
surprisingly large (100s of m vertically and of unknown dimensions
horizontally) for rather small (3-5 mm long) zooplankters. Zooplankton ultimately interact with one another on a one-to-one
basis (i.e., individuals, not larger units, mate, feed, or die):
perhaps this is an appropriate scale to examine if possible. Chlorophyll, a measure of potential food for herbivores, is distributed
in layers as thin as one meter or less, and most of the density changes
at the pycnocline occurs over short distances; any system must at
least resolve those scales.

In addition, biology faces some restraints a bit peculiar to
it. Biology is not a heavily funded field, and it is by and large
very conservative. The best chances for success lie with relatively
inexpensive techniques or gear, probably involving designs built
about new configurations or combinations of existing (although state-
of-the-art) technologies. Truly massive, original attempts to find
new ways to investigate the biology of the oceans are just not very

likely to succeed financially, if for no other reason. The systems
must be such that they can become common tools once developed. Bio-·
logical parameters in the ocean are so variable that unique and/or
high cost systems are not appropriate. We need many of the devices
operating in many locations, not a few more very expensive, one-copy
devices with long lists of potential users (e.g., the submersible
ALVIN).

POSSIBLE APPROACHES

   Where does one look for candidate devices or taxonomies?
Appropriate areas to consider are fields which are plagued with
noise problems. In many cases, "noise" in one field may be "information" in another. Conductivity-Temperature-Depth (CTD) units
record glitches that may be organisms. Physical oceanographers
using instruments with sensitive accelerometers record transient
signal spikes believed to be organisms hitting the sensors. Underwater optical work is plagued by scattering and absorption problems.
Sound bounces off compressibility discontinuities, most of which are
organisms. Optical densitometers have their records speckled with
interference rings believed to be caused by particles. All these
"noise" problems may really be important biological information not
yet perceived as such.
   A brief listing of the requirements for any new system will be
useful.

   1. One must begin with questions important to the organisms
      being investigated.
   2. One will almost certainly have to operate in a "new
      taxonomy".
   3. This new taxonomy must be built upon biologically reasonable
      assumptions.
   4. It must yield interpretable, meaningful biological data.
   5. It must sample very rapidly, in order to counter the variabil·
      ity in the ocean and allow us to partition it into signal
      and noise components.
   6. The data must be amenable to automatic data processing *with-
      out* human manipulation of each datum individually.
   7. It must collect environmental information fitted in partic-
      ular to criterion no. 1 (above).
   8. We must eventually be able to integrate the "new taxonomy"
      data with that from other systems (e.g., traditional taxo-
      nomic analyses).
   9. It should be able to (eventually) provide 3- or 4-dimen-
      sional data.
   10. It should be inexpensive and robust, so that it may become
      a common tool.

   Of these criteria, only numbers 9 and 10 are very difficult.
Three dimensional capabilities are likely to be far in the future,
and significant improvements in 2-D capabilities are badly needed.
"Inexpensive" is a relative term; enormous amounts of time and

money have already been expended on these problems, to little
avail. Against that sum, the probable cost of any one such new
system may be relatively insignificant. Certainly, new ideas and
attempts to implement them are badly needed if we are to make progress on the two basic questions: what?, followed by why?

SUMMARY

1. *What are the questions?* What are the patterns of distribution of zooplankton and other organisms in time and
   space, and what drives or maintains those patterns?
2. *What have we accomplished so far?* We have a reasonably
   good knowledge of global zoogeography. We have tried
   numerous variations on net-and-microscope methods of
   describing smaller-scale distributions with little
   success. "Species" plus "nets" seem to be fundamentally
   improper tools with which to answer many questions about
   spatial distributions.
3. *What can we do at present?* Some new systems under development, especially acoustical systems, appear likely to provide much-improved descriptive capabilities on small
   scales. Improved net systems are available but are hin-
   dered by dependence on the human eye and microscopic analysis of samples, plus extremely low data acquisition
   rates.
4. *What must we do in the future?* We need to break out of
   the "species paradigm", establish new taxonomies, and
   develop the tools for them. These tools must collect
   good data at very high rates so that we may describe
   organisms' distributions (no. 1, above) and partition
   their variability into signal and noise. Tools are
   needed not only for investigation of zooplankton, but
   also for other groups known to be both ecologically
   important and unreliably reported. We must ultimately
   meld the new data with the old.
5. *What are some possible approaches?* This question is,
   properly, the subject of this entire volume.

REFERENCES

Backus, R. H., J. E. Craddock, R. L. Haedrich, and B. H. Robison, 1977.
Atlantic mesopelagic zoogeography. Part III. In: *Fishes of
the Western North Atlantic. Memoirs of the Sears Foundation
for Marine Research* 7: 266-287.

Brinton, E., 1962. The distribution of Pacific Euphausiids. *Bull.
Scripps Institution of Oceanography, University of California
at San Diego* 8: 51-270.

Cody, M. L., 1974. *Competition and the Structure of Bird Communities*. Princeton Univ. Press, Princeton, New Jersey. 318 pp.

Dagg, M., 1977. Some effects of patchy food environments on copepods. *Limnol. Oceanogr.* 22: 99-107.

Darwin, C., 1859. *On the Origin of Species. Facsimile of First Edition.* Atheneum, New York (1967), xxvii + 502 pp.

Ekman, S. P., 1953. *Zoogeography of the Sea.* Sidgewick, London, xiv + 417 pp.

Elton, C., 1927. *Animal Ecology.* (Reissue by Science Paperbacks, 1966). 207 pp.

Elton, C., 1966. *The Pattern of Animal Communities.* Methuen, London. 432 pp.

Fager, E. W., 1963. Communities of organisms. In: *The Sea,* (M. N. Hill, ed.), pp. 415-437. Wiley Interscience, New York.

Fager, E. W. and J. A. McGowan, 1963. Zooplankton species groups in the North Pacific. *Science* 140: 453-460.

Haury, L. R., J. A. McGowan, and P. H. Wiebe, 1978. Patterns and processes in the time-space scales of plankton distributions. In: *Spatial Pattern in Plankton Communities,* (J. H. Steele, ed.), pp. 277-327. NATO Conference Series, series IV, Vol. 3. Marine Sciences. Plenum Press, New York. 470 pp.

Kuhn, T., 1968. *The Structure of Scientific Revolutions.* Univ. Chicago Press, Illinois. 411 pp.

Levins, R., 1968. *Evolution in Changing Environments.* Princeton Univ. Press, Princeton, New Jersey. 120 pp.

Longhurst, A. R., A. D. Reith, R. E. Bower, and D. L. R. Seibert, 1966. A new system for the collection of multiple serial plankton samples. *Deep-Sea Res.* 13: 213-222.

MacArthur, R. H. and E. O. Wilson, 1967. *The Theory of Island Biogeography.* Princeton Univ. Press, Princeton, New Jersey. 203 pp.

May, R. M., 1973. *Stability and Complexity in Model Ecosystems.* Princeton Univ. Press, Princeton, New Jersey 235 pp.

McGowan, J. A., 1974. The nature of marine ecosystems. In: *The Biology of the Oceanic Pacific,* (C. B. Miller, ed.), pp. 9-28. Oregon Univ. Press, Corvallis, Oregon.

Mullin, M. M. and E. R. Brooks, 1976. Some consequences of distributional heterogeneity of phytoplankton and zooplankton. *Limnol. Oceanogr.* 21: 784-796.

Shulenberger, E., 1978. Vertical distributions, diurnal migrations, and sampling problems of hyperiid amphipods in the North Pacific central gyre. *Deep-Sea Res.* 25: 604-625.

Scripps Institution of Oceanography, 1966. A new opening-closing paired zooplankton net. *Scripps Institution of Oceanography Reference #66-23.* Unpublished document. 56 pp.

Scripps Institution of Oceanography, 1974. Data report, physical, chemical, and biological data. Climax I Expedition, 19 Sept. - 28 Sept. 1968. *Scripps Institution of Oceanography Reference #74-20.* 41 pp.

Steiglitz, K., 1974. *An Introduction to Discrete Systems.* Wiley, New York. 318 pp.

Sverdrup, H. U., M. W. Johnson, and R. H. Fleming, 1942. *The Oceans, Their Physics, Chemistry, and General Biology.* Prentice Hall, New Jersey. x + 1087 pp.

Taylor, L. R., 1977. Megamouth: a new family of shark. *Oceans* 10: 46-47.

Venrick, E. L., 1971. Recurrent groups of diatom species in the North Pacific. *Ecol.* 52: 614-625.

Whittaker, R. H., 1975. *Communities and Ecosystems* (Second Edition). MacMillan, New York. xvii + 385 pp.

Wiebe, P. H., 1971. A computer model study of plankton patchiness and its effects on sampling error. *Limnol. Oceanogr.* 16: 29-38.

Shallow Water Marine Biological Research

Bruce C. Coull

ABSTRACT: Shallow-water marine biological research is restricted by labor intensive sorting and identification of collected organisms and the inability to monitor continuously the relevant environmental parameters affecting organisms. Thirty respondents to a questionnaire mailed to 90 shallow-water marine biologists provided a variety of suggestions as to their research needs. These suggestions, which in general call for increased automated instrumentation, are summarized and the priority instrumentation needs for research advancement in shallow-water research are discussed.

INTRODUCTION

Shallow-water marine biologists, those working in water depths less than 100 m, traditionally have been concerned with accurate sampling to represent correctly either the organism, population, or community level of interest. Historically, much of shallow-water marine biological research involved collecting, counting, identifying, and compiling. Indeed, such work is prerequisite before attempting to understand the dynamic aspects of any system and, in

Contribution Number 290 from the Belle W. Baruch Institute for Marine Biology and Coastal Research, University of South Carolina.

many unstudied regions of the world's oceans, it is still a necessary first phase of marine biological investigations. The experimental or manipulative phase of shallow-water research has come of age primarily within the past decade, often after the descriptive phase has been completed. From such an hypothesis-testing approach, we have learned much about ecosystem inter-actions and functions and, also, there has been an increased number of shallow-water marine biologists involved. In the following sections of this paper, I shall attempt to synthesize the "state-of-the-art" in shallow-water marine biological research and present the needs as envisioned by myself and thirty colleagues who responded to a questionnaire circulated by mail. Of 90 questionnaires mailed (recipients were researchers presently studying shallow water marine biological problems) 30 were returned. A sample questionnaire is appended (Appendix I).

STATE-OF-THE-ART IN MARINE BIOLOGICAL RESEARCH

Field Collections

Since sample collection varies considerably with the type of assemblage to be studied, I have divided the responses to the questionnaire into three general categories: 1) plankton (zoo- and phyto-); 2) benthos (including hard and soft bottom); and 3) nekton.

Sampling for plankton is regularly done with nets, pumps, or various types of collection bottles (e.g. Nansen, Niskin, VanDorn) not much differently from what was done 21 or more years ago (Hedgpeth, 1957). What has become more sophisticated during the past 2-3 decades has been the level of collecting accuracy with these standard devices. Plankton nets, for example, can now be opened and closed at a desired depth to collect a discrete "layer" of plankton. Opening/closing mechanisms of "water sample" bottles have become much more durable, precise, and dependable. Further-more, recording the volume of water passed through a towed net has necessitated the development of accurate, lightweight flow meters suspended in the net's mouth. At present several inexpensive ones are commercially available. Continuous plankton recorders, available since the early 1930's (Hardy, 1939) and subsequently modified, are currently not in wide use in the United States or elsewhere. They are expensive, require continual monitoring, and often render collected specimens unrecognizable which makes them unusable for low level identification. More recently, direct hand collections and observation of marine plankters (Alldredge, 1972; Bé et al., 1977) have provided a more complete understanding of how these organisms live *in situ*.

Soft bottom (sand and mud) infaunal benthos are traditionally collected by a variety of grabs or corers which collect the bottom sediments and associated organisms. Depending upon the water depth and/or organisms to be collected, the size and weight of the grab or corer will vary. Collecting devices are used remotely in deeper waters, but should be hand-held in intertidal or shallow subtidal

systems.   The efficiency of these different devices varies according
to the substrate, water depth, lowering speed, and nature of the
sediment-water interface.   Such devices may not be as quantitative
as benthic ecologists would like to think (Flanagan, 1970; McIntyre,
1971; Ankar, 1977).   Carey and Heyamoto (1972) have presented an
overview of the available methods for collecting soft bottom benthos.
On hard substrates (rocky shores, coral reefs) organisms are usually
removed individually by scrapping or their densities and position are
quantified by photographing them *in situ.*

Nekton (particularly fish, large crustaceans, and swimming
mollusks) are regularly collected by nets or trawls.   There are,
of course, a variety of net sizes and trawl widths available for
collecting nekton; but all seem to fail in their ability to capture
rapidly moving taxa and do not provide true quantitative estimates
of the sampled fauna.   True "epibenthic" (natant) forms are best
collected by towing epibenthic sleds (none of which is available
commercially).   Spotter planes, recording fathometers, and acoustical devices are regularly used to find schools of nekton, particularly fish, but collection again becomes a problem of using the
appropriate net or trawl.

Collection of physical-chemical data, an often necessary
correlate of the marine biological collection, is usually more
automated than the collection of traditional "biological" data.
Continuous recording of temperature, salinity, and tidal height,
for example, are, of course, a regular occurrence on many marine
biological collecting trips and the analysis of water column micro-
nutrients can be made directly aboard ship.   The actual methods of
collection of biological samples has not seen great changes over
the past 20 years.

Table I summarizes the primary methods available and the needs
as perceived by the 30 questionnaire responders and myself.

TABLE 1: Present methods and needs for field
sampling in shallow water marine biological
research (summary of 30 questionnaires).

Biological Category	Present Primary Methods	Needs
Phytoplankton	nets, bottles, rosettes, pumps, fluorometer, C_{14}	continuous environmental data; moored fluorometer; light-weight pump; opening/closing array with sensor
Zooplankton	nets, bottles, echo-sounders, flow meters	continuous environmental data; *in situ* particle counter; efficient nets (detritus); telemetered open/close nets; temperature-salinity sensors on open/close nets; cheap *in situ* flow meter readout
Benthos	grabs, corers, dredges, digging, photographs	continuous environmental data; quantitative epibenthic sampler; large hard sand grab; gentle mud siever
Nekton	nets, trawls, seines, traps	continuous environmental data; quantitative trawls, nets; micro-telemetry

Laboratory Analyses

In the laboratory much of the work still requires time-consuming hand sorting of samples, usually under a microscope. In all categories (i.e., plankton, benthos, nekton) sorting was the one laboratory problem which was consistently cited in the questionnaires returned. Many phytoplankton researchers, however, have mechanized their sorting procedure through use of electronic particle counters, but not once was such a counter mentioned as being in use by benthic or nekton researchers. Further, since much phytoplankton research is aimed at quantification of primary production (see review of Vollenweider, 1969), various automated devices, such as scintillation counters for C_{14} measurement, O_2 probes to measure changes in photosynthetic oxygen production, and fluorometers to measure fluorescence due to chlorophyll changes, have come into general laboratory use in the past 20-30 years. Marine biologists working with animals also use some automation in attempts to evaluate secondary production (Edmonson and Winberg, 1971) and common in-use instruments are bomb calorimeters, chemostats, carbon analyzers, and various O_2 probes and/or respirometers to measure energy loss due to respiration. Even so, most laboratory work in present day shallow-water marine biological research entails a great amount of labor-intensive sample sorting, whether that be to enumerate the taxa for population/community studies or to extract live organisms for experimental purposes. Table 2 summarizes the primary laboratory methods and needs of the 30 questionnaire respondents and myself.

TABLE 2: Present primary laboratory methods and priority needs of shallow-water marine biologists (summary of 30 questionnaires).

Biological Category	Present Methods	Needs
Phytoplankton	electronic counters, autoanalyzers, spectrophotometers	increased sensitivity in present instrumentation; image analysis
Zooplankton	hand-sorting, autoanalyzers, electrodes	automatic sorter; image analysis
Benthos	hand-sorting; autoanalyzers, x-ray	automatic sorter; image analysis
Nekton	hand-sorting	automatic sorter; image analysis

Field Experimentation

Overall field experimentation has not been the forte of marine biologists. Of course, the *in situ* C_{14} experiments pioneered by Steemann-Nielsen (1952) are indeed field oriented, but it was not until recently that larger scale manipulative experiments were attempted. Pioneered by west coast rocky shore ecologists (Paine, 1966; Dayton, 1971; Connell, 1972), benthic ecology is presently in a "caging" frenzy. This technique of excluding potential predators from a site by erecting an enclosure has provided significant insights into how marine benthic communities are structured. Microcosm studies, where a "typical" piece of the marine environment is monitored and then altered (e.g., polluted or over-stocked) also have been extremely valuable to look at eco-system response to some particular perturbation (see Bell and Coull, 1978). Multidisciplinary, multi-personnel studies at the Controlled Environmental Pollution Experiment (CEPEX) project in British Columbia and the Marine Ecosystem Research Laboratory (MERL) project in Rhode Island are examples of enclosing an "eco-system" with subsequent disturbance and monitored responses. Total metabolism studies of the water column and/or benthic systems have been accomplished also. Usually the entire "system" is enclosed and subsequent changes in metabolism monitored *in situ* (see Pamatmat, 1977, for a review of techniques for benthic systems).

INSTRUMENTATION NEEDS OF SHALLOW-WATER MARINE BIOLOGISTS

Again the instrumentation needs are identified for the three large categories, plankton, benthos, and nekton. In all three categories an appeal for an efficient, inexpensive, continuous recording data logger for microenvironmental param-eters is recurrent. Obviously different disciplines desire that this recorder track different variables. Phytoplankton researchers argue that continuously resolving qualitative changes in phytoplankton is a major problem and a moored, self-contained fluorometer is the primary piece of hardware needed. This instrument needs to be relatively low in cost so that vertical arrays can be deployed on transects across the region of study. Dr. R. T. Barber (in his questionnaire response) contends that such equipment will permit a breakthrough in biological oceanography comparable with the breakthrough provided in physical oceanography by the development of a dependable moored current meter. Barber cautions, however, "...not...that such an instrument will solve all our problems; the moored current meter has some severe deficiencies but it is a salutary advance over its predecessor".

Benthic ecologists' ability to measure physical variables continuously *in situ* is also restricted by expense and lack of adequate instrumentation. Dr. R. T. Paine states that "what is needed are real advances in measuring the physical forces 'seen' by organisms (plant and animal)...The devices will be electronic, miniaturized to some degree, and have not yet been developed adequately". Almost all the responses from benthic ecologists echoed Paine's sentiments on the need to have continuous monitor-ing devices for a variety of sediment and overlying water column parameters. Similarly the nekton researchers also requested continuous monitoring devices for the relevant environmental

variables. Thus this need expressed by a consensus of the marine
biologists surveyed would appear to be a major priority development in shallow-water marine biological research.

Besides the need for the continuous environmental data, each
particular sub-set surveyed had more specific needs. The needs
of phytoplankton investigators were primarily centered around
more accurate assessment of environmental variables and the afore-
mentioned *in situ* chlorophyll measurements. A distinct need of
several phyto- and zooplankton workers, particularly those who
operate in shallow water, was for lightweight, inexpensive,
dependable collecting apparatus. Suggestions included: small
bongo nets for use off skiffs, a lightweight battery-operated
waterproof pump, and opening/closing nets that can be operated
manually in shallow water. A specific need in zooplankton
research is for the development of multiple samplers that can
be used simultaneously at different depths with direct flowmeter readouts appearing aboard so that nets can be monitored
before retrieving the samples.

Benthic investigators have few suggestions to improve field
sampling of organisms, but the nekton researchers have innumerable
problems adequately sampling their quarry. All presently available
nekton samplers are inherently biased to organisms that do not swim
away. Suggestions for new technological devices included a finely
tuned bioacoustical monitor that can accurately determine size of
the school and size classes of the particular school and telemetered
release devices that would allow capture (particularly at depth) of
a discrete nektonic assemblage. The overall plea was for quantitative collecting devices so truly appropriate estimates of standing
stocks can be made. The recurring problem of sampling an actively
swimming assemblage quantitatively remains one of the major problems
in shallow-water marine biology today.

Much laboratory work in marine biology has been automated,
particularly when the measurement is one of rate or chemical composition. Very clearly the major and most time-consuming problem in
any laboratory work with marine organisms is sorting, identifying,
enumerating, and removing specimens. Except for those phytoplankton
researchers who have devices to alleviate much of this tedium
through the use of particle counters (e.g., Coulter counters), there
was a recurrent and persistent cry for some type of automatic sorter
and identifier. In zooplankton research, H. P. Jeffries and
colleagues (University of Rhode Island) are presently developing a
pattern recognition system with electronic image analysis. Such a
system would be a major breakthrough in zooplankton research and
obviate the need for the time-consuming microscopic sorting
presently required to determine zooplankton community and population
dynamics.

Benthic biologists have the same identification problems as
those of zooplankton researchers, but additionally benthic researchers
have the problem of quantitatively removing their taxa from the sediment. Standard sieving techniques are time-consuming and tend to
break many animals apart. Even after animals are removed from the
sediment, a tedious identification procedure must follow. Sorely
needed is an optical scanner which can adequately distinguish the

great variety in size and shape that benthic organisms exhibit, and classify these to at least major taxa. Subsequently, if such a device could further distinguish subtle differences between congeners, it would be ideal. Benthic research progresses at a relatively slow pace primarily because of the logistic, mechanical, and labor-intensive problems involved in faunal extraction and identification. To solve this problem would be *the* giant breakthrough in benthic research.

An "automatic sorter" was mentioned by every soft-bottom benthic ecologist as the primary need. Almost equal importance was given to the ability to trace benthic organisms in the sediment. The ability to follow sediment dwellers would provide significant insight into their natural history and biology. Hard-bottom benthic ecologists urge the development of a 3-dimensional digitizer for stereo pairs of photographs to determine growth of epifaunal organisms.

The nekton researchers also echoed the time-consuming process of sorting and identification in their laboratory work. They too urge the development of automated sorting and counting devices.

Experimentation

Experimentation in the shallow-water marine environment has obviously been limited by the constraints listed above to more traditional laboratory and field measurements. *In situ* field manipulations of total assemblages has been the forte of the benthic ecologists through the use of exclosures (see Paine, 1966; Dayton, 1972; Woodin, 1974; and Virnstein, 1977). Although no suggestions were forthcoming for modifying the present exclosure design, there was the recurrent theme of building such exclosures with supplemental automatic recording devices that would allow continuous monitoring of various environmental variables (e.g. Eh, O_2, pH, temperature, salinity, sediment compaction). The same theme is expressed in the collection of unmanipulated field samples. Several benthic investigators also expressed interest in the development of efficient and quantitative methods for transplanting sediment and its associated organisms to microcosms with as little disturbance as possible. The use of microcosms for analyzing total ecosystem energetics is becoming more fashionable, and again, as in all "field" work, there is a need for accurate continuous measurements of environmental variables, as well as a solution to the sorting, identification procedure which hampers all marine ecology. An extremely promising avenue of benthic research is the interaction of macrofaunal larvae with adults of the same or different species and the interaction of macrofaunal larvae with meiofauna. At present there is no suitable mechanism for holding such larvae *in situ*, i.e., one with controlled conditions which allows free communication between benthic and pelagic systems so that larvae can settle and interactions between the size classes be noted.

Plankton systems are extremely difficult to experiment with *in situ*. Of course, phytoplankton researchers have pioneered in the field of using isotopes to measure primary production, heterotrophic uptake, nitrogen release, etc. The stable isotope measurements of the paleoclimatologists (e.g., Williams et al., 1978) where different carbon isotopes are taken up by organisms with a carbonate test in relation to the temperature at the time of test formation, can be adapted to extant species. E. B. Haines (personal communication) has used this technique to distinguish different sources of carbon in detritus based on the premise that different plants have different carbon isotopes depending on basic physiology, genetics,etc. This technique obviously has strong potential for further application in shallow-water experimentation and should certainly be pursued.

Remote sensing experimentation is also an area of great interest to marine biologists which needs to be more fully utilized. Extremely valuable monitoring of vegetation distributions, amount and types of chlorophyll in shallow waters, sediment distributions, and suspended sediment loads has been done. The question now is whether such remote sensors can be used for uninterrupted monitoring of smaller scale events. Refinement of such sensors should continue to the point where small scale monitoring is feasible.

Development of hardy, inexpensive micro-telemetry devices would be extremely valuable in behavioral studies of many marine organisms. The problem of where species "X" goes during a particular season is often a great puzzle. Present "mark and recapture" tagging techniques are inadequate to measure migration patterns and success of a species after migration. Of particular relevance is the movement of commercially important species (e.g., penaeid shrimp, crabs, fish) which are thought to regularly migrate inshore/offshore. We do not know exactly where the species go nor do we know what they do while they are gone. If, for example, the penaeid shrimp, *Penaeus setiferous*, an important commercial crop in the southeastern and Gulf states, could be followed during its suspected winter offshore migration utilizing a micro-radio technique, one would certainly be better able to predict potential commercial catch and to improve economic forecasting in the coastal zone. Such a device needs to be tiny so as to not impede movement or harm such a relatively small animal, and yet it should provide a strong, persistent and long-lived signal to allow careful tracking. Cost must be minimal since many of the micro-telemeters will most likely be lost or damaged. No such devices are available, but their development would be extremely valuable to marine biologists.

Shallow-water marine biological research entails many labor intensive tasks and the lack of development of techniques to shorten this time-consuming procedure appears to be a major obstacle in advancement of present-day experimentation. This lack of techniques coupled with the need for continuous, accurate monitoring of environmental variables, but remembering that the accuracy of the measurement need not be greater than the particular situation demands, appear to be the critical needs of shallow-water marine biologists in 1978.

ACKNOWLEDGMENTS

The following responded by returning the questionnaire and to each I express my sincere thanks: D. M. Allen, H. Austin, R. T. Barber, J. M. Bishop, T. Brown, M. A. Buzas, N. R. Cooley, J. M. Dean, R. H. Gore, E. B. Haines, D. R. Heinle, S. S. Herman, H. P. Jeffries, B. A. Kask, J. J. Lee, R. J. Livingston, J. V. Merriner, E. L. Mills, W. J. North, R. T. Paine, C. H. Peterson, T. T. Polgar, J. W. Porter, M. Reeve, J. Sibert, S. E. Stancyk, J. P. Sutherland, W. H. Sutcliffe, R. W. Virnstein, and T. Whitledge.

REFERENCES

Alldredge, A. L., 1972. Abandoned larvacean houses: a unique food pelagic environment. *Science* 177: 885-887.

Ankar, S., 1977. Digging profile and penetration of the Van Veen grab in different sediment types. *Contr. Asko Lab.* 16: 1-22.

Bé, A. W. H., C. Hemleben, O. R. Anderson, M. Spindler, J. Hacunda, and S. Tuntivate-Choy, 1977. Laboratory and field observations of living planktonic Foraminifera. *Micropaleontology* 23: 155-179.

Bell, S. S. and B. C. Coull, 1978. Field evidence that shrimp predation regulates meiofauna. *Oecologia* (Berl.) 35: 141-148.

Carey, A. G. and H. Heyamoto, 1972. Techniques and equipment for sampling benthic organisms. In: *The Columbia River Estuary and Adjacent Ocean Waters: Bioenvironmental Studies*, (A. T. Pruter and D. L. Alverson, eds.), pp. 378-408. University of Washington Press.

Connell, J. H., 1972. Community interactions on marine inter-tidal shores. *Ann. Rev. Ecol. Syst.* 3: 179-192.

Dayton, P. K., 1971. Competition, disturbance, and community organization: the provision and subsequent utilization of space in a rocky intertidal community. *Ecol. Monogr.* 41: 351-389.

Edmonson, W. T. and G. G. Winberg, 1971. A manual on methods for the assessment of secondary production in fresh waters. *IBP Handbook No. 17.* Blackwell Sci. Publ., Oxford. 368 pp.

Flannagan, J. F., 1970. Efficiencies of various grabs and corers in sampling fresh water benthos. *J. Fish. Res. Bd. Can.* 27: 1691-1700.

Hardy, A. C., 1939. Ecological investigations with the continu-ous plankton recorder. *Hull Bull. Mar. Ecol.* 1: 1-57.

Hedgpeth, J. W., 1957. Obtaining ecological data in the sea. In: *Treatise on Marine Ecology and Paleoecology*, Vol. 1, (J. W. Hedgpeth, ed.), *Geol. Soc. Amer. Mem.* 67: 53-86.

McIntyre, A. D., 1971. Deficiency of gravity corers for sampling meiobenthos and sediments. *Nature* 231: 260.

Paine, R. T., 1966. Food web complexity and species diversity. *Amer. Natur.* 100: 65-75.

Pamatmat, M. M., 1977. Benthic community metabolism: a review and assessment of present status and outlook. In: *Ecology of Marine Benthos*, (B. C. Coull, ed.), pp. 89-112. Univer-sity of South Carolina Press, Columbia.

Steemann-Nielsen, E., 1952. The use of radioactive carbon (14C)
    for measuring organic production in the sea. *J. Conseil* 18:
    117-140.
Virnstein, R. W., 1977. The importance of predation by crabs
    and fishes on benthic infauna in Chesapeake Bay. *Ecology* 1199-
    1217.
Vollenweider, R. A., 1969. A manual for measuring primary pro-
    duction in aquatic environments. *IBP Handbook, No. 12,*
    Blackwell Sci. Publ., Oxford. 224 pp.
Williams, D. F., R. C. Thunell, and J. P. Kennett, 1978.
    Periodic flooding and stagnation of the eastern Mediterranean
    Sea during the late Quaternary. *Science* 252-254.
Woodin, S. A., 1974. Polychaete abundance patterns in a marine
    soft-sediment environment: the importance of biological
    interaction. *Ecol. Monogr.* 44: 171-187.

APPENDIX 1: Questionnaire on Instrumentation
Needs in Shallow Water Marine Biological Research*

1. Primary organism group of interest (e.g. phytoplankton, fish,
    benthos, etc.).

2. Primary habitat of your research organism (e.g., rocky intertidal;
    estuarine water column, shelf, soft bottom, coral reef, etc.).

3. a. How do you collect your field samples?

    b. What "gadget", either mechanical or electronic, would
        allow you to collect your samples more efficiently?
        Is it presently available? If not available, do you
        think it could be engineered? What needs do you have
        to collect more efficiently?

4. a. Do you do field manipulations in your research?

    b. If so, would mechanical or electronic instruments aid
        in your manipulations? How? What would you like to
        see developed to aid your field manipulations?

5. a. Is your laboratory work mechanized?

    b. If so, how? (what instruments, etc.)

    c. If not, can you envision an instrument that would
        significantly reduce your lab time? What is it?
        And what suggestions do you have for its design?

6. What specific problems does your research have that yet-
    to-be-developed instrumentation could solve?

7. Additional comments-suggestions etc.

*Questionnaire sent to 90 shallow-water marine biologists.
Much of this paper is based on the 30 responses received
to this questionnaire.

Deep Ocean Microbiology

J. W. Deming
P. S. Tabor
R. R. Colwell

ABSTRACT: A major impetus behind fundamental
research in baromicrobiology is the growing
need to understand biological processes as
they occur under the hyperbaric conditions of
the deep ocean environment. Biogeochemical
cycles in the marine environment remain in-
completely described due to a paucity of information on metabolic rates of microorganisms
in the deep sea, even though marine bacteria
have been shown to be ubiquitous, even in the
deepest regions of the world oceans. Marine
microbiologists have recognized the importance
of a multi-disciplinary approach in clarifying
the role of marine bacteria in the transformation of matter in the deep sea. Taxonomic
studies of the structure and diversity of deep-
sea bacterial populations, physiological studies,
in which information concerning metabolic rates
of natural populations under *in situ* conditions,
and the recognition of pressure-sensitive sites
of enzymes in biochemical reaction sequences examined *in vitro*, have been done. The essence
of these findings is reviewed in this paper.

The diverse experiments carried out in many laboratories have had the unifying, underlying objective of characterizing barophilism, the obligate requirement by organisms for elevated hydrostatic pressure for optimal growth. Barophilic
responses have been recorded, but only in a few
instances. Investigators have concluded that
there may be unique molecular characteristics
circumscribing barophiles, based on kinetic
studies in which barosensitive and barotolerant
bacteria were compared. Technological impasses
have prevented the isolation and study of true
barophiles in pure culture, although barotolerant bacteria isolated from deep ocean water and
sediment samples have provided useful physiological systems for investigating effects of hydrostatic pressure.

Recent research in deep ocean microbial ecology
has yielded improvements in samplers and culturing methods that permit the collection and maintenance of water samples under conditions of *in
situ* deep-sea environmental pressure. This paper
outlines the design and operation of pressure-retaining samplers currently in use. In addition
rates of utilization, obtained from recent experiments and from previously published reports
concerning deep-sea populations, measured under
in situ conditions of pressure, are summarized,
and effects of elevated pressure levels, nutrient
concentrations, and efficiency of available substrate utilization are discussed.

INTRODUCTION

    It has been known since the voyages of the TRAVAILLEUR and
TALISMAN (1882-1883) that bacteria are present in ocean sediment,
even at depths of ≥5100 m (Certes, 1884). Although microorganisms
are considered to be ubiquitous in the ocean environment, their
distribution is not homogeneous. Occurrence and population densities of marine bacteria are affected by association with other
organisms and aggregation and attachment to particulates, as well
as by temperature, salinity, and other environmental parameters
(Ashby and Rhodes-Roberts, 1976). Having established the occurrence
and distribution of bacteria, marine microbiologists now are examing
the questions of rates of activity and community structure in the
deep sea. The role of deep-sea bacteria in energy transfer, nutrient regeneration, and trophic cycles in the marine environment
remains to be clarified and quantitated.
    The average depth of the ocean is estimated to be *ca.* 3800 m.
In deeper regions, such as the abyssal plains and deep-sea trenches,
where depths may reach 11,500 m, as in the Challenger Deep, environmental parameters are essentially constant. Because of water tem-

peratures in the range -1 to 4°C low nutrient concentrations, 0.35
to 0.70 mg dissolved organic carbon per liter (Menzel and Ryther,
1970), salinities of 34-35 o/oo, and long residence times for deep
water masses, *ca*. 350 to 950 years (Broeker, 1963), the deep ocean
environment is considered to be stable, relative to coastal waters
and estuaries.

Psychrophilic bacteria are readily isolated from deep ocean
water and sediment samples, psychrophiles generally being defined
as those bacteria capable of growth at temperatures of 0°C to 20°C,
with an optimal growth temperature of <15°C. In fact, the incidence of psychrophilic marine bacteria that can be isolated from
water samples appears to increase with depth (Wirsen and Jannasch,
1975). The relative geologic stability of the deeper regions of
the sea may be a factor in the evolution of psychrophilic bacteria
(Brock et al., 1973; Morita, 1976).

A unique parameter of the deep ocean is hydrostatic pressure,
which increases approximately one atmosphere for every ten meters
increase in depth. Historically, it has been an intriguing question
for marine microbiologists as to whether populations of deep-sea
bacteria exist which are uniquely adapted, not only to cold temperatures, but also to the hydrostatic pressure of the deep ocean. Barophilic responses have been observed, i.e., it has been shown that
deep-sea bacteria proliferate under *in situ* pressures in the laboratory, but, in comparison, fail to grow, or grow poorly,at atmospheric
pressure (ZoBell and Morita, 1957; Mitskevitch and Kriss, 1966;
Yayanos et al., 1979; Deming, Tabor, and Colwell, unpublished data).
Studies of barophiles, i.e., bacteria growing in pure culture and
demonstrating a requirement for elevated hydrostatic pressure, have
not been reported yet. Strictly defined, psychrophiles expire at
temperatures >20-22°C (Morita, 1976). By analogy, barophilic bacteria, if present in a deep-sea sample, may not survive decompression
during sample retrieval. The technological difficulties of maintaining elevated hydrostatic pressure, during and after sample collection,
have severely restricted attempts to isolate barophilic organisms.
Attention has been focused, until recently, on those bacteria surviv-
ing decompression that can be isolated in pure culture at ambient
pressure for subsequent studies under defined conditions in the laboratory, usually simulated *in situ* conditions.

Standard methods for obtaining deep-sea sediment cores subject
the samples to decompression during retrieval. Total, viable, aero-
bic, heterotrophic bacterial counts for such sediment samples may be
as high as 1000 colony forming units per gram (cfu/gm) wet weight
when the sediment sample is aseptically diluted and spread-plated on
a seawater agar medium. Similarly, deep-sea water samples yield, on
average, a total number of viable bacteria per ml, 2-3 orders of magnitude less than sediment collected at the same station. In both
cases, the bacterial populations are small, compared to the abundance
of bacteria in water and sediment samples collected from shallow,
coastal waters, i.e., approximately 10^6 cfu per ml and 10^8 cfu/gm wet
weight, respectively. A similar relationship for total direct counts
of deep ocean samples, relative to coastal water and sediment samples,
also has been observed (Jannasch and Jones, 1959; Tabor, Deming, and
Colwell, unpublished data).

Standard microbiological methods for isolation of pure cultures do not incorporate pressure as a culturing condition, thus may not provide complete data on the occurrence and distribution of viable bacteria in the deep sea. Viable bacteria obtained using routine sampling and isolation methods may, in fact, be metabolically inactive under deep-sea conditions, considering that many of these isolates, when tested in the laboratory, are inhibited by elevated, i.e., *in situ*, hydrostatic pressure (ZoBell and Oppenheimer, 1950; ZoBell and Morita, 1957; Tabor, Ohwada, and Colwell, unpublished data). Rather than comprising a true sampling of indigenous deep-sea bacteria, such isolates may represent the opportunistic, baroduric bacteria sinking to the deepest parts of the world oceans, either free-living or attached to particulates or animal/plant debris, and managing only to survive deep-sea conditions.

Despite the fact that studies of deep-sea bacteria have been restricted to those strains capable of growth at atmospheric pressure, interesting patterns of microbial adaptation to the deep ocean environment have been described. Taxonomic studies, in which hundreds of isolates were subjected to detailed phenetic and genetic analyses, have shown that bacteria present in sediment collected from the deep sea are predominantly Gram-negative rods that demonstrate a wide diversity of morphological and biochemical characteristics (Liston, 1968; Quigley and Colwell, 1968). In contrast, bacteria in water and sediment from shallower regions show greater biochemical specialization. Liston (1968) suggested that the high degree of metabolic diversity expressed by most deep-sea sediment isolates, when examined in the laboratory, can be ascribed to a versatility required to deal with nutritional and other forms of stress in deeper water. Specific niches within the deep ocean, however, have been shown to harbor distinct populations of microorganisms. For example, bacteria isolated from deep-sea sediment, water, and intestinal tracts of benthic fauna tend to group separately, when examined by numerical taxonomy (Tabor, Ohwada, Kuchinsky, and Colwell, unpublished data).

Barophilism or barotolerance may, indeed, confer a selective advantage on deep-sea bacteria, and play an important role in the ecology of the deep ocean environment (Colwell and Kettling, 1974). Unfortunately, physiological studies designed to measure bacterial response to increased hydrostatic pressure have included, in general, only a small number of deep-sea or marine isolates. The molecular basis for barotolerance, thus, has been deduced largely from studies of barophobic organisms, in particular, terrestrial strains of *Escherichia coli*. Pressure inhibition of bacterial metabolism appears to be highly complex, and a universally accepted explanation remains to be formulated.

Hydrostatic pressure is known to effect volume changes normally incurred during formation of enzyme-substrate complexes (Johnson et al., 1954; Zimmerman, 1970). Where a positive volume change is required, enzyme reaction rates will be inhibited by elevated pressures. Regulation of gene expression, which involves delicate allosteric influences by inducer and/or repressor control molecules, will also be affected by the application of hydrostatic pressure (Hedén, 1964). In the late 1960's, Pollard and coworkers established the relative order of anabolic pathways that are sensitive to pressure: protein,

DNA, and RNA synthesis (Pollard and Weller. 1966). For example, RNA synthesis in *E. coli* was demonstrated to occur in the absence of protein synthesis at 600 atm (Yayanos and Pollard, 1969). Work done with whole cell suspensions of *E. coli* confirmed that protein synthesis is inhibited at pressures which do not markedly affect the synthesis of nucleic acids (Arnold and Albright, 1971). When marine bacteria were used as experimental systems, as, for example, a psychrophilic strain of *Vibrio marinus*, salinity was found to play an important role in overall resistance to pressure effects: maximum hydrostatic pressure permitting cell division increased when concentrations of salts in the growth medium were also increased (Albright and Henigman, 1971).

Pressure effects on protein synthesis have been examined in considerable detail at the molecular level. Studies with cell-free extracts, employing both terrestrial and marine isolates, including *Pseudomonas bathycetes*, a deep-sea bacterium isolated by ZoBell and Morita from deep-sea sediment collected from the Marianas Trench during the Dodo Expedition of 1964, and identified by Quigley and Colwell (1968), indicate that ribosome conformation, and, more specifically, the ability of the ribosome to complex with transfer RNA, are key elements in barotolerant protein synthesis (Schwarz and Landau, 1972a, 1972b; Swartz et al., 1974; Pope et al., 1975; Smith et al., 1975). Furthermore, as has been demonstrated for cell division, specific ion concentrations can affect relative barotolerance of protein-synthesizing systems in cell-free extracts, at least in the case of *P. bathycetes* (Smith et al., 1976).

Determination of pressure inhibition sites at the molecular level in the bacterial cell comprises an important step towards understanding the metabolic capabilities of bacteria existing under elevated hydrostatic pressure. Unfortunately, it is difficult to explain the behavior of intact cells in the deep sea, based on results obtained with cell-free extracts (Arnold and Albright, 1971), especially when such studies do not simulate deep-sea conditions of low temperature and nutrient concentration.

Pressure and temperature cannot be considered independently since increased hydrostatic pressure and decreased temperature act synergistically on metabolic activity, both changes resulting in a decrease in the molecular volume of the system (Morita, 1976). The amino acid transport systems of marine psychrophiles grown at *in situ* temperature were strongly inhibited under conditions of elevated hydrostatic pressure, i.e., 380 atm, the pressure encountered at the average depth of the ocean (Paul and Morita, 1971). Activity of respiratory enzymes required for subsequent amino acid metabolism were not affected significantly. The initial sites of pressure damage, under these conditions, appear to be related not to specific intracellular enzymes, but rather to the membrane itself.

It has been speculated that bacteria in the deep sea are subjected to starvation due to an inability to obtain energy, i.e., to transport nutrients into the cell (Paul and Morita, 1971). Interestingly, a marine psychrophile starved of nutrients prior to pressure application, i.e., grown at nutrient concentrations similar to that in the deep sea, survived pressure effects better than cells growing in logarithmic phase with an adequate supply of nutrient(Novitsky and Morita, 1978).

When *in situ* conditions are re-created faithfully, physiological studies of barotolerant marine isolates provide valuable insight into survival mechanisms of bacteria in the deep sea (Morita, 1978). Nevertheless, all of the studies accomplished to date suffer the same handicap: deep-sea isolates available for experimentation have undergone decompression at some step during sampling, and, therefore, may not be entirely representative of the deep-sea microflora. If a population of true barophiles exists which is capable of significant metabolic activity in the deep sea, how can the presence and activity of such a population be detected and measured?

With the aid of the research submersible ALVIN, a variety of *in situ* incubation experiments have been conducted at permanent bottom stations in the North Atlantic, at depths less than 2000 m (Jannasch et al., 1971; Jannasch and Wirsen, 1973). In these experiments, decompression effects were avoided by inoculating and incubating sterile media on the ocean floor. Utilization of selected substrates over incubation periods as long as 15 months was monitored. No barophilic response was detected. Growth and metabolism of natural microbial populations, as well as pure and mixed cultures, appeared to be reduced *in situ* by a factor of 100 when compared to laboratory controls held at one atm (Jannasch et al., 1971). *In situ* experiments of this type suffer two major drawbacks: dependence on a submersible vessel with limited depth range; and necessity to extrapolate metabolic rates from end-point measurements.

To measure metabolic rates occurring in the deep sea, the baromicrobiologist requires sampling equipment capable of maintaining *in situ* conditions of pressure and temperature from sample collection through completion of time course experiments conducted on board ship and, subsequently, in the laboratory. Two laboratories, one located at Woods Hole Oceanographic Institution and the other in the Department of Microbiology at the University of Maryland, have developed and operated deep ocean water samplers providing this capability. Limitations of classical sampling methods employed in deep-sea studies have led to the development of these unique, pressure-retaining samplers. As a result, data are now available from deployment of the samplers (*vide infra*).

PROBLEMS OF MICROBIOLOGICAL SAMPLING IN THE DEEP SEA

Sophisticated deep ocean biological sampling poses a wide range of exceptional problems, the least of which is that samples to be collected lie 4.5-6.5 miles beneath the ocean surface. Sampling must be accomplished within vertical limits of 10-100 cm in the sediment and 10-20 m in the water column. Since vertical positioning of samplers is, at best, uncertain, exact positioning is not feasible as a routine operation on a minimum cost basis. The density of organisms in the water column is usually low and often not known prior to sampling. Sampling events, such as bottom contact, can be signalled to the surface by acoustic transponders but *in situ* operation of samplers is monitored only rarely. None of the samplers presently available can be interrogated so that sampling can be modified while in progress. In addition to these universal problems, microbiological sampling must overcome other difficulties, including problems of contamination. Populations of bacteria in surface waters are, on average, 2-4 orders of magnitude larger than in deep water beyond the continental slopes. Collection of samples without contamination becomes a major problem since samplers must pass through the surface waters on deployment and retrieval. Furthermore, *in situ* temperature and pressure are significantly different from those at the surface, severely affecting natural populations during retrieval (ZoBell, 1968; Seki and Robinson, 1969). Maintaining pressure and temperature is a technical problem that has recently been overcome for water sampling, but still remains in the case of sediment sampling.

The J-Z and Niskin samplers (ZoBell, 1941; Niskin, 1962) have been used for microbiological studies in the deep sea. However, external pressures in the deep ocean prevent these samplers from functioning properly. Air in tubing, sample collection bulb, and Niskin bag is compressible and the samplers remain collapsed, even after triggered to open. The air can be replaced by injecting sterile water into the bulbs and tubing, but the suspicion is that deep ocean water samples thus collected become contaminated during retrieval ascent, as pressure diminishes (ZoBell, 1954; Wirsen and Jannasch, 1976).

Sediment cores and grabs are subject to wash-out during retrieval, resulting in removal of bacteria from inorganic particles by percolation of sea water through the sediment or total loss of the upper centimeters of the sediment material. Percolation of the sediment occurs even in samplers designed to seal the water-sediment interface during retrieval and poses a serious sampling problem, since numbers of microorganisms, as well as microbial activity, have been shown to decrease by several orders of magnitude from the sediment-water interface to a depth of 5 cm or less in deep-sea sediments (Wirsen and Jannasch, 1976; Norkans and Stehn, 1978; Tabor, Ohwada, and Colwell, unpublished data).

Marine microbiologists rely on samplers designed by macrobiologists, i.e., biological oceanographers, to study the microflora of deep-sea animals. Unfortunately, without pressurized traps, the animals are often damaged during retrieval. In addition, contamination is introduced via the sediment in the case of benthic animals, as well as from surface water, and from contact with other organisms in the sample. Intestinal flora of deep-sea fauna can be obtained by aseptic dissection but only when the gut is intact and not damaged.

DEVELOPMENT OF DEEP OCEAN MICROBIOLOGY SAMPLERS

Marine baromicrobiologists have, in the main, taken two experimental approaches to study microbial activity without sample decompression. One is *in situ* study and the other involves retrieval of samples for subsequent examination in the laboratory. *In situ* baromicrobiology experiments require deep-sea moorings, manned and unmanned submersibles, or free-fall vehicles to place experimental apparatus on the deep-sea floor.

Such experiments have been, of necessity, relatively simple in
design. Both deep-sea moorings and the research submersible
ALVIN have been employed for *in situ* experiments designed to
quantitate utilization of soluble and solid substrates incubated at ocean depths of 1830-5000 m by natural populations
collected from various depths, including surface water, water
200 m off-shore, the water-sediment interface, and shallow
depths in the sediment (Jannasch et al., 1971; Jannasch and
Wirsen, 1973; Tuttle and Jannasch, 1976; Wirsen and Jannasch,
1976). Parallel studies with aliquots of the same inocula were
conducted at one atm for shorter incubation times to provide
control ambient pressure reactions.

The surface area of the substrates exposed to the surrounding water and the susceptibility of solid substrates to attack
by macroorganisms have been recognized as the most significant
factors in determining rate of utilization (Sieburth and Dietz,
1974; Wirsen and Jannasch, 1976). The Remote Underwater Manipulator (RUM) has also been used to collect and fix samples at
1200 m with minimal decompression for subsequent morphological
examination of deep-sea bacteria (Carlucci et al., 1976).

Fine particles of sediments can prevent valves of pressure-
retaining samplers from sealing, causing loss of pressure both
at the time of sampling and during transfer manipulations. To
date, no sampler has been designed which will overcome this
valve closure problem. Thus, *in situ* experiments are the only
available methods for conducting microbiological studies of
deep-sea sediment.

In situ experiments yield "end-point" data, since they cannot
be monitored at short intervals over long periods of time. Such
experiments have limitations for assessing and interpreting *in situ*
microbial activity at significant depths. Water samplers have been
designed that eliminate this restriction and also overcome some of
the sampling problems (*vide infra*).

Pressure-retaining samplers and transfer systems permit studies
of natural deep ocean populations without changing *in situ* conditions of pressure and temperature. A preliminary description of a
sampler with capabilities for sampling seawater at depths of 2000 m
has been described by Jannasch et al. (1973), and results of utilization of labeled soluble substrates by undecompressed samples, and
a description of a sampler capable of sampling seawater at depths
of 6000 m have been reported by Jannasch et al. (1976). The samplers were used as incubation chambers and subsamples (13 ml) could
be removed without decompression of the original sample by adding a
sterile solution of equivalent displacement volume, resulting in
dilution of the remaining sample. A concentrating sampler and
transfer/storage units have also been developed (Jannasch and
Wirsen, 1977), whereby the organisms in 3 l of seawater can be
concentrated during sampling by filtration with a Nuclepore filter,
with the final sample volume reduced to approximately 13 ml. Bacteria are dislodged from the filter by a magnetic stirring bar and
scraper. The undecompressed 13 ml concentrate can be transferred
to a pressurized unit, fitted with gas accumulator, for storage,
thereby allowing the sampler to be prepared for repeated deployment.

On return to the laboratory, the samples in the storage units must
be transferred and diluted into pressurized 1 l incubation vessels
for measurement of microbial activity. If all of the concentrated
sample is transferred, the final concentration of the suspension
in the pressurized vessel is approximately two-fold.

Preliminary descriptions of pressure-retaining samplers developed in our laboratory at the University of Maryland, in collaboration with the National Bureau of Standards, have also been reported (Tabor and Colwell, 1976; Colwell and Tabor, 1979). Samples
have been collected from 3,450 m, bacteria in the seawater sample
enumerated, and cell types observed by scanning electron microscopy.
Respiration of labeled glutamate has also been measured, using undecompressed subsamples of the original seawater samples. Only limited subsampling was possible in the early stages of the sampler
development and measurements were end-point experiments, from which
in situ rates of activity were estimated.

Improvements and modifications have been made on this sampler,
including development of a transfer and sub-sampling system. Two
samplers are presently in use and both have the capability of retaining the *in situ* pressure of seawater collected from any depth.
The loss in pressure of the sampler upon retrieval is only 7%,
arising from expansion of the sampler during retrieval. The
first of two deep ocean sampler units to be constructed (DOS
No. 1) is equipped with a latex bladder in the sample chamber,
eliminating dilution by the displacement volume during sub-
sampling. When the transfer system is employed, subsamples
are transferred without decompression or exposure to shear
forces, since pre-pressurized reactors are connected to the
sampler for transfer and subsequent incubation. The transfer
reactors have been modified by addition of a valve to the base
of the vessel which serves to regulate the displacement volume.
An incubation chamber is connected to the sampling valve and
extends into the vessel of the reactor. Incubation chambers are
latex bladders of 100 ml capacity or 50 ml plastic syringes.
When in operation at sea, the sample chamber of the DOS can be
washed, autoclaved and pressurized for repeated use. A second
sampling unit, DOS No. 2, does not contain a bladder and the
sample is diluted during successive sub-sampling. Sterile salts
solution provides displacement volume when a sub-sample is taken.
In this respect, DOS No. 2 is similar to the two incubation samplers constructed by Jannasch and Wirsen (1976). A schematic of
the sub-sampling system developed at the University of Maryland
is shown in Figure 1.

STUDIES TO DETERMINE RATES OF MICROBIAL ACTIVITY IN THE DEEP OCEAN

A large number of deep-sea samples, predominantly sediment,
have been used to measure microbial activity under conditions of
pressure and temperature simulating those of the original sample,
i.e., samples were decompressed upon retrieval and subsequently
recompressed for study. The general experimental design has been
to add substrate or reagents to the samples, subsample at given

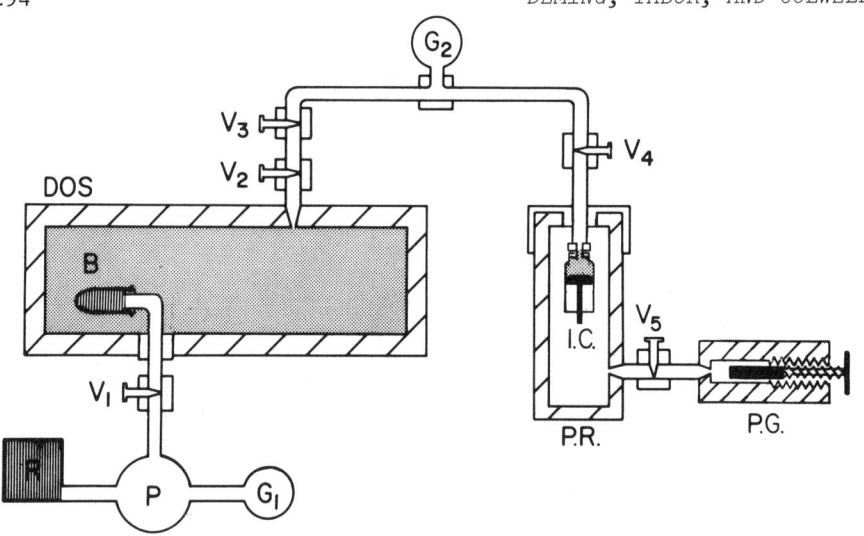

FIGURE 1: Deep Ocean Sampler (DOS) schematic
diagram. The diagram for DOS No. 1 shows the
bladder arrangement for maintaining volume
without dilution of the sample during transfer
of sample from one pressure unit to another
without loss of pressure. In preparation of
the sampler for sample collection, the bladder
(B) is collapsed, retaining only enough water
to prevent the walls of the bladder from stick-
ing together. Upon retrieval, care is taken to
maintain temperature of the sampler at <5°C.
Internal pressure of the sampler is measured by
attaching a pump-gauge (P-G) assembly. Substrate is added to the sample by one of two methods: 1) substrate is preloaded in the incubation chamber (I.C.) receiving a subsample as
diagrammed. Transfer of sample is accomplished
after the receiving pressure reactor (P.R.) is
pressurized to a pressure equivalent to that of
the DOS by means of a pressure generator (P.G.);
and 2) substrate is preloaded into a high pressure, sterile reservoir connected to the pump
and the DOS. Substrate is pumped into sampler.
Method (2) is used with DOS No. 2 which has no
bladder. During subsampling, the sample volume
is replaced with a sterile salts solution, pH
7.8, via the sterile reservoir.

time intervals, and measure activity. Aliquots are repressurized
and incubated under pressure, with aliquots also incubated at one
atm to serve as decompressed controls. Results obtained from pressurized samples, when compared with those carried out at one atm,
indicate inhibitory effects of pressure. Barophilic responses have

TABLE 1: Simulated *in situ* measurements of deep-
sea populations.

Sample Type	Depth (m)	Measurement	Reference

Sediment	7,020-10,210	*Enumeration of sulfate reducers
		*Aerobic enumeration using the
		  MPN method
			ZoBell and Morita,
			  1957; 1959

Sediment	?	*Growth at 1000 atm without sub- strate addition	ZoBell, 1968

Sediment	7,750 and 8,130	Incorporation, respiration, and
		  amino acid pool measurement
		  using ^{14}C-amino acids
			Schwarz and Colwell,
			  1975

Amphipod gut flora	7,050	*Growth
		*Respiration and incorporation
		  of ^{14}C-starch
			Schwarz et al.,
			  1976

*Results interpreted by authors cited as indicative of a barophilic response.

been observed, however. Types of measurements that have resulted
in observations of barophilic responses from simulated *in situ* experiments are given in Table 1. *In situ* experiments, using undecompressed water samples, also suggest inhibitory effects of deep-
sea pressure, with the exception of an observation indicating increased growth of microorganisms in seawater held under *in situ*
conditions of pressure (Seki et al., 1974). In general, results
obtained from simulated experiments involving deep-sea samples
have not answered the question of whether a portion of the deep-
sea microbial population is, indeed, adapted to *in situ* conditions
of pressure and temperature. However, interest in barophilic bacteria continues (Colwell and Morita, 1974; Dietz and Yayanos, 1978;
Morita, 1978; Yayanos et al., 1979).

Because metabolic activities of microbial populations may indicate rates of conversion and mineralization of organic matter in
the deep sea, they have been of principal interest in deep-sea ecology. *In situ* experiments and studies of undecompressed samples
accomplished to date are considered pioneering efforts in this area
of marine microbiology. Microbial activities in sediment at 1830 m
have been measured using unlabeled substrates at sample concentrations of 2-1,000 µg/liter. Approximately 1.3-16.5% of the substrates used in the experiments were utilized after 51 weeks, with
in situ incorporation ranging from 0.26-4.08% of that demonstrated
by control cultures held at one atm. Total substrate utilized was
calculated to be 0.50-3.08% of that of control cultures held at one
atm (Jannasch and Wirsen, 1973).

Microbial activity in undecompressed water samples collected
from less than 50 m above the sea floor at depths of 3450-7730 m
in the Puerto Rico Trench has been measured in our laboratory
(Tabor, Ohwada, and Colwell, unpublished data). Some results are
cited in Table 2. Utilization of soluble amino acids was measured
as radioactivity incorporated into the particulate fraction (>0.22
µm) and as $^{14}CO_2$ evolved from respiration for samples incubated in
the chamber of the DOS or transferred as subsamples to reactors
without decompression. Microbial activity was found to be surprisingly low. In sufficient data have been obtained to correlate total
utilization with depth of the sample, or utilization with measure-

TABLE 2: Utilization of organic compounds by deep ocean microbial populations.

Method	Location	Depth (m)	Day at Measurement	Substrate Concentration µg/ml	Total Utilization µg/ml	% Added Substrate Utilized	% Respired
Undecompressed sample incubated with ^{14}C-glutamate	P. R. Trench	3,450	161 301	0.12	5.5×10^{-2}	.046 no increase	only respiration measured
Undecompressed sample incubated with ^{14}C-amino acids	Gilliss Deep	6,040	43 111	0.16	4.0×10^{-2} 2.4×10^{-2}	.025	58 73
Undecompressed sample incubated with ^{14}C-glutamate	Gilliss Deep	6,040	326	0.18	10.4×10^{-2}	.058	91
Undecompressed sample incubated with ^{14}C-glutamate	Brownson Deep	7,730	159	0.50	103.1×10^{-2}	.206	80
Undecompressed sample incubated with ^{14}C-glutamate	P. R. Trench	7,350	48	0.18	46.6×10^{-2}	.259	97
In situ incubation with starch*	N. W. Atlantic	1,830	357	1,000.00	30.0×10^{1}	.030	--

*Jannasch and Wirsen, 1973.

ments of *in situ* dissolved organic carbon concentration. However, it has been observed that utilization is dependent upon the concentration of substrate added. A 2.5-fold increase in amino acid concentration resulted in a significant increase in total utilization. *In situ* experiments carried out by Jannasch and Wirsen (1973) employed substrate concentrations 1,000-fold greater than the concentrations used with undecompressed samples, and total utilization was at least three orders of magnitude higher. The percent of added substrate utilized did not vary significantly for the samples reported here. The *in situ* sample was neither more nor less efficient than the undecompressed samples. Comparisons between data obtained with *in situ* samples (Jannasch and Wirsen, 1973) and undecompressed samples employed in our study must take into account difference in methods. No clear relationship between increased respiration with decreasing substrate was observed from results of studies with the undecompressed samples. The initial rate of activity of samples held in the DOS decreased significantly before 16 weeks, perhaps before 7 weeks.

Measurements of respiration and incorporation, using subsamples taken from undecompressed deep water samples incubated in pressure-retaining samplers have been reported (Jannasch et al., 1976), whereby measurements taken at intervals of a few days, over a period up to five weeks, showed an initial lag, a period of increasing activity, ranging from 4-16 days, followed by a plateau in activity, with little or no increase in utilization thereafter. Corresponding measurements taken on decompressed samples showed total utilization during the period of increasing activity 2.1-4-fold greater than that of undecompressed samples.

Results of several samples reported by Jannasch et al (1976) were analyzed and compared with data for which similar calculations were reported and units were converted for uniformity of presentation (Table 3). Similar measurements for concentrated, undecompressed samples have been reported by Jannasch and Wirsen (1977).

Measurements of incorporation and respiration of subsamples from undecompressed samples have also been obtained using our DOS. Samples were collected on Cruise No. 44 of the R/V OCEANUS at a depth of 3,600 m (ALVIN Station) and subsamples (50 ml) were withdrawn for analysis. Radioactivity in the particulate, CO_2, and particulate filtrate fractions was determined for the subsamples, with activity corrected for dilution occurring when subsamples were removed (Fig. 2). Utilization could be measured in the sample for about 60 days, with the rate of activity increasing up to 12 days. In the parallel, decompressed sample, incorporation was 1.6 times greater at two days incubation. Respiration of the undecompressed sample was greater than that of the sample held at one atm. Cells were enumerated by direct counts of subsamples, using epifluorescent microscopy, at 8 hr, 60 days, and 130 days.

In Table 3, data for utilization are given for five undecompressed deep ocean water samples. It should be noted that concentrations of substrates used by Jannasch et al. (1976) and Jannasch and Wirsen (1977) were two orders of magnitude greater than in the experiments reported here. Percentage of respiration, with respect to total substrate utilized, i.e., respiration and incorporation,

TABLE 3: Utilization of organic compounds by deep ocean microbial populations.

Method	Location	Depth (m)	Concentration μg/ml	Rate of Utilization μg/l/day	% Added Substrate Utilized	% Respired
Undecompressed samples incubated with ^{14}C-glutamate*	N. W. Atlantic	1,800	5.58	133.00	2.38	84
Undecompressed samples incubated with ^{14}C-glutamate*	N. W. Atlantic	3,000	5.78	69.00	1.19	89
Undecompressed samples incubated with ^{14}C-casamino acids*	N. W. Atlantic	3,130	1.40	75.00	5.36	71
Undecompressed samples incubated with ^{14}C-casamino acids**	N. W. Atlantic	2,600	5.00	94.00	1.88	63
Undecompressed sample incubated with ^{14}C-glutamate	N. W. Atlantic	3,550	0.04	0.14	0.35	94

*Jannasch et al., 1976.
**Jannasch and Wirsen, 1977.

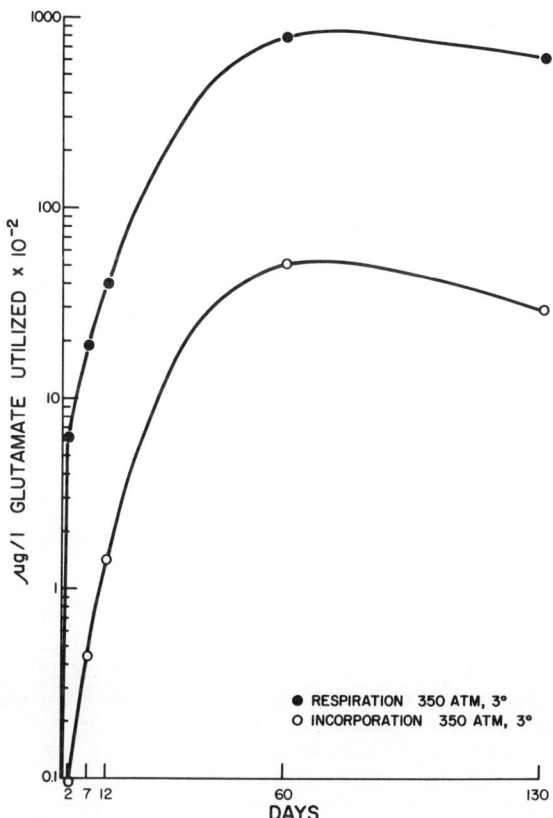

FIGURE 2: Uptake and respiration of C^{14}-labeled
glutamate by a deep-sea water sample maintained
undecompressed in the DOS (see text for details).

was higher than that reported by Jannasch et al. (1976). For samples with comparable substrate additions, sample utilization indicated a relationship with sample depth, i.e., more rapid rates were
measured for samples held at lower pressure levels.

High concentrations of dissolved substrate added to samples
for measurement of stimulated response may provide erroneous results
with respect to actual conditions in the deep ocean. Williams and
Carlucci (1976) compiled results of experiments measuring endogenous
activity, as well as stimulated activity, arguing that the latter
may not represent actual deep-sea microbial activity. The types of
measurements they presented were not entirely comparable, since
some assumptions were required in order to convert the results to
a carbon-utilization format. The methods and utilization rates
compiled by Williams and Carlucci (1976) and the measurement of
O_2 consumption reported for the station sampled by Jannasch and
Wirsen (1973) are listed in Table 4. Rates given are approximately
an order of magnitude lower than total utilization of undecompressed

TABLE 4: Utilization of organic compounds by deep
ocean microbial populations.

Method	Location	Depth (m)	Total Utilization µgC/1/day
Calculated from vertical diffusion-advection models*	Central Pacific	1,000-4,000	4.9×10^{-3}
	S. Pacific	1,000-4,000	1.8×10^{-3}
In situ incubation with ^{14}C-labeled substrate*	N. W. Atlantic	5,000	$5.4\text{-}35.6 \times 10^{-3}$
	W. Pacific	<1,000	$3.1\text{-} 9.2 \times 10^{-2}$
O_2 uptake by electron transport system*	E. Tropical Pacific	1,000-6,000	1.9×10^{-3}
		6,000	1.1×10^{-3}
ATP content of cells*	N. E. Pacific	1,000-4,000	1.2×10^{-3}
Growth of a repressurized bacterial culture*	N. Central Pacific	2,000	$1.1\text{-}16.7 \times 10^{-4}$
In situ O_2 consumption in sediment**	N. W. Atlantic	1,850	5.28 µgC/m^2/day

*Williams and Carlucci, 1976.
**Smith and Teal, 1973 (with O_2 conversion from (*)).

samples given in Table 2. These rates are two to four orders of
magnitude lower than rates obtained for the undecompressed measurements presented in Table 3. If O_2 consumption of a liter of sediment slurry can be considered to comprise an area of 100 cm by 10 cm
by 1 cm deep, and having a utilization rate of 5.28×10^{-1} µg C/liter/
day, the actual, total utilization in the sediment after 357 days can
be estimated to be approximately 60% of the total utilization reported
by Jannasch and Wirsen (1973) for their *in situ* water sample.

CONCLUSIONS

    In general, extrapolating from results of investigations pre-
sented here, as the concentration of substrate is increased, utilization will increase, as reported by Wirsen and Jannasch (1975) in
their metabolic studies of psychrophilic bacteria subjected to elevated pressure. Thus, substrate concentration is an important factor for *in situ* growth of marine bacteria.
    Utilization of substrates by bacteria in seawater at one atm
pressure is 1.6-4 times greater than that of bacteria in seawater
held under pressure. A portion of a given natural bacterial population is very probably inhibited by elevated *in situ* pressure, although the existence of barophilic organisms cannot be ruled out.
Results of substrate utilization studies, therefore, indicate the
expected response for natural bacterial populations under *in situ*
deep-sea conditions, based on studies comparing undecompressed and
decompressed activity of the total microbial populations in samples.
Further studies designed to observe and quantitate barophilic activity must focus on autecological responses of the individual cells
using the pressure-retaining samplers.
    The importance of maintaining deep ocean pressure and tempera-
ture conditions, from the time of sampling until completion of the

experiment and sample analysis, cannot be over-emphasized. Obligate psychrophiles have been shown to comprise a significant portion of the microbial population in the deep sea (Wirsen and Jannasch, 1975; Norkans and Stehn, 1978). Maintenance of *in situ* pressure may prove equally important in preventing loss of the pressure-adapted portion of the active deep-sea bacterial population. Investigations employing pressure-retaining samplers can, therefore, best address the question of the importance of barophilic bacterial populations in deep-sea biogeochemical cycles.

A critical and accurate description of growth characteristics of obligate barophiles under elevated pressure may require initial isolation in pure culture. The isolation of bacteria from undecompressed, concentrated samples using solid media in a helium-oxygen atmosphere is in progress (Taylor and Jannasch, 1976; Taylor, 1979). Observation of the effects of decompression on deep-sea populations using an autecological approach, for example, microautoradiography, would also be useful (Brock and Brock, 1966; Peroni and Lavarello, 1975; Hoppe, 1976).

If barophiles are found to survive at one atm, even for short periods of time after retrieval from the deep ocean, isolation in pure culture will be possible. Silica gel media has been shown to be useful for isolation, growth, and characterization of barophiles under pressure (Dietz and Yayanos, 1978). Pure cultures thus obtained, combined with autecological methods for direct observation of substrate-responsive cells (Kogure et al., 1979) will be valuable in elucidating the distribution and functional role of bacteria adapted to deep-sea conditions.

ACKNOWLEDGMENTS

Deep ocean research at the University of Maryland is supported by National Science Foundation Grant No. OCE 76-82655 and taxonomic studies of deep-sea bacteria by National Science Foundation Grant No. DEB 77-14646 A01.

REFERENCES

Albright, L. J. and J. F. Henigman, 1971. Seawater salts-hydrostatic pressure effects upon cell division of several bacteria. *Can. J. Microbiol.* 17: 1246-1248.

Arnold, R. M. and L. J. Albright, 1971. Hydrostatic pressure effects on the translation stages of protein synthesis in a cell-free system from *Escherichia coli*. *Biochim. et Biophys. Acta* 238: 347-354.

Ashby, R. E. and M. E. Rhodes-Roberts, 1976. The use of analysis of variance to examine the variations between samples of marine bacterial populations. *J. Appl. Bacteriol.* 41: 439-451.

Brock, T. D. and M. L. Brock, 1966. Autoradiography as a tool in microbial ecology. *Nature* 209: 734-736.

Brock, T. D., F. Passman, and I. Yoder, 1973. Absence of obligately psychrophilic bacteria in constantly cold springs associated with caves in southern Indiana. *Amer. Mid. Nat.* 90: 240-246.

Broecker, W., 1963. Radioisotopes and large-scale oceanic mixing. In: *The Sea*, Vol. 2, (M. N. Hill, ed.), pp. 88-108. Interscience Publishers, New York.

Carlucci, A. F., S. L. Shimp, P. A. Jumars, and H. W. Paerl, 1976. *In situ* morphologies of deep-sea and sediment bacteria. *Can. J. Microbiol.* 22: 1667-1671.

Certes, A., 1884. Sur la culture, a l'abri des germes atmospheriques, des eaux et des sediments rapportes par les expeditions du "Travailleur" et du "Talisman", 1882-1883. *C. R. Acad. Sci., Paris* 98: 690.

Colwell, R. R. and R. C. Kettling, Jr., 1974. Isolation and characterization of some deep sea bacteria. In: *Effects of Ocean Environment on Microbial Activities*, (R. R. Colwell and R. Y. Morita, eds.), pp. 227-241. University Park Press, Baltimore.

Colwell, R. R. and R. Y. Morita, 1974. Pressure effects. In: *Effects of the Ocean Environment on Microbial Activities*, (R. R. Colwell and R. Y. Morita, eds.), pp. 131-241. University Park Press, Baltimore.

Colwell, R. R. and P. S. Tabor, 1979. Microbiological studies of decompressed and undecompressed water samples collected with a deep ocean sampler. In: *The Dynamic Environment of the Ocean Floor*, (K. A. Fanning and F. T. Manheim, eds.), D. C. Heath & Co., Lexington. (in press).

Dietz, A. S. and A. A. Yayanos, 1978. Silica gel media for isolating and studying bacteria under hydrostatic pressure. *Appl. Environ. Microbiol.* 36: 966-968.

Hedén, C. G., 1964. General effects of pressure at physiological temperatures. *Bacteriol. Rev.* 28: 14-29.

Hoppe, H. -G., 1976. Determination and properties of actively metabolizing heterotrophic bacteria in the sea, investigated by means of microautoradiography. *Mar. Biol.* 36: 291-302.

Jannasch, H. W., K. Eimkjellen, C. O. Wirsen, and A. Farmanfarmaian, 1971. Microbial degradation of organic matter in the deep sea. *Science* 171: 672-675.

Jannasch, H. W. and G. E. Jones, 1959. Bacterial populations in seawater as determined by different methods of enumeration. *Limnol. Oceanogr.* 4: 128-139.

Jannasch, H. W. and C. O. Wirsen, 1973. Deep-sea microorganisms: *in situ* response to nutrient enrichment. *Science* 180: 641-643.

Jannasch, H. W. and C. O. Wirsen, 1977. Retrieval of concentrated and decompressed microbial populations from the deep sea. *Appl. Environ. Microbiol.* 33: 642-646.

Jannasch, H. W., C. O. Wirsen, and C. D. Taylor, 1976. Undecompressed microbial populations from the deep sea. *Appl. Environ. Microbiol.* 32: 360-367.

Jannasch, H. W., C. O. Wirsen, and C. L. Winget, 1973. A bacteriological pressure-retaining deep-sea sampler and culture vessel. *Deep-Sea Res.* 20: 661-664.

Johnson, F. H., H. Eyring, and M. J. Polissar, 1954. *The Kinetic Basis of Molecular Biology*. John Wiley, New York. 874 pp.

Kogure, K., U. Simidu, and N. Taga, 1979. A direct microscopic method for counting living marine bacteria. *Can. J. Microbiol.* 25: 415-420.

Liston, J., 1968. Distribution, taxonomy, and function of heterotropic bacteria on the sea floor. *Bull. Misaki Marine Biol. Inst.,Kyoto Univ.* 12: 97-104.

Menzel, D. W. and J. H. Ryther, 1970. Distribution and cycling of organic matter in the oceans. In: *Organic Matter in Natural Waters*, (D. W. Hood, ed.), pp. 31-54. University of Alaska: Institute of Marine Science Occasional Publication No. 1.

Mitskevitch, I. N. and A. E. Kriss, 1966. High-pressure tolerance of *Pseudomonas* sp., strain 8113, isolated from the bottom of a deep-water basin of the Black Sea. *Doklady Akademii nauk SSSR, Biological Science Section* 171: 822-824.

Morita, R. Y., 1976. Survival of bacteria in cold and moderate hydrostatic pressure environments with special reference to psychrophilic and barophilic bacteria. In: *The Survival of Vegetative Microbes*, (T. G. R. Gray and J. R. Postgate, eds.), pp. 279-298. Cambridge University Press, Cambridge.

Morita, R. Y., 1979. Current status of the microbiology of the deep-sea. *Ambio* (in press).

Niskin, S. J., 1962. A water sampler for microbiological studies. *Deep-Sea Res. Oceanogr. Abst.* 9: 501-503.

Norkans, B. and B. O. Stehn, 1978. Sediment bacteria in the deep Norwegian Sea. *Mar. Biol.* 47: 201-209.

Novitsky, J. A. and R. Y. Morita, 1978. Starvation-induced barotolerance as a survival mechanism of a psychrophilic marine vibrio in the waters of the Atlantic convergence. *Mar. Biol.* 49: 7-10.

Paul, K. L. and R. Y. Morita, 1971. Effects of hydrostatic pressure and temperature on uptake and respiration of amino acids by a facultatively psychrophilic marine bacterium. *J. Bacteriol.* 108: 835-843.

Peroni, C. and O. Lavarello, 1975. Microbial activities as a function of water depth in the Ligurian Sea: an autoradiographic study. *Mar. Biol.* 30: 37-50.

Pollard, E. C. and P. K. Weller, 1966. Effect of hydrostatic pressure on synthetic processes in bacteria. *Biochem. Biophys. Acta* 112: 573-580.

Pope, D. H., W. P. Smith, R. W. Swartz, and J. V. Landau, 1975. Role of bacterial ribosomes in barotolerance. *J. Bacteriol.* 121: 664-669.

Quigley, M. M. and R. R. Colwell, 1968. Properties of bacteria isolated from deep-sea sediments. *J. Bacteriol.* 95: 211-220.

Schwarz, J. R. and R. R. Colwell, 1975. Heterotrophic activity of deep-sea sediment bacteria. *Appl. Microbiol.* 30: 639-649.

Schwarz, J. R. and J. V. Landau, 1972a. Hydrostatic pressure effects on *Escherichia coli*: site of inhibition of protein synthesis. *J. Bacteriol.* 109: 945-948.

Schwarz, J. R. and J. V. Landau, 1972b. Inhibition of cell-free protein synthesis by hydrostatic pressure. *J. Bacteriol.* 112: 1222-1227.

Schwarz, J. R., A. A. Yayanos, and R. R. Colwell, 1976. Metabolic

activities of the intestinal microflora of a deep-sea inverte-
brate. *Appl. Environ. Microbiol.* 31: 46-48.

Seki, H. and D. G. Robinson, 1969. Effect of decompression on activity of microorganisms in sea water. *Int. Rev. Gesamten Hydrobiol.*54: 201-205.

Seki, H., E. Wada, I. Koike, and A. Hattori, 1974. Evidence of high
organotrophic potentiality of bacteria in the deep ocean. *Mar.
Biol.* 26: 1-4.

Sieburth, J. McN. and A. S. Dietz, 1974. Biodeterioration in the sea
and its inhibition. In: *Effects of the Ocean Environment on
Microbial Activities,* (R. R. Colwell and R. Y. Morita, eds.), pp.
318-326. University Park Press, Baltimore.

Smith, W. P., J. V. Landau, and D. H. Pope, 1976. Specific ion concentration as a factor in barotolerant protein synthesis in bacteria. *J. Bacteriol.* 126: 654-660.

Smith, W. P., D. H. Pope, and J. V. Landau, 1975. Role of bacterial
ribosome subunits in barotolerance. *J. Bacteriol.* 124: 582-584.

Smith, K. L. and J. M. Teal, 1973. Deep-sea benthic community respiration: an *in situ* study at 1850 meters. *Science* 179: 282-
283.

Swartz, R. W., J. R. Schwarz, and J. V. Landau, 1974. Comparative
effects of pressure on protein and RNA synthesis isolated from
marine sediments. In: *Effects of Ocean Environment on Microbial
Activities,* (R. R. Colwell and R. Y. Morita, eds.), pp. 173-179.
University Park Press, Baltimore.

Tabor, P. S. and R. R. Colwell, 1976. Initial investigations with a
deep ocean *in situ* sampler. *Proc. MTS/IEEE OCEANS '76,* Washington, D. C., pp. 13D-1-13D-4.

Taylor, C. D., 1979. Growth of a bacterium under a high-pressure
oxyhelium atmosphere. *Appl. Environ. Microbiol.* 37: 42-49.

Taylor, C. D. and H. W. Jannasch, 1976. Activity of bacteria at high
pressure in an oxy-helium atmosphere. *Abstr. Annu. Meet. Amer.
Soc. Microbiol., 1976, N42:* 177.

Tuttle, J. H. and H. W. Jannasch, 1976. Microbial utilization of
thiosulfate in the deep sea. *Limnol. Oceanogr.* 21: 697-701.

Williams, P. M. and A. F. Carlucci, 1976. Bacterial utilization of
organic matter in the deep sea. *Nature* 262: 810-811.

Wirsen, C. O. and H. W. Jannasch, 1975. Activity of marine psychrophilic bacteria at elevated hydrostatic pressures and low temperatures. *Mar. Biol.* 31: 201-208.

Wirsen, C. O. and H. W. Jannasch, 1976. Decomposition of solid
organic materials in the deep sea. *Environ. Sci. & Technol.* 10:
880-886.

Yayanos, A. A., A. S. Dietz, and R. Van Boxtel, 1979. Isolation of
a deep-sea barophilic bacterium and some of its growth characteristics. *Science* (in press).

Yayanos, A. A. and E. C. Pollard, 1969. A study of the effects of
hydrostatic pressure on macromolecular synthesis in *Escherichia
coli. Biophys. J.* 9: 1464-1482.

Zimmerman, A. M., ed. *High Pressure Effects on Cellular Processes.*
Academic Press, New York.

ZoBell, C. E., 1954. Some effects of high hydrostatic pressure on
apparatus on the Danish Galathea Deep-Sea Expedition. *Deep-
Sea Res.* 2: 24-32.

ZoBell, C. E., 1968. Bacterial life in the deep sea. *Bull. Misaki Mar. Biol. Inst., Kyoto Univ.* 12: 77–96.

ZoBell, C. E. and R. Y. Morita, 1957. Barophilic bacteria in some deep sea sediments. *J. Bacteriol.* 73: 563–568.

ZoBell, C. E. and C. H. Oppenheimer, 1950. Some effects of hydrostatic pressure on the multiplication and morphology of marine bacteria. *J. Bacteriol.* 60: 771–781.

ZoBell, C. E. and R. Y. Morita, 1959. Deep-sea bacteria. *Galathea Report* 1: 139–154. Dansk Videnskabs Forlog, A/S, Copenhagen.

ZoBell, C. E., 1941. Apparatus for collecting water samples from different depths for bacteriological analysis. *J. Mar. Res.* 4: 173–188.

Measurement and Instrument Needs Identified in a Case History of Deep-Sea Amphipod Research

A. A. Yayanos

ABSTRACT: The identification of measurement and instrument needs in deep-sea biology is approached by reviewing the techniques that have been most important in the development of our research with deep-sea amphipods. Such techniques include deep-sea photography, retrieval of live animals with pressure vessels and microbial cultivation. Refinements in the use of these methods are discussed. Future research, it is felt, would benefit from the use of image analysis technology and from the use of traps which catch rising particles.

INTRODUCTION

Amphipods can be caught on sandy beaches, in the Mariana Trench at a 10,500 m depth and in most other parts of the oceans. This ubiquitous distribution and the predictability of catching amphipods make them attractive as research organisms. Two milestones achieved in our laboratory studies are the cultivation of both deep-sea amphipods and deep-sea microbes associated with live or dead amphipods. This paper will discuss key instruments and techniques that have assisted our efforts and will underscore measurement methodology which could assist continued work.

307

FIGURE 1: A 35 mm deep-sea camera was focused on
bait (dead fish) for over 35 h. This picture was
taken at 20 h 50 min and reveals hundreds of amphipods on and around the bait. The free vehicle
camera was deployed at 11° 21.3'N, 142° 13.8' E
and recovered at 11° 20.6'N, 142° 4.7'E. Water
depth was approximately 10,599 m.

DEEP-SEA PHOTOGRAPHY IN AMPHIPOD RESEARCH

Our work has focused on deep-sea amphipods because evidence both
from photographs as well as from trapping indicates that they live at
all depths of the ocean and can be captured easily. Isaacs (see
Hessler et al., 1978) deployed cameras many times into the Peru-Chile
trench in 1972. Along with these camera drops, R. Wisner and R.
McConnaughey (personal communication) deployed baited traps and hooks.

TABLE 1: An analysis of deep-sea photography.

Camera Variables:

    Moored or towed
    Still or motion pictures
    Stereo or two-dimensional views
    Color or black and white film or video tape

Experimental Design Variables:

    Bait type
    With or without bait
    Duration of deployment
    Trapped or free animals
    Deploy with other measuring equipment:
        electrodes
        hydrophones
        photometers
        sonar
        nets

Data Handling and Analysis:

    Visual inspection
    Mandated archiving of data
    Analyze with computer-assisted image analysis

At depths greater than approximately 7,000 m, fish were never seen in the photographs, but amphipods were always present in great numbers. The baited traps also returned only amphipods from these great depths. However, at shallower depths of approximately 5,000 m, both fish and amphipods were abundant in photographs (Hessler et al., 1978), but the latter were far easier to catch. These and other observations prompted Hessler et al. (1972) to point out the potential significance of amphipods for physiological studies. Subsequent photographic research has shown the presence of amphipods in the Philippine Trench (e.g., Hessler et al., 1979) and in the Mariana Trench (Fig. 1).

DEEP-SEA PHOTOGRAPHY IN PROSPECT

    Table 1 is an attempt to summarize the research potential of deep-sea photography and to list some of the variables that can be controlled. The monograph edited by Hersey (1967) includes articles dealing with many of these variables.
    Towed cameras provide considerable amounts of information covering sizeable areas of the sea floor. Over 4,000 pictures taken with towed cameras at seven stations with depths between 6,758 and 8,930 m

on a Scripps expedition in the Pacific Ocean have been studied by
Lemche et al. (1976) for a faunistic survey. Pictures of the Palau
Trench floor revealed a few amphipods possibly feeding on a piece of
presumed organic debris. Quite apparent in the pictures is the disturbance of the bottom sediments by organisms, a process of interest
to geologists. For example, Pequegnat et al. (1972) studied photographs of the floor of the Gulf of Mexico to determine the influence
of organisms on the sediment.

Leaving a camera on the sea floor for a long period of time has
only recently been done with some fascinating results. Paul et al.
(1978) photographed the sea floor at a single locus without added
bait for more than 200 days. They saw (among other things) a holo-
thurian deposit a fecal cast which appeared to decay during their
observations over 137 days. They also saw a hemichordate leave a
mucous sheath which was nearly entirely decomposed after 12 days.
The rapid rates of decay that are inferred from these observations
are claimed to be more rapid than processes found in other kinds of
experiments (Grassle, 1977; Jannasch and Wirsen, 1977). The processes of recolonization studied by Grassle (1977) may possibly not
be comparable to those of decay.

Whereas the kind of bait which is sufficient to attract amphipods is known, the attractants for many other organisms are not.
Perhaps as a result of studying pictures taken as a function of
type of bait, we could learn the basis for swarming by holothurians
(see Heezen and Hollister, 1971) or how to attract other animals.
Lemche et al. (1976) believe they were able to detect bacterial
slicks on the sediment. Could bacterial slicks attract deposit
feeders?

Cameras have been deployed with other instruments to provide
complimentary information. Menzies et al. (1963) designed and deployed a camera-grab device which photographed the sediment and
then sampled it. The number of animals in the sample was compared
to that in the photographs. Such an approach can help delimit the
accuracy of a sampling method as a function of species (Menzies et
al., 1963), sediment type, water depth, sea state and other variables.

The use of sonar with cameras has been of some value but has
not been done with baited cameras. The latter offers interesting
possibilities for acoustic studies: the density of organisms approaching bait would (or might) change with time; the kinds of organisms attracted might change with time; their three-dimensional
distribution as a function of time could be known; and the physical
properties of trapped individuals could be determined. Such determinations would appear to provide an unusual amount of information
to test sound scattering predictions. The knowledge thereby gained
could be used to assess populations as well as to test theories.

Cameras can also be deployed with photometers, such as was
done by Breslau et al. (1967). The rigging was such that photographs were taken only when a luminescent signal of a prescribed
intensity was noted. Such simultaneous observations would appear
to be necessary to understand the origin of luminescence in the
sea to a fuller extent.

The animals in the above observations were free. Photography
of trapped animals could reveal, among other things, their be-

havior, the length of intermolt periods, feeding habits, swimming
speed, and effects of confinement in traps of various sizes.
These observations could corroborate results (e.g., on behavior and
pleopod beats) from laboratory studies of deep-sea animals, such as
those reported by Yayanos (1978).

Other aspects of deep-sea photography that are listed in Table
1 are the handling and analysis of data. As scattered collections
are consolidated and augmented through, hopefully, mandated archiving
then the use of image analysis to scan thousands of frames will be
useful and necessary.

CULTIVATION OF DEEP-SEA ANIMALS AND MICROBES

Life can be studied in the deep-sea with submersibles (Grassle
et al., 1975; Jannasch and Wirsen, 1977) and with instruments placed
in the deep ocean (Smith and Hessler, 1974; Smith et al., 1976). Organisms from the deep sea can be investigated with insulated pressure-retaining samplers and traps (Macdonald and Gilchrist, 1969;
Jannasch and Wirsen, 1973; Colwell and Tabor, 1977; Jannasch and
Wirsen, 1977; Yayanos, 1977; Macdonald and Gilchrist, 1978; Yayanos,
1978; and others discussed in Brauer, 1972). Field and laboratory
studies provide overlapping sets of results. *In situ* studies will
clearly be necessary for the study of populations, communities and
the deep-sea ecosystem.

A pressure-retaining amphipod trap (PRAT) (Fig. 2) has been
developed (Yayanos, 1977) and successfully used on several deep-sea
expeditions to recover live amphipods from depths as great as 5,900 m
(Yayanos, 1978). On recovery of a PRAT, it is used as a high pressure aquarium by attaching it to a high pressure pumping system
(Yayanos, 1978; Yayanos and Van Boxtel, personal communication). The
effluent from the PRAT can be analyzed for products of metabolism,
for oxygen, and for microbes. One problem encountered in this study
is that small samples of sea water must be analyzed because the
volume of the PRAT is small (about 400 ml). (Analyses which use 2
ml or less of sea water are being developed.)

We have been able (unpublished data) to demonstrate that an amphipod becomes immobile when decompressed from 580 bars (the pressure at the depth where it was captured) to 200 bars. Also, immobi-
lity occurs when the animals are warmed at 580 bars to 10°C. These
observations lead to the definition of bounds for the permissible
temperature and pressure changes within which the retrieval and
maintenance of live deep-sea amphipods will be possible. To date
animals have been maintained for about 17 days, but longer periods
of maintenance should be possible.

Amphipods live in association with a variety of microbes
(Yayanos, 1976; Hessler et al., 1978). Although attempts to isolate some of these microbes have been an on-going project in this
laboratory for several years, only recently have we met with success.

A requirement of our methods has been that the cultivation for
the isolations be done only at a high pressure equal to that where
an inoculum originated. Some of our initial experiments were done
with microbes that were on or in the bodies of amphipods that were

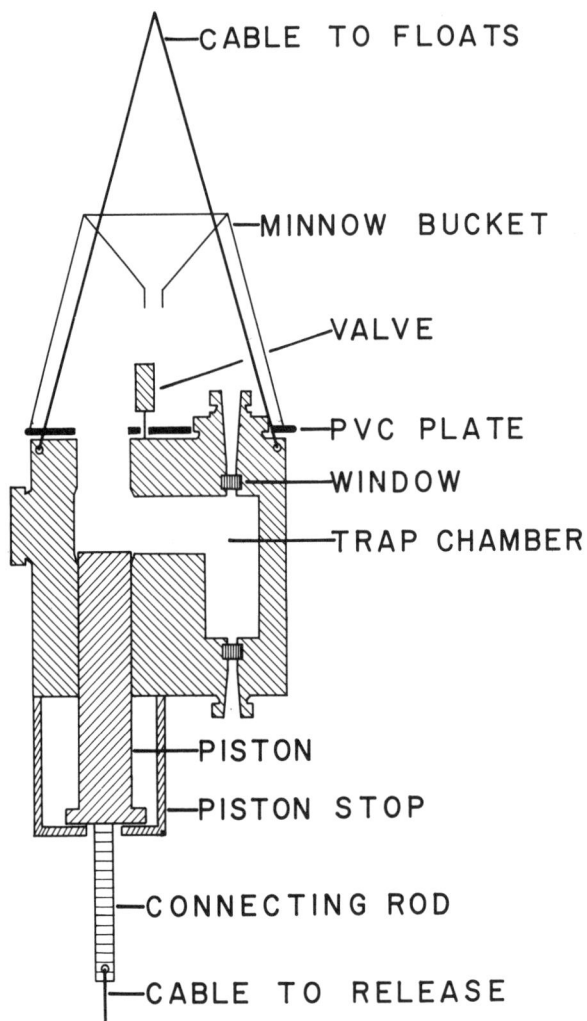

FIGURE 2: A schematic of a PRAT (pressure-retaining amphipod trap) showing how it might appear on the sea floor. The piston is anchored
to the sea floor by ballast attached to the release mechanism. Floats are attached to the
trap body thereby pulling it over the piston
until the piston abuts the piston stop. When
the ballast is released, springs (not shown) pull
the piston into the trap body and seal the trap
chamber and any animals in it. Details are given
in Yayanos (1977, 1978).

retrieved dead either with decompression or with warming. We
reasoned that bacteria should be hardier than amphipods. For reasons which are not entirely known, it proved impossible to obtain
growth of turbid suspensions of bacteria in liquid cultures at high
pressures. We tried to grow colonies of bacteria on agar at high
pressures. The surface of agar was covered with a fluorocarbon (FC-

FIGURE 3:   Tubes of silica gels made with nutrient
medium and innoculated with a bacterium, *Micrococcus
euryhalis*.   The colonies grew at 101 bars.

70) liquid (suggested by J. Hauxhurst, personal communication) to
separate the agar surface from the hydraulic fluid.   But deep-sea
bacteria did not colonize the agar surfaces at high pressures.   The
unsuccessful attempts with this technique have been due to the presence of non-viable microbes.   This circumstance could have arisen
from the thermal shock during retrieval of the amphipods or from
warming encountered during an unfortunate power failure in our
laboratory.
     Silica gels can be used instead of agar to immobilize cells
in a manner permitting their growth into colonies.   Silica gels
with nutrients can be made at $2^{\circ}C$,   inoculated prior to gelation
and then incubated at high pressures (Dietz and Yayanos, 1978)
(Fig. 3).   This technique has resulted in the recovery of several
psychrophilic bacteria and one barophilic one.   Barophilic bacteria
are defined as those that grow preferentially or exclusively at

high pressures. There are possibly other barophilic bacteria among
the isolates in our pressure vessels, which remain to be tested for
a barophilic response.

The barophilic bacterium, CNPT-3, which we have studied,
(Yayanos et al., 1979) has a generation time of 6 to 9 hours at
580 bars and 2 to 4°C. These conditions approximate closely those
prevailing at 5,700 m in the central North Pacific Ocean. Most
work reported in the literature on deep-sea marine bacteria has
been done with organisms that do not grow optimally under deep-sea
conditions and that do not grow as rapidly as CNPT-3 does. It remains as an active area of research to find out what the normal
spectrum of microbial inhabitants and their growth rates in the
deep-sea are.

In the brief time available, it is best to summarize our ex-
periences by saying that microorganisms do not appear to be suscep-
tible to decompression inactivation but are very sensitive to thermal
inactivation. Whether this pattern of sensitivity extends to the
depths of the sea greater than 5,700 m remains to be determined.
Thus we recommend that warming of samples be avoided and that cultivation be done under the high pressures at the origin of the samples. We have not addressed the problem of the axenic nature of
samples and will not do so here beyond saying that the deep sea is
an open system into which intruders from shallow waters are not ex-
cluded.

Image analysis techniques will undoubtedly assist microbiological research. First, such instruments can automatically count and
size bacteria in a light microscopic view of a culture. Second, bacteria differing from each other in size or shape can be separately
enumerated in mixed cultures. Mixed culture studies are otherwise
difficult. Finally, image analysis instruments will augment and accelerate the development of autofluorescent methods (e.g., Mink and
Dugan, 1977) and fluorescent labelling techniques (e.g., Young and
Smith, 1964; Schmidt, 1973; Fliermans and Schmidt, 1975).

RISING PARTICLES GENERATED BY AMPHIPODS AND OTHER ORGANISMS

Lipids are present in organisms for a variety of functions.
In the marine environment, some lipids are accumulated in mass
quantities in a striking fashion by a variety of creatures to
serve functions peculiar to life in a water atmosphere. Lipid-
containing structures are used for buoyancy devices in organisms
ranging in size from microscopic plankton to large whales (e.g.,
Denton, 1961; Smayda, 1970; Alexander, 1972; Bone, 1973; Clark,
1978a,b,c; Yayanos et al., 1978). Accumulations of low density
lipids in eggs of larvae of many organisms also provide buoyancy
and thereby allow for dispersal by mass water movements. Russell's
(1976) monograph contains many examples illustrating the role of
lipids in fish egg dispersal. Fulton (1973) documented the rise
of benthically produced lipid-containing eggs of the copepod, *Calanus plumchrus*.

Deep-sea amphipods were observed by Hessler et al. (1972) to
have oily bodies. Yayanos and Nevenzel (1978) found that an amphi-

TABLE 2: Rising Particle Hypothesis.

Processes Generating Rising Particles:

Inefficient predation leading to lipids being released
Laying of lipid-containing eggs
Consumption of fatty meals by scavengers which then
become buoyant
Lipids released from animals by their autolytic enzymes
and by decomposer microorganisms

Measurement Needs:

Design and deployment of rising particle traps
Methods to analyze rapidly the contents of net tows for
rising particles

pod from the Philippine Trench had 26% of its dry weight as lipid, demonstrating that large amounts of lipid are present in an ubiquitous benthic organism, the amphipod. If released from the body of an amphipod or of any other organism, such lipid-rich material would rise upward in the water column. Table 2 summarizes most of the processes which could lead to the production of lipid-containing particles. The magnitude of such an upward flux of matter in the ocean has not been determined since only sediment traps and not rising particle traps have been deployed. A situation in the deep ocean that would cause mass mortality of mobile animals could result in an amount of feeding and predation that might make the generation of rising particles sizeable.

Lipid-containing particles will not only rise, but also do so rapidly because of their low density (compared to water). Calculations by Yayanos and Nevenzel (1978) show that such objects could rise 5,000 m in less than a day. They have referred to such a flux as the rising particle hypothesis.

Two approaches (Table 2) appear possible for testing the importance of this rising particle hypothesis. One is to use traps (similar to sediment traps) that passively collect particles moving upwards. Calculations based only on the abundance of fish eggs in parts of the North Atlantic suggest that such a passive collection of rising particles should be feasible. A second approach would be to collect actively samples with nets or water samplers (large volume bottles or pumps) and then to analyze the particulates present for buoyancy at temperatures and pressures identical to those at the depth of the samples. An instrument would need to be designed for such analyses. The collection of buoyant particles by means of traps should be the easiest approach. Someday such devices should contribute to our knowledge of vertical mixing processes and to our ability to catch eggs of deep-sea organisms. Eggs of rattails, deep-sea eels and some brotulids are buoyant (Marshall, 1970) with development taking place as the eggs rise through the

water column (Merrett, 1978). The ability to catch early developmental stages of deep-sea organisms would allow the study of em-
bryos in pressure-retaining traps. These stages in the life cycle
of an organism are likely the ones most sensitive to pollutants
whose effects could be determined with captive, viable eggs.

Clearly, measurements of this upward flux are needed to better
understand the pathway of carbon in the sea, to provide us with the
developmental stages of deep-sea animals and to evaluate the feasibility of placing industrial, urban or nuclear wastes on the sea
floor. Toxic compounds or elements that could bind to lipid-containing particles might be rapidly returned to the upper portion of the
oceanic water column.

REFERENCES

Alexander, R. McN., 1972. The energetics of vertical migration by
    fishes. *Symp. Soc. Exp. Biol.* 26: 273-274.
Bone, Q., 1973. A note on the buoyancy of some lantern-fishes
    (Myctophoidei). *J. Mar. Biol. Assoc. U. K.* 53: 619-633.
Brauer, R. W., ed., 1972. *Barobiology and the Experimental Biology
    of the Deep-Sea.* University of North Carolina Sea Grant Pro-
    gram, Chapel Hill, North Carolina. 428 pp.
Breslau, L. R., G. L. Clarke, and H. E. Edgerton, 1967. Optically
    triggered underwater cameras for marine biology. In: *Deep-
    Sea Photography*, (J. B. Hersey, ed.), pp. 223-228. The Johns
    Hopkins Press, Baltimore, Maryland.
Clarke, G. L. and C. J. Hubbard, 1959. Quantitative records of
    the luminescent flashing of oceanic animals at great depths.
    Limnol. Oceanogr. 4: 163-180.
Clarke, M. R., 1978a. Structure and proportions of the spermaceti
    organ in the sperm whale. *J. Mar. Biol. Assoc. U. K.* 58: 1-17.
Clarke, M. R., 1978b. Physical properties of spermaceti oil in the
    sperm whale. *J. Mar. Biol. Assoc. U. K.* 58: 19-26.
Clarke, M. R., 1978c. Buoyancy control as a function of the sperma-
    ceti organ in the sperm whale. *J. Mar. Biol. Assoc. U. K.* 58:
    27-71.
Colwell, R. R. and P. Tabor, 1977. Heterotrophic activity in de-
    compressed and undercompressed water samples collected with a
    deep ocean sampler. In: *Benthic Boundary Layer and Geochemistry
    of Interstitial Waters.* Proc. Joint Oceanog. Assembly, Edinburgh,
    Scotland.
Denton, E. G., 1961. The buoyancy of fish and cephalopods. *Prog. in
    Biophys.* 11: 177-234.
Dietz, A. S. and A. A. Yayanos, 1978. Silica gel media for isolating
    and studying bacteria under hydrostatic pressure. *Appl. Environ.
    Microbiol.* 36: 966-968.
Fliermans, C. B. and E. L. Schmidt, 1975. Fluorescence microscopy:
    Direct detection, enumeration and spatial distribution of bac-
    teria in aquatic systems. *Arch. Hydrobiol.* 76: 33-42.
Fulton, J., 1973. Some aspects of the life history of *Calanus
    plumchrus* in the Strait of Georgia. *J. Fish. Res. Bd. Canada* 30:
    811-815.

Grassle, J. F., 1977. Slow recolonization of deep-sea sediment. *Nature* 265: 618–619.

Grassle, J. F., H. L. Sanders, R. R. Hessler, G. T. Rowe, and T. McLellan, 1975. Pattern and zonation: a study of the bathyal megafauna using the research submersible ALVIN. *Deep-Sea Res.* 22: 457–481.

Heezen, B. C. and C. D. Hollister, 1971. *The Face of the Deep*, Oxford University Press, London. 659 pp.

Hessler, R. R., J. D. Isaacs, and E. L. Mills, 1972. Giant amphipod from the abyssal Pacific Ocean. *Science* 175: 636–637.

Hessler, R. R., C. Ingram, A. A. Yayanos, and B. Burnett, 1978. Scavenging amphipods from the floor of the Philippine Trench. *Deep-Sea Res.* 25: 1029–1047.

Jannasch, H. W. and C. O. Wirsen, 1973. Deep-sea microorganisms: *In situ* response to nutrient enrichment. *Science* 180: 641–643.

Jannasch, H. W. and C. O. Wirsen, 1977. Microbial life in the deep sea. *Sci. Amer.* 236: 42–52.

Lemche, H., B. Hansen, F. J. Madsen, O. S. Tendal, and T. Wolff, 1976. Hadal life as analyzed from photographs. *Vidensk. Meddr. Dansk. Naturh. Foren.* 139: 263–336.

Macdonald, A. G. and I. Gilchrist, 1978. Further studies on the pressure tolerance of deep-sea crustacea, with observations using a new high-pressure trap. *Mar. Biol.* 45: 9–21.

Macdonald, A. G. and I. Gilchrist, 1969. Recovery of deep seawater at constant pressure. *Nature* 222: 71–72.

Marshall, N. B., 1953. Egg size in Arctic, Antarctic and deep-sea fishes. *Evolution* 7: 328–341.

Marshall, N. B., 1966. *The Life of Fishes*, pp. 272–273. Universe Books, New York.

Marshall, N. B., 1971. *Exploration in the Life of Fishes*, Harvard University Press, Cambridge, Massachusetts. 204 pp.

Menzies, R. J., L. Smith, and K. O. Emery, 1963. A combined underwater camera and bottom grab: a new tool for investigation of deep-sea benthos. *Int. Revue ges. Hydrobiol.* 48: 529–545.

Merrett, N. R., 1978. On the identity and pelagic occurrence of larval and juvenile stages of rattail fishes (family Macrouridae) from 60°N, 20°W and 53°N, 20°W. *Deep-Sea Res.* 25: 147–160.

Mink, R. W. and P. R. Dugan, 1977. Tentative identification of methanogenic bacteria by fluorescence microscopy. *Appl. Environ. Microbiol.* 33: 713–717.

Paul, A. Z., 1976. Deep-sea bottom photographs show that benthic organisms remove sediment cover from manganese nodules. *Nature* 263: 50–51.

Paul, A. Z., E. M. Thorndike, L. G. Sullivan, B. C. Heezen, and R. D. Gerard, 1978. Observations of the deep-sea floor from 202 days of time-lapse photography. *Nature* 272: 812–814.

Pequegnat, W. E., B. M. James, A. H. Bouma, W. R. Bryant, and A. D. Fredericks, 1972. Photographic study of deep-sea environments of the Gulf of Mexico. In: *Texas A&M University Oceanographic Studies*, Volume 3, (R. Rezak and V. J. Henry, eds.), pp. 67–128, Gulf Publishing Company, Houston.

Russell, F. S., 1976. *The Eggs and Planktonic Stages of British Marine Fishes*. Academic Press, New York. 524 pp.

Schmidt, E. L., 1973. Flourescent antibody techniques for the study of microbial ecology. *Bull. Ecol. Res. Comm. (Stockholm)* 17: 67–76.

Smayda, T. J., 1970. The suspension and sinking of phytoplankton in the sea. *Oceanogr. Mar. Biol. Ann. Rev.* 8: 353–414.

Smith, K. L., Jr., C. H. Clifford, A. H. Eliason, B. Walden, G. T. Rowe, and J. M. Teal, 1976. A free vehicle for measuring benthic community metabolism. *Limnol. Oceanogr.* 21: 164–170.

Smith, K. L., Jr., and R. R. Hessler, 1974. Respiration of benthopelagic fishes: *In situ* measurements at 1230 meters. *Science* 184: 72–73.

Yayanos, A. A., 1976. Recovery of amphipods from deep ocean trenches at near *in situ* pressures: description of the instrument and scanning electron microscopy of animal-associated microbes. *Biophys. J.* 16: 180a.

Yayanos, A. A., 1977. Simply actuated closure for a pressure vessel: design for its use to trap deep-sea animals. *Rev. Sci. Instrum.* 48: 786–789.

Yayanos, A. A., 1978. Recovery and maintenance of live amphipods at 580 bars pressure from 5700 m depths of the central North Pacific Ocean. *Science* 200: 1056.

Yayanos, A. A., A. A. Benson, and J. C. Nevenzel, 1978. The pressure-volume-temperature (PVT) properties of a lipid mixture from a marine copepod, *Calanus plumchrus*: Implications for buoyancy and sound scattering. *Deep-Sea Res.* 25: 257–268.

Yayanos, A. A., A. S. Dietz, and R. Van Boxtel, in press. Isolation of a deep-sea barophilic bacterium and some of its growth characteristics. *Science*.

Yayanos, A. A. and J. C. Nevenzel, 1978. Rising particle hypothesis: Rapid ascent of matter from the deep ocean. *Die Naturwissenschaften* 65: 255–256.

Young, M. R. and A. V. Smith, 1964. The use of euchrysine in staining cells and tissues for fluorescence microscopy. *J. Roy. Microscop. Soc.* 82: 233–244.

In Situ **Studies of Deep-Sea Communities**

J. Frederick Grassle

ABSTRACT: Submersibles are likely to be increasingly
important in studies of deep-sea benthic communities.
To use the submersible efficiently, future studies
will be done at permanent bottom stations where geologists, chemists, biologists, and physical oceano-
raphers can cooperate to study the factors controlling the major fluxes ot organic and inorganic constituents in the benthic boundary layer.

By perturbing deep-sea communities in various ways
it is possible to sort out how individual life cycles of populations and interrelationships among
species are related to patterns of environmental
variation. The relatively constant physical regime
of the deep sea makes it an excellent laboratory
for studying the mechanisms by which populations
are adapted to the environment and each other.

INTRODUCTION

Until recently deep-sea biologists have been at a disadvantage
relative to colleagues working on shallow-water systems. Sampling

Contribution Number 4329 from the Woods Hole Oceanographic Institution.

the deep-sea benthos from surface ships does not provide the close-
up view of the system which is a prerequisite for the study of
the more accessible benthic habitats. The relatively sessile organisms of benthic habitats are particularly suited for field experimentation; field experiments on rocky shores, coral reefs, and soft
bottom benthos have provided data pertinent to a number of fundamental issues in ecology (Connell, 1974; Paine, 1977). Ideally, simple
yes or no answers to basic questions can be achieved by designing
experiments where all factors vary naturally except one, which is
controlled. Experiments of this sort are most successful in areas
where the physical environment is sufficiently uniform to provide
adequate experimental controls and where populations do not fluctuate erratically.

In contrast with the detailed experimental studies in accessible
areas, even the most up-to-date techniques for working on deep-sea
communities from surface vessels are like trying to study forest
ecology from an airship. Large trawls and nets give snapshot views
of a remarkably high diversity of life integrated over transects
that may be a mile or more long. Box cores deployed from surface
vessels provide a more accurate sampling method but cannot be positioned with sufficient accuracy to obtain precisely spaced samples.
Large quantitative box core samples are useful in describing what
is there but do not enable us to dissect the processes operating in
the complex and diverse deep-sea ecosystem. To understand organisms'
relationships to the environment it is necessary to sample repeatedly
for periods longer than the generation time of species and on spatial
scales smaller than the area encountered by single individuals
through an entire life cycle. *In situ* studies are required, and the
ability to return to the same spot is of utmost importance. This can
be achieved with manned or unmanned vehicles equipped with precise
navigational capabilities and visually operated manipulators.

Some of the work that is accomplished with submersibles can also
be accomplished with free vehicles (e.g., Smith et al., 1976). Free
vehicles equipped with transponders, timers, or a combination of these
have the advantage that they cost little more than a single submersible dive and can be reused. They are also not limited by depth or
location. However, existing designs have a number of disadvantages:

1) The operation of the instrument cannot be observed.
2) Manipulation is limited to release commands.
3) There is no control of placement on the bottom.
4) There are no back-up systems if a major component fails.
5) The risk of not retrieving the instrument increases with
    increasing length of deployment.

It is not possible to deploy a free vehicle for an entire year
with confidence that all the electronic and mechanical equipment will
still function at the end of a year. Even well-designed instruments
are prone to failure unless the same instrument has been repeatedly
deployed. One of the most important uses of submersibles is to observe the initial operation of free vehicle instruments.

Unmanned submersibles are likely to be important in the future
but none is presently available for biological work. A somewhat less

FIGURE 1:   The ALVIN basket attached to the front
of the submersible.   Each instrument has a T-
shaped handle so that it can be picked up and
operated by the mechanical manipulator seen behind the basket.   The equipment in the basket
includes three box cores, tube cores, a mud
tray, a specimen retrieval box, and wood panels.

versatile unmanned vehicle could be designed to do the same work for
most of the tasks assigned to a manned submersible.   However, the
main advantage of submersibles is to increase the resolution of
sampling and observation through continuous visual observation.
For example, submersibles are the only way to

1)   Sample precisely small-scale features such as sediment forms,
     rocks, or individual organisms.
2)   Sample repeatedly with respect to specific experiments or
     features of the bottom over time spans up to several years.
3)   Push sampling devices and other instruments into the bottom
     without disturbance of the sediment-water interface.
4)   Locate objects and sample in complex rocky topography where
     tethered devices could not move over the bottom without en-
     countering obstacles.
5)   Sample specific layers in the water column.

FIGURES 2 and 3 show a free vehicle with
four mud trays used to study rates of colonization of deep-sea animals. The column in the center is an AMF acoustic release transponder. The
trays are deployed and recovered with the lids
closed.

BOTTOM STATIONS

     The availability of submersibles led to the establishment of
permanent bottom stations at 1800 m and 3640 m in the Atlantic off
New England and at 2000 m in the Tongue of the Ocean. At these and
other stations to be established in the future a number of processes
can be studied simultaneously.

Figure 1 illustrates an ALVIN basket with tube cores, a box core, specimen box, wood panels, and a sediment tray ready for deployment at a permanent bottom station at 3640 m depth. We have used the trays with azoic deep-sea sediment to demonstrate low rates of colonization in the deep sea (Grassle, 1977). The wood panels have been used for studies of wood borers and studies of the role of wood in deep-sea communities (Turner, 1973, 1977). Figures 2 and 3 illustrate mud trays, similar to those deployed with ALVIN, set up as a free vehicle. The lids release when magnesium bolts holding the lids closed dissolve and the shock cords running to the top of the transponder snap the lids open. When the transponder drops the weight, rods running through the ends of the shock cords slide out and the lids close. The lids are held shut against a latex tubing gasket by ceramic magnets. The first time the free vehicle was

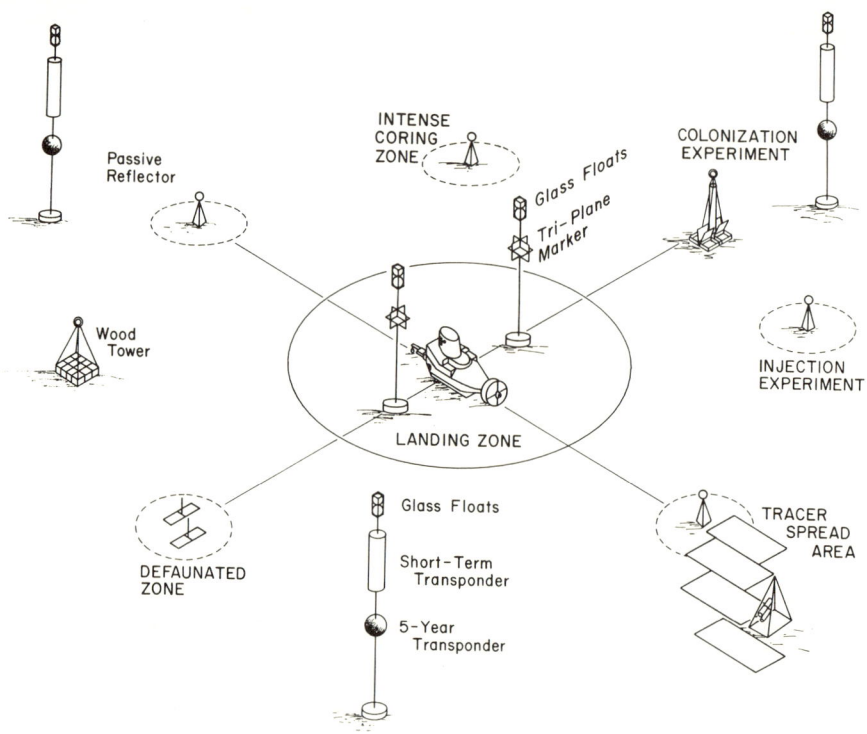

FIGURE 4: A plan for a permanent bottom station.
Various experiments involving either tracers or
perturbations of benthic communities are shown
at positions that can be relocated in the navigation net consisting of three long-term and
three short-term transponders.

deployed some of the flotation was lost and it had to be recovered
by the ALVIN. The second time everything worked well and we observed the lids closing with no disturbance of the sediment surface.
     The first markers were simple metal sonar targets placed on the
bottom in a line at the 1800 m permanent bottom station. These reflectors show up on a sonar screen at horizontal distances of 400-
500 m (Bland et al., 1976). In areas of uneven bottom terrain or
where rocks present false targets, considerable dive time is lost
searching for the station. The next generation of bottom station
markers were triplane reflectors attached to floats and 37 kHz
pingers on the bottom. These navigational aids could only be used
by the submersible as it neared the bottom. Navigational errors
at the surface have also occasionally resulted in loss of dive time.
The best results have been obtained using acoustic transponders that
can reply to signals sent from the surface vessel. Acoustic transponders have been used to mark bottom stations for periods of up to
one year, and with the recent availability of transponders using
lithium batteries we expect to mark stations for periods up to five
years without replacing the markers. By using two or more transponders it is possible to navigate the submersible on the bottom
with an accuracy of ±10-20 m and a repeatability of ±1-3 m (Bland et
al., 1976). Figure 4 illustrates a bottom station with a variety of

experiments positioned with a long-term navigation net consisting of both short-term and long-term transponders.

To use the submersible efficiently, the participants in future bottom station studies should include geologists, chemists, biologists, and physical oceanographers. Cooperation among specialists in a number of disciplines is necessary if we are to understand the factors controlling the major fluxes of organic and inorganic materials in the benthic boundary layer.

Benthic organisms depend on the flux of large particles from the surface for food. Sediment trap moorings are needed to understand the spatial and temporal variation of these fluxes. Fine-scale studies of currents are needed to understand patterns of deposition and resuspension of particles. These patterns both influence, and are influenced by, the activities of animals. Within the sediment the activities of individual organisms are important both in determining rates of burial of materials and exchanges across the sediment-water interface. Studies of the flux of materials and rates of activity in the benthic boundary layer will involve long-term experiments with both inert particles and radioisotope tracers.

Although less accessible, the deep sea has some advantages for the study of fundamental ecological questions. At least over short distances the bottom is not very patchy; the patchiness that does occur is on a scale of individuals (Jumars, 1975, 1976, 1978). The relative uniformity of the bottom means that it is easier to obtain adequate control samples. However, the environmental changes that occur are infrequent disturbances on the scale of the ambit of individual organisms. These disturbances usually result from activities of individual organisms or material settling from the surface and may be very important in the development of community structure (Dayton and Hessler, 1972; Grassle, 1978). Many shallow-water environments are so variable on several time scales that it is very difficult to resolve different sources of disturbance. In the deep sea it is possible to refer to disturbance events and their effects on the population biology of individual species. Because of the great range of life histories observed in deep-sea species (Grassle, 1978) and the uniformity of the environment, the deep-sea is an excellent laboratory for studying the relationship between individual life cycles and biotic interrelationships and the pattern of environmental change. So many variables are constant in deep-sea experiments that perturbation experiments more nearly approach a laboratory situation.

The disadvantages of the deep-sea environment for experimentation are the low density of organisms and the relatively long time spans required to measure significant changes. Initial colonization can be measured in a matter of months with a $\frac{1}{4} m^2$ experimental tray, but in one experiment, after two years, adults of only two species had become established in the trays (Grassle, 1977).

PREVIOUS STUDIES

Rate measurements of a number of biological processes in the deep sea have been accomplished with experiments at permanent stations using ALVIN. Rates of microbial activity are an order of magnitude

lower than in shallow water (Jannasch and Wirsen, 1973; Jannasch et
al., 1976; Wirsen and Jannasch, 1976). Rates of oxygen consumption of
deep-sea communities are 2-3 orders of magnitude lower than in shallow water (Smith and Teal, 1973). Rates of growth and maturation of
benthic organisms range from two months for wood-boring bivalves
(Turner, 1973) to two years for the faster growing infaunal bivalves
(Grassle, 1977), to 50 years for the slow growing long-lived species
of infaunal bivalves (Turekian et al., 1975).

Three current explanations for the relatively low rates of biological processes include 1) physiological limitation resulting from
high pressure, 2) low levels of food reaching the bottom, and 3) the
relatively constant environmental conditions (more particularly the
pattern of environmental variation relative to the life span of deep-
sea species). These explanations relate to the factors determining
rates of life processes that have evolved in different environments.
Studies on a molecular level indicate that enzymes have evolved to
function at rates normal for shallow water despite high pressures
(Hochachka, 1975; Low and Somero, 1975). There is no reason to suppose that animals with high rates of activity could not evolve in
the deep sea and observations of behaviour of individual animals
clearly indicate at least occasional high rates of activity (Cohen,
1977; Paul et al., 1978).

The relatively low rate of food input to the deep sea limits the
number of animals (Sanders et al., 1965; Rowe et al., 1974), which
taxa are abundant (Hessler and Jumars, 1974; Rex, 1976) and the for-
aging strategy of deep-sea species (Jumars and Fauchald, 1977).
Species with at least a good short-term supply of food, such as scavengers on large bait falls, might be expected to have high rates of
activity. The macrourid fish (*Coryphaenoides armatus*) that is normally attracted to bait is capable of at least short periods of rapid
movement and has access to occasional large concentrations of food.
Smith (1978) has suggested that *Coryphaenoides* is well adapted to
these occasional windfalls by maintaining energy stores in the form
of lipid and glycogen. Each individual travels over a large enough
area so that enough food is found to support a large motile animal.
Even so, the respiration of these animals is low in comparison with
shallow-water marine fish that have been studied (Smith, 1978).

In practice it is virtually impossible to separate the effects
of amount of food from the frequency and spatial pattern of the food
supply reaching the bottom. High metabolic rates, high rates of
growth, early maturation, rapid increase in population size are all
adaptations to changing environments. Individuals with high metabolic rates are better able to maintain normal bodily functions independent of changes in the environment. Some species maintain
their independence of environmental change by the movement of individuals to new habitats and sources of food when existing resources
are depleted. Omnivorous species such as *Coryphaenoides armatus*
maintain their food supply in this way.

The more sessile benthic species rely on life cycle adaptations
to insure that some portion of the species population finds a suitable habitat. Some species in the deep sea can maintain populations
from generation to generation in a single place and it is likely that
these species will have long life spans (e.g., *Tindaria callistifor-*

mis, that may have an extended life span of 100 years) (Turekian et al., 1975). Wood-boring species in the genus *Xylophaga* live in a substratum where both the habitat and food supply may be depleted in a generation. The adults are not motile enough to find new pieces of wood, and the planktonic larvae serve as the dispersal phase in the life cycle. Such species require the ability to increase rapidly and produce large numbers of young so that some portion of the population will find a suitable substratum. A simple way to rank species in terms of life cycle adaptations is by the ratio of the generation time to the length of time the habitat remains suitable for reproduction (Southwood et al., 1974).

In addition to being important in the evolution of life histories, the pattern of habitat regeneration is important in maintaining the high diversity of species (Grassle, 1967; Hessler and Sanders, 1967; Sanders, 1968, 1969; Coull, 1972; Hessler and Jumars, 1974; Jumars, 1976; Rex, 1976; Gage, 1977) . The importance of disturbance in marine communities was first demonstrated by Dayton (1971). The initial assumption that the deep sea was spatially uniform prevented the direct application of this idea to the deep sea (Dayton and Hessler, 1972) even though disturbance was known to be important in other species-rich communities such as coral reefs (Grassle, 1973; Connell, 1976; Sale, 1977; Connell, 1978) and rain forests (Janzen, 1970; Grubb, 1977).

The combination of long life cycles and infrequent small-scale disturbance are likely to result in a nonequilibrium micromosaic. This does not necessarily mean that considerable specialization of animals in terms of biotic interaction is unlikely. Just as some animals may adapt to particular kinds of abiotic disturbance, others may adapt to the presence of other living organisms. The individuals of the longer-lived species may provide a particularly suitable microhabitat for other species.

Upon further experimentation it may be possible to understand whether the source and extent of disturbance are important in the evolution of particular life history features. Major unanswered questions involve understanding how the kinds of interrelationships between species relate to patterns of environmental variation and the importance of genetic mechanisms in life cycle adaptations (Grassle and Grassle, 1978).

SUMMARY

Our preliminary information indicates that the deep sea may be a very fragile ecosystem. The species of the deep sea have evolved in a relatively unchanging physical environment. Environmental changes of such magnitude as to have little or no effect on land or in shallow water may severely affect deep-sea communities. Measurements of rates of response of populations to disturbance can be of practical value in predicting the effects of such disturbances as dumping of wastes or deep-sea mining.

ACKNOWLEDGMENTS

I thank Linda Morse-Porteous for a number of helpful suggestions on the manuscript. This material is based on research supported by the National Science Foundation under Grant Nos. OCE-7621968 and OCE-7819820.

REFERENCES

Bland, E. L., J. D. Donnelly, and L. A. Shumaker, 1976. ALVIN users manual. *Woods Hole Oceanogr. Inst. Tech. Memorandum 3-76*. 36 pp.
Cohen, D. M., 1977. Swimming performance of the gadoid fish, *Antimora rostrata* at 2400 meters. *Deep-Sea Res*. 24: 275-277.
Connell, J. H., 1974. Ecology: field experiments in marine ecology. In: *Experimental Marine Biology*, (R. N. Mariscal, ed.), pp. 21-54. Academic Press, New York.
Connell, J. H., 1976. Competitive interactions and the species diversity of corals. In: *Coelenterate Ecology and Behavior*, (G. O. Mackie, ed.), pp. 51-58. Plenum, New York.
Connell, J. H., 1978. Diversity in tropical rain forests and coral reefs. *Sci*. 199: 1302-1310.
Coull, B. C., 1972. Species diversity and faunal affinities of meio-benthic copepoda in the deep sea. *Mar. Biol*. 14: 48-51.
Dayton, P. K., 1971. Competition, disturbance and community organization: the provision and subsequent utilization of space in a rocky intertidal community. *Ecol. Monogr*. 41: 351-389.
Dayton, P. K. and R. R. Hessler, 1972. The role of biological disturbance in maintaining diversity in the deep sea. *Deep-Sea Res*. 19: 199-208.
Gage, J. D., 1977. Structure of the abyssal macrobenthic community in the Rockall Trough. In: *Biology of Benthic Organisms*, (B. F. Keegan, P. Ó. Céidigh, and P. J. S. Boaden, eds.), pp. 247-260. Pergamon Press, Oxford.
Grassle, J. F., 1967. Influence of environmental variation on species diversity in benthic communities of the continental shelf and slope. Ph.D. Dissertation, Duke University, *No. 68-2723 University Microfilms Inc.*, Ann Arbor, Michigan.
Grassle, J. F., 1973. Variety in coral reef communities. In: *The Geology and Biology of Coral Reefs*, (R. Endean and O. A. Jones, eds.), pp. 247-270. Academic Press, New York.
Grassle, J. F., 1977. Slow recolonization of deep-sea sediment. *Nature* 265: 618-619.
Grassle, J. F., 1978. Diversity and population dynamics of benthic organisms. *Oceanus* 21: 42-49.
Grassle, J. F. and J. P. Grassle, 1978. Life histories and genetic variation in marine invertebrates. In: *Marine Organisms*, (B. Battaglia and J. Beardmore, eds.), pp. 347-363. Plenum Press, New York.
Grubb, P. J., 1977. The maintenance of species richness in plant communities: the importance of the regeneration niche. *Biol. Rev*. 52: 107-145.

Hessler, R. R. and P. A. Jumars, 1974. Abyssal community analysis from replicate box cores in the central North Pacific. *Deep-Sea Res.* 21: 185-209.

Hessler, R. R. and H. L. Sanders, 1967. Faunal diversity in the deep sea. *Deep-Sea Res.* 14: 65-78.

Hochachka, P. W., 1975. Why study proteins of abyssal organisms. *Comp. Biochem. Physiol.* 52B: 1-2.

Jannasch, H. W. and C. O. Wirsen, 1973. Deep-sea microorganisms: *in situ* response to nutrient enrichment. *Science* 180: 641-643.

Jannasch, H. W., C. O. Wirsen, and C. D. Taylor, 1976. Undecompressed microbial populations from the deep sea. *Appl. Environ. Microbiol.* 32: 360-367.

Janzen, D. H., 1970. Herbivores and the number of tree species in tropical forests. *Amer. Natur.* 104: 501-528.

Jumars, P. A., 1975. Environmental grain and polychaete species diversity in a bathyal benthic community. *Mar. Biol.* 30: 253-266.

Jumars, P. A., 1976. Deep-sea species diversity: does it have a characteristic scale? *J. Mar. Res.* 34: 217-246.

Jumars, P. A., 1978. Spatial autocorrelation with RUM (Remote Underwater Manipulator): vertical and horizontal structure of a bathyal benthic community. *Deep-Sea Res.* 25: 589-604.

Jumars, P. A. and K. Fauchald, 1977. Between-community contrasts in successful polychaete feeding strategies. In: *Ecology of Marine Benthos*, Belle W. Baruch Library in Marine Science, No. 6, (B. C. Coull, ed.), pp. 1-20. University of South Carolina Press.

Low, P. S. and G. N. Somero, 1975. Pressure effects on enzyme structure and function *in vitro* and under simulated *in vivo* conditions. *Comp. Biochem. Physiol.* 52B: 67-74.

Paine, R. T., 1977. Controlled manipulations in the marine intertidal zone, and their contributions to ecological theory. In: *The Changing Scenes in Natural Sciences 1776-1976*, Academy of Natural Sciences (Philadelphia), Special Publication 12. pp. 245-270.

Paul, A. Z., E. M. Thorndike, L. G. Sullivan, B. C. Heezen, and R. D. Gerard, 1978. Observations of the deep-sea floor from 202 days of time-lapse photography. *Nature* 272: 812-814.

Rex, M. A., 1976. Biological accommodation in the deep-sea benthos: comparative evidence on the importance of predation and productivity. *Deep-Sea Res.* 23: 975-987.

Rowe, G. T., P. T. Polloni, and S. G. Hornor, 1974. Benthic biomass estimates from the Northwestern Atlantic Ocean and northern Gulf of Mexico. *Deep-Sea Res.* 21: 641-650.

Sale, P. F., 1977. Maintenance of high diversity in coral reef fish communities. *Amer. Natur.* 111: 337-359.

Sanders, H. L., 1968. Marine benthic diversity: a comparative study. *Amer. Natur.* 102: 243-282.

Sanders, H. L., 1969. Benthic diversity and the stability-time hypothesis. *Brookhaven Symposia in Biology* 22: 71-81.

Sanders, H. L., R. R. Hessler, and G. R. Hampson, 1965. An introduction to the study of deep-sea benthic faunal assemblages along the Gay Head-Bermuda transect. *Deep-Sea Res.* 12: 845-867.

Smith, K. L., 1978. Metabolism of the abyssopelagic rattail *Coryphaenoides armatus* measured *in situ*. *Nature* 274: 362-364.

Smith, K. L., Jr., and J. M. Teal, 1973. Deep-sea community respiration: an *in situ* study at 1850 meters. *Science* 179: 282-283.

Smith, K. L., Jr., C. H. Clifford, A. H. Eliason, B. Walden, G. T.
Rowe, and J. M. Teal, 1976. A free vehicle for measuring benthic
community metabolism. *Limnol. Oceanogr.* 21: 164-170.

Southwood, T. R. E., R. M. May, M. P. Hassell, and G. R. Conway, 1974.
Ecological strategies and population parameters. *Amer. Natur.*
108: 791-804.

Turekian, K. K., J. K. Cochran, D. P. Kharkar, R. M. Cerrato, J. R.
Vaisnys, H. L. Sanders, J. F. Grassle, and J. A. Allen, 1975.
Slow growth rate of a deep-sea clam determined by ^{228}Ra chronol-
ogy. *Proc. Nat. Acad. Sci. U. S. A.* 72: 2829-2832.

Turner, R. D., 1973. Wood-boring bivalves, opportunistic species in
the deep-sea. *Science* 180: 1377-1379.

Turner, R. D., 1978. Wood, molluscs and deep-sea food chains. *Bull.
Amer. Malacol. Union for 1977*: 13-19.

Wirsen, C. O. and H. W. Jannasch, 1976. Decomposition of solid organic materials in the deep sea. *Environ. Sci. Technol.* 10: 880-
886.

Description of a Discrete-Depth Multisampler for Midwater Trawling

Roderick Mesecar

ABSTRACT: Trawl net systems can provide useful information on the distribution and abundance of small animals. This paper recognizes, however, that different net configurations catch different types and sizes of animals. Therefore, the major emphasis of this paper is to describe a modular net system that can be reconfigured for multiple sampling at discrete-depths down to 1000 m.

INTRODUCTION

Since this workshop is dedicated to advanced concepts for biological measurement techniques, or the lack thereof, it is appropriate that the topic of trawl nets be included. Some marine biologists are pursuing an understanding of the nature and role of nekton (free swimmers) in the marine ecosystem. Where actual specimens are required for positive identification and data on age, growth, and other life-history factors, the trawl net appears to be the only sampling instrument available.

Different net systems catch different types and sizes of animals, and, generally, for the many net configurations, good evaluations of their catch efficiencies have not been done. The animal catch com-

parisons that have been done suffer from inadequate sample size, differences in net mesh, and flow characteristics, more precise estimates
of filtration efficiency, and a general lack of experimental design
(Harrison, 1967; Aron and Collard, 1969; Atsatt and Seapy, 1974).

Because of the lack of quantitative net intercomparisons, there
is no rational basis for specifying the net type and size for sampling
the broad range of nektonic animals. Biological oceanographers have
traditionally used midwater trawls with mouth openings less than 8 m^2.
These nets have provided useful data on the distribution and abundance
of small animals, but they appear to be inadequate in sampling more
agile specimens. The larger animals with generally better developed
sensory systems and strong swimming ability seem adept at avoiding
nets. If one can conclude that bigger nets can catch bigger animals,
then a contribution to the biological community would be the development of a midwater net system with discrete-depth sampling capability
and a mouth opening greater than 250 m^2.

The major portion of this paper describes a modular net system
that has been used for multiple samplings at discrete depths down to
1000 m, and which has the adaptive flexibility for mouth openings of
approximately 1, 5, 8, and 80 m^2. The essence of a discrete-depth,
multisampling net system is the mechanization for monitoring environmental parameters about the net and the ability to command the en-
trapment and isolation of specimens at a chosen depth without contamination by specimens from surrounding depths.

MIDWATER TRAWL SYSTEM COMPONENTS

The essential components of the net and command/monitoring system
are illustrated in Figure 1. The midwater trawl system consists of an
Isaacs-Kidd Midwater Trawl (IKMT) with a 5.4 m^2 mouth-opening and a
1 m^2 multiple sampler (MS) with five codend nets. An 11 mm armored
coaxial cable is used for towing the trawl system and conducting
electronic commands and data. The cable is terminated at a commu-
tator swivel. A two-conductor armored cable extends along the bridle and safety lines from the swivel to a pressure housing containing
the *in situ* electronics located on top of the MS. Data on sampling
depths, temperatures, and water flow through the MS are displayed
aboard ship on a digital-panel meter and a strip-chart recorder.
Upon command from the surface, an electric motor is actuated in the
pressure housing, which causes an external shaft to rotate one revolution. This shaft is coupled to a series of five cams that sequentially release hinged bars under tension from elastic shock
cords. With the release of each bar, one codend net is closed and
another opened. Positive actuation of net release is confirmed at
the surface recorder.

The IKMT net preceding the MS is 4 m long; the five nets aft
of the MS are 4.6 m long. A 5 mm mesh net is used for the cylindrical portions of the MS nets and for a liner in the IKMT net.
The terminal 1.8 m sections of the MS nets are made of 0.57 mm
mesh netting. They are conical, with 22 cm diameter collecting
buckets of 0.57 mm mesh. These buckets are mounted along the inside of a 0.8 m diameter aluminum ring connected to the MS by
safety cables (Fig. 1). The lengths of the safety lines are long

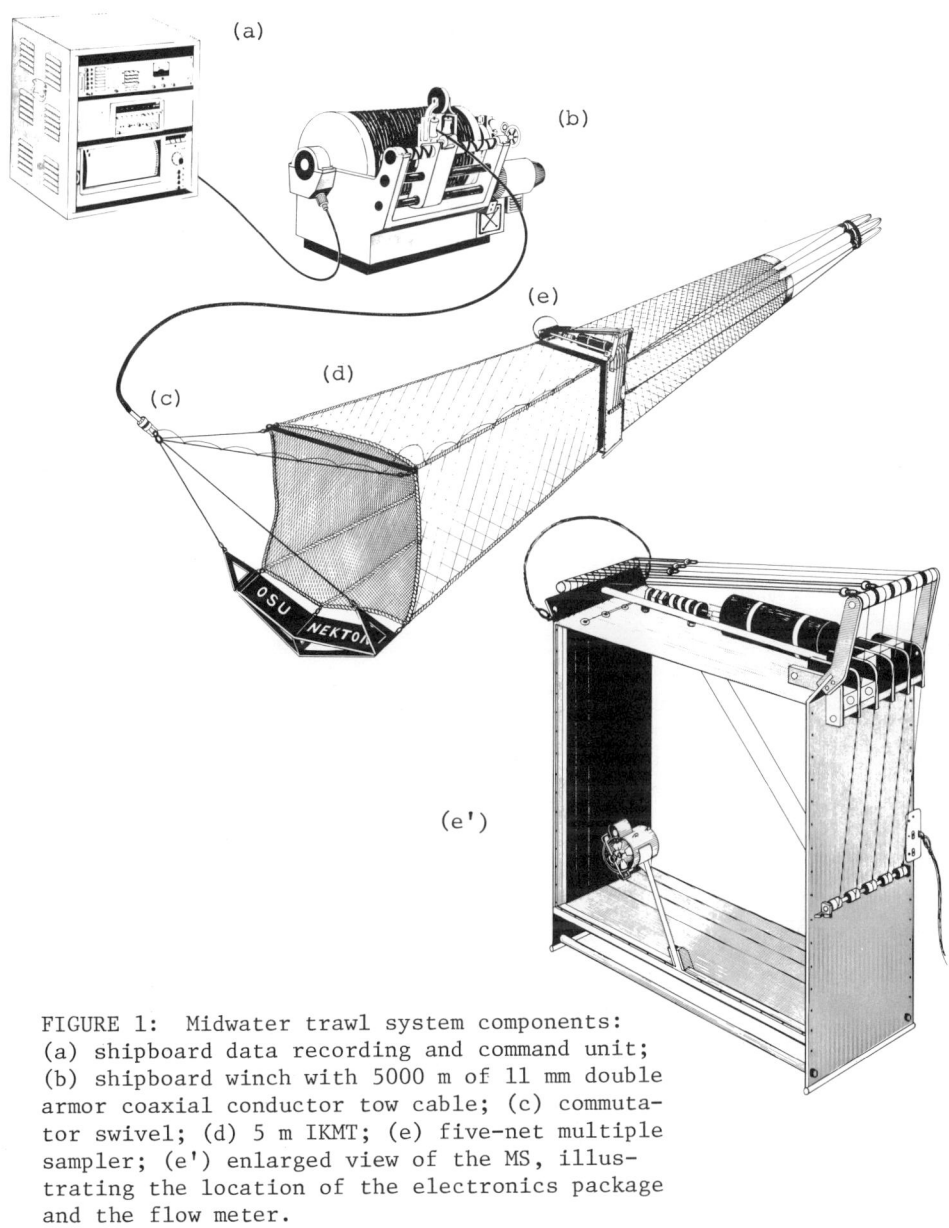

FIGURE 1: Midwater trawl system components:
(a) shipboard data recording and command unit;
(b) shipboard winch with 5000 m of 11 mm double
armor coaxial conductor tow cable; (c) commuta-
tor swivel; (d) 5 m IKMT; (e) five-net multiple
sampler; (e') enlarged view of the MS, illustrating the location of the electronics package
and the flow meter.

enough to prevent the net from bagging, but short enough to relieve
strain on the net.

The MS box (Fig. 1 e'), constructed from 6 mm thick aluminum,
is 101 cm wide, 101 cm tall, and 38 cm deep. Stainless steel rods
2 cm in diameter are used for the net bars. Fully assembled, excluding nets, the MS weighs 41 kg in air.

The codend net opening and closing sequences are similar to

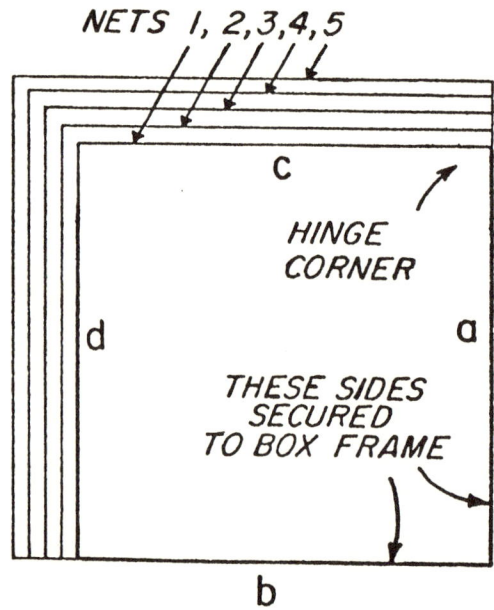

FIGURE 2: Illustration of the codend multiple
net opening and closing sequence.

those described by Bé (1962). Figure 2 is a frontal schematic lay-
out for the interleaving of the five nets. Net sides a and b which
do not move are attached to comparable sides of the codend box.
Side c is attached to one of five pivoting solid bars that are
hinged as noted. When the bar is released, side c will pivot until
it is flat against side a and side d is against b. This operation
closes the first net and opens the second. Subsequent bar releases duplicate this sequence until all the nets are used.

In practice, it is important to keep the net sides tight
against the box sides to prevent animals from slipping around them
and into another sample net. To provide the net side seal needed
for this large codend unit, a 1.5 mm diameter steel cable is sewn
into the leading edge of side d of each net and kept very taut. The
pivoting bar is sewn in the leading edge of side c. As shown in
Figure 3, the cable is an integral part of the net opening and closing
mechanism and has tension equivalent to that exerted by the elastic
cord. When the net bars are in a cocked position, the energy stored
in the elastic cords is used to make the net transitions.

During the bar transition from a cocked to released position, the
cable momentarily goes slack. While this is happening, two small
cable stops prevent the cable from changing its length. The taut cable length is self-adjusting and eliminates dependence on the dimensions of the fabric collar remaining stable to maintain the sealed
edges. The five codend nets are totally isolated from each other aft
of the MS box. Transition time from one net opening to another is
less than two seconds.

A general block diagram of the command/monitor electrical system
is illustrated in Figure 4. Signals to and from the MS are trans-

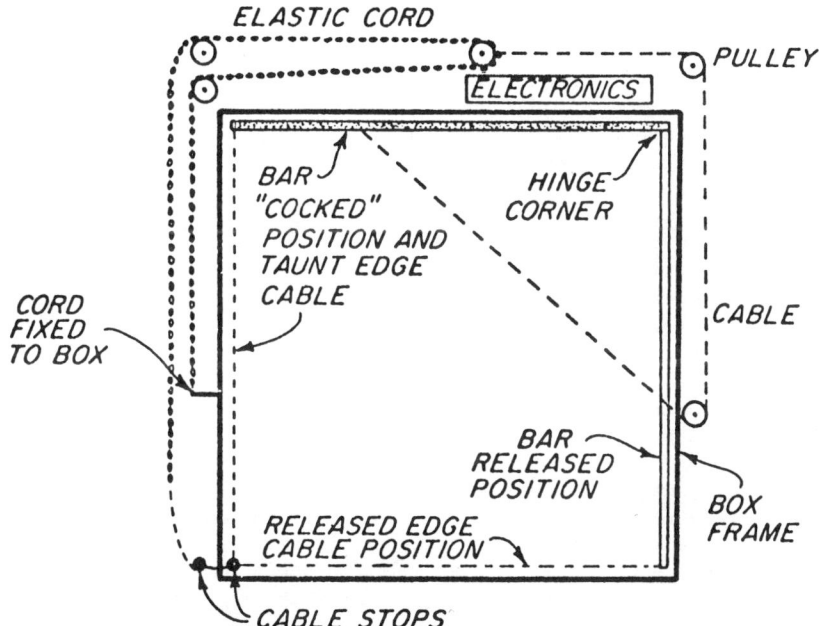

FIGURE 3: Schematic illustration of the pulley, elastic cord and cable linkage system used on the codend sample controller box to maintain the net seal integrity around the codend net edges.

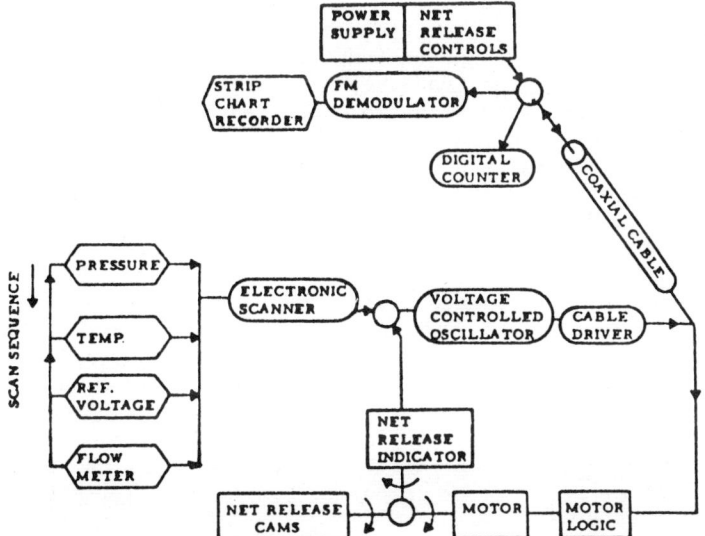

FIGURE 4: Block diagram of the command/monitor electronics system.

mitted down the 4600 m of 11 mm diameter coaxial conductor tow cable,
through the electric swivel, and into the pressure housing on the MS.
This pressure housing contains the net bar actuators, some transducers, and scanning and signal transmission electronics.

Tone frequencies, similar to the telephone system, are used for
the command designation to the MS. Three basic command functions
are used: one to change the MS net that is sampling, one to start
an automatic monitoring sequence of the MS net status and transducers, and one to lock selectively on any monitoring channel, i.e.,
temperature, pressure, and other parameters. In the automatic mode,
pressure (depth), temperature, flow (revolutions), and net opening
are scanned sequentially and transmitted as frequency-modulated (FM)
signals from the transducers on the MS to the recording unit on deck.

When a net release command is actuated, a 2 rpm motor in the MS
pressure housing makes one shaft revolution. Cams located on top of
the MS are coupled directly with the motor shaft in the electronics
package. During one motor-shaft revolution, one cam turns 360°, re-
leasing one lever bar that holds the net bar in a cocked position.
This operation is repeated five times for release of five nets.
During the motor operating period, an FM signal that identifies which
net is opened is transmitted to the surface. Actuation of the net-
release motor interrupts the automatic scan sequence of the transducer outputs.

Between net actuations, the electronic scanner sequentially con-
nects the transducer outputs for discrete periods of time to a voltage-
controlled oscillator (VCO) generating FM signals. The VCO output is
coupled through an electronic driver stage to the coaxial cable. Signals are displayed aboard ship in two ways: on an analog strip-chart
recorder, and on a digital-panel meter. The recorder offers a quick
observation of a tow pattern of the trawl. The digital readouts,
which are periodically written on the strip-chart record, give the
greatest resolution. The maximum resolving capability in the monitoring system is one part in one thousand of transducer output signal.
Depth is monitored with a potentiometric type Servonic model H-172-5
pressure transducer. The transducer is calibrated in the laboratory
with a secondary-pressure standard. The depth resolution is ±1 m and
it is transducer-limited.

Water temperature is sensed by a 10 kΩ thermistor (Yellow Springs
Instrument Corporation) at 25°C. It is calibrated to ±0.02°C in an
ice bath with a Hewlett-Packard quartz thermometer and referenced by a
platinum thermometer and Mueller bridge. The thermistor time constant
is 30 sec.

A flowmeter, with a reed switch to produce a signal every 1000
revolutions of the impeller, is mounted near the center of the MS
frame and connected by electrical cable to the electronic unit on top
of the MS. During some tows, a flowmeter is mounted on a stanchion
of the trawl depressor as well as in the MS opening. The distance
recorded by the MS flowmeter is 77.4% (s.d. = 6.7%, n = 7) of the distance recorded by the meter in front of the mouth of the trawl. Therefore, the volume filtered is calculated as (1/0.774) [MS flowmeter distance (m)] x (5.4 m^2).

A voltage reference is used to excite the pressure and temperature
transducer and to act as a reference value. This reference is moni-

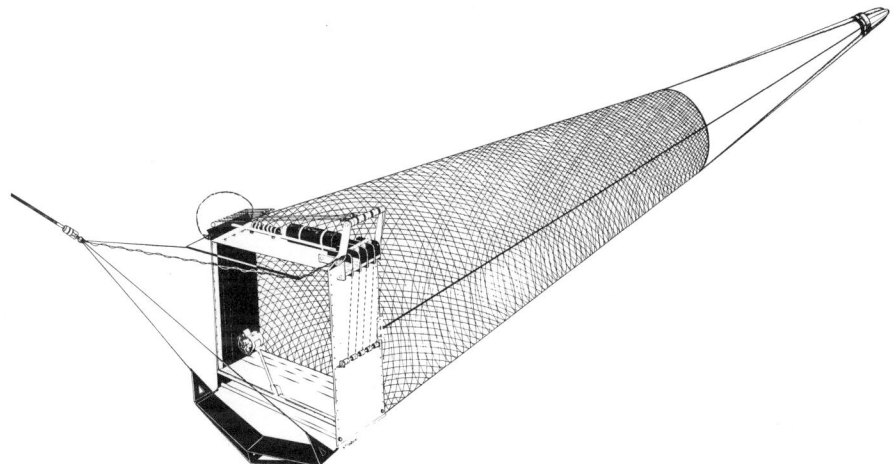

FIGURE 5: Multiple sampling net with 1 m^2 opening.
The net is 7 m long with the initial two-thirds
fabricated from 0.62 cm mesh and the final one-
third from 0.57 mm mesh.

tored each scan cycle along with the transducer signals. If the
reference has changed during a tow, it indicates not only an error in
data but an electrical malfunction in the transmitting electronics.

By removing the collector portion of the IKMT net configuration
(Fig. 1), the remaining net can be towed as a 1 m^2 mouth opening, as
shown in Figure 5. In both cases, blade-type depressors are used to
acquire the depressing force necessary to get the net to sampling
depth. These steel depressors are effective at tow speeds of 2-3
knots.

The net-release mechanism shown with the previous net configurations has also been fitted to a rectangular meter net that uses a
different opening and closing concept, as seen in Figure 6. The
outer frame dimensions of the net are about 3 m x 5 m, which, when
being towed at a 45° mouth angle, provides an opening of 7.5 m^2.
The top of the uppermost net (shown in Figure 6) is attached to the
upper support bar. The bottom of this same net is common to the top
of the net below and is attached to a moveable bar. One to five nets
can be configured in this fashion so that, initially, all of the nets
go into the water nestled against the upper support structure. When
a net release command is given, one of the moveable bars is dropped,
thus opening a net. Upon the next release command, the next dropping
bar will close the first net and open a second net. This sequencing
can be continued until all nets have been used. Care has been taken
to insure that the bifolds along each side of the net and between
moveable bars are properly nestled to prevent animals from getting
into the nets when they are not being fished. These nets are roughly
10 m long and are fabricated from 0.62 cm mesh. A multiplane kite-
otter design depressor, which is fabricated from aluminum for handling
purposes, is used with this net system.

The fourth net configuration (Fig. 7) has a much larger mouth opening (80 m^2) than any presented so far. When in use, it is about 10 m in diameter and 50 m long. The net is fabricated from 1.6 cm mesh netting the first 30 m back from the mouth, with the next 10 m made with 0.62 cm mesh netting. This net also has the option of discrete-depth, multiple sampling because the last 7 m of it consists of the MS net shown in Figure 5. However, in this application of the MS, the net closure commands are done with a preset timer unit mounted on top of the MS box frame. Depth of the net and local water temperature are presently monitored on shipboard via an acoustic link to a transducer electronics package on the net head rope. Coupling the MS net-release mechanism to the acoustic-command link is scheduled for the near term.

This net system requires a vessel with dual trawl winches to accommodate the otter-boards used in maintaining the net-mouth opening. If one of the winches were outfitted with conducting tow cable, the standard command/monitor electronics could be used with this net system as well. In applications, to date, this net system has been used with commercial fishing trawlers.

SUMMARY

A certain amount of net design detail has been deleted in an attempt to cause the reader to focus on the net systems as a sampling instrument and their implicit engineering problems. The net hardware is difficult to disguise. Its presence, through sight, sound, pressure waves, and other physical parameters, can be sensed by the animals and the more agile animals can avoid capture. These factors, coupled with the mechanics of filtering large volumes of water, make capturing animals that vary in size ratio by more than a million-to-one a formidable sampling task.

To meet the need for accurate and precise nekton sampling, net systems should be implemented with additional technology to develop more knowledge on net tow characteristics and on animal avoidance. The expanded use of high-frequency acoustics to monitor animal activity in front of the net would be helpful. The incorporation of low-light level video cameras with high-bandwidth filter-optics cable and pattern requisition technology may also be useful in evaluating the animal avoidance problem.

ACKNOWLEDGMENTS

Some of the equipment described in this paper was developed with funding from the Office of Naval Research contract N00014-79-C-0004.

REFERENCES

Atsatt, L. H. and R. R. Seapy, 1974. An analysis of sampling variability in replicated midwater trawls off southern California. *J. Exp. Mar. Biol. Ecol.* 14: 261-273.

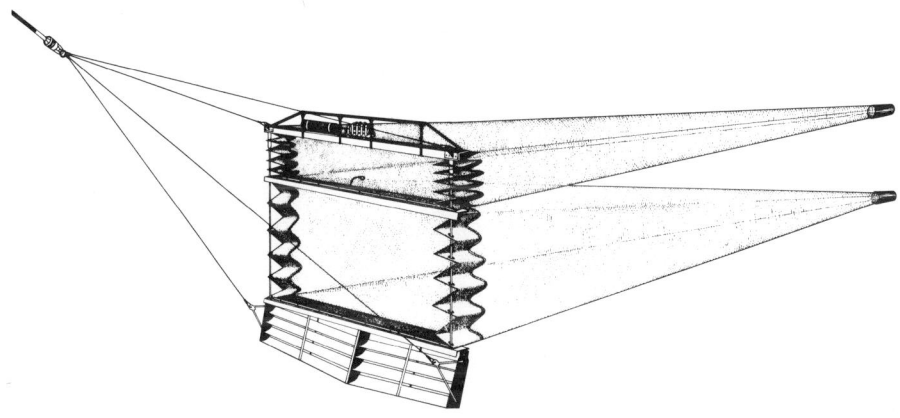

FIGURE 6: A rectangular meter net that is configured for multiple 7.5 m^2 open and closing nets.

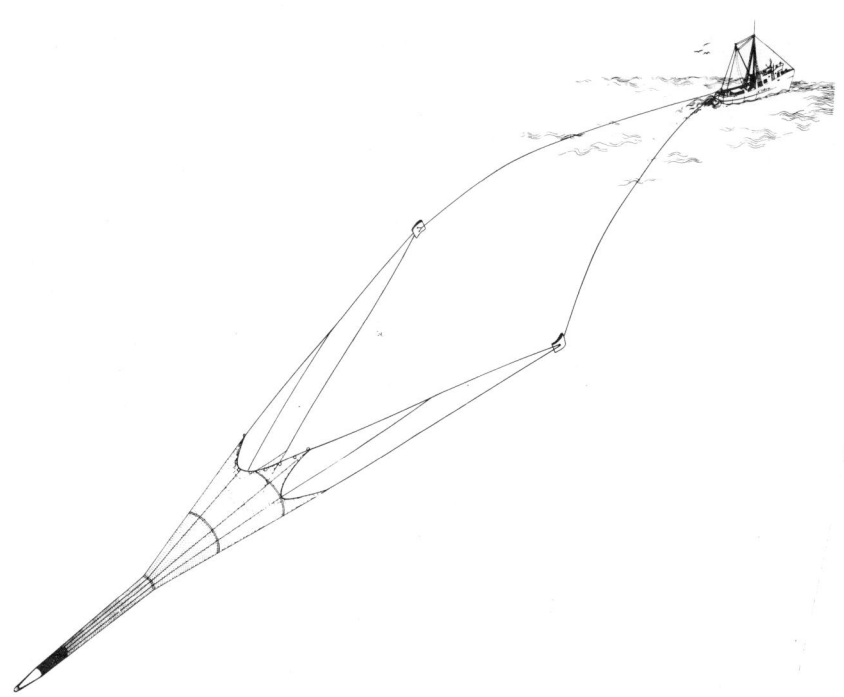

FIGURE 7: An 80 m^2 midwater trawl adapted with
a five-net multiple sampler codend unit.

Aron, W. and S. Collard, 1969. A study of the influence of net speed
    on catch. *Limnol. Oceanogr*. 14: 242-249.
Bé, A. W. H., 1962. Quantitative multiple opening and closing plank-
    ton samplers. *Deep-Sea Res*. 9: 144-151.
Harrison, C. M. H., 1967. On methods of sampling mesopelagic fishes.
    Symp. Zool. Soc. London 19: 71-126.

Microprocessors at the Woods Hole Oceanographic Institution 1978

Earl E. Hays

ABSTRACT: Microprocessors are being used at
Woods Hole in three instrumentation developments. The systems are surface current measurements, long range float tracking, and low frequency acoustic propagation fluctuation studies.
The functions performed by the microprocessor
are described, and the effort required to develop
is presented.

Marine biology measurements can use so many different sensors
that I only wish to make a short comment on the development of them.
The most attractive is a new and rare invention which allows the
measurement of some variable in a simple direct way. Typically
sensors have required a long development period and require careful
calibration and maintenance. As technology advances, I am sure that
sensors for many specifics will continue to become available. I
think that microprocessors, about which I am going to talk, could
have a significant influence on the use and development of sensors,
but having had no direct experience I will not discuss them.

I think the systems part of measurement programs is about to
go through a significant if not major revolution because of the
development of the large scale integrated (LSI) circuits, which in

certain forms are known as microprocessors. These started to appear in the early seventies, and are now a standard product of many companies. The literature on these devices is abundant; there are several books offered as textbooks for senior level engineering students, as well as a variety of short courses offered on their use. *Scientific American* had an issue within the past year devoted to the subject of microelectronics, which gives a good overview of the present status. According to some people, the microprocessors will be almost running the world within a generation.

A microprocessor usually contains a control/timing section, a memory, an arithmetic/logic section and an input/output section. It is therefore the nucleus for a computer, and with the addition of peripherals can become one. The control/timing and arithmetic/logic sections are not large enough that complicated data handling and processing can be carried out efficiently, but they are large enough that many of the desires of a measurement system can be met. The desires, as I see them, are to log in time the amplitudes of a number of variables, and to cause something to happen when the values of some of the variables reach certain levels, or even to do something that depends upon a fairly complicated function of combinations of variables. This is where the subject that I lightly passed by is so important. The sensors for the variables must produce a voltage or a frequency that can be recognized by the processor. It is possible to use the processor to detect signals in a poor signal-to-noise situation.

At Woods Hole we have used microprocessors in several ways. Unfortunately none of these have been in marine biology. There has recently been one experiment in marine biology that could have benefitted from their use, but it was conceived and completed before we were using them. This experiment involved the capture of a small bottom environment in a container on the sea floor, with gentle circulation of the water and introduction of oxygen and aromatic hydrocarbons, and subsequent study of the effect upon the animals. One can now imagine the experiment running under the control of the processor, where it measures and stores the concentration of oxygen, carbon dioxide, the aromatics and the circulation, adding oxygen and the aromatics or nutrients according to the experimental design. It is not too hard to imagine, but much harder to implement, the measure of the animal concentrations and status by recorded acoustic means. This simple example should illustrate the power of these tiny devices. Now I will give some examples of what we have done with microprocessors at Woods Hole and what effort was required for their implementation.

The deep float tracking program uses a microprocessor (MP) in its remote listening stations. The deep floats (called Webb or SOFAR floats) are neutrally buoyant drifting instruments, whose depth is set by ballasting. These floats periodically emit a low frequency signal (270 Herz [Hz]) of such intensity it can be received at long ranges (2000 km). By recording arrival times at several stations the motion of the float can be determined and so the deep currents. The original floats pinged on a schedule of a few minutes with a one second long pulse. After some experience with these, the pulse was lengthened to 80 sec, being a frequency modulated (FM) slide with a 1.5 Hz bandwidth. The signal is detected after filtering and clipping by a correlation against a replica of the outgoing signal. This procedure allows a lower signal level at the transducer face for the same signal-to-noise at the receiver. In practice, the received signals were tape-recorded at the receiving station and sent to a central processing computer. The use of such floats in areas where no shore stations are available would have posed a problem in the days before MPs. We (A. Bradley, D. Webb) have built and used an autonomous listening station to record signal arrivals. The MP takes the incoming signal after it has been filtered and clipped and correlates it with a replica of the emitted signal. It selects the four largest signals in each 10 min interval and stores the amplitude and time of arrival in a digital tape recorder. Using an eight hr ping repetition rate and setting the firing times within 20 min blocks allows one frequency band to listen to 24 floats without confusion (10 min represents about 1200 km so the motion of the floats will usually not cause overlap.) The MP does more than this, however; it causes the instrument to transpond to a coded signal from a nearby ship, and also release when it hears another special signal. It also creates a telemetry capability so that the instrument can be checked after it is initially launched. We are examining the ability to store data for a few days which can be telemetered to a ship in a short time so that the performance and position of the floats can be checked without recovering the mooring. This development represents somewhat over a year of professional time and $10,000-$12,000 worth of hardware.

We have a program in which we are studying the fluctuations in amplitude and phase of low frequency sound (200 Hz) in the ocean over fairly long ranges (25-1000 km) caused by temporal variations in the water column (internal waves, eddies, currents). From a communication viewpoint it is important to know the character of the fluctuations so that systems can be made optimal. From an oceanographic viewpoint, it may be possible to tell what is happening in the ocean. We started out using continuous wave (CW) which lumped all the multipaths into one signal. Now we (R. Spindel) are attempting to sort out the effects along the various paths by using a signal that allows us to separate the arrivals in time and an array that separates up and downgoing signals. As the sources and receivers are on simple one-legged moorings they can move around with the currents, so we also want to keep track of their positions. To do all of this we use an instrument called the Digital Buoy (DIBOS).

The signal put out by the source, for example, is a "phase shift keyed" pseudorandom sequence 6.3 sec long, repeated 10 times in a sequence every 10 min. The phase shift base rate is 10 Hz. The MP collects the quadrature components of the incoming signal every 0.1 sec, averages the corresponding ones in the 6.3 sec sequences for the 10 repetitions and stores the average on the digital tape. The next data processing is done after instrument recovery. The MP has several other tasks besides this primary one. It collects and stores temperature and navigation data. The navigation data are somewhat complex. CW beacons (about 12 kHz) are deployed on the bottom about a water depth away from the DIBOS. As the instrument moves around in the currents, there is a Doppler shift or phase change in the received 12 kHz signal. The MP accumulates the total phase change between each reception of the low frequency signal and

records it on the digital tape. With two beacons and some geometry
the motion of the instrument can be determined to within ten centimeters. Timekeeping and putting the 12 bit data information onto
the four channel digital tape recorder is also handled by the MP.
This represents about eight months of professional time and $9,000
(with deep ocean pressure case).

Figure 1 shows the "boards" of the DIBOS laid out on a bench
before assembly. The microprocessor chip is the light colored piece
in the second board from left and bottom. Figure 2 shows the DIBOS
assembled ready to be placed in the pressure case. The inside diameter of the pressure case is 15 cm and its length 1.5 m.

Another use of the MP is in an instrument which observed surface
currents from ships in transit or otherwise. With the coming of new
navigation systems, the ship's motion will be determined precisely
and by measuring the relative current, the surface currents can be
obtained. A fish has been built that is towed behind the ship, and
it senses speed, heading, roll, pitch, temperature and pressure and
contains two calibration voltages. These outputs are serially sensed,
digitized in the fish and sent up the wire to the ship where the MP
is. The shipboard equipment consists of a serial in, parallel out
converter, the MP, a display unit, a keyboard, output ports tape
recorder and oscillator (clock). The MP controls the whole system,
putting the data on the digital tape, putting any of the variables
on display when queried via the keyboard, and keeping time. This
package also serves as the playback system to whichever computer is
used to combine the ship's navigation with the relative currents to
obtain the true currents. This is about seven man months and $6500.

Recently an application arose that demonstrates the versatility
and effectiveness of the microcomputer in solving problems with instrumentation. A LORAN C navigation system was about to be sent to
sea for several months of field work. This system consisted of a
LORAN C receiver, a computer terminal, a precision real-time clock
and an interface to interconnect these units. The receiver processes
signals from a chain of LORAN C stations, providing a reduced lati-
tude/longitude position and time delays from the slave stations within the chain. This data is sent to the terminal where it is recorded
on cassette for later processing and printed for operator viewing.
The real-time clock was to provide time information (day-hour-minute-
second) to accompany the recorded positional data. At the last moment the manufacturer of the real-time clock delayed delivery for a
period of up to six months. This forced a search for an alternative
device that could simulate or replace this function. After a brief
engineering meeting it was decided to do this with a microcomputer.
A kit was purchased from a local supplier and constructed. A program was written using a support package for the microprocessor on
a larger computer system. This program simulated the functions of
the real-time clock. Within five days the navigation system was
sent to sea in full operational order.

The microcomputer performed the real-time clock simulation by
using an internal one second interrupt to maintain time information
in computer memory. This information was displayed on the microcomputer front panel and transmitted to the navigation system at 10-
sec intervals. The transmission format was identical to that origi-

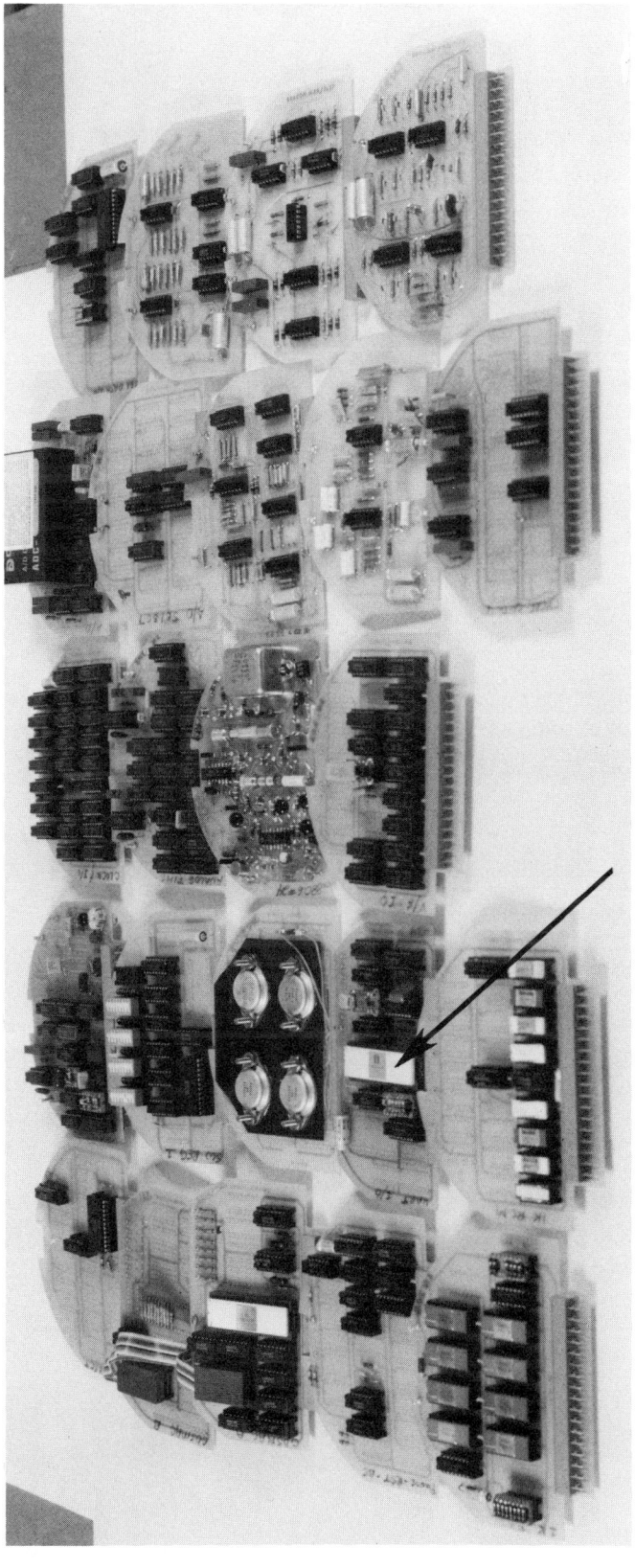

FIGURE 1: DIBOS unit major components. The microprocessor is the light colored chip in the second column, second board from bottom.

FIGURE 2: DIBOS unit assembled to go in case.
Diameter 15 cm, length 1.5 cm.

nally specified for the real-time clock, and the navigation system
noticed no differences. However, the accuracy of the computer-im-
plemented clock was not equal to the original specifications. This
situation has since been rectified by the use of a precision time
base for the computer interrupts.

The computer implementation of the real-time clock cost approximately one-third of the price projected for the purchase of the original clock. It was accomplished in only a few days and has proven more
versatile than the original would have been. This is a clear demonstration of how the microcomputer and its associated periphery can
solve instrumentation problems. This required one week of professional
time and $1,200.

I believe these examples cover the range of complexity that many
marine biological measurements will require. The man months, except
for the LORAN fix were those required for very competent engineers,
who had not used MPs, but were conversant with digital logic and circuits. This probably resulted in a slightly higher time requirement.
Of some interest was our experience in modifying a DIBOS from the
operation as described, to one of listening for a high frequency trigger signal, sampling and recording a low frequency signal for a given
length of time and then waiting for the next trigger. The engineer/
programming for this change was one month. I suppose that each biologist who wishes to make measurements that require such systems will
find that his situation has unique characteristics in the availability
of engineering and programming help, funding restraints, and testing
capability. I recommend a contact with the university EE or physics
departments. You will find that they will have junior, senior and
graduate students, who, with some faculty help and advice, will be
able to give you the desired system. It is important that you understand how the system works to a reasonable extent, so stay in close
touch with the progress. Building flexibility in sounds desirable,
but often results in more difficulty than the creation of two separate
systems.

HYDROACOUSTICS AND MARINE BIOLOGY

Hydroacoustics and Fisheries Biomass Estimations

J. B. Suomala, Jr.
J. B. Lozow

BACKGROUND DISCUSSION

Hydroacoustical equipment has been employed for more than 40 years to aid in the capture of fish in nearly all the oceans and seas. This equipment, which can perform both echo-sounding and echo-ranging functions, is currently produced in over 50 variants by approximately 20 enterprises and companies throughout the world. It is commercially available at costs ranging from less than $3,000 to over $300,000, depending upon the complexity.

Echo sounding is generally defined as the propagation of hydroacoustical energy in a direction which is nominally vertical. Echo ranging is propagation roughly parallel to the surface of the sea. Nearly all hydroacoustic devices operate in a similar manner. The echo sounder or echo ranger produces an electrical signal of a specific carrier frequency, amplitude and duration at the terminals of a transducer. The transducer converts this electrical signal into a dynamic pressure wave of a corresponding frequency, amplitude and duration. This hydroacoustical signal, called a pulse, is radiated from the transducer and spreads spherically in water. The radiated energy is greatest along its axis and varies off axis as a function of its directivity. When this pulse encounters an object whose acoustical properties are different from those of water, a portion of the hydroacoustical energy is reflected or scattered back toward the source. This reflected or back-scattered energy also spreads spherically and varies with aspect angle as a function

353

of the directivity of the object. When the back-scattered energy
impinges on the transducer, an electrical signal is produced which
is handled by various methods to indicate the presence of the
detected target.

The most common medium employed to display the electrical
signal is a paper chart called an echogram. The echogram is
generated by activating an electrical stylus (or stylii) which
produces a mark on electrosensitive paper. Depending upon the
position, size, and the density of the marks, an experienced
observer can deduce the approximate range and composition of the
target signal, e.g., bottom and/or aquatic targets. In addition,
a mark-to-mark correlation can sometimes yield information
concerning the number of targets. Variations of the basic echogram display include modifications to increase the probability of
discrimination between the desired target signals and the signal
from the sea bottom, the addition of expanded range displays,
special cathode ray tube displays to examine complex echo forma-
tions, and computer-generated displays which may, under certain
conditions, provide relative-position plots of the vessel and the
insonified targets. In addition, there are audible outputs
available to aid further in the decision process to deploy or
maneuver a net, the capture device universally employed in active
fishing. Further, there are net-mounted hydroacoustical transducers available which can provide indications of the position of
the net relative to the sea bottom or surface, and the presence
of fish targets which have a high probability of capture.

The principal use of the described equipment (there are many
other variants) is to detect and locate the fish desired for
capture. It should be realized that the results obtained can vary
considerably, because it is the interpretation of the displays,
along with other factors, which determines the success of the
fishing activity. In certain intensive single-species fisheries,
in well-defined spatial locations, identification of the fish may
be surmised by noting specific characteristics of the data display.
To a certain extent, quantification of insonified aggregations can
be inferred by a subjective comparison of the display and experience
relating to previous capture rates in similar circumstances.

It is clear that successful use of hydroacoustical equipment
for fish detection, location, and subsequent capture depends
ultimately upon human factors such as concentration, perception,
intelligence, etc., coupled with training and experience. It is
also clear that professional fishermen have learned to employ
hydroacoustical equipment with significant success in the effective
exploitation of the fish resources of the world.

In parallel with the improvements in the detection methods, a
number of devices to derive the quantity of detected targets were
produced. In the decade that has elapsed, there has been considerable international activity which has resulted in the assembly and
deployment of specialized (and generally expensive) apparatus which
handle and process received echo signals. The subsequent outputs
are meant to provide direct and speedy estimations of aquatic biomass which, in turn, are intended to help provide a rationale for
the decisions of various fisheries exploitation and management
organizations.

Currently, there are more than 12 medium-to-large fisheries research vessels conducting hydroacoustical activities in various seas and oceans. The annual cost for these activities approaches $15 million. The purpose of the following is to discuss the problems which confront the hydroacoustical practitioner in the pursuit of aquatic biomass measurements.

FUNDAMENTAL CONSIDERATIONS

Earlier, the process by which a pulsed hydroacoustical transmission is converted into an electrical signal at the terminals of the receiving transducer was described. Information in this electrical signal may be contained in phase variations, amplitude, frequency, and time duration. Also, the time delay between signal transmission and reception is an important source of information. For purposes of target quantification, the physics of sound propagation in water and the limited practical knowledge of hydroacoustical scattering from fish targets inhibits phase processing. Therefore, the information concerning the number of targets which produce an echo may be contained in the remaining characteristics, e.g., amplitude, frequency, and time duration. For example, if a single fish target is insonified, the echo signal would have (in addition to a particular amplitude) a time duration on the order of the transmitted pulse duration. If several targets were packed tightly together, the echo-signal envelope would have a larger amplitude compared to a single target, but still have a time duration on the order of the transmitted pulse. Conversely, if the targets were uniformly distributed within an insonified volume, the echo-signal envelope could have a time duration considerably longer than the transmitted pulse, since the target echos would be arriving in a random sequence.

In essence, all fish target quantification schemes employ pulsed hydroacoustical transmissions and scattering which behave according to linear hydroacoustical theory. The identification of the type of quantification has evolved from the postulated density of individual fish targets of interest in the insonified volume of the sea. When the fish targets can be resolved, the quantification technique is called echo counting. When the fish targets cannot be resolved, the quantification technique is called echo integration.

ECHO COUNTING

The fundamental requirement for counting is the joint capability of the combination of the hydroacoustical apparatus and the received echo-signal processor to resolve or distinguish between the objects to be counted, and then to perform the counting function.

The resolution of targets refers to two coordinates, range and angle. Range resolution refers to the separation of targets at the same angle referenced to a coordinate origin, and is expressed in terms of the distance between them. Angle resolution

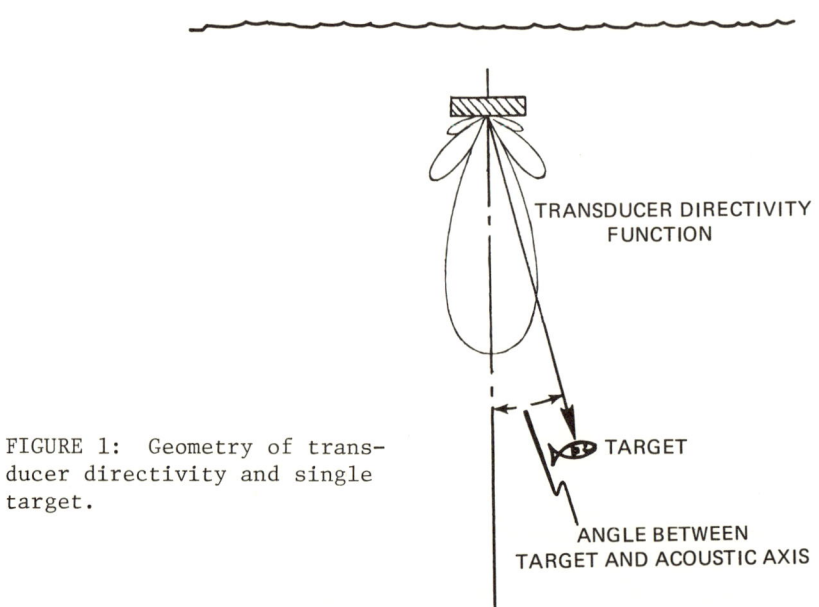

FIGURE 1: Geometry of transducer directivity and single
target.

TRANSDUCER DIRECTIVITY
FUNCTION

TARGET

ANGLE BETWEEN
TARGET AND ACOUSTIC AXIS

refers to the angular separation of targets at the same range and
is expressed in terms of the angle between them. In addition to
the basic considerations of target resolution, there are additional
vital practical factors that must be recognized, especially when
fish are the targets of interest.

Consider the situation depicted in Figure 1, which schemati-
cally shows two fundamental elements in a hydroacoustical environment, the transducer and a single fish target. The directivity of
the transducer determines the hydroacoustical energy projected and
received as a function of the angle between the target fish and
the acoustic axis of the transducer. The maximum energy is on the
acoustic axis and diminishes off axis. If there is little or no
knowledge of the back-scattering properties of the fish target
(except perhaps that a single large fish returns a larger echo
signal than a single small fish), then it is possible that a
single large fish target off axis may produce an echo signal
smaller or equal to a small fish directly on axis. Indeed, the
differences in the echo signals are related to the size and
position of the fish target.

If the fish target has a buoyancy regulating organ (most of
the fish that inhabit the midwater regions of the world's major
fishing areas do), otherwise known as a swim bladder or simply a
bladder, a further complication must be considered. There have
been numerous experiments conducted to measure the back-scattering
characteristics of individual bladder fish. The results of these
experiments showed that for a small head-down attitude, approximately 5 to 10°, the back scattering increased by as much as a
factor of 4 (Nakken and Olsen, 1973).
These data should be regarded approximate since the fish were

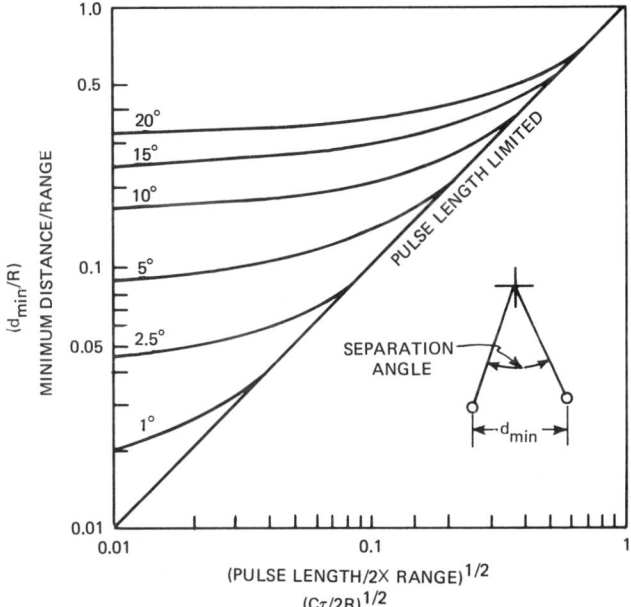

FIGURE 2: Geometric constraints for target resolution.

either dead or in a distressed state during the measurement procedure which may have introduced inaccuracy in the results.

At this point, assume there were other fish in the vicinity of the target fish and that a number were at approximately the same range, but at arbitrary angles relative to the acoustic axis of the transducer. The received echo signal would then be indistinguishable in duration from that of a single fish target and the echo level would depend upon the back-scattering characteristics of the various fish.

Now assume there were other fish in the vicinity of the target fish, but neither at the same range nor at the same angle relative to the transducer axis. Through simple geometrical considerations, it can be shown that an unresolvable echo signal is obtained if the fish in the insonified region of interest are separated by a distance which is a function of the transmitted pulse, the angle between them, and their ranges from the transducer. Figure 2 conveys the geometric constraints, which, in turn, affect the resolution capability of the hydroacoustical equipment. Two hypothetical targets are envisioned to be within the detection zone of the transducer. If the distance between the targets is decreased, the received echos from the individual targets will merge and eventually the distinction between them will disappear.

Practically, the maximum separation angle between two scatterers is roughly equivalent to the beamwidth angle of the

directivity function (most investigators visualize the transducer
directivity as a conical beam with sharply defined limits at the
half-power beamwidth angle). For example, if the transmitted
pulse length was 0.5 millisecond, the average range to the targets
30 meters, and the angle between targets 10°, then $(C\tau/2R)^{\frac{1}{2}} = 0.1$
(C in this expression is the speed of sound in water, approximately
1.5 meters/millisecond). In Figure 2, an ordinate value of 0.1
projected vertically to an angle of 10° projects horizontally to a
value of 0.2 on the abscissa. From this value, a minimum distance
between targets of six meters is calculated. Note that if the
included angle is 0°, the minimum separation distance is $C\tau/2$.

Finally, assume the target fish was near the sea bottom. From
the preceding discussion, it is evident that the fish must be above
the sea bottom at a distance greater than the range resolution of
the hydroacoustical apparatus. If the sea bottom is sloping or
undulating, the effective range and angle resolution is a matter
of considerable uncertainty.

Numerous investigations by various groups and individuals
worldwide have recognized, in one way or another, the complexities
just described. There have been a myriad of technical variations
to hydroacoustical equipment and the received echo-signal processes
to increase resolution capability and the probability of discrimination of a single fish target. It is reasonable to state that, in
the past decade, there has not been a variant or a combination of
variants that have not been assembled and deployed for trials
onboard fisheries research vessels.

ECHO INTEGRATION

This form of echo-signal processing is, technically speaking,
echo envelope integration. In effect the procedure obtains the
energy of the echo signal in a selected interval of time (Lozow
and Suomala, 1971). Echo integration is applied to circumstances in
which the signal is scattered from an aggregation of unresolved
targets. This situation is depicted in Figure 3. Unlike Figure 1,
the targets of interest are visualized as an aggregation in a
range interval or layer.
The mean spacing between the insonified targets is so small that
resolution is beyond the capability of the hydroacoustical equipment. It is assumed that the targets are uniformly and independently dispersed in the insonified volume and that the density
may be interpreted as the mean rate at which the targets occur
per unit volume. It is also postulated that these occurrences
are a form of a random phenomenon which can be modeled as a
three-dimensional (Poisson) distribution. Figure 4 illustrates
the mean distance between targets versus the mean density for a
Poisson distribution. This figure, in conjunction with Figure 2,
can provide the reader a useful guide to estimate the transition
point between the echo-counting and echo-integration methods. In
addition, it is assumed that the scattering from each point
scatterer is not affected by the proximity of any adjacent point.

It has been shown mathematically, and observed in the

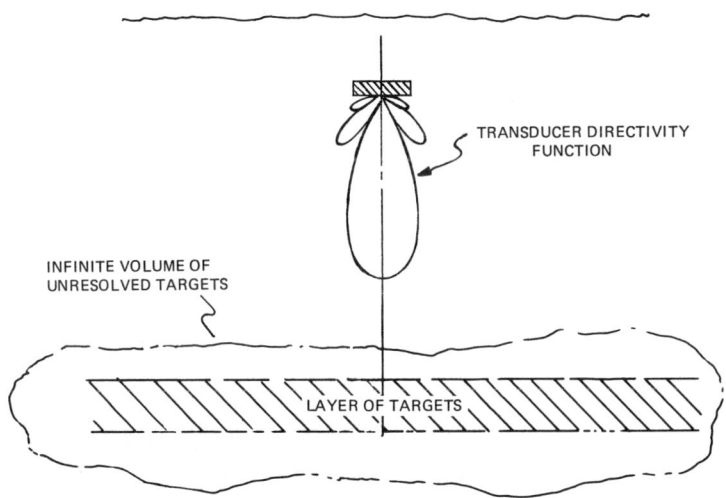

FIGURE 3: Geometry of transducer and aggregation of unresolved targets.

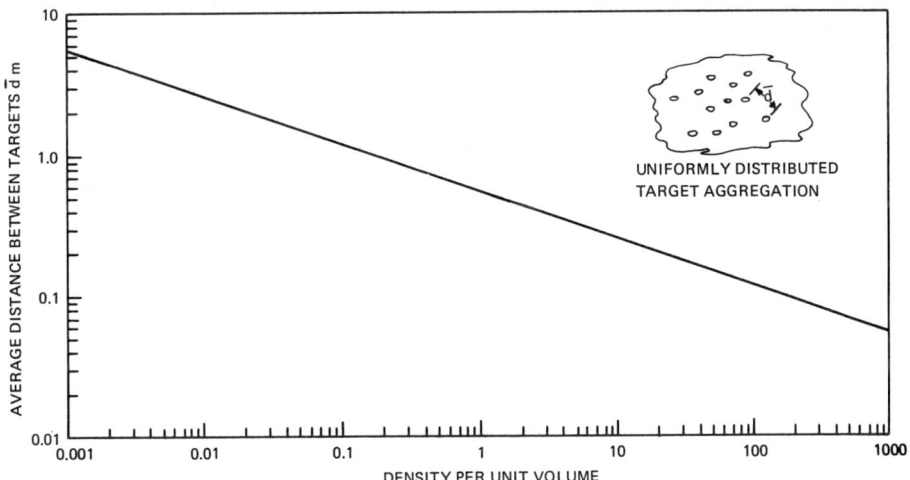

FIGURE 4: Average distance (m) between targets versus density for Poisson distribution.

scattering of hydroacoustical energy from air bubbles in water,
that the received signal fluctuates in an erratic or random
manner. These random fluctuations can be described in several
mathematical forms, depending upon the density of the targets
contributing to the received echo signal at any instant of time
(Middleton, 1967; Ol'Sheveskii, 1967; Weston, 1967). Considerable analytical applications have been made with the scattering
model previously described. Based upon the simultaneous phase
incoherent echos from aggregations of point scatterers, this
model is almost universally accepted as representative of the
environment encountered during fish abundance measurements
employing hydroacoustical methods.

In theory, the mean hydroacoustical echo signal is the
product of the average target density and the average independent
back-scattering characteristics of the insonified targets.
Serial estimations of target density should, therefore, be proportional to the number of the insonified targets. At this
point, it is appropriate to address the crucial element in the
application of hydroacoustical techniques for fisheries biomass
estimations, the scattering characteristics of the fish of
interest.

SCATTERING CHARACTERISTICS OF FISH

Although the principal interest here is the quantification
of fish, there have been a considerable number of investigations
concerning the identification of such. These investigations
have confirmed that the bladder is the major contributor to the
scattering mechanism of fish over a wide range of frequencies
(McCartney and Stubbs, 1971). Various techniques have been
employed to discriminate between fish with bladders and those
without. Other investigations have sought to identify the
species of bladder fish by the frequency at which the bladder
resonates. Further application of the resonance phenomenon
seeks to derive the size of the fish by the scattered energy
at resonance (Holliday, 1973a). Finally, other investigations
measure the frequency shift (Doppler effect) between transmitted
and received signals to derive target velocity information
(Holliday, 1973b). Most of these investigations have demonstrated that, under carefully controlled conditions, the identification of bladder fish versus nonbladder fish is qualitatively
feasible. Furthermore, the swimming velocity and acceleration
of certain fish, such as tuna, can be determined with reasonable
accuracy. In fact, it seems unlikely that the scattering
characteristics of fish, associated with identification, have
not been thoroughly investigated to satisfy a military requirement, if not for purely scientific purposes.

The scattering characteristic of a single fish is of
primary concern, in particular, the equivalent scattering cross
section of the fish as a function of aspect angle, from above or
in a dorsal view. The cross section, normalized to a standard
reference, is called the target strength. There have been

numerous investigations of the scattering cross section of fish
of similar morphology and physiology, but the scatter of the
data prevents selection of a practical value for a given fish
length. Figure 5 shows the approximate statistical limits of
the measurements of the scattering cross section of fish (Love,
1971; Goddard and Welsby, 1975). This figure depicts the scatter-

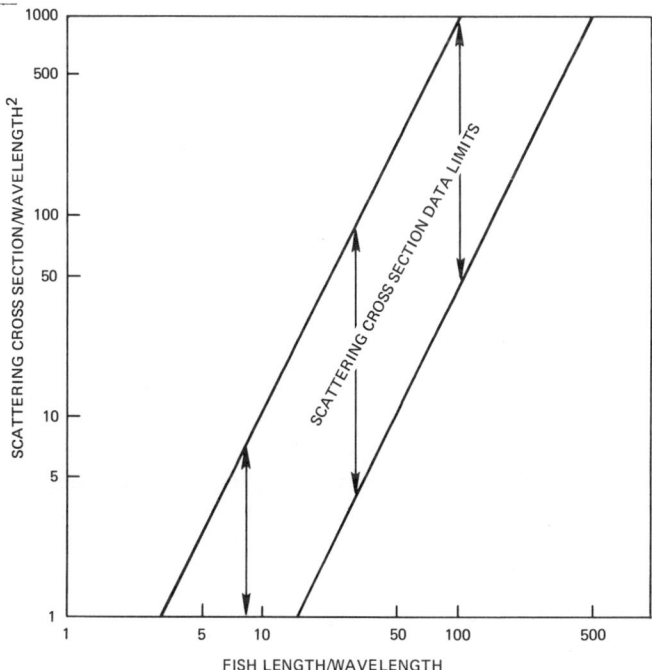

FIGURE 5: Scattering cross section versus fish
length.

ing cross section versus fish length, normalized to the hydroacoustical transmitted carrier frequency.
It can be seen that the limits of the data are greater than an
order of magnitude over a wide range of size and frequencies.
Since the mean hydroacoustical echo signal is assumed to be
the product of the average target density and the average
independent back-scattering characteristic of the insonified
targets, the resulting biomass estimation would be uncertain
by this same order of magnitude.

In recent years, investigators have sought to obviate the
uncertainty in scattering cross section by performing *in situ*
measurements. This method assumes that if a statistically
significant number of measurements of the back scatter from
resolved individual fish can be obtained, an average scattering
cross section can be derived (Traynor, 1975). Assuming that it
is possible to automate the recognition of single resolved
targets, doubts persist concerning the utility of such an
activity until some further questions can be answered. Princi-

pally, is there a significant change in the scattering cross
section related to the change of depth which the fish has
experienced? There have been numerous observations confirming
the vertical migrations of various fish species, but to date,
there is no evidence of reliable measurements which have
quantified this effect. In addition, there is no reliable data
concerning the possibility of variation in scattering cross
section versus rate-of-depth adaptation of the fish (i.e., the
time period over which the fish achieves neutral buoyancy). Of
course, if the scattering cross section varies with changes in
attitude, then this further complicates the situation.

Another *in situ* method for measuring the scattering
characteristics of fish involves the insonification of fish
confined in a net (Burczynski and Azzali, 1977). In this
situation, the average density of the fish is such that
individuals cannot be resolved and echo envelope integration
is performed on the received signal. Explicit in the measurement of the received echo signal from an insonified aggregation
of fish confined in a net of finite size is the selection of an
effective transducer directivity function angle. Implicit in
the measurement of this signal to derive target density, or
biomass, is the knowledge of the average individual backscattering characteristics of the insonified fish. This *in
situ* method merely approaches the problem from another
direction and does not, in any way, obviate the need to address
and quantify the critical factors discussed and to be discussed.

CALIBRATION

The importance of calibration of the hydroacoustical
apparatus is stressed by nearly every investigator. In order
to scale correctly the level of the received echo signals, the
level of the transmitted hydroacoustical intensity and the
sensitivity of the transducer to back-scattered energy must be
known. These two parameters are known technically as source
level and receiving voltage response, respectively. The source
level specifies the amount of sound emitted by the transducer
and is defined as the intensity of the dynamic sound field
(pressure) at a specified distance along the acoustic axis of
the transducer. The receiving voltage response of the transducer
is specified by the amplitude of the voltage produced across its
terminals by an impinging acoustic plane wave. Alternate methods
for determining the source level and/or voltage response of
transducers are possible. They may be measured individually or
coupled as a lumped parameter, depending upon the selected
calibration method.

In addition to source level and voltage response determination,
the spatial integration of the two-way transducer directivity
function, as well as the two-way acoustic path attenuation losses,
must be accurately estimated in order to scale the received echo
signal. Naturally, uncertainty in the calibration constants leads
to systematic errors in subsequent biomass estimates. That is,

inaccuracies in estimates of the system (and environmental parameters) inevitably lead to inaccurate scaling and thus affect the quality of any biomass estimate. Clearly, the final accuracy of a biomass estimate is limited by the lack of accurate knowledge of the various systems and environmental quantities necessary to properly scale the received echo signal.

FINAL COMMENTS

For the moment, assume that the positive identification of a single resolved fish target can be achieved either by automatic or semiautomatic methods. It may be possible that this can be achieved under carefully selected and controlled conditions. Under these same conditions, the calculation of the average scattering cross section of the insonified fish targets is also possible. If there were concurrent information concerning the behavior of the fish, it is reasonable to expect that the numerical value is indeed the true average value of the scattering cross section of the insonified fish. Obviously, if the fish targets were of a single species of nearly uniform size, counting would be feasible and an approximate estimate of the density and biomass could be derived.

Assume that there are several morphologically and physiologically similar species in the insonified region and they are approximately the same size. Under these conditions, a detailed knowledge of the species distribution would be mandatory in order to give credibility to either a count or density estimate by species.

Now assume that an aggregation of unresolved fish targets of the same size and species is insonified. Can the echo-integration process derive an estimate of the true density of the insonified fish? The answer to this question is dependent upon how the hypothetical scattering model relates to the true environment.

Presently, it can only be suggested that theoretical models permit the calculation of apparent density. However, this calculation should be considered with reservation since there is only general knowledge of the fish in terms of social groupings, prey-predator relationships, and physiological processes as they affect the aggregate echo signal. For example, at times, do the fish come close enough to one another so that the scattering of hydroacoustical energy is no longer independent and multiple scattering occurs? What is the scattering mechanism under these conditions? There is also the possible occurrence of a form of scattering which is not the result of the incoherent summation of the various echos, but contains a coherent component. In theory, the effect of a coherent component can seriously distort the estimate of target density.

All these and other effects can only be surmised at present since there appears to be no rigorous experimental evidence to establish their quantitative levels. It is suggested that hydroacoustical methods may provide reasonable estimates of relative aquatic biomass compared to other relative methods of estimation,

e.g., nets and larvae surveys, under certain prescribed conditions.
It is also suggested that estimations of the absolute abundance of
a specific fish species may eventually be feasible, under severely
restricted circumstances. These restricted circumstances demand
an intimate knowledge of the behavior of the insonified fish as it
affects the scattering coefficient and the resulting echo signals.
It is therefore of crucial importance that considerable emphasis
must be directed to the detailed understanding of specific fish
behavior *in situ* if the application of hydroacoustical methods for
aquatic biomass and, finally, fish abundance is to provide credible
estimates for fisheries management decisions.

ACKNOWLEDGMENTS

The preparation of this document was sponsored by the United
States National Marine Fisheries Service Contract No. 03-5-043-311
(CSDL Project No. 70278). The publication of this document is only
for the exchange and stimulation of ideas.

REFERENCES

Burczynski, J. and M. Azzali, 1977. *Report to the Government of
Italy on the Quantitative Acoustic Estimation of Sardine Stock
and Distribution in the Northern Adriatic*, Rome, FAO/ITA/TF-
ITA-3. 59 pp.
Goddard, G. and V. Welsby, 1975. *Statistical Measurements of the
Acoustic Target Strength of Live Fish*, Univ. of Birmingham,
Dept of Electronic and Electrical Engineering, Memorandum
No. 456. 37 pp.
Holliday, D., 1973a. The use of swimbladder resonance in the
sizing of schooled pelagic fish. *ICES/FOA/ICNAS Symposium
on Acoustic Methods in Fisheries Research*, Bergen, Contri-
bution No. 9. 6 pp.
Love, R., 1971. Measurements of fish target strength: A review.
U. S. Fish. Bull. 69: 703-715.
Lozow, J. and J. Suomala, Jr., 1971. *The Application of Hydro-
acoustic Methods for Aquatic Biomass Measurements, A Note on
Echo Envelope Sampling and Integration*. Report R-712.
Massachusetts Institute of Technology, Charles Stark Draper
Laboratory, Cambridge, MA. pp. 1-91.
McCartney, B. and A. Stubbs, 1971. Measurements of the acoustic
target strengths of fish in dorsal aspect, including swim
bladder resonance. *J. Sound Vib.* 15: 397-420.
Middleton, D., 1967. A statistical theory of reverberation and
similar first-order scattered fields, parts I and II.
*Institute of Electrical and Electronic Engineers Trans-
actions on Information Theory IT-13*. pp. 372-414.
Nakken, O. and K. Olsen, 1973. Target strength measurements of
fish. *ICES/FAO/ICNAF Symposium*, Contribution No. 24. 18 pp.
Ol'Sheveskii, V., 1967. *Characteristics of Sea Reverberation*

(Translated from Russian), Consultants Bureau, NW. pp. 1-159.
Traynor, J., 1975. Studies on indirect and direct methods of *in situ* fish target measurement. *Acoustic Surveying of Fish Populations*, Institute of Acoustics, Univ. of Birmingham.
Weston, D., 1967. Sound propagation in the presence of bladder fish. In: *Underwater Acoustics*, (V. M. Albers, ed.), Vol. 2, pp. 55-88. Plenum Press, NY.

Volume Reverberation—Help or Hindrance?

Charles L. Brown
Thomas M. Fitzgerald

ABSTRACT: The effectiveness of active sonar systems is often limited by volume reverberation
caused by biological scatterers, chiefly, fish
possessing gas-filled swimbladders. Volume scattering strength is a function of frequency, time
of day, season, and geographical location. For
an accurate prediction of sonar system performance, volume scattering strength data are required. The general sonar equation is discussed
in terms of optimum frequencies and limitations
imposed by volume reverberation. Some non-linear
acoustic techniques are also presented as a means
to reduce reverberation. From discrete-depth
collection of mid-water fish, in an ocean area
off Bermuda, scattering strength profiles were
predicted successfully at frequencies ranging
between 3.5 kHz and 15.5 kHz. Subsequent applications of this information allowed the development of a model to predict scattering strength
profiles of a Gulf Stream warm core eddy at any
frequency between 2 and 20 kHz.

INTRODUCTION

The theme of this paper may be embodied in the phrase, "frequency diversity". The United States Navy operates a wide variety of active sonar systems including: large scale submarine and surface ship sonar systems; intermediate size systems, such as torpedoes; smaller size systems found in sonobuoys; and special purpose systems, such as high resolution side-scanning sonar systems. These systems operate over a wide bandwidth of frequencies in a variety of environments. The acoustical properties of marine biological systems are also highly frequency dependent. Consequently, it would be expected that various sonar systems are influenced to a greater or lesser degree by the biological content of a particular geographical location. In general, sonar systems are designed to operate under an ambient noise limited condition. In practice, sonar systems must also contend with self-noise problems which are generated by the platform moving through the water. Ideally, by proper design and the use of various noise reduction techniques, it is possible to keep self-noise of a system below ambient noise. In many cases, however, volume reverberation can dominate the noise background of an active sonar system, which will limit its predicted performance. It is the objective of this paper to show in general terms how volume reverberation can limit the expected performance of a system and to show the frequency diversity associated with volume reverberation produced by biological scatterers.

SONAR EQUATION

The performance of a sonar system can be predicted using the generalized active sonar equation as shown in the following equation:

$$N_E = L_S + N_{TS} - 2N_W - N_{SD} - L_N + N_{DI} - N_{BW} \qquad (1)$$

The various terms in the sonar equation are defined in dB as follows:

N_E = signal excess

L_S = sound pressure level (dB//μPa/m)

N_{TS} = target strength

N_W = propagation loss (spreading loss and absorption)

N_{SD} = detection threshold (indicates how far above or below the noise level an echo must be in order to be recognized as an echo consistent with some probability of false alarm rate)

L_N = omnidirectional ambient sea state noise in a one Hertz band (function of sea state and frequency)

SONAR EQUATION

$$N_E = L_S + N_{TS} - 2N_W - N_{SD} - L_N + N_{DI} - N_{BW}$$

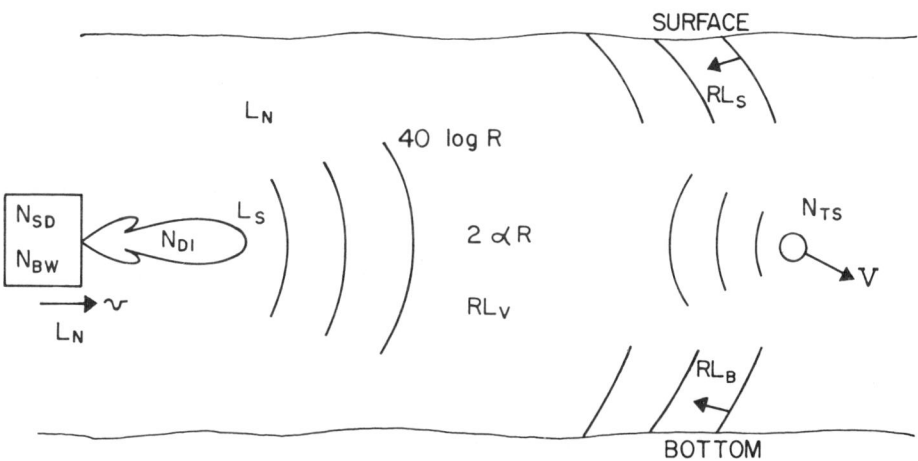

FIGURE 1: Sonar equation and schematic representation of each term.

N_{DI} = directivity index of the sonar transducer which is related to the aperture size

N_{BW} = bandwidth of receiving sonar

By knowing each term in the sonar equation, the expected performance in terms of acquisition range can be predicted. The general situation for a sonar system is shown in Figure 1. A given amount of acoustic energy is emitted from a transducer in a directional beam having some directivity index N_{DI}. The directivity index (N_{DI}) is given by 20 log ka for a circular transducer of radius a where k = $2\pi/\lambda$ and λ is the acoustic wavelength. This energy propagates through the medium and undergoes spherical spreading (20 log R) or cylindrical spreading (10 log R) and absorption (2 α R) where α is the attenuation coefficient. It strikes some target which can be moving with the velocity, v, such that the returning energy undergoes a Doppler shift proportional to the velocity of the moving target. Upon returning to the transmitter, the signal is received and processed through the receiver which has a detection threshold N_{SD} and a bandwidth given by N_{BW} = 10 log bandwidth. This returning energy constitutes the echo level which must be recognized as an echo from the target against the ambient noise background which is given by $L_N - N_{DI}$.

In addition to the ambient noise, it is also possible for biological species to scatter energy back towards the receiving transducer. This backscattered energy constitutes volume reverberation as

VOLUME REVERBERATION

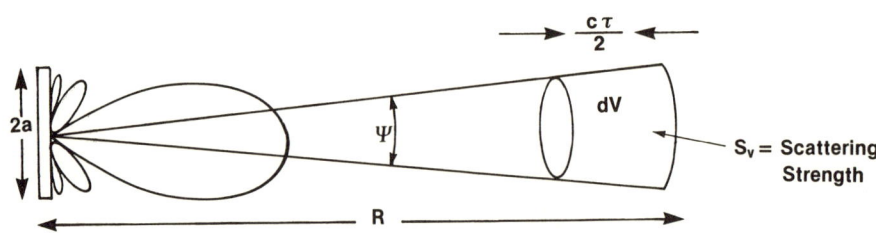

$$RL_v = L_s - 40 \log R + S_v + 10 \log V$$

$$\text{where: } V = \frac{c\tau}{2} \, \Psi \, R^2$$

FIGURE 2:   Origin of volume reverberation.

shown in Figure 2. Volume reverberation level is given by the following equation:

$$RL_V = L_S - 40 \log R + S_V + 10 \log V - 2 \, \alpha \, R \tag{2}$$

where L_S is the source level, R is the range to some scattering volume V, and S_V is the scattering strength. The scattering volume is given by:

$$V = \frac{c\tau}{2} \, \psi \, R^2 \tag{3}$$

where τ = pulse length, c = velocity of sound (yards/sec) and ψ = the isonofied solid angle in radians, which is given by

$$10 \log \psi = -20 \log ka + 7.7 \text{ dB} \tag{3a}$$

A sample calculation for predicting the range of a sonar system is shown in Figure 3. The target strength is assumed to be +10 dB and the frequency is equal to 12 kHz which is the usual echo sounder frequency. The transducer diameter is assumed to be 1 ft. The amount of electrical power available is equal to 2kW and a nominal 50% efficiency for the transducer is assumed. In Figure 3 the echo level in dB is plotted vs range in kyds. The echo level decays until at a range of approximately 9 kyds, the echo level is equal to the ambient noise associated with a sea state 1 condition assuming a 10% bandwidth. Under these conditions, the active acquisition range would be approximately equal to 9 kyds, where a detection threshold of 0 dB is assumed. If ambient sea state noise were higher, as indicated on the figure for a sea state 2 condition, the acoustic range would be

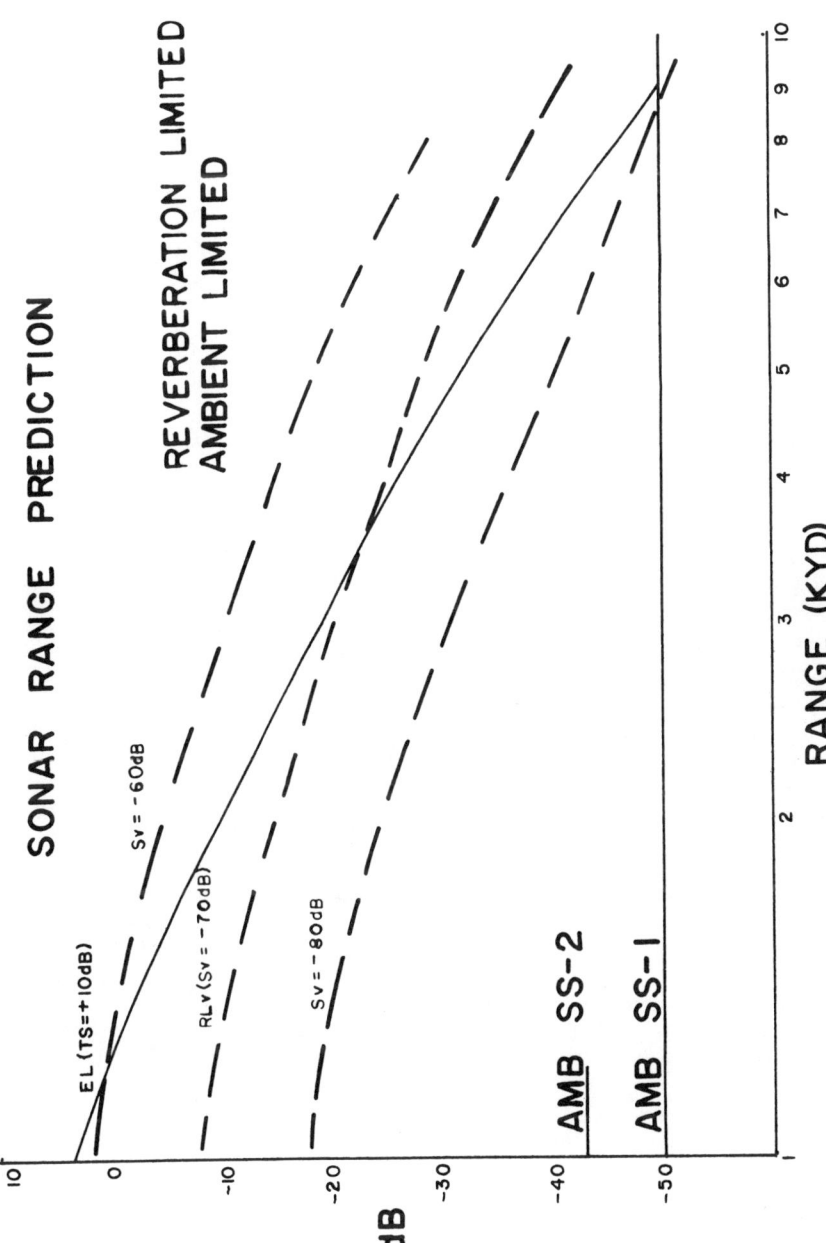

FIGURE 3: Echo level and volume reverberation level as a function of range.

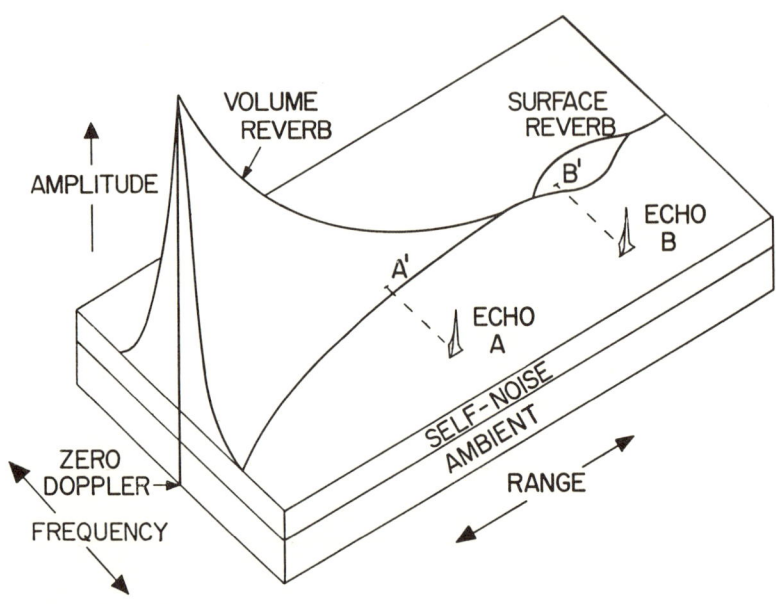

FIGURE 4:   Range–Doppler map.

subsequently reduced to about 7 kyds. The curve for the echo level
can be moved up or down, depending upon the particular target strength
assumed and where this curve crosses the noise level curve determines
the expected range for this particular sonar system.

The dashed curves in Figure 3 represent the expected reverberation levels as a function of range for three different values of the
scattering strength. For the case of the lower scattering strength
of −80 dB the reverberation levels are below the echo levels at all
ranges. In this case, the sonar system is said to be ambient limited.
If the scattering strength is 10 dB higher, i.e., −70 dB, the reverberation level and the echo level are approximately equal at a range
of about 4 kyds. In this case, the system is said to be reverberation
limited and the expected range would be approximately 4 kyds. In the
worst condition (S_V = −60 dB) assumed for this calculation the reverberation level is above the echo level at a range of about 1 kyd.
Thus, depending on the scattering strength, the expected acquisition
range of this sonar system could extend from a value of 9 kyds to a
minimum value somewhere below 1 kyd. Thus the particular value of
the scattering strength determines the expected range of this particular sonar system.

For an operational system, the situation can be better than as
depicted in Figure 3 because a target of interest can be moving at a
sufficiently high velocity such that the Doppler shift associated
with the echo level can be significantly outside the reverberation
bandwidth. This is depicted in Figure 4 on a range-Doppler map. For
example, the echo labeled A has a high enough Doppler such that it
has to be detected against a self-noise or an ambient limited back-

ground and as shown in Figure 4, could be identified as a true echo. If, however, the velocity were much smaller, such that the echo moved to the position labeled A', the echo could become buried in the volume reverberation and not be easily identified as a true echo. The echo labeled B in Figure 4 is of a high enough Doppler value to be outside the surface reverberation bandwidth, and, if it were to move to a position B', it could in this case be buried in surface reverberation. In general, the reverberation bandwidth is determined by the pulsewidth and the beamwidth used with a particular sonar system, and depends on the velocity of the scatterers, e.g., if the scatterers were moving with a relatively high velocity, the reverberation bandwidth would also be spread proportionately.

SONAR LIMITATIONS

In general, sonar systems are limited by two primary parameters: 1) the electrical power available; and 2) the physical size of the aperture available. In addition to these parameters, properties of the medium must also be considered. For example, given unlimited amounts of power, a system will eventually undergo cavitation, so that cavitation becomes a limiting parameter. Cavitation is a function of pulsewidth, frequency, and water depth. In general, short pulse widths and higher frequencies reduce cavitation. Near the surface, cavitation takes place more readily than at much deeper depths.

In addition to cavitation, non-linear effects in the ocean medium also limit the amount of power that can be put into the water. At very high power levels, the ocean behaves non-linearly such that the energy put into the ocean at a given frequency occurs not only at the given frequency of interest, but at higher harmonic frequencies.

Also, the ocean is not at a uniform temperature, but exhibits strong vertical and horizontal temperature gradients which cause bending of the sound waves, thereby introducing additional losses to the system. In addition, the ocean is inhomogeneous, such that biological scatterers can produce an apparent increased absorption loss again degrading the expected performance of the system. In spite of these limitations, it is possible to provide some rationale for a judicious choice of operating frequencies.

OPTIMUM FREQUENCIES

The design of any sonar system usually involves a trade-off analysis between what is desired and what can be realistically obtained within acoustic limitations. For example, the requirements for a search sonar, a communication sonar, and a side scan sonar can be quite different. These different requirements are usually not compatible and are often diametrically opposed, e.g., long range usually means low frequencies and wide beamwidths to overcome absorption losses and provide a wide area search while high resolution usually means high frequency (short range) and narrow beams (small area coverage) to provide the desired spatial discrimi-

nation. The function of the sonar designer is to match these requirements with the limitation imposed by the sonar platform (e.g.,
hydrodynamics, power, aperture), the component technology available
(e.g., transducer elements, power amplifiers, signal processors),
and the medium (e.g., noise, absorption, reverberation). While the
detailed analysis of any particular system is beyond the intent of
this paper, a case of general interest will be considered.

The performance of a particular sonar system against a given
target is often measured in terms of the expected acquisition range
as determined from the sonar equation. The calculation of this
range presupposes the existence of a sonar system such as the postulated 12 kHz echo sounder used to generate the curves of Figure 3.
The inverse problem can also be posed: "given a requirement to detect a particular target at a given range, is there a "best" frequency to use?". It will be noted that most of the terms in the sonar equation (Equation 1) are frequency dependent. Thus, in principle, it should be possible to optimize the signal excess by differentiating the sonar equation with respect to frequency, setting
it equal to zero and solving for the optimum frequency.

In order to make the calculation more tractable, some assumptions
are warranted. The most variable term in the sonar equation with respect to frequency is the ambient noise term. At low frequency (below 1 kHz) ambient noise is dominated by shipping noise. At high
frequencies (above 100 kHz) ambient noise increases with frequency
and is attributed to the thermal molecular noise of the ocean. In
the intermediate range (approximately 1 kHz to 100 kHz) ambient
noise is attributed primarily to surface sea state noise, which depends on the sea state and varies approximately –5 to –6 dB/octave.
Based on Knudsen sea state ambient noise curves, the omnidirectional
sea state noise for sea state zero can be expressed as

$$L_N = -55.3 \text{ dB} = 16.7 \log f \text{ (kHz)} \tag{4}$$

assuming a 5 dB/octave decrease as the frequency is increased.
Confining the frequency range to be investigated to between 1 kHz
to 100 kHz, Equation 4 is assumed to represent the frequency dependence of sea state ambient noise.

Within this frequency range, the wavelength varies between 5
feet and .05 ft. Assuming that the targets of interest (e.g., a
submarine) are large compared with an acoustic wavelength, then
ka >> 1 and the scattering strength is approximately independent
of frequency (geometric scattering as opposed to Rayleigh). This
assumption would not be valid when small biological scatterers are
considered. Indeed, as will be shown later, biological scattering
strengths are highly frequency dependent.

While the receiver bandwidth must be chosen wide enough to encompass relative Doppler shifts between own platform and target, the
frequency dependence of the bandwidth will be linear. The detection
threshold is assumed independent of frequency although this is valid
only for specific cases depending on the type of processing used after
the transducer.

The source level in dB// 1 μPa @ 1 m is given by

$$SL = 171 \text{ dB} + 10 \log W + N_{DI} \qquad (5)$$

where W is the acoustic power projected into the water. The acoustic power in the water will depend on the efficiency of the transducer and the electrical power available.

The absorption coefficient is a function of temperature and frequency and over the frequency range of interest (1 kHz to 100 kHz) can be expressed as

$$\alpha = K(T) \ f^2 \qquad (6)$$

where f is the frequency in kHz and K(T) is a constant depending on the temperature T. Representative values of K(T) are: $K(5^\circ C) = .0071$, $K(15^\circ C) = .004$, and $K(22.5^\circ C) = .0024$.

Collecting only those terms in the sonar equation that are frequency dependent (under the above assumptions) yields:

$$46.7 \log f - 2 \ K(T) \ R \ f^2 = N_E \qquad (7)$$

Differentiating equation 7 with respect to frequency, setting the derivative equal to zero and solving for the optimum frequency yields:

$$f_o = \frac{2.252}{\sqrt{K(T) \ R(kyd)}} \qquad (8)$$

It can be seen from equation 8 that the optimum frequency (i.e., that frequency that will provide a maximum echo level to ambient noise ratio) depends on both range and K(T) which is related to the absorption coefficient at a particular temperature T.

The relatively strong dependence on the temperature is illustrated in the following table for the case of a target at a range of 9 kyd.

Range = 9 kyd

Temperature (°C)	Optimum Frequency (kHz)
5.0	08.91
15.0	11.87
22.5	15.32

It can be seen that the optimum frequency at 9 kyds varies over a relatively wide band of frequencies depending on the water tempera-

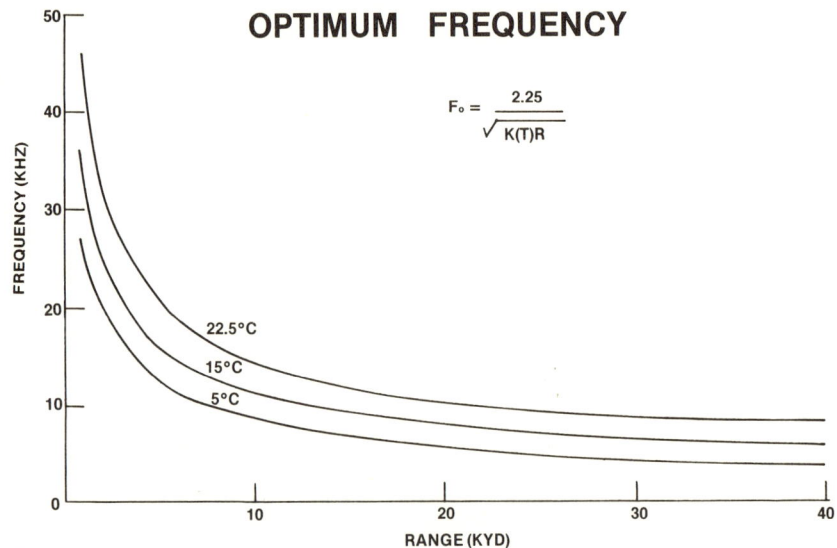

FIGURE 5: Optimum frequency for the ambient
noise limited case as a function of range.

ture. At 15°C the optimum frequency is about 12 kHz which corresponds to the frequency used in the earlier echo sounder example.

    A plot of equation 8 is shown in Figure 5 for three different
temperatures. Two major features are to be noted from Figure 5:
1) at a given range, the lower the temperature, the lower the operating frequency; and 2) at a given temperature, the shorter the range,
the higher the operating frequency. Thus a sonar system that operates
over a wide bandwidth of frequencies would be capable of adapting to
the particular environment of interest.

    The above discussions were based on the assumption that the tar-
get strength was independent of frequency over the frequency range
considered. In practice the target strength can vary with frequency
depending on target size, aspect, target material, and hull construction techniques. If the frequency dependence of target strength for
a particular target is known, a similar equation to equation 8 can be
obtained by minimizing the signal excess for the sonar equation. Although biological scattering strengths also vary with frequency, a
particular species of interest could be considered to have a constant
target strength over a limited band of frequencies and a similar optimization could be done.

REVERBERATION LIMITATIONS

    In terms of active sonar systems, biological scatterers are con-
sidered a hindrance in detecting targets of interest. In an active
sonar system volume reverberation can mask a target of interest as
was shown earlier in Figure 3. In terms of acoustics as a tool for

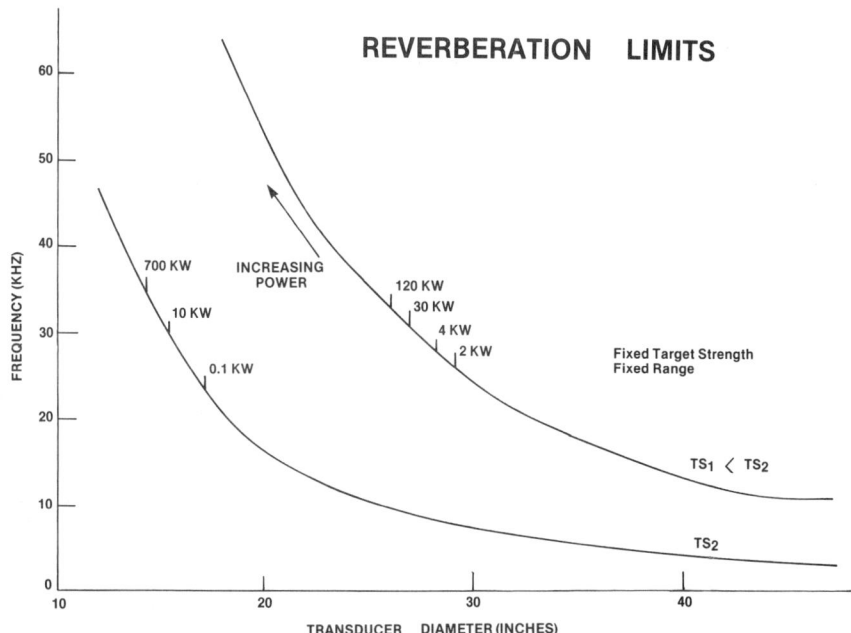

FIGURE 6: Sample calculation for frequency,
power and transducer diameter required to
maintain reverberation level equal to the
noise level for an unspecified target at an
unspecified range.

the marine biologist, the biological scatterers are the targets of
interest.

In the reverberation limited case, an optimum frequency calculation to minimize reverberation is not possible in a practical sense.
A tradeoff is required among three parameters: frequency, size of the
aperture available, and power available. (Strictly speaking, an optimum frequency does exist for the reverberation limited case, but it
turns out to be infinitely high.) Ideally, for a fixed target strength
and a fixed range, the reverberation level should be below the echo
level at all ranges. In order to have this happen, a choice must be
made between frequency, aperture size, and power. A plot of frequency
vs. aperture size is shown in Figure 6, for a fixed unspecified target strength and a fixed unspecified range. The points labeled on
the curve in terms of kilowatt show the amounts of electrical power
that must be made available in order to meet the above stated condition, i.e., to keep the reverberation level below the echo level out
to the desired range of operation. For example, at approximately 25
kHz, if there are 2 kW of acoustic power available, then a system
with an aperture diameter of approximately 30 in would suffice. If
the aperture available is smaller, then both the frequency and the

WAVE EQUATION

$$\nabla^2 P = \frac{1}{c^2} \frac{\partial^2 P}{\partial t^2} + \frac{1}{c^2} \frac{\partial^2}{\partial t^2} F(A, B, p^2, u^2)$$

$$P = A\varsigma \qquad\qquad + \qquad B\,p^2$$

LINEAR   NON-LINEAR

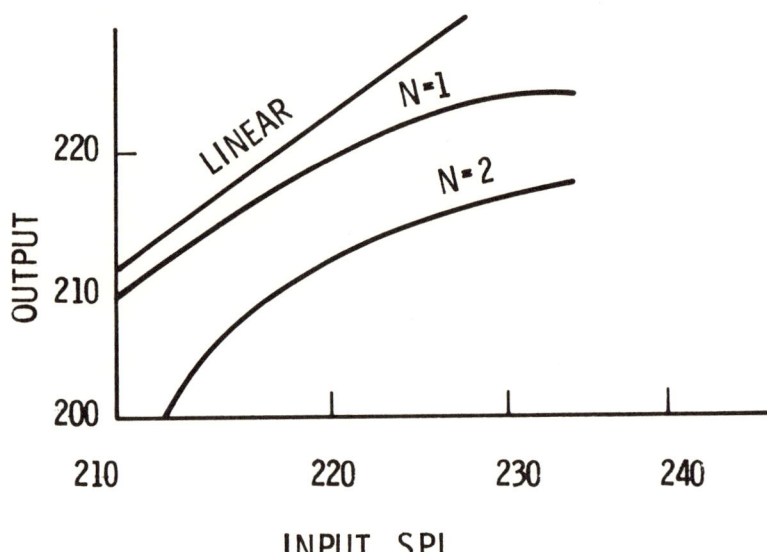

2f, 3f, • • • •

HARMONICS

FIGURE 7: Harmonic generation for nonlinear
(finite amplitude) sound waves.

power must increase. Eventually, as the aperture becomes smaller,
the frequency becomes much higher and the power level becomes prohib-
itive. The second curve in Figure 6 labeled TS_2 is a curve calculated for a fixed target strength higher than TS_1, again it can be
seen that as the aperture size gets smaller, frequency and power
levels must increase. In most operational systems, the aperture
size available and the power available are fixed so that a trade off
is required between the performance wanted in the ambient limited

case and the performance wanted in the reverberation limited case. If the system is designed for optimum performance in the ambient limited case, the limitations which are imposed by reverberation associated with conventional acoustic systems must be accepted.

NON-LINEAR ACOUSTICS

Propagation of sound waves is usually described by the wave equation which is derived for infinitesimal displacements, i.e., the pressure is assumed to vary linearly with small changes in the density. In dealing with high-powered sonar systems, however, non-linear or finite amplitude effects must be considered. In the finite amplitude case, the pressure is no longer linearly related to the density changes, but higher order terms quadratic in the density must be included. This leads to a non-linear wave equation containing higher order terms in the pressure as shown in general terms in Figure 7.

The non-linear wave equation leads to generation of harmonics. For example, if a single frequency f is put into the water at very high amplitudes, then harmonics at frequencies 2f, 3f, etc., are generated in the water. Figure 7 shows a plot of the input sound pressure level expected vs the actual output obtained assuming the transducer remains linear and taking into consideration the efficiency of the transducer. In the linear case, if the input sound pressure level is increased by 10 dB, the output would be expected to increase by 10 dB as well. This is true at low sound pressure levels. As the sound pressure levels are increased, however, departure from the linear case occurs. In other words, for a 10 dB increase in the input, much less is observed at the output. The missing energy goes into harmonic generation. As the input sound pressure is increased, the harmonics start to grow in the water so that non-linear behavior represents the limitation of a conventional sonar system. As more and more energy is put in at the fundamental frequency, a smaller disproportionate amount at the fundamental frequency appears in the water. The missing energy occurs at the harmonic frequencies.

Another representation of harmonic generation is also shown in Figure 8 where the signal near the transducer is more or less a pure sine wave and the signal spectrum is represented by a single frequency. As the sound wave propagates in the medium, it begins to distort into a saw-toothed type pattern as shown and the signal spectrum shows an increase in the harmonic frequencies.

One of the properties of the second harmonic is that its beamwidth is approximately one-half as wide as that associated with the fundamental (first harmonic). In addition, the sidelobes associated with the harmonics are greatly suppressed. Thus, a better signal-to-reverberation ratio is expected with the harmonics than with the fundamentals in a highly reverberent environment. An experimental observation of this is shown in Figure 9.

The top photo in Figure 9 is a conventional echo from a transducer driven at a fundamental frequency f with a pulsewidth of 25 msec.

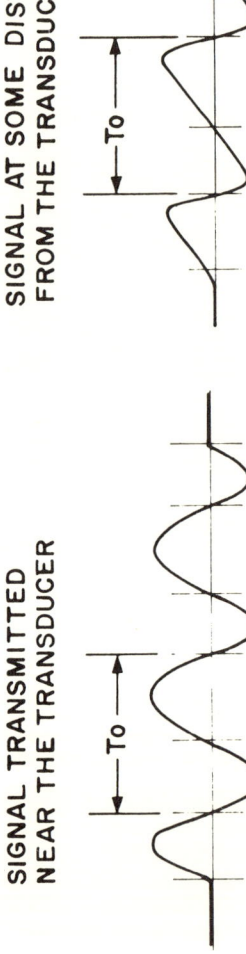

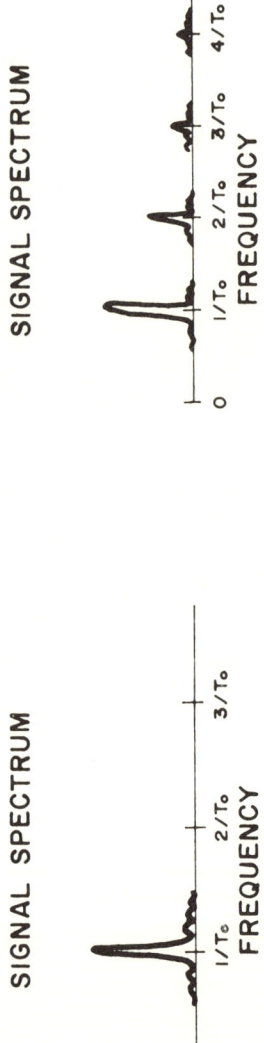

FIGURE 8: Wave form distortion and spectrum changes for finite amplitude sound waves.

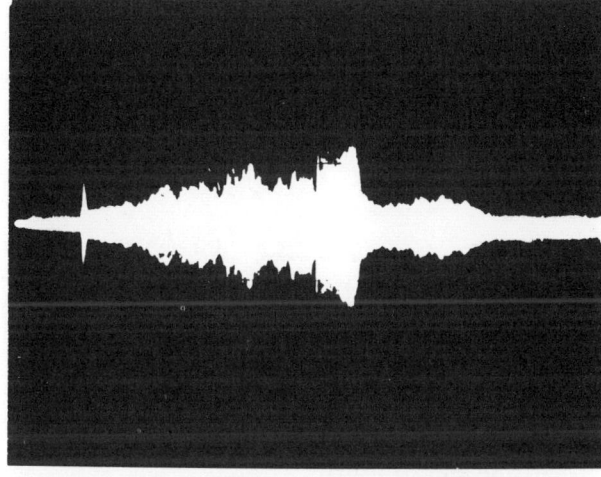

CONVENTIONAL
PULSE WIDTH:
25 msec

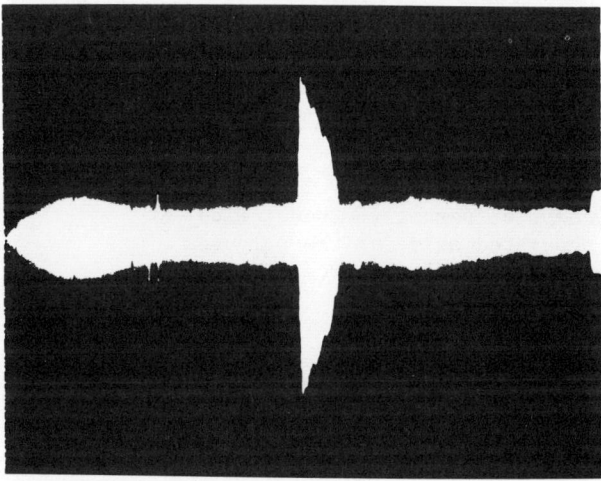

HARMONIC
PULSE WIDTH:
25 msec

FIGURE 9:   Conventional versus harmonic echo.

What is seen is a large amount of sidelobe and surface reverberation
with the target imbedded in the reverberation. The lower photo in
Figure 9 is the echo from the same transducer for the same pulse and
for the same target, only now the output has been filtered so that
only the second harmonic at frequency 2f is being observed. It can
be seen than in the lower figure the signal-to-noise or the signal-
to-reverberation ratio is better than the conventional case. Thus,
the harmonic signals which are generated by the non-linear properties
of the medium can provide a better signal-to-noise or signal-to-reverberation ratio when the conventional system is reverberation limited.

Another consequence of the non-linear behavior of the ocean at
high amplitudes is the parametric sonar system (Konrad and Carlton,
1973; Konrad, 1975). A parametric sonar system is shown in Figure 10

PARAMETRIC SONAR

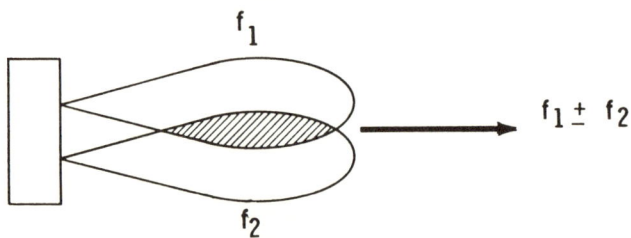

f_1

$f_1 \pm f_2$

f_2

PROPERTIES OF PARAMETRIC SONAR

NARROW BEAMWIDTH

REDUCED SIDE LOBES

WIDE BANDWIDTH

REDUCED APERTURE

LOW EFFICIENCY (?)

FIGURE 10: Concept of parametric sonar (The
two beams are separated only for the purposes
of illustration. In actual practice both
beams overlap.)

where two high frequencies f_1 and f_2 are propagated from the same
transducer and allowed to mix or interact in a volume of water in
front of the transducer. This interaction produces a sum and difference frequency $f_1 \pm f_2$. The sum frequency $f_1 + f_2$ is a very high
frequency and attenuates rapidly. The difference frequency $f_1 - f_2$
is a much lower frequency, and once generated, propagates as a conventional sound wave. The properties of the parametric sonar are
such that the beamwidths associated with the difference frequency
are much narrower than could be generated conventionally from a
transducer of the same size (Moffett et al., 1976). The beamwidth
of the difference frequency is approximately the same as the higher
primary frequencies. Also, the sidelobes at the difference frequency are greatly reduced.

The parametric sonar system is also an inherently wide bandwidth system. For example, if the two primary frequencies were
100 kHz and 110 kHz, the difference frequency would be 10 kHz.
If the primary frequencies were 100 kHz and 120 kHz then the difference frequency would be 20 kHz. Thus a relatively small change
at the higher primary frequencies is translated into a wide bandwidth frequency at the difference frequency, e.g., a 1% modulation
at the primary frequencies could be translated into a 10% modulation at the difference frequencies.

One of the drawbacks most often mentioned in the case of a para-

ECHO-RANGING PERFORMANCE

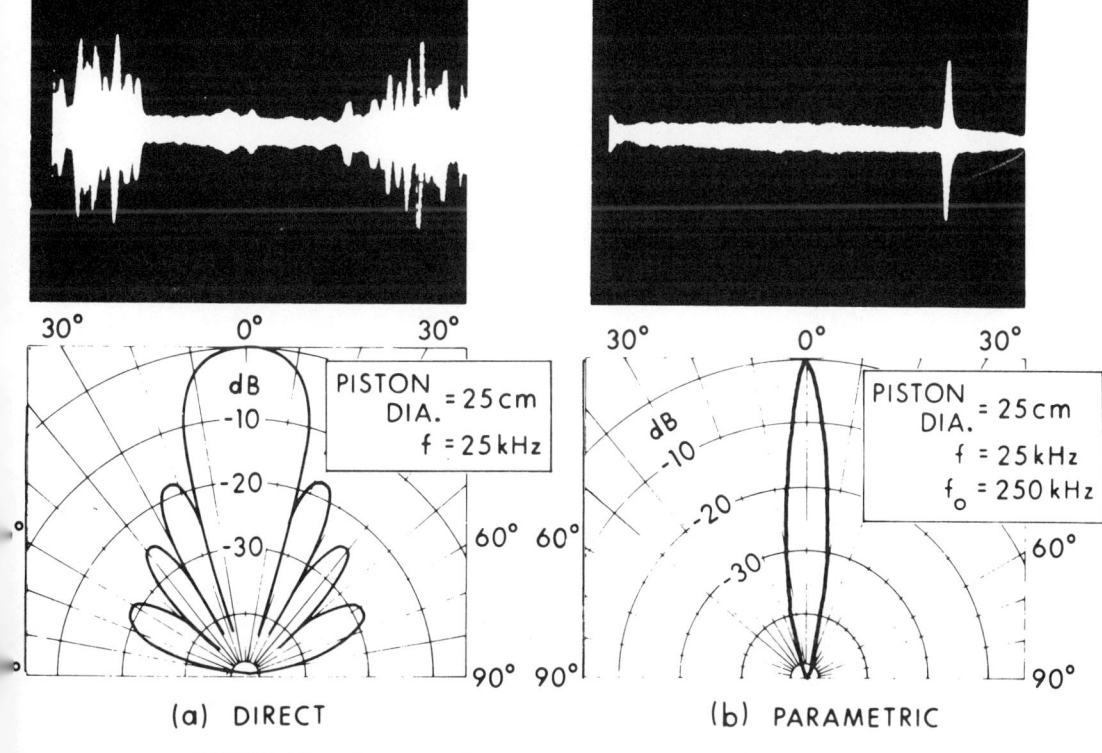

FIGURE 11: Parametric versus conventional echo
ranging (photo courtesy of W. Konrad).

metric sonar is its relatively low efficiency. However, high-
powered sonar systems, which may be efficient at modest power levels,
become very inefficient at very high power levels. Also, in many
cases where the aperture size is limited, it is simply impossible to
generate such narrow beams via conventional acoustics.

An example of an echo ranging experiment with a parametric sonar system is shown in Figure 11 (Konrad, 1973). A piston transducer of 25 CM diameter was driven directly at a conventional frequency of 25 kHz. The relatively wide conventional beam pattern
with its sidelobe is shown in Figure 11a. The first part of the
conventional echo photograph shows sidelobe reverberation. Later
in time the mainbeam surface reverberation is shown with the target
imbedded in the reverberation. The same transducer was then driven
parametrically at a center frequency of 250 kHz, i.e., it was driven
at two frequencies, 250 kHz ± 12.5 kHz in order to generate a 25 kHz
difference frequency. The beam pattern associated with the parametrically generated difference frequency shows a much narrower beam
pattern with sidelobes down almost 50 dB (Figure 11b). The upper
echo trace taken with the parametric system shows no sidelobe reverberation, no surface reverberation, and the target is clearly
distinguishable from the noise background.

ACOUSTIC LENS

The above sections have dealt primarily with active sonar systems. There are also applications (e.g., biological sampling via
broad band explosive sources) for receiving systems that require
beamwidths that are constant with frequency (Sternberg et al., 1978).
Figure 12 shows a technique to generate beamwidths that remain constant over a wide bandwidth of frequencies. The upper figure in
Figure 12 shows an acoustic lens where the beamwidth normally varies
with frequency. The lower figure in Figure 12 shows essentially the
same acoustic lens where a filter plate has been placed in front of
the lens. The filter plate is a conical tapered section of varying
thickness becoming thinner towards the center. The filter plate acts
as a variable aperture which depends on the frequency such that the
beamwidth remains essentially constant over a wide bandwidth of frequencies. Figure 13 shows measured beamwidths taken from 25 kHz to
300 kHz at MRA, i.e., along the maximum response axis, and for two
other cases where the beam has been tilted or steered 15° off axis
and 30° off axis. The beamwidth remains relatively constant over a
very wide band of frequency.

RAY BENDING

Figure 14 is a ray trace diagram for a source at 500 ft where
the sound waves or sound rays are plotted in one degree increments
as a function of range and depth. The curve on the left labeled
SVP7 is the sound velocity profile varying from the surface to a
depth of 1500 ft. The curves show that in a non-isovelocity medium,
sound waves do not travel in straight lines and additional losses
are introduced into the sonar equation which must be taken into
account in predicting the performance of any sonar system or in
determining the distribution of biological scatters.

VOLUME REVERBERATION

A major cause for the loss of sound energy during an acoustic
transmission is the re-radiation of sound, called scattering.
Losses may occur at the ocean surface, bottom, or within the ocean's
volume (Fig. 15). Discontinuities formed by temperature and/or salinity gradients and inorganic and organic particles all contribute
to differences in density which cause acoustic energy to be scattered; the sum total of all scattering is called reverberation. Comprehensive and detailed information on underwater sound is given in
Albers (1965, 1967), Urick (1975), and Clay and Medwin (1977).
At the commonly used active sonar frequencies, biological organisms are the major source of volume reverberation, overshadowing
inorganic sources such as dust, density microstructure, turbulence,
ship's wakes, etc. A type of volume reverberation called backscattering, where the direction of acoustic energy is scattered back toward the source, places a major limitation on the performance of active sonar systems used by the military. On the positive side, back-

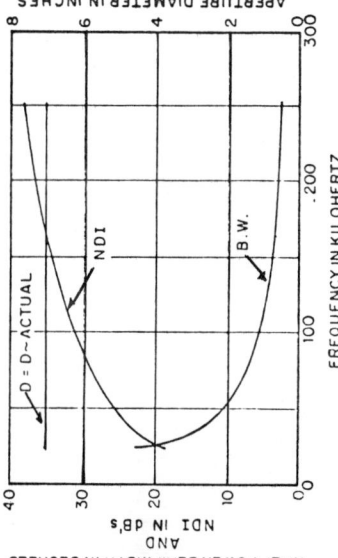

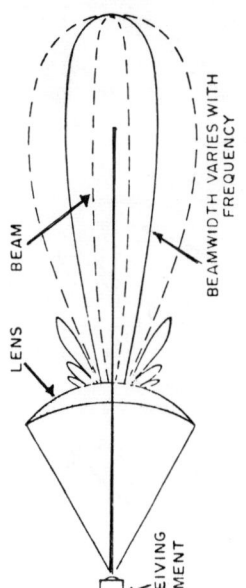

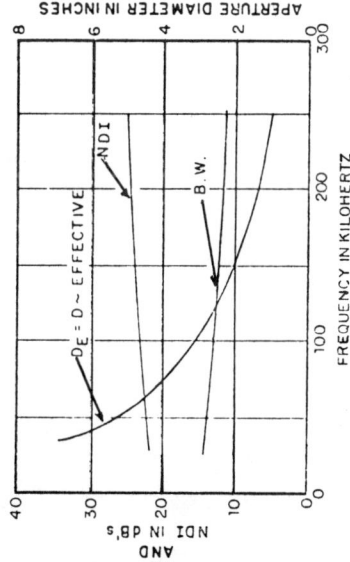

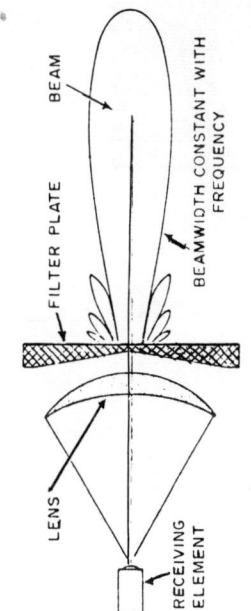

FIGURE 12: Conventional versus acoustic lens/
filter plate directivity indices and beam—
widths.

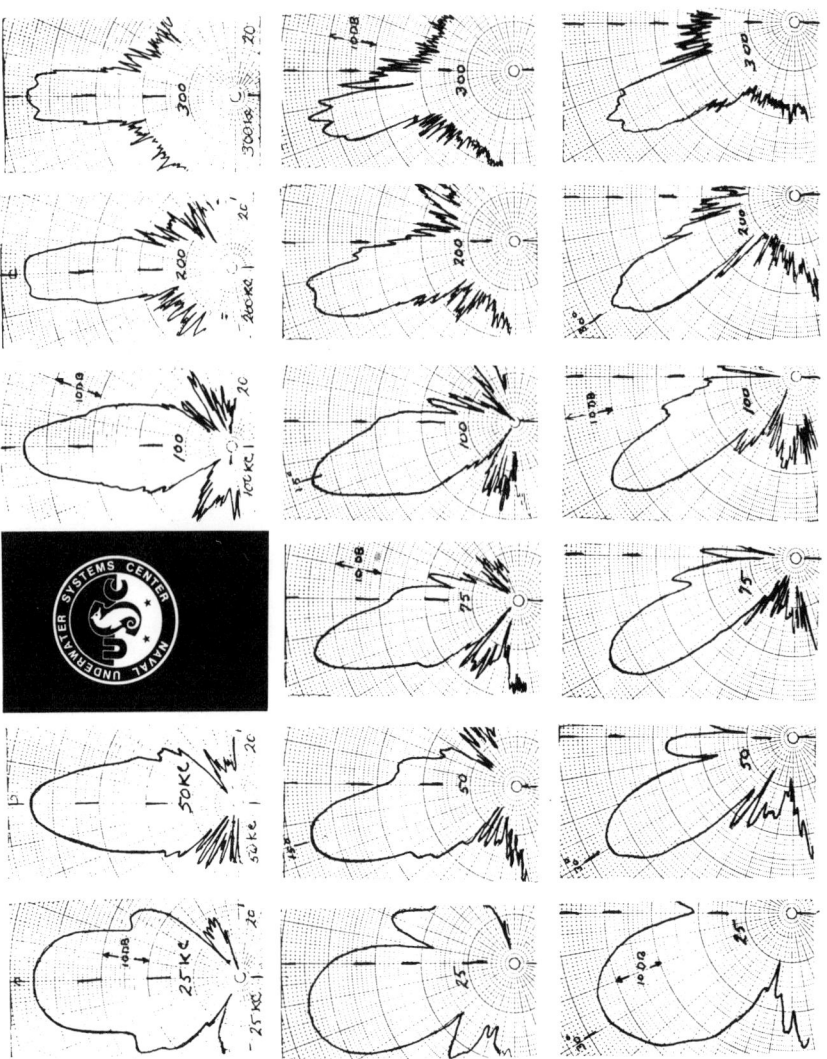

FIGURE 13: Beam patterns for acoustic lens with filter plate.

SENSOR SEARCH EVALUATOR

SD = 500. OFT   RAY INCR = 1.000 DEGS

PA = 1.0 DEGS   HALF BM = 10.00 DEGS

RANGE (KYD) = 7.0

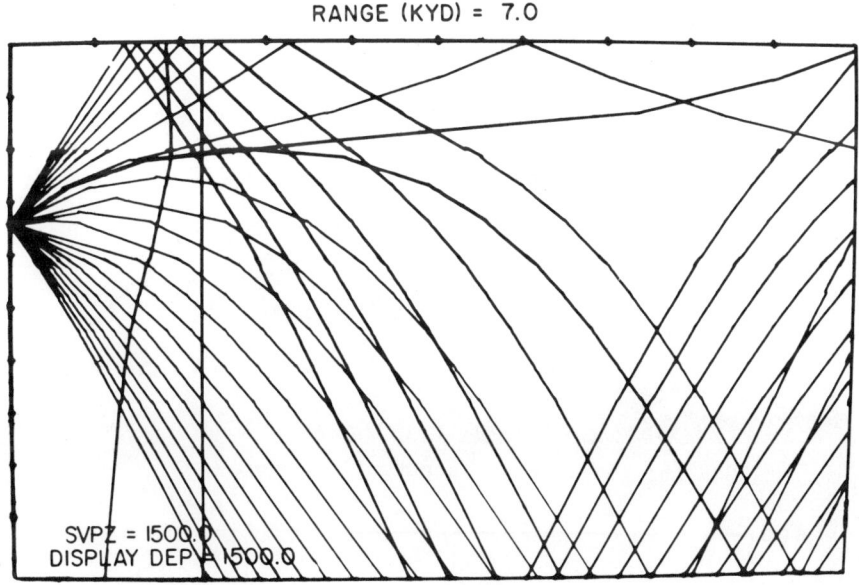

SVPZ = 1500.0
DISPLAY DEPTH 1500.0

FIGURE 14: Sound velocity profile and ray
trace diagram.

scattering records are utilized to map inhomogeneities in the ocean,
to locate and map biological populations, and to count and describe
individual species or assemblages of species (see Harden Jones,
Holliday, and Suomala, this volume).

Between 1 and 20 kHz, the major contributors to biological reverberation are fish possessing gas-filled swimbladders, whereas at
higher frequencies plankton generally become the dominant scatters
(Farquhar, 1970; Andersen and Zahuranec, 1977). Swimbladders (which
are usually shaped like a prolate spheroid) are treated as spherical
bubbles with equivalent radii which maintain a constant volume with
depth. Bubbles become resonant at a particular depth depending upon
their size and the incident frequency. At the depth of resonance,
the maximum effective acoustic scattering cross-section is orders of
magnitude larger than the actual geometric cross-section of a bubble
or bladder (Fig. 16). Thus, a few gas-bladdered fish ($\sim$1/1000 m^3 of
water) near or at resonance can account for considerable scattering.

Changes in resonant frequency with resulting changes in scattering strength occur when members of aquatic animal populations
undergo diel migration from deeper daytime depths to shallower night-

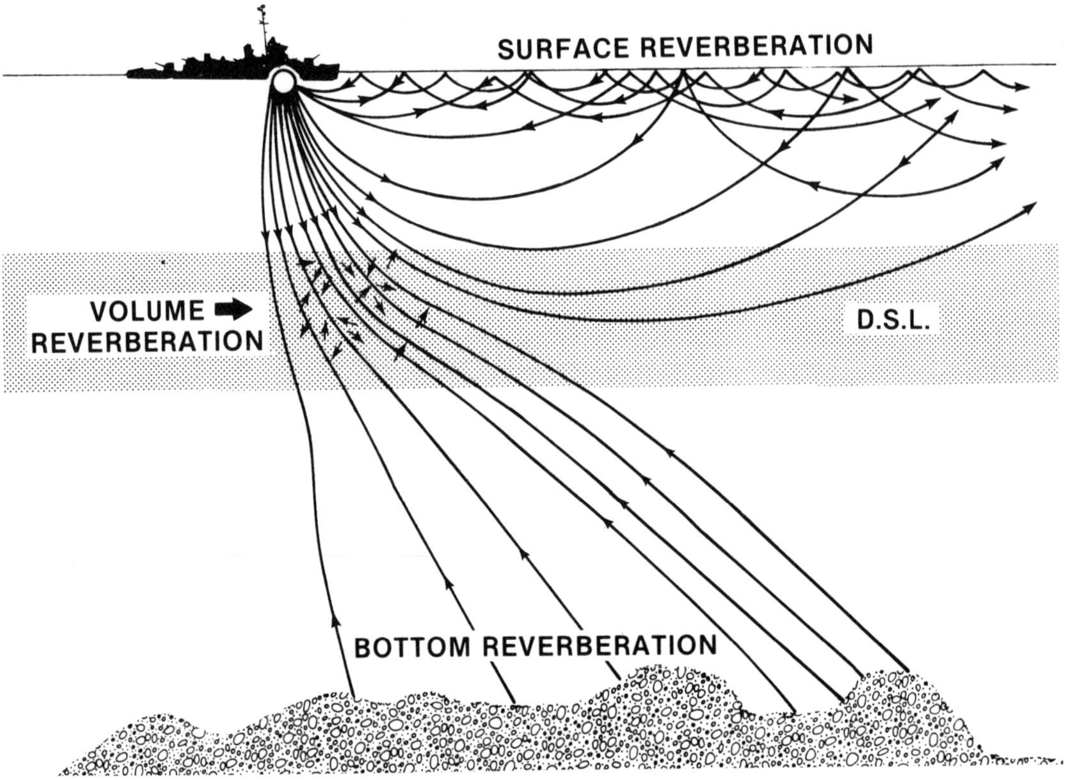

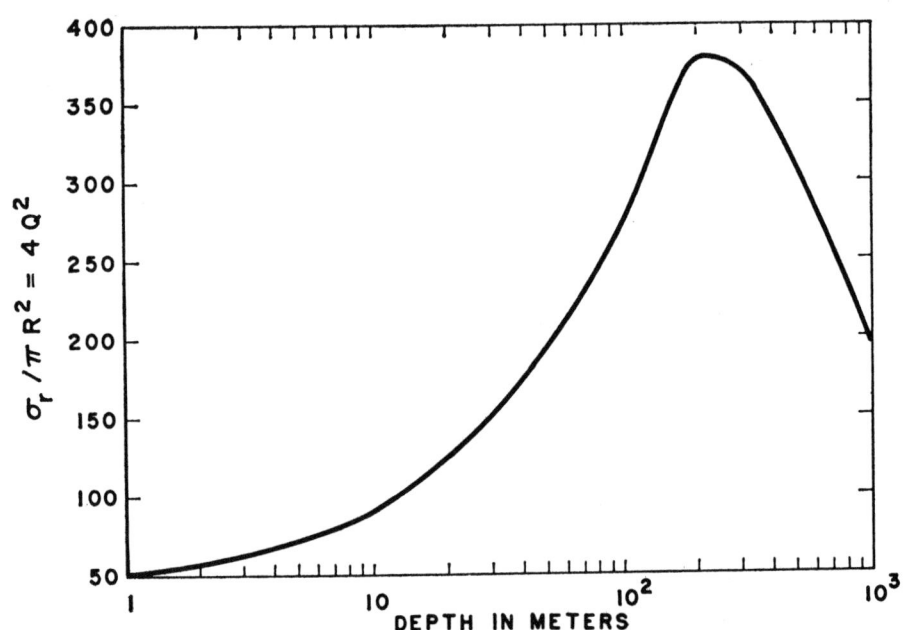

SURFACE REVERBERATION

VOLUME ➡
REVERBERATION

D.S.L.

BOTTOM REVERBERATION

FIGURE 15: Three major scattering areas.

FIGURE 16: Ratio of the resonant to the
geometric cross section of a fish swimbladder
as a function of depth. (At 5 kHz/ after
Chapman and Marshall, 1966.)

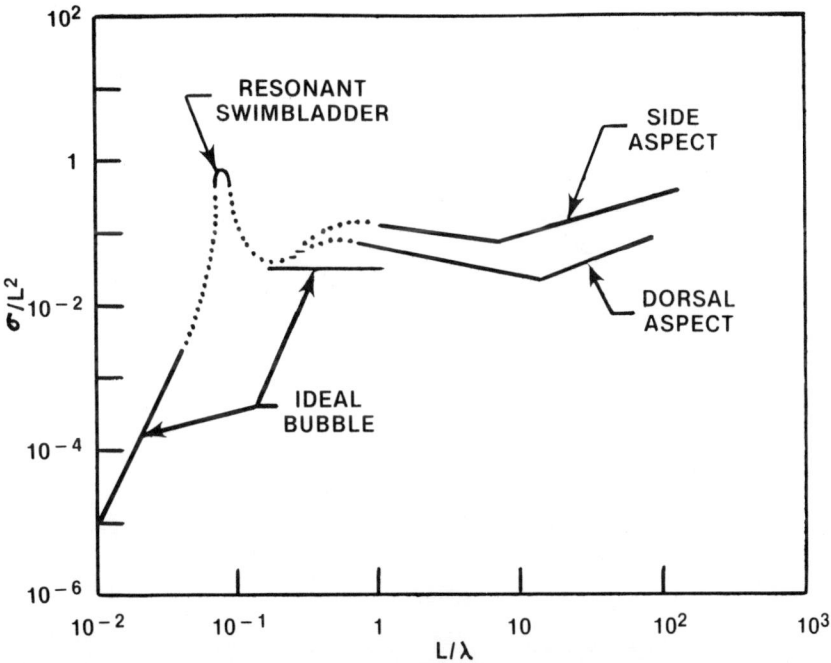

FIGURE 17: Estimated dorsal-aspect and
maximum side-aspect acoustic cross-section
of an individual bladder fish, assuming D =
20 ft, Q = 5, and R/L = 1/20. (After Love,
1970). Also shown are three regions of scattering.

time levels (i.e., the Deep Scattering Layers). Gas-filled swimbladders of a given size resonate at higher frequencies during the
day when the animals are at depth, and at lower frequencies at night
when they are shallower. Therefore, a single, migrating swimbladdered
fish scatters sound over a relatively wide frequency band in the
course of the day.

As a function of frequency, an assemblage of fish or other scatterers of varying sizes will scatter acoustic energy over three different regions of acoustic wavelengths depending upon the ratio of
the length of the scatterer to incident acoustic wavelength (Fig. 17).
Where the length of the scatterer is much less than the acoustic wavelength, Rayleigh scattering takes place; near or at resonance, in the
case of a bladder, resonant scattering occurs; and when the wavelength is about equal to or greater than scatterer length, geometric
scattering exists. In addition, the aspect of a scatterer with respect to the incident wavefront is also a factor that determines
scattering level (Love, 1970).

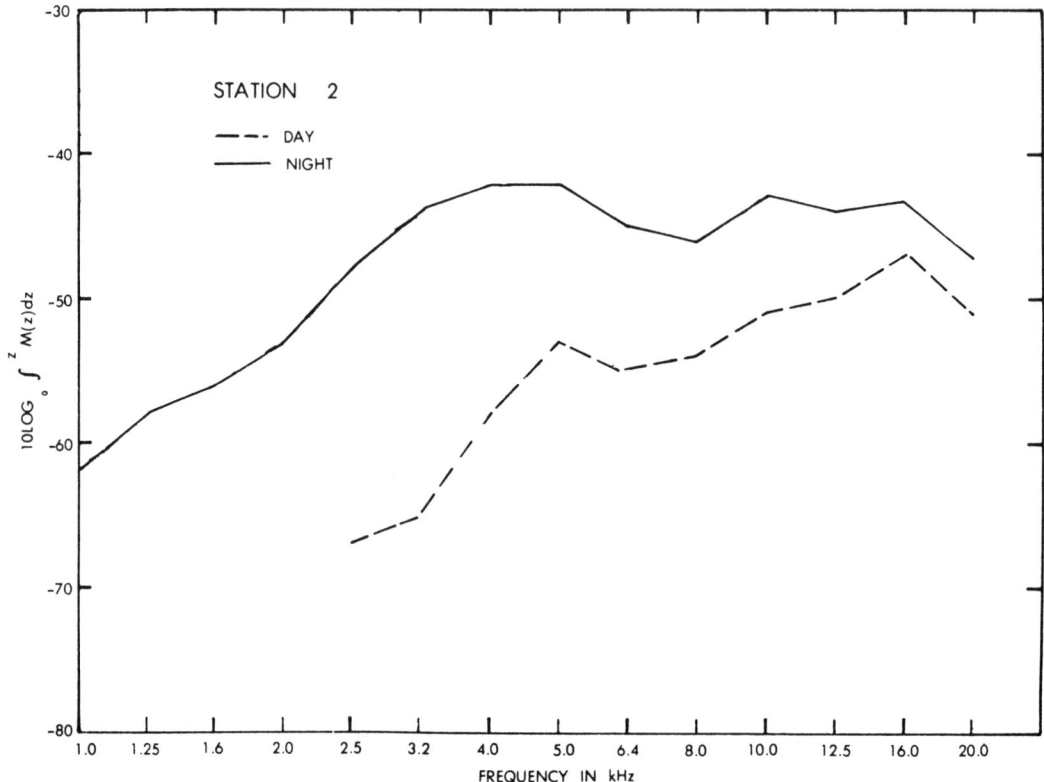

FIGURE 18: Daytime and nighttime scattering
strengths versus frequency for station 2
(After Gold and Renshaw, 1978).

MEASUREMENT OF SCATTERING STRENGTH

Small explosive charges ($\sim$0.5 - 2.0 lbs TNT) are used as broadband acoustic sources for estimating reverberation levels over a frequency range of about 1-20 kHz. By this method, the scattering
strength of the water column is commonly integrated over the depth
interval of near surface to 1000 m.

The magnitude of a water column scattering strength (s.s.) for
day and night over a frequency range of 1-20 kHz can be seen in Figure 18. Nighttime values are considerably higher than day values at
all frequencies; interestingly, no reverberation was measured in the
day below approximately 2.5 kHz, presumably because no "large" scatterers were present to scatter sound at this low frequency, but at
night as scatterers migrated to shallower depths their resonance
changed and 2.5 kHz data were obtained. Peaks in magnitude of s.s.
shift in frequency between day and night also; a peak is observed
between 4.0 and 5.0 kHz both day and night with the daytime high
about 16.0 kHz, while at night peaks are seen at about 9.0 and 16.0

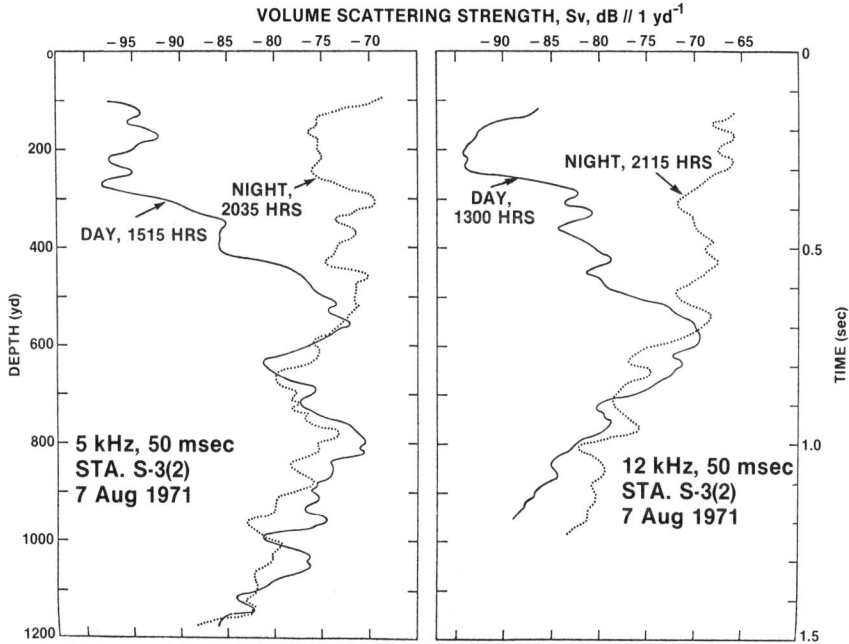

FIGURE 19:   Comparison of day and night profiles of volume scattering-strength, during
winter, 1971 (After Batzler, 1975).

kHz as well. The advantage of integrated column strength measurement is the wide frequency coverage; the disadvantage is the lack of
depth information.

Another technique used for measuring volume reverberation em-
ploys single-frequency transducers in a downward looking mode, very
similar to the manner in which echo sounders operate. Depth profiles of s.s. for day and hight at 5.0 and 12.0 kHz are given in
Figure 19. Like the column strength data shown in Figure 18, significant day/night differences in values are shown for the two frequencies, but in contrast to column strength data, scattering
strength as a function of depth is depicted. Here, it is noted
that s.s. below about 550 yds for both day and night is similar,
while nighttime values are markedly higher at both frequencies in
the upper 500 yds of the water column. Moreover, s.s. levels were
generally greater at 12 kHz than at 5 kHz, which was measured to a
deeper depth for both day and night. The shapes of the profiles
also provide a rough indication of the distribution of scatterers
for a given frequency and would be expected to change with changes
in frequency.

PREDICTING SCATTERING STRENGTH FROM BIOLOGICAL DATA

Scattering strength for a given frequency can be calculated from data on numbers of swimbladdered fish, their bladder size, its resonant frequency, and the depth of capture (Andreeva, 1964). Conversely, from broad-band acoustic volume reverberation measurements, effective fish bladder radii and fish abundances have been estimated (Chapman and Marshall, 1966). Since fish distribution is a function of geographical location, season, and time of day, it follows that prediction of scattering strength should take these variables into account.

Love (1975), utilizing swimbladder volume data and acoustic data taken in the Mediterranean Sea, reported essentially good agreement between predictions from biological trawl data and measured reverberation between 6.3 and 20.0 kHz, but less satisfactory agreement below 6.3 kHz. However, he compared his trawl data to derivatives of the measured scattering strength rather than to the directly measured data. In addition, he did not report any statistical variability about the mean values of the differences between biologically predicted and the derivatives of the acoustic data.

In contrast, scattering strengths calculated from bladdered fish data collected with a mid-water trawl off Bermuda, were always lower than concurrently acoustically measured s.s. taken at six frequencies between 3.85 kHz and 15.5 kHz (Brooks and Brown, 1978). This result was attributed mainly to the inability of sampling methods to furnish the necessary precise inputs on difficult-to-obtain swimbladder volume information.

The inconsistently low scattering strength values calculated from bladder information led to additional research which showed that numbers of swimbladdered fish alone could be compared to concurrently measured scattering strength profiles. As a result, fish numbers were linearly correlated with acoustic scattering strength for specific depths, frequencies, and seasons (Table 1). Correlation coefficients (r) ranged between 0.64 and 0.96.

In an effort to expand the capabilities of the relationships found between fish numbers and measured scattering strengths for predictive purposes, equations were obtained from linear regression analyses from the combined data on number of scatterers per 10^4 m^3 (X) and scattering strength in dB (Y) measured during summertime cruises 10, 12, and 14. At 15.5 kHz, the equation resulting from this analysis is

$$Y = -74.1 + 0.17 \ X \ , \quad r = 0.68 \qquad\qquad (9)$$

This equation was used to predict scattering strengths at 15.5 kHz from biological trawl data in the following manner. Acoustical data collected during the spring of 1970 (Fisch and Dullea, 1973; Fisch, 1977), for which no suitable biological data are available, were combined with biological data collected in the spring of 1972 (for which no suitable acoustical data are available). The assumption was made that the biological populations and the resultant volume reverberation levels were essentially similar for spring-

TABLE 1: Results of regression analyses of number
of scatterers versus measured scattering strengths

	Frequency (kHz)	Correlation Coefficient (r)	Regression Equation*
Night cruise 10	03.85	0.96	Y = -79.8 + 0.72X
29 May - 11 June 1970	15.50	0.87	Y = -75.9 + 0.26X
Night cruise 12	03.85	0.87	Y = -79.6 + 0.14X
20 August - 8 September 1971	15.50	0.64	Y = -75.4 + 0.25X
Night cruise 14	03.85	0.90	Y = -76.5 + 0.22X
4 June - 12 June 1972	05.00	0.88	Y = -74.2 + 0.24X
	07.00	0.75	Y = -67.8 + 0.18X
	09.00	0.76	Y = -65.8 + 0.19X
	13.50	0.64	Y = -69.3 + 0.08X
	15.50	0.73	Y = -70.3 + 0.11X

*Y = scattering strength in dB; X = number of swimbladdered fish/$10^4 m^3$.

time (February - March) during these two different years. However,
it was realized that population density, species composition, and
vertical distribution of scatterers were somewhat different between
spring and summer months and probably contributed to disparities
between the predicted (summertime) scattering strength profiles and
measured springtime (1970) profiles.

Predicted scattering strength profiles were calculated for 15.5
kHz by substituting for "X" in equation 9 the number of gas-filled
swimbladder fish per $10^4 m^3$ captured at each of the depths sampled
during the spring of 1972. These predicted values in Figure 20
(open triangles) are compared with scattering strengths measured
during the spring of 1970 (closed circles). It is evident that
very good similarity in magnitude and shape was achieved. The
maximum difference between predicted and measured scattering
strengths means is 3.9 dB, which occurs at 175 m. At most depths,
however, the two 15.5 kHz profiles differ by less than 2 dB. Moreover, only three of the predicted values exceed the limits described
by ±1 standard deviation on either side of the acoustically measured
scattering strength. Similar results were obtained at 3.85 kHz
(Brooks and Brown, 1978).

For the most part, the degree of agreement between the predictions and the measured scattering strengths especially at 15.5 kHz,
is very good and decidedly superior to the values predicted by the
utilization of swimbladder information. The success of the fish
abundance model is based on the supposition that the numbers of
swimbladdered fish used serve as a reliable index to the total population of potential scatterers and that the distribution of bladder volumes is invariant in this ocean area.

To provide estimates of scattering strength at other discrete

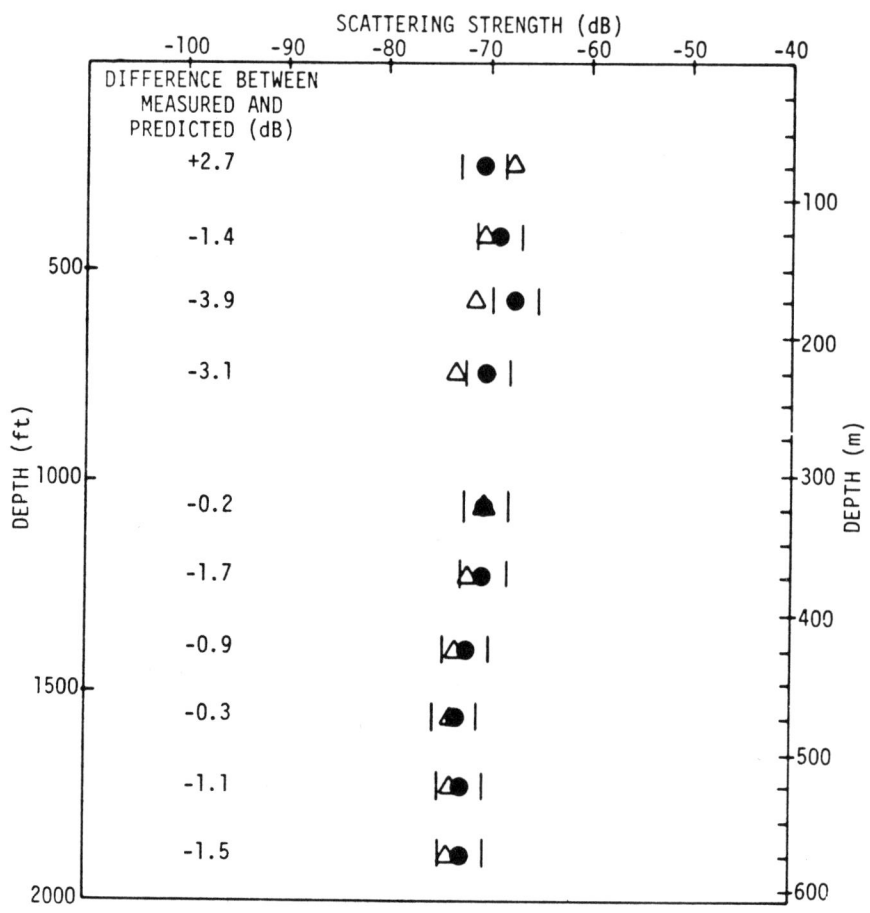

FIGURE 20: Comparison of predicted scattering
strengths with scattering strengths measured
at 15.5 kHz.

frequencies and for other areas of oceans where vertical fish population density data are available, multiple regression techniques
were applied to the present set of data to yield prediction equations over the frequency range of 2.0 to 20.0 kHz (Brown and Brooks,
unpublished data). For example, utilizing data collected on the
depth distribution and abundances of swimbladdered lanternfish
(family Myctophidae) in and around a Gulf Stream warm core eddy
(Krueger et al., 1977) a multiple regression equation with fre-

WARM CORE EDDY IN GULF STREAM

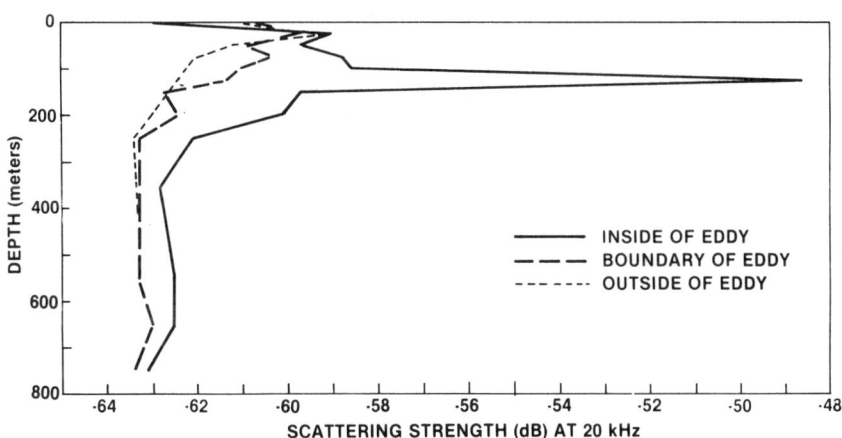

FIGURE 21: Variation in predicted nighttime scattering strengths associated with a Gulf Stream warm core eddy, summer, 1975.

quency in kHz and numbers of fish/10^4 m^3 as variables (r = 0.84) was used to predict 20.0 kHz nighttime scattering strength profiles in slope water, the eddy boundary and in the eddy core. In Figure 21 a marked increase in scattering strength is noted in the eddy core as compared to the slope and boundary water which have similar scattering levels. This increased s.s. in the eddy core corresponds to the large increase in numbers of lantern fish sampled at 125 m in the core. No other large abundances of lantern-fish were reported in the slope or eddy-boundary water during the sampling period (Krueger et al., 1977). Importantly, however, verification of the model is required for the expanded frequency coverage and for geographical locations other than the Bermuda area.

CONCLUSION

Volume reverberation is a major hindrance to active sonar systems designed for and used by the military. Conversely, volume reverberation caused by biological scatterers is utilized as an aid by marine biologists to map biological populations. Both these applications extend over a wide frequency range that must be recognized and treated differently for each application.

REFERENCES

Albers, V. M., 1965. *Underwater Acoustic Handbook - II*. The Pennsyl-
    vania State University Press, University Park. 356 pp.
Albers, V. M., ed., 1967. *Underwater Acoustics, Volume 2*. Plenum
    Press, New York. 416 pp.
Andersen, N. R. and B. J. Zahuranec, eds., 1977. *Oceanic Sound Scat-
    tering Prediction*. Plenum Press, New York. 859 pp.
Andreeva, I. B., 1964. Scattering of sound by air bladders of fish
    in deep sound-scattering ocean layers. *Akust. Zh.* 10: 20-24.
    English Translation Soviet Phys. *Acoust*. 17-20.
Batzler, W. E., 1975. Deep-scattering-layer observations off New
    Zealand and comparison with other volume scattering measurements.
    J. Acoust. Soc. Am. 58: 51-71.
Brooks, A. L. and C. L. Brown, 1978. Ocean acre final report: A
    comparison of volume scattering prediction models. *NUSC Tech.
    Rpt. 5619*. Naval Underwater Sys. Ctr., New London, Connecticut.
    40 pp.
Chapman, R. P. and J. R. Marshall, 1966. Reverberation from deep
    scattering layers in the western North Atlantic. *J. Acoustic.
    Soc. Am.* 40: 405-411.
Clay, C. S. and H. Medwin, 1977. *Acoustical Oceanography*. John
    Wiley and Sons, New York. 544 pp.
Farquhar, G. D., ed., 1970. Proceedings of an international sympo-
    sium on biological sound scattering in the ocean. *Maury Center
    Rpt. 005*. Department of the Navy, Washington, D. C. 642 pp.
Fisch, N. P., 1977. Acoustic volume scattering at the Bermuda Ocean
    acre site (Cruise 14 and related earlier studies). *NUSC Tech.
    Rpt. 5365*. Naval Underwater Sys. Ctr., New London, Connecticut.
    12 pp.
Fisch, N. P. and R. K. Dullea, 1973. Acoustic volume scattering at
    the Bermuda Ocean acre site during spring and summer 1970 and
    summer 1971 (Cruises 9, 10, and 12). *NUSC Tech. Rpt. 4469*. Naval
    Underwater Sys. Ctr., New London, Connecticut. 24 pp.
Gold, B. A. and W. E. Renshaw, 1978. Joint volume reverberation and
    biological measurements in the tropical western Atlantic. *J.
    Acoust. Soc. Am.* 63: 1809-1819.
Konrad, W. L., 1973. Application of the parametric source to bottom
    and sub-bottom profiling. *NUSC Tech. Memo. TDIX-43-73*. Naval
    Underwater Sys. Ctr., New London, Connecticut.
Konrad, W. L., 1975. Design and performance of parametric-sonar
    system. *NUSC Tech. Rpt. 5227*. Naval Underwater Sys. Ctr., New
    London, Connecticut.
Konrad, W. L. and L. F. Carlton, 1973. A parametric source for deep-
    scattering layer investigation. *NUSC Tech. Memo. TDX-32-73*.
    Naval Underwater Sys. Ctr., New London, Connecticut.
Krueger, W. H., R. H. Gibbs, Jr., R. C. Kleckner, A. A. Keller, and
    M. J. Keene, 1977. Distribution and abundance of mesopelagic
    fishes on Cruises 2 and 3 at deepwater dumpsite 106. In: *Base-
    line Report of Environmental Conditions in Deepwater Dumpsite 106,
    Volume II*, pp. 377-421. U. S. Department of Commerce, National
    Oceanic and Atmospheric Administration, Rockville, Maryland.

Love, R. H., 1970. Dorsal-aspect target strength of an individual fish. *J. Acoust. Soc. Am.* 49: 816-823.

Love, R. H., 1970. Predictions of volume scattering strengths from biological trawl data. *J. Acoust. Soc. Am.* 57: 300-306.

Moffett, M. B., R. H. Mellen, and W. L. Konrad, 1976. Parametric sources of rectangular aperture. *NUSC Tech. Memo. 306-331-76.* Naval Underwater Sys. Ctr., New London, Connecticut.

Sternberg, R. L., W. A. Anderson, and G. T. Stevens, 1978. Log-periodic acoustic lens-acoustic filter plate study. *J. Acoust. Soc. Am.* 63: 1617-1621.

Urick, R. J., 1975. *Principles of Underwater Sound, Second Edition.* McGraw Hill Book Company. 384 pp.

High Resolution Sonars

H. L. Warner

ABSTRACT: A brief description of the major mine
hunting sonars developed by the Navy over the
past 25 years is given in terms of their search
geometries. Sonar information equations which
determine information capacity, density, and
rate of a sonar in terms of spatial and temporal bandwidths are presented and can be used
to determine the potential capability of a sonar to detect and classify targets. Possible
means of enhancing detection of targets whose
characteristics are known are discussed with
emphasis on increasing temporal bandwidths when
size constraints limit the spatial bandwidth of
a sonar.

INTRODUCTION

The main purpose of this paper is to discuss briefly some of
the aspects of searching for small underwater objects using high
resolution sonars and how the target detection problem affects
sonar design. In mine countermeasures applications two general
types of sonars are recognized: search sonars and classification

399

sonars. Search sonars are those which are designed to detect and
localize targets with emphasis given to searching as large an area
as possible in a given time. Classification sonars are designed
to increase the possibility of identifying a target from its acoustic attributes while eliminating most other echoes from consideration. The targets are categorized by their expected location with
respect to the sea boundaries. The simplest detection problem for
a given target occurs when its position is away from any boundaries
and only the sea volume has to be searched. Increased difficulty
is encountered as target position shifts nearer the sea surface or
bottom, and the most difficult detection problem occurs, as one
would expect, if the target becomes buried in the sea bottom.

SEARCH SONARS

The earliest search sonars had only a single receiver beam and
information channel and required mechanical movement of the transducer to scan out the search area or sector. As the need for improved search rates arose, means were found to increase the information rates of sonars by increasing both the number of beams and
the signal bandwidth.

The ability of a sonar to transfer information about the distribution of acoustic reflectivity in a region of interest can be
determined by calculating the sonar information capacity, density,
and rate. The number of possible messages the sonar can present to
an observer can be derived directly from the sonar resolving powers
(number of range elements, bearing elements, and the dynamic range
of the displayed signal) or by starting with general concepts used
in the field of information theory.

The information capacity of a sonar for a single look is

$$N_c = n_r \cdot n_b \cdot \Gamma = \frac{r}{\Delta r} \cdot \frac{\phi}{\theta} \cdot \Gamma \text{ (bits/look)}$$

where $\Gamma = \text{LOG}_2$ (displayed levels), n_r is the number range increments,
n_b the number of beams, r the search range, ϕ the search sector, and
θ the beamwidth.

The information density of a sonar is obtained by dividing the
information capacity by an area or by a sector. Thus,

$$N_\rho = \frac{1}{\Delta r} \cdot \frac{1}{\theta} \cdot \Gamma = \frac{2}{c} W_T \cdot W_s \cdot \Gamma \text{ (bits/radian-metre)}$$

where W_T and W_s are the temporal and spatial bandwidths of a pulsed
sonar. N_ρ is a good indicator of the relative classification or
imaging ability of sonars.

If the information capacity is divided by the time per look, the
information rate of a sonar is obtained. For a pulsed sonar $T_\ell = 2$
r/c and

$$N_r = \frac{1}{\tau} \cdot \frac{\phi}{\theta} \cdot \Gamma \text{ (bits/sec)}$$

where

$$\tau = \frac{2\Delta r}{c} = \text{pulse duration}$$

For a continuous transmission frequency modulated (CTFM) sonar the time per look is the reciprocal of the filter bandwidth and

$$N_r = n_f W_f \cdot \frac{\phi}{\Theta} \cdot \Gamma \quad (\text{bits/sec})$$

where n_f is the number of filters of bandwidth W_f in the range analyzer. It can be seen that when the analyzed bandwidth of a frequency modulated (FM) sonar equals the effective bandwidth ($1/\tau = n_f W_f$) of a pulsed sonar, the information rates of the two sonars will be the same.

It should be noted that information rate is a direct indicator of the potential capability of a sonar, but, producing a sonar having large bandwidths without providing related processing to make use of the available information does not result in improved performance. In other words, one has to do more than vary the frequency of transmission and increase the size of the transducer to build a better sonar. The search sector ϕ of a sonar is that area or volume covered by the receiver beam patterns and is determined by the size of the independent samples taken of the array during beamforming (i.e., usually the element size). Of the many possible configurations, the more useful search geometries have been the following.

1. Ahead-Look -- Where the beam patterns are narrow horizontally and broad in the vertical dimension with the center beam directed in the direction of sonar motion. These sonars can be hull-mounted or located in a towed body. The ahead-looking sonar is very useful in shallow water because it can detect targets anywhere between the surface and the bottom in time to allow object avoidance maneuvers by the sonar craft.

2. Side-Look -- Usually a single beam to each side of a towed body. Each beam is very narrow in the direction of motion and broad in the other dimension. Side-look sonars are very useful in bottom mapping applications and because the largest transducer dimension lies along the length of the body, the beampattern can be made quite narrow. This is very useful in target classification based on size and shape considerations. The main disadvantage of this type of sonar is that it must move in order to generate a display.

3. Radial-Look -- Pencil beams are formed in a plane perpendicular to the direction of motion of the sonar body. The radial-look sonar provides the most efficient way (in terms of numbers of beams required) to search a given volume.

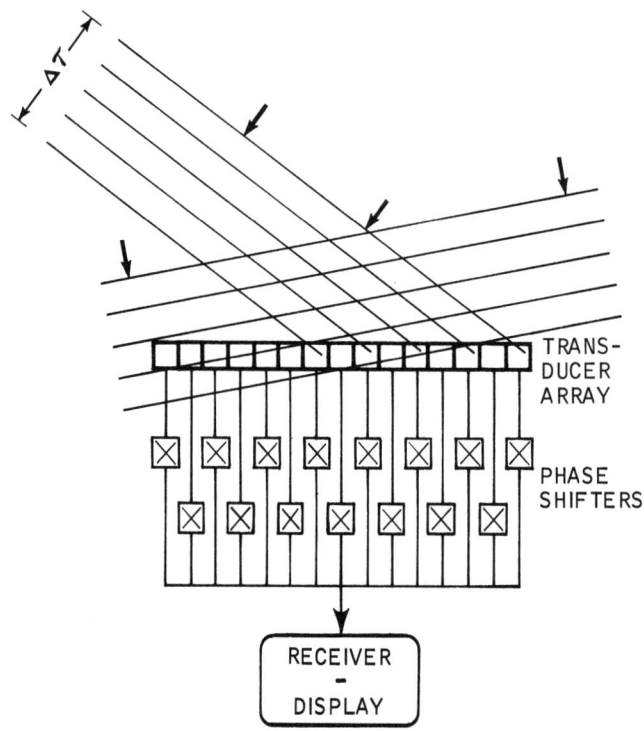

FIGURE 1: Reception of short duration echo
pulses by a phase-shift beamformer showing
the decrease in effective aperture for off-
axis targets.

     In order to maximize the use of the available spatial bandwidth
of a sonar, processing is done either to scan a single beam across
the search sector or to form enough simultaneous beams to cover the
sector. In both methods, the beamforming can be accomplished by
either phase shift techniques or by true time delays. Phase shift
beamforming is usually the simplest and least costly to implement,
but it has a disadvantage in some applications where very short
pulses are used to obtain good range resolution. Figure 1 illustrates the arrival of a short pulse from an off-axis target. Since
the instantaneous signals from each hydrophone element are only
phase shifted to form the off-axis beams, the short pulse results
in an effective shortening of the transducer array; hence, the beampattern is broadened in direct proportion. Because of this problem,
sonars using cylindrical shaped arrays must be made larger than
would be required if delay line beamforming were used. True delay
beamformers can be implemented using delay lines, acoustic lenses,
or any of the many types of electronic memories now available.
     Figure 2 illustrates the primary sonars developed by the Navy
over the last 25 years. The earliest major shipboard sonar was the

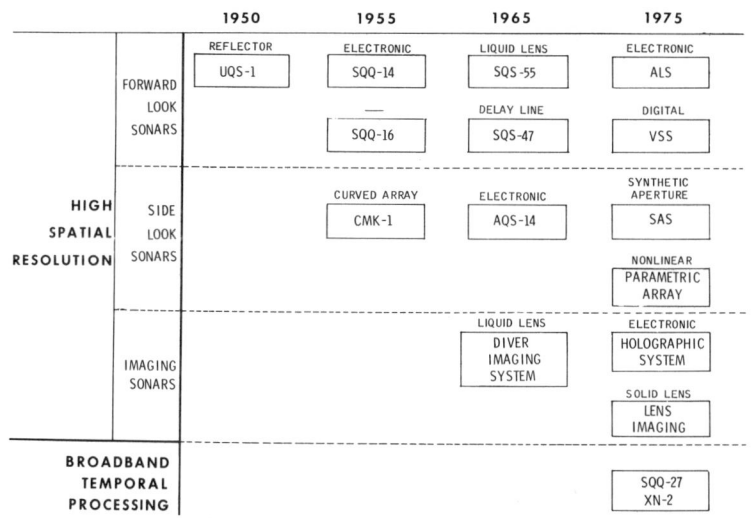

		1950	**1955**	**1965**	**1975**
	FORWARD LOOK SONARS	REFLECTOR UQS-1	ELECTRONIC SQQ-14	LIQUID LENS SQS-55	ELECTRONIC ALS
			SQQ-16	DELAY LINE SQS-47	DIGITAL VSS
HIGH SPATIAL RESOLUTION	SIDE LOOK SONARS		CURVED ARRAY CMK-1	ELECTRONIC AQS-14	SYNTHETIC APERTURE SAS
					NONLINEAR PARAMETRIC ARRAY
	IMAGING SONARS			LIQUID LENS DIVER IMAGING SYSTEM	ELECTRONIC HOLOGRAPHIC SYSTEM
					SOLID LENS LENS IMAGING
BROADBAND TEMPORAL PROCESSING					SQQ-27 XN-2

FIGURE 2:   Primary sonars developed by the
Navy intended for detection of small underwater targets.

AN/UQS-1 which used a reflector to form 23 beams covering a 20-degree
sector. The sector could be mechanically oriented in different directions. The sonar was hull-mounted and its performance suffered
when sound propagation near the surface was less than perfect. This
sonar design was the first successful attempt to increase performance
by increasing the number of simultaneous beams. The UQS-1 reflector
provided a true delay beamformer whereas the next shipboard sonar,
the AN/SQQ-14, used a phase shift technique. This sonar, based on a
British design, electronically scanned a single beam through a larger
sector during a time equal to the transmitted pulse duration. This
sonar also had a classify mode and a capability to lower the sound-
head down below near-surface sound propagation problems. The AN/SQQ-
16 was a CTFM sonar of similar capabilities to the SQQ-14 but whose
single beam was mechanically rotated. It was intended for use from
a small boat and provided variable depth operation from its over-the-
stern mount. The AN/SQS-55 and the AN/SQS-47 were the first sonars
to form many simultaneous beams over large search sectors (up to 120
degrees) using true time delay techniques. These sonars provided
significant increases in information rate over previous designs and
were much smaller in size due to the use of solid state electronics.
Both sonars were column-mounted for over-the-side use on small boats
and were designed to detect swimmers and small mines in a river environment. The 1975 sonars shown are still under development.
      The side-look sonars listed in Figure 2 are primarily classifi-
cation sonars. They operate at relatively high frequencies and because of their relatively large spatial and temporal bandwidths have
the highest information densities of any sonars. The information
rates of these sonars are low due to the small number of beams formed
at a time. The imaging sonars have both high information density and

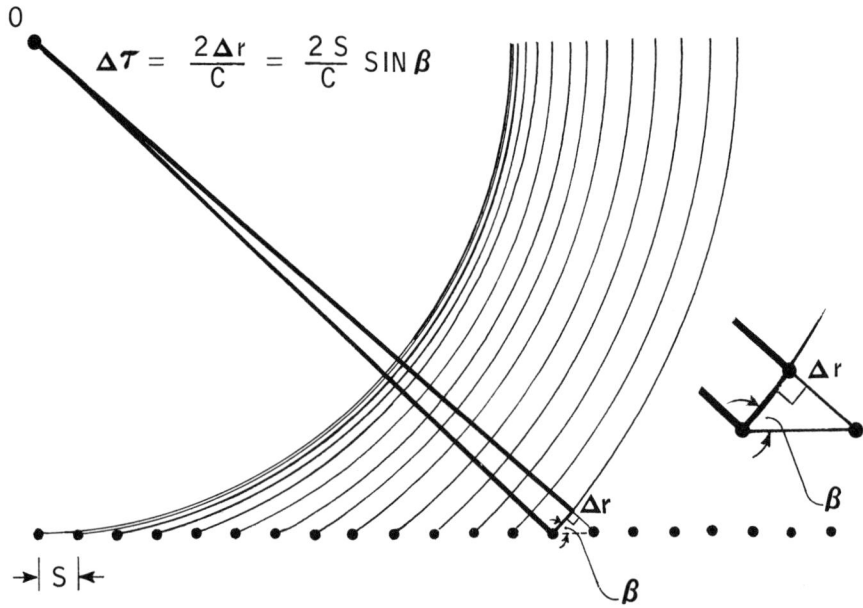

FIGURE 3.   Frequency sweep resulting from a
set of equally spaced reflectors due to an
obliquely incident impulse.

information rates, but due to size-wavelength constraints they operate only at very short ranges.

DETECTION OF TARGETS

In the acoustic detection of targets nearly every physical
characteristic of the object has some effect on its acoustic properties. Assuming the target of interest to be passive (i.e., not
a sound source), the interaction of the insonifying signal with the
target is the only basis for target detection. In most search problems the target and its characteristics are not within the control
of the designer or user of search sonars. The designer can only
measure test targets and choose some detection method based on the
results of the measurements. Most sonars rely on target reflectivity for detection so that the effective cross-section of the target
must exceed the background reverberation cross-section from the sea
bottom, surface, and volume. Thus, detection improves as sonar res-
olutions are made to approach target size; however, this ideal design can only be achieved for fairly large targets at ranges less
than 100 metres because of sonar size constraints. Ideally, for
any given target, one could design a transmitted signal which fits
the target exactly, resulting in some chosen characteristic of the
received signal being maximized only when the target is present.
However, a number of conditions occur in the practical application
of this concept which, so far, have limited its success. As a simple example, consider the array of point reflectors shown in Figure 3
with a single acoustic impulse transmitted from position 0. The arri-

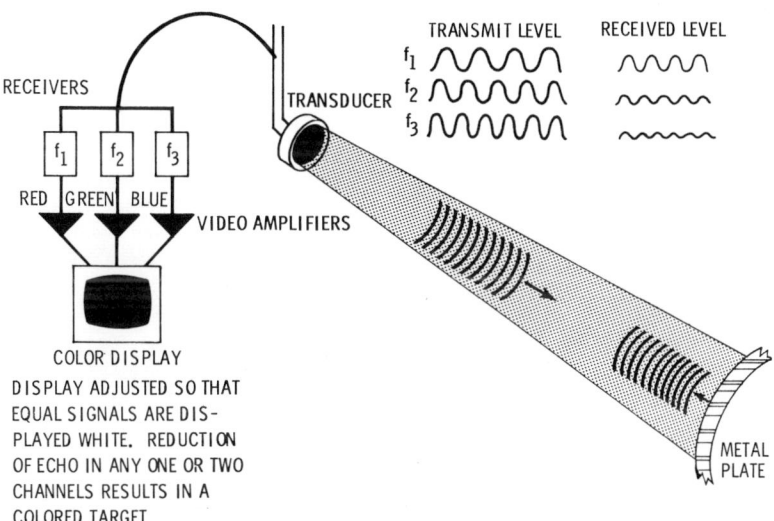

FIGURE 4. Technique for representing sonar
target bandwidth characteristics in terms of
displayed color.

val time of the pulse at each reflector varies with the distance and
the angle so that the difference in arrival times at consecutive reflectors increases as the pulse travels outward and eventually would
approach a time period given by the spacing divided by the speed of
sound. If the reflected signal is recorded at point 0, the time interval between pulses is given by

$$\Delta\tau = \frac{2\Delta r}{C} = \frac{2S}{C} \sin\beta$$

which means that the received signal would start at some high frequency and sweep down to a frequency approaching C/2S impulses per
second. Frequency sweeps representing expected geometries of reflectors could be stored and compared with signals received from
unknown reflector arrays as a detection technique. The difficulty
with this method is that the catalog of expected signals gets large
very quickly and processing time goes up accordingly. One is then
tempted to transmit the time-reversed signal of the one received
after transmitting the impulse. This signal would start at the
limiting frequency, C/2S, and sweep up with a duration and rate determined by the number and extent of the reglectors. If done properly, this process would result in the reception of a single impulse at point 0 with an amplitude N times as large as any of the
N cycles in the reflected frequency sweep. Unfortunately, the
transmitted signal would have to be different for every change in
target and target-sonar geometry for the technique to work. This
design is not impossible, but not yet practical for most applications.
    Another approach which may be useful when sonar spatial resolu-
tion is inadequate may be implemented against targets which reflect
some frequency better than others. Such is the case with metal
plates which exhibit an interference between signals reflected from
the front and back surfaces. Figure 4 shows a simplified version of

a detector that does not require specific information on the material or its thickness, only that the targets of interest be selectively reflective; i.e., have an "acoustic color". Temporal bandwidth requirements should be less than an octave (as in the visual detection of color) since interference effects repeat for each frequency doubling. This technique may also be applicable in the classification of fish -- especially those having swim bladders -- where one might expect the bubble resonance or the rapid decrease in target strength at frequencies below resonance to result in the target acquiring an acoustic color when the echo is compared with the transmitted spectrum.

If Rayleigh scattering occurs at frequencies below swim-bladder resonance, the target strength should vary inversely as the fourth power of the wavelength. This decrease would result in a rapid loss of reception of echoes if the operating frequency of a sonar could be varied downward on successive transmissions. In the case where the spatial resolution of the sonar is not adequate to resolve individual fish, the slope of the echo frequency versus amplitude curve would be even more sensitive to the presence of different sized fish and we are faced, again, with having more unknowns than measurables. It may be that by measuring the echo strength at some frequency where the wavelength is quite short compared with bubble circumference, and by comparing this value with simultaneous measurements made around resonance, the number and size of fish in a given resolution cell could be determined. Such a technique suggests the use of a parametric array sonar, where high frequencies can be used on transmission and the non-linear products provide low frequency signals of relatively broad bandwidths upon reception.

As mentioned earlier, the science of designing sonars is based on a prior knowledge of the target and its background. In the absence of such measurements one is forced to use available sonars to verify that fish are detectable and then to try to correlate sonar measurements with some physical sampling of the fish population. Under such conditions, calibration of the sonar is very important. A passive target which is useful for such purposes was developed as a by-product of Naval Coastal Systems Center's (NCSC's) acoustic lens work. The target is a stainless steel sphere filled with a fluid having a low sound velocity compared with sea water (Folds, 1971). This target has the property of having a large effective cross-section in all directions and can be made to have a constant target strength over a wide frequency range. A 15 cm sphere, filled with a mixture of fluids having a density of $1.8 - 1.9$ grams/cm^3 will have a target strength of -10 dB compared with -28 dB for a similar air-filled sphere.

SUMMARY

Sonar information equations describing the information capacity, density, and rate can be used to estimate the potential capability of a sonar to detect and classify targets. Since the equations are derived from fundamental concepts of spatial and temporal bandwidths, any sonar can be specified regardless of the complexity of its processors if the bandwidth limitations are determined. Also, the relationships show that expected sonar performance can be maintained by increasing one bandwidth when the other has to be decreased in a given application. A look at high resolution sonars developed by the Navy over the last 25 years shows a steady increase in sonar information rate capability. This increase has usually been accomplished by increases in the spatial bandwidth of the sonars. It is suggested that in applications where spatial resolution of targets is not feasible, increased temporal bandwidth with associated signal processing is the most promising approach.

REFERENCES

Folds, D. L., 1971. Target strength of focused liquid-filled spherical reflectors. *J. Acoust. Soc. Amer.* 49: 1596-1599.

Acoustics and the Fisheries: Recent Work with Sector-Scanning Sonar at the Lowestoft Laboratory

F. R. Harden Jones

ABSTRACT: A high resolution 300 kHz modulation
scanning sonar has been installed in the Ministry of Agriculture, Fisheries and Food's research vessel CLIONE. The equipment is stabi-
lized against roll, pitch, and yaw and can be
used in horizontal (plan) or vertical (elevation) mode to scan a 30^0 sector. The equipment
has been used to determine the efficiency of
the Granton otter trawl for catching plaice.
In these experiments plaice fitted with acoustic transponding tags were released individually
and kept under surveillance by sonar. A second
research vessel, towing an otter trawl, was directed to catch the fish, the whole catching
process being followed by sonar and recorded on
16 mm film and video tape.

For 166 valid attacks made on individual fish
lying between the otter boards, the overall efficiency of the gear was 44% $\pm$ 8%. For fish lying
between the otter boards and wing ends of the net,
the efficiency was 22% $\pm$ 10%, and for those in
the path of the net 61% $\pm$ 10%. The addition of

a board-to-board tickler chain increased the overall efficiency of the net to 67%, and to 48% and
79% for fish in the positions 'boards to wing
ends' and 'path of the net' respectively.

The high resolution and exceptional picture-forming qualities of the sector-scanning sonar have
been applied to: wreck and bottom surveys; measurements of the parameters of fishing gear; fish
tracking; the structure and behaviour of fish
shoals; and the study of fish target strengths.
Modulation sector-scanning sonar has proved to be
a remarkable and versatile invention and very reliable in operation.

INTRODUCTION

    The Fisheries Laboratory, Lowestoft, United Kingdom, is one of
several controlled by the Ministry of Agriculture, Fisheries and
Food. As might be expected for an establishment supported by public funds, our research and development objectives are clearly
defined:

        Research on marine fish and shellfish stocks and
    their environment is designed to provide the Fisheries
    Ministers with the scientific information needed to
    discharge their responsibilities to the fishing indus-
    try and consumers. It provides the basis for the for-
    mulation of domestic policy and the discharge of inter-
    national obligations, particularly those related to
    conservation and environmental protection (Anon., 1974).

Within this framework there are three broad aims of marine research:
to ensure the rational exploitation and conservation of the stocks
of fish and shellfish, to promote the efficiency of the industry by
technological development, and to protect the marine environment. I
am going to tell you how we have used acoustic techniques to tackle
a problem that is relevant to the first two of the three aims: it
concerns the efficiency of the otter trawl.

THE OTTER TRAWL

    In the beam trawl the net is kept open by a beam which is
attached to the headline and supported on two iron shoes which
ride along the bottom. But, in the otter trawl, the net is kept
open by two spreading boards which were, in the first otters,
attached directly to the wings of the net.
    The bottom otter trawl is probably the most important fishing
gear in the world and, in the United Kingdom, it accounts for at
least half of the value of all the fish landed by British vessels.
It was invented and patented by an engineer, James Robert Scott
(Scott, 1894), whose father was the manager of the General Steam

Fishing Company of Granton, near Edinburgh in Scotland. The
Company was struggling to survive a steady decline in earnings (a
trading loss of 3600 pounds sterling was reported for 1894) and the
first trials of the new beamless trawl were made on S. T. ATHOLE in
June 1894 (M'Intosh, 1894). Within six months it was clear that the
trial was a success and in 1895 several Scottish and English vessels
were fitted out to work the gear. By 1896 the new trawl was generally adopted by British steam trawlers, and, in December of the same
year, the High Court upheld the validity of Scott's patent (Anon.,
1897). With subsequent modifications and additions in the 1920's
and 1930's - such as Vigneron's bridle arrangements, Phillips' patent floats and bobbins, and Brighouse's patent footropes - Scott's
original invention provided the basic design for many of the bottom
trawls now in use.

When first introduced, Scott's trawl caught 50% more roundfish,
such as cod and haddock, than the traditional beam trawl (Cunningham,
1895; Garstang, 1900) and it fished as well by day as it did by
night, in contrast to the beam trawl whose catches were low in daylight. This small but interesting detail appears to have been for-
gotten by fishery biologists. Vessels, working the new patent gear,
outfished all others, and the duration of voyages was cut by half
with a consequent improvement in the quality of the catch and its
sale price: small wonder that the beamers complained that "there
can be no competing with a machine that sweeps the sea as cleanly
as a mowing machine reaps a lawn" (Anon., 1895).

The spread of Scott's trawl from wing to wing was unlikely to
have been much greater than that of the larger beam trawls of the
day - about 14 to 15 m - so that the area of bottom swept clear by
both gears was probably much the same. It was never claimed that
the new trawl caught more flatfish than the old, and the Scottish
catch statistics reported by Garstang (1900) confirm this point.
Cunningham (1895) attributed the increased catch of roundfish by
Scott's trawl to a somewhat higher headline height and its ability
to catch fish during the day to the absence of the beam which was
presumed to scare fish while they still had time to avoid the net.
Headline floats are not mentioned in Scott's patent but the gain
should have followed from the freedom of the headline.

Vigneron's patent (1920) introduced the use of bridles between
the otter boards and the wing ends of the trawl. One of the reasons
for this measure was to allow silt and stones stirred up by the
doors to pass outside the net and so improve the quality and thus
the value of the catch by reducing damage to the fish in the codend. The cut of the net was also altered to allow the headline to
fly higher. The introduction of bridles and the so-called Vigneron-
Dahl (V-D) gear was followed by a marked increase in catch-rate,
particularly in roundfish (Bagenal, 1958). The increase appeared
to be proportional to the increased spread between the otter boards.
This suggested that the net was now catching fish which were originally lying between the otter boards and the wing ends and had subsequently moved into the path of the net, possibly in response to
the presence of the otter boards or to the bridles themselves. The
increased catch-rates following the adoption of the V-D gear went a
little way to relieve the problems of the industry in the middle

1920's when the benefits of the increased stock following the reduction in fishing effort during the 1914-1918 war were exhausted.
A somewhat similar situation arose after the 1939-1945 war and the
decline in catch-rate - particularly among the distant-water vessels-
led to the so-called "SARO" project which was a United Kingdom attempt
to improve the otter trawl (Margetts, 1974). This acronym is derived
from SAunders-ROe Ltd, the agents for the development work. The project, which was sponsored by the British Government, the British
Trawlers' Federation and the White Fish Authority, had as its ultimate
objective the design of a trawl with greater spread and headline
height which would require no more power than that normally used in
trawling (White Fish Authority, 1960). The larger net was expected
to catch more fish and while the engineering objective was achieved-
the new trawl having a mouth area about 2½ times that of the standard Granton trawl - trials did not show any substantial increase
in catch-rates (White Fish Authority, 1963).

     The fishing power of a gear can be assessed in terms of the
catch taken from a given density of fish per unit of fishing time
and Gulland (1969) argues that it can be considered in two parts.
Firstly, there is the area or volume over which the influence of
the gear extends and within which the fish are liable to be caught;
secondly, there is the proportion of the fish within the influence
of the gear that are actually caught. The distinction is between
fishing intensity and gear efficiency. Fishing intensity is measured in terms of area or volume swept per unit time and can be
increased by towing the same trawl faster or by towing a larger
trawl at the same speed. But the efficiency of the gear must also
be considered. If the fishing intensity is increased by towing
faster, or by making the trawl larger, the gear may become less
efficient and fishing power and catch-per-effort could decrease.
One suspects - with the clarity of vision so characteristic of
hindsight - that this is what happened with the SARO trawl.

     For many years little was known about the efficiency of a
trawl in terms of the proportion of the fish that came within its
influence and were caught. The problem is relevant to two of the
three broad aims of our research programme: efficiency is important when the catches of research vessels are used to estimate
stock densities in assessment studies, and in any research programme to improve trawls. For example, does the trawl catch 50%
or 90% of the fish that pass between the otter boards? This simple question has been impossible to answer until recently and the
change in circumstance has arisen because we have been able to make
use of high resolution sector-scanning sonar and acoustic transponding tags. While an account of some of our work has already
been published (Harden Jones et al., 1977), I suspect that the
particular problem and the equipment and methods we used to solve
it may be new to many investigators.

THE SECTOR-SCANNING SONAR

     The Fisheries Laboratory, Lowestoft, has been associated with
the application of sector-scanning sonar to fisheries problems for

TABLE 1: Some parameters of the sector-scanning
sonar. The operating frequency is 300 kHz;
pulse length 100 µs; transmission rate 4/sec on
the 183 m range and 2/sec on the 366 m range.

Scanning of Receiver Beam	
Scans/sec	10,000
Scanning sector	30°
Independent channels	75

Transmitter Characteristics	
Array form	curved
Horizontal coverage	30°
Vertical beam width	5°
Nominal range resolution	8 cm

Receiver Array Characteristics	
Array form	linear
Horizontal beam width	0.33°
Vertical beam width	10°
Number of elements	75

over 15 years (Harden Jones and McCartney, 1962; Cushing and Harden
Jones, 1966). In 1969 the British Admiralty Research Laboratory's
300 kHz sector-scanning sonar was fitted in the Ministry of Agriculture, Fisheries and Food's research vessel CLIONE. The modulation scanning system was invented by Voglis and subsequently patented (Voglis et al., 1957). Further details have been given by
Voglis (1971, 1972). Voglis and Cook (1966) described the parameters of the equipment originally installed in R.V. CLIONE which
are summarized in Table 1. I believe that the AN/SQQ-14 mentioned
at this conference by Dr. Warner is the American counterpart to
this sonar. Mitson and Cook (1971) described the shipborne installation and stabilization system which allows the transducer to be used
in the horizontal or vertical mode to present a plan or elevation
picture of the target or area under surveillance. The facility for
horizontal and vertical scanning allows the range, bearing, and
depth of a target to be determined and the high resolution of the
system gives a very detailed picture on the B-scan display. Thus
an otter trawl can be clearly observed at ranges out to 150 m, and
the main warps, otter boards, eddy trails, dan lenos, headline and
floats, footrope, selvedges, and cod-end of the net are clearly
visible.

The sector-scanner can detect 30-40 cm roundfish out to a
range of 150 m. We hoped that it might be possible to detect and
count individual fish in front of the trawl and, with the catch
on the deck, determine what proportion of them were caught. But
it proved impossible to follow, in the horizontal mode, the echoes
from a single fish against the bottom reverberation and this led

to the development of an acoustic transponding tag which returned
a powerful and unambiguous signal when insonified by the outgoing
pulse of the sonar. The tag was invented and described by Mitson
and Storeton-West (1971) and, when used in conjunction with the
sector-scanner provided a novel approach to the problem of gear
efficiency.

THE METHOD

Our objective was to determine the efficiency of the Granton
trawl for plaice (*Pleuronectes platessa* L.) in quantitative terms
with an error limit of $\pm$ 10%. The plaice does not have a swimbladder and the work was restricted to this species because substantial problems relating to pressure adaptation were foreseen
with roundfish, such as cod, which has a closed swimbladder and
a slow rate of gas secretion equivalent to an adaptation rate of
1 m hr^{-1} at 10°C (Harden Jones and Scholes, unpublished data).
It was hoped that the results would show the extent, if any, to
which the efficiency of the trawl could be improved. In principle
the method was simple. Plaice previously caught by trawl in the
North Sea were held in the laboratory and tagged with Petersen
disc or button tags. The fish (size range 30-50 cm) for a particular cruise were carried on board R.V. CLIONE and kept in large
deck tanks provided with a copious flow of sea water and 10-15 cm
layer of clean sand. A single plaice carrying an acoustic transponding tag attached to the eye of the Petersen disc tag was released from R.V. CLIONE and kept under surveillance, at a range
of 120-140 m, for a 2-4 hr settling-down period. Then a second
research vessel, R.V. CORELLA, towing an otter trawl, was directed
by radio link to catch this particular fish. The whole fish-cap-
ture process was followed by sector-scanning sonar on R.V. CLIONE
from the moment when R.V. CORELLA passed near or over the plaice
until either the fish was seen to be caught in the trawl or was
left behind in the event of an unsuccessful attack. The behaviour
of the fish in relation to the gear was observed and subsequently
analyzed from film records of the sonar display.

THE SCALE OF THE WORK

At the onset we were concerned with the research effort that
would be required to reach our objective in terms of the error
limit. The number of valid attacks required to determine the
efficiency of the trawl at a 95% confidence level was estimated
as $n = 4pqL^{-2}$ where p and q are the binomial proportions or percentages and L the limit of error (Snedecor and Cochran, 1967).
With an efficiency of 50% - the worst possible case - and an error
limit of $\pm$ 10%, 100 valid attacks would be required to reach the
objective. We soon found out that the average attack rate was 2.5
attacks per day's absence from port when steaming time, bad weather,
breakdowns, mistakes, and the plain misfortunes that attend
all sea-going exercises were taken into consideration. Because it

TABLE 2: Some details of R.V. CORELLA's otter trawl.

Otter boards	3.05 x 1.27 m 700 kg
Backstrop	7.3 m
Bridle	18.3 m
Dan leno	Scuttle bobbin with butterfly
Legs	3.0 m (lower of 25 mm chain)
Headline	23.8 m
Floats	25 x 20 cm diam.
Groundrope	36.6 m (6 m bosom with 14 x 20 cm diam. bobbins and 2 x 15.3 m wing pieces with 9.0 cm diam. rubber discs on 22 mm diam. wire, loosely wound with 6 mm chain)
Tickler chains	Board-to-board 61 m, mid-wing 22 mm, bunt 13.7 m, all 16 mm chain

was clear that the work would require a substantial effort involving two ships, the load was spread over several years so that the research vessels could continue to meet the laboratory's commitment to other programmes.

R.V. CORELLA'S OTTER TRAWL

R.V. CORELLA used a small Granton otter trawl of 23.8 m headline length. The meshes of the net were normal for North Sea trawling, ranging from 140 mm in the forward parts to 80 mm in the codend. The bridles were 18.3 m long and the otter board back-strops 7.3 m, giving a total distance from the otter board to the dan leno of 25.6 m. The trawl was rigged for general purpose use and not specially for plaice fishing. Some details of the trawl are given in Table 2. Compared with the most efficient commercial rig for plaice fishing on good ground, R.V. CORELLA's trawl had longer bridles, fewer and shorter tickler chains, less groundrope chain, and more floats, and was also fitted with dan leno scuttle bobbins and bosom bobbins. The work was divided into two phases, and in the first a base line for the efficiency of the trawl was determined when it was not rigged with a board-to-board tickler chain. This chain was left off to enable the possible herding effect of the bridles to be determined: careful examination of the gear

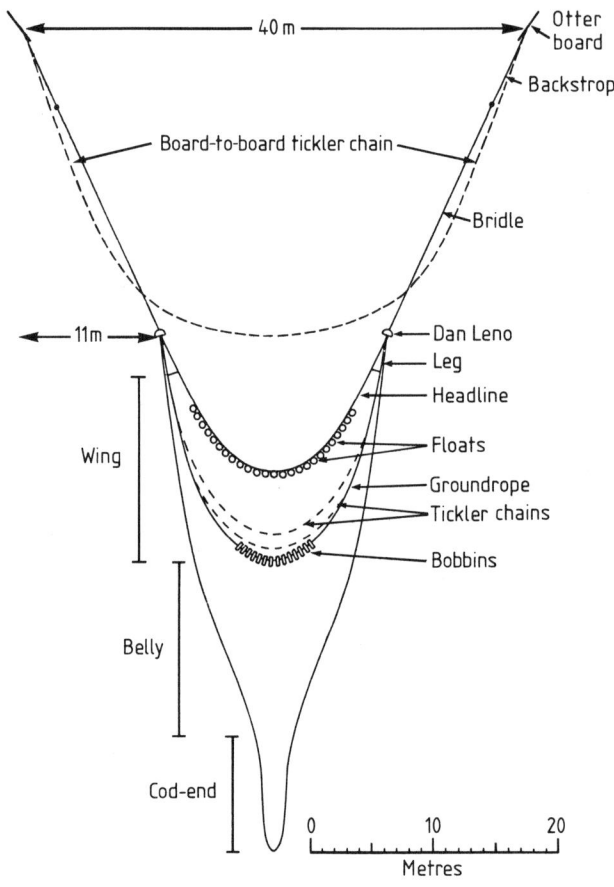

FIGURE 1:   Diagram to show, in plan view, the main
parameters of R.V. CORELLA's otter trawl.   The
board-to-board tickler chain was only rigged in
the second series of trials.

after hauling showed that the backstrop-bridle complex never
dragged on the bottom.   In phase 2 a limited number of attacks
were made with the board-to-board tickler chain on the gear, to
determine the effect, if any, that this chain had on the efficiency of the trawl.

A plan view of the geometry of the trawl is shown in Figure 1.
The maximum headline height was estimated as 1.5–2.0 m.   Most
attacks were made in 26 m of water when the length of warp out
(water surface to otter board) was 260 m and the towing speed
through the water 1.8 m s^{-1} (3.5 knots).   Under these conditions
the spread between the otter boards was 35–40 m of which the
middle 18 m was covered by the path of the net.

TABLE 3: The efficiency of the Granton otter trawl
for plaice as determined by the capture of fish
fitted with acoustic transponding tags. The data
relating to the efficiency of the gear without a
board-to-board tickler chain has been published
(Harden Jones et al., 1977) while that relating to
the gear with such a chain is new and should be regarded as preliminary only (Harden Jones, Arnold
and Greer Walker, unpublished data).

Position of Fish	Board-to-Board Chain Not Fitted			Board-to-Board Chain Fitted		
	Valid Attacks	Number of Fish Caught	Efficiency %	Valid Attacks	Number of Fish Caught	Efficiency %
Within boards	166	73	44	48*	32	67
Between boards and wing ends	72	16	22	21	10	48
In path of net	94	57	61	24	19	79

*The position of three of these fish with respect to the wing ends of the net was uncertain.

WORKING ARRANGEMENTS

The response of plaice to the trawl might be different by day
and by night and this could lead to differences in the efficiency
of the gear. It seemed wise to concentrate the effort on day or
night work rather than to attempt both. Day work was chosen because plaice are thought to come off the bottom at night when they
would not be accessible to the gear; and while two vessels can
work close together at night under good conditions it seemed only
prudent - in the first instance - to master the technique by day.
So the work was kept to daylight hours and restricted to one area
in the Southern Bight of the North Sea - the Black Bank-Brown Bank
grounds, 140 km north-east of Lowestoft - where the bottom was
smooth, relatively free from noise and fastners, and there was
little interference from traffic.

THE RESULTS

Table 3 shows that the overall efficiency of the gear without
the board-to-board tickler chain was 44% and that there was a marked
increase in efficiency when this chain was added to give an overall
figure up to 67%.

The board-to-board tickler chain increases the efficiency of
the gear for fish in the path of the net and, to a greater degree,
for fish lying between the wing ends and the otter boards. Detailed
study of the films of the sonar display suggest that some plaice, as
judged by the weak signal from the acoustic transponding tag were
well buried in the sandy bottom, and it was these fish that were
often left behind by the trawl even when they were directly in the
path of the net. One plaice obviously buried in the bottom was undisturbed by three successive attacks in which the trawl, which was
not fitted with the board-to-board tickler chain, passed straight
over the fish. The net was hauled up, the board-to-board tickler

shackled into position and a fourth attack made on the recalcitrant
beast. When the catenary of the board-to-board tickler chain passed
over the fish, the signal from the transponding tag appeared in full
strength with many multiple signals and the fish was immediately
caught and subsequently recovered alive and strong in the cod-end on
hauling. There can be little doubt that part of the increased efficiency which accompanies the use of the board-to-board tickler chain
reflects its ability to 'dig' fish out of the bottom. But this
chain brings about an even greater increase in efficiency for those
fish lying between the boards and wing ends and here part of its
role would appear to be in herding those fish into the path of the
net which would otherwise be undisturbed by the bridle which is 50 cm
or more off the bottom. But there is more work to be done in analyzing the films, frame by frame, before we can add much to the bare
figures summarized in Table 3.

DISCUSSION

     This account shows how two new acoustic techniques have been
brought to bear on a hitherto intractable problem of fisheries
research. The results allow one to estimate the efficiency of the
trawl quantitatively and the films of the sonar display also provide the basic material from which certain gear parameters can be
estimated directly (such as the spread of the otter boards and the
wing ends) and the behaviour of fish studied in relation to the gear.
     The overall estimate of 67% for the efficiency of the trawl
bears comparison with figures quoted for a range of species and
gears (Edwards, 1968; Kjelson and Colby, 1977). If the trawl is
indeed as efficient as our results suggest, further improvements
could well concern other aspects of efficiency in terms of reducing
drag (and thus fuel cost) or vulnerability to damage (and thus
avoiding loss of fishing time in mending and replacing gear).
     Changes have been made to the sector-scanner in the last five
years. The original equipment has been returned to the Admiralty
and replaced by a solid state counterpart with improved reliability
and performance (Holley et al., 1975). One of the most useful additions to the equipment was an instrumentation video recorder which
allows the details of a particular attack to be repeatedly examined
on a second display so that all concerned would agree as to what had
happened while events were fresh in the mind. The detailed notes
and comments made at the time are invaluable when working up the
film in the laboratory.
     The sector-scanner has been applied to a range of problems in
the marine field including wreck and bottom surveys (Voglis and
Cook, 1966), the location of anisotropic sound sources (Voglis and
Cook, 1970), environmental protection in relation to gravel extraction (Dickson and Lee, 1973), surveys of dredged channels (Dickson
et al., 1978), studies on fishing gear (Cook, 1966; Cushing and
Harden Jones, 1966; Margetts, 1971; Harden Jones et al., 1977),
fish shoals (Cushing, 1977), the tracking of individual fish (Greer
Walker et al., 1978), the movements of seabed drifters (Harden Jones
et al., 1973; Dickson, 1976), and the tail beat frequency of a

large shark (Harden Jones, 1973). The transponding acoustic tag, originally developed for fish tracking, has been produced in a 'long life' form and as an 'artificial pebble' or stationary bottom marker has been used in sediment transport studies. The 'long life' tag has also been used to position, locate and recover current meter arrays and other underwater devices. All these applications have made use of the picture-forming qualities of the equipment which are excellent, but recent studies (Cook and Rushworth, 1978) have shown that the sector-scanner can also be applied to the study of fish target strengths which could lead to important advances in this field.

There can be little doubt that modulation sector-scanning sonar has proved to be a remarkable and versatile invention and at Lowestoft, at least, we are much indebted to its inventor.

REFERENCES

Anon., 1895. *Fish Trades Gz.*, 2, 13 (634): 12.

Anon., 1897. Scott V. Hamling and Co., Ld. *Rep. Pat. Des. Trade Mark Cases, Lond.* 14: 123-142.

Bagenal, T. B., 1958. An analysis of the variability associated with the Vigneron-Dahl modification of the otter trawl by day and by night and a discussion of its action. *J. Cons. Int. Explor. Mer.* 24: 62-79.

Cook, J. C., 1966. Scanning sonar and the fishing industry. *Ultrasonics* 8: 139-144.

Cook, J. C. and A. Rushworth, 1978. Measuring fish target strength and shoaling density using a high resolution scanning sonar. *Symposium on Acoustics in Fisheries*, Hull, United Kingdom. Univ. Bath, U. K. (unpaginated).

Cunningham, J. T., 1895. North Sea investigations. *J. Mar. Biol. Ass. U. K.* 4: 97-143.

Cushing, D. H., 1977. Observations on fish shoals with the ARL scanner. *Rapp. Cons. Int. Explor. Mer.* 170: 15-20.

Cushing, D. H. and F. R. Harden Jones, 1966. Sea trials with modulation sector scanning sonar, with an appendix by J. A. Gulland. *J. Cons. Int. Explor. Mer.* 30: 324-345.

Dickson, R. R., 1976. Field tests of seabed·drifters, using sector-scanning sonar techniques. *J. Cons. Int. Explor. Mer.* 37: 3-15.

Dickson, R. R., D. N. Langhorne, R. S. Millner, and E. G. Shreeve, 1978. An examination of a dredged channel using sector scanning sonar in side-scan mode. *J. Cons. Int. Explor. Mar.* 38: 41-47.

Dickson, R. R. and A. J. Lee, 1973. The effects of marine gravel extraction on the topography of the sea bed. *Offshore Serv.* 6 (6): 32-36 and 39: 6(7): 56, 58, and 61.

Edwards, R. L., 1968. Fishery resources of the North Atlantic area. In: *The Future of the Fishing Industry of the United States*, (de Witt Gilbert, ed.), pp. 52-60. Univ. Wash. Publs. Fish., NS 4: 346 pp.

Garstang, W., 1900. The impoverishment of the sea. *J. Mar. Biol. Ass. U.K.* 6: 1-69.

Greer Walker, M., F. R. Harden Jones, and G. P. Arnold, 1978. The

movements of plaice (*Pleuronectes platessa* L.) tracked in the
open sea. *J. Cons. Int. Explor. Mer.* 38: 58-86.

Gulland, J. A., 1969. Manual of methods for fish stock assessment.
Part 1. Fish population analysis. *FAO Man. Fish Sci.* No. 4,
154 pp.

Harden Jones, F. R., 1973. Tail beat frequency, amplitude and
swimming speed of a shark tracked by sector scanning sonar. *J.
Cons. Int. Explor. Mer.* 35: 95-97.

Harden Jones, F. R., M. Greer Walker, and G. P. Arnold, 1973. The
movements of a Woodhead seabed drifter tracked by sector scanning
sonar. *J. Cons. Int. Explor. Mer.* 35: 87-92.

Harden Jones, F. R. and B. S. McCartney, 1962. The use of electronic
sector-scanning sonar for following the movements of fish shoals:
sea trials on RRS DISCOVERY II. *J. Cons. Int. Explor. Mer.* 27:
141-149.

Harden Jones, F. R., A. R. Margetts, M. Greer Walker, and G. P.
Arnold, 1977. The efficiency of the Granton otter trawl deter-
mined by sector-scanning sonar and acoustic transponding tags.
Rapp. Cons. Int. Explor. Mer. 170: 45-51.

Holley, M. L., R. B. Mitson, and A. R. Pratt, 1975. Developments in
sector scanning sonar. *I.E.R.E. Conf. Proc. No. 32:* 139-153.

Kjelson, M. A. and D. R. Colby, 1977. The evaluation and use of
gear efficiencies in the estimation of estuarine fish abundance.
In: *Estuarine Processes, 2,* (M. Wiley, ed.), pp. 416-424. Academic Press, New York.

M'Intosh, W. C., 1894. Remarks on trawling. *12th Ann. Rep. Fish.
Bd. Scotl. Part III:* 165-195.

Margetts, A. R., 1971. Sector-scanning sonar used for observing
deep-sea trawling. In: *Modern Fishing Gear of the World, 3,*
(H. Kristjonsson, ed.), pp. 137-140. Fishing News (Books) Ltd.,
Lond.

Margetts, A. R., 1974. Modern developments of fishing gear. In:
Sea Fisheries Research, (F. R. Harden Jones, ed.), pp. 243-
259. Elek Science, Lond.

Mitson, R. B. and J. C. Cook, 1971. Shipboard installation and
trials of an electronic sector-scanning sonar. *Radio Electron.
Engr.* 41: 339-350.

Mitson, R. B. and T. J. Storeton-West, 1971. A transponding acoustic
fish tag. *Radio Electron. Engr.* 41: 483-489.

Scott, J. R., 1894. Improvements applicable to trawl nets. *British
Patent (3187).* 2 pp.

Snedecor, G. W. and W. G. Cochran, 1967. *Statistical Methods,* 6th
Edition. Iowa State University Press. 593 pp.

Vigneron, J. -B. J. A., 1920. Improvements in and relating to
trawling gear for deep sea fishing. *British Patent (175824).*
5 pp.

Voglis, G. M., 1971. A general treatment of modulation scanning as
applied to linear arrays. Part 1. *Ultrasonics* 9: 142-153.
Part 2, 9: 215-223.

Voglis, G. M., 1972. Design features of advanced scanning sonars
based on modulation scanning. Part 1. *Ultrasonics* 10: 16-25.
Part 2, 10: 103-113.

Voglis, G. M. and J. C. Cook, 1966. Underwater applications of an advanced acoustic scanning equipment. *Ultrasonics* 4: 1-9.

Voglis, G. M. and J. C. Cook, 1970. A new source of acoustic noise observed in the North Sea. *Ultrasonics* 8: 100-101.

Voglis, G. M., H. F. Willis, and J. Buckingham, 1957. Direction finding apparatus. *British Patent (926850)*. 6 pp.

White Fish Authority, 1960. *Annual Report and Accounts for Year Ended 31 March 1960*. 50 pp.

White Fish Authority, 1963. *Annual Report and Accounts for Year Ended 31 March 1963*. 55 pp.

Use of Acoustic Frequency Diversity for Marine Biological Measurements

D. V. Holliday

ABSTRACT: The conclusions one may draw upon
looking into the ocean with sound of a particular frequency are severely colored by the
frequency chosen. The dependence of acoustic
scattering from marine biological targets on
frequency has been known since early observations of the deep scattering layer. It is
only in recent years, however, that the electronics associated with complex signal and data
processing of acoustic information have become
sufficiently small, reliable and inexpensive to
incorporate in broadband or multi-frequency
sensors for use at sea. This technological advance, which appears to be in its infancy, has
already made possible the design of acoustic
instruments that make the use of frequency as
valuable in some bioacoustic applications as
time or space. A simplified review of frequency dependence of sound scattering from
marine organisms is followed by two examples
of the use of frequency diversity taken from
the author's recent work.

INTRODUCTION

More than 30 years have elapsed since investigators at the University of California Division of War Research discovered that sound scattering in the open ocean could not be explained by a uniform distribution of reflectors (Duvall and Christiansen, 1946; Eyring et al., 1948). These investigators discovered that a vertical stratification of scatterers was required to explain their echo sounding observations. The layers, initially termed the "ECR layers" after the initials of their discoverers, were soon correctly linked to a biological origin (Johnson, 1948) and became widely known as the deep scattering layers (DSL). When trawls conducted in the acoustically detected scattering layers revealed only low densities of midwater nekton, numerous non-biological hypotheses were advanced in lieu of a biological origin for the scattering. The apparent inconsistency between net hauls and the intensity of sound scattering from the deep scattering layers was based in part on use of the Rayleigh model of sound scattering from an object which is very small compared to the wavelength of the sound being scattered. After a number of detailed investigations, physical and chemical explanations proved no more satisfying than did the original impressions of the match between biological theories and the acoustic observations.

Recognizing that Rayleigh scattering was not sufficient to describe scattering from many marine organisms which were comparable in size with the acoustic wavelengths in use, Anderson (1950) noted that some components of the typical plankton community might be approximated by a fluid sphere. The fluid sphere scattering model, though approximate both in shape and in its neglect of appendages and carapace, is still widely used even though a comprehensive experimental verification has never been accomplished. Even with this substantial theoretical contribution, the acoustic scattering levels at low frequencies, e.g., 12 and 18 kHz, could not be explained in terms of the biomass from net hauls in the DSL.

There are both fishes and zooplankton species which cannot be adequately described by either the Rayleigh or the Anderson models. These organisms are characterized by an included gas bubble, either a swimbladder or a pneumatophore. Minnaert (1933) is generally credited with the original work in modeling acoustic radiation from an ideal spherical gas bubble enclosed by a liquid. When scattering is enhanced at one or more frequencies relative to that at nearby frequencies, the scattering is termed resonance scattering. This is the case for scattering of sound from a gas bubble. For a free bubble, scattering at the resonance frequency can be enhanced many times, e.g., 100–1000 times above that at a few hundred Hertz away. Since the frequency at which this enhancement occurs depends on both bubble size and the ambient pressure (depth), the choice of acoustic frequency and the target location may make the difference between an indication of a dense layer or group and not observing the target at all for this class of scatterers. While complicating the choice of operating frequency, the strong variation in backscattering with frequency also admits the possibility of acoustic target classification and sizing, a subject to be discussed in more detail later. Strasberg (1953), Devin (1959), Andreeva (1964), and Weston (1967) modified the free bubble model to include first order corrections for shape, a variety of energy losses which occur in non-ideal bubble, and the effects of a swimbladder and surrounding tissue.

Two points in the history of marine acoustics are worthy of special emphasis. First, the entire quantitative history of sound scattering from marine targets dates from the end of World War II. The dependence of sound scattering on frequency was noted by the earliest investigators (Physics of Sound in the Sea, 1946) but the levels varied with geographic location and no theory was advanced to explain the differences. Hersey and Backus (1954) were among the first to attempt an interpretation of observed frequency and temporal dependence of sound scattering in terms of the behavior of biological organisms, i.e., the migration of swimbladder-bearing midwater fishes. Fewer than 25 years have elapsed since this pioneering work, a relatively short period in the history of most scientific disciplines. Secondly, the acoustically motivated division of marine organisms into those with gas bubbles and those without occurred relatively early but has not been vigorously pursued. The necessity for different mathematical treatments of these two types of targets reflects basic differences in the character of sound scattering from each, an observation which can sometimes be used to distinguish acoustically between co-occurring species.

Many marine biologists are familiar with the ubiquitous 12 kHz echo sounder. Some have used other frequencies as high as 200 kHz (Pieper, 1977) or as low as 3.5 kHz. A very few have had multiple frequencies available. In most cases, the choice of frequency has been dictated by existing, installed hardware. The electromagnetic analog of frequency in the visible range is color. If one were restricted to illuminating an object he wishes to describe with light of only one color, the color would be carefully chosen. For many problems, given the choice, one would wish to use all of the colors he could see, i.e., white light. In acoustics, broadband frequency coverage is analogous to white light.

In an attempt to lead the reader to an appreciation of the importance of his choice of frequency and the potentials of using multiple frequencies, we will concentrate on the use one can make of the frequency dependence of sound scattering from living marine organisms. We will touch briefly on the basic physics of echo formation, and describe the frequency dependence of two of the sound scattering models most frequently used in work with marine organisms. With that background, we will examine two specific examples in which we have attempted to exploit variations in sound scattering at different frequencies. One of these examples involves multiple surveys in which fish size distributions were estimated for schooling, epipelagic zone, swimbladder-bearing fish. The other involves detailed studies of small scale layering of millimeter-sized plankton.

THE BASIC PHYSICS OF ECHO FORMATION

Scattering is defined as a change in the direction of a wave

due to an encounter with an inhomogeneity in the medium in which
the wave is propagating. This change in direction is often accompanied by a change in the intensity of the wave field around the
scattering object. The difference between the wave pattern which
would have existed had the inhomogeneity not been there and the
waves which are observed with the inhomogeneity present is called
the scattered wave. The energy which is reflected 180°, i.e., toward the original wave source is termed the backscattered wave.

   Whether dealing with surface waves on a body of water, electromagnetic waves (light, x-rays, etc.) or acoustic waves, describing scattering quantitatively inevitably requires a working
familiarity with advanced mathematics. The exact solution of
most practical problems, such as the scattering from a copepod
or euphausiid, is beyond current technology.

   For those interested in an introduction to the mathematics
of acoustic scattering, the classic book by Lord Rayleigh (1945)
and two excellent papers by Freedman (1962a,b) originally published
in *Acustica* are recommended. Reprints of these papers can be
found in Albers (1972). Additionally, two books, one by Skudrzyk
(1971) and the other by Morse and Ingard (1968) are recommended for
an in-depth treatment of the subject. For our purposes, we will
only introduce the subject of scattering in sufficient detail to
discuss the two common mathematical models for sound scattering
from marine organisms.

   The speed with which sound travels in a medium, c, depends
on the adiabatic compressibility, κ, and the density, ρ, (Equation 1)
of the medium:

$$c^2 = 1/\rho\kappa. \tag{1}$$

Though sound speed may be measured directly, it is much more common
to measure temperature and salinity at a particular depth in the
ocean. Precise empirical relations between temperature, salinity,
depth (pressure), and the sound speed allow us to bypass the more
difficult problems of measuring sound speed or κ and ρ directly.
A detailed review of empirical relationships by Mackenzie (1971)
was reprinted in Albers (1972).

   The frequency, f, of a pure tone is related to the speed of
sound through a parameter called wavelength (Equation 2). For
ocean surface waves, the wavelength is the distance between wave
peaks. For acoustic waves, the distance between peaks in local
pressure can be used to define wavelength.

$$c = f\lambda \tag{2}$$

   Visualize a sound wave propagating from sea water on the left,
(Fig. 1) into a volume filled with protoplasm (perhaps a salp) on
the right. An incident wave, traveling in the direction of the
arrow labeled p_i impinges on a surface, its angle θ_i measured from
the normal to the surface. The reflected wave, p_r, leaves the surface of angle θ_r. When $\theta_i = -\theta_r$, the reflection is termed specular.
Some of the sound penetrates the material on the right of the sur-
face, setting up waves in that medium. The penetrating ray, p_t, trav-

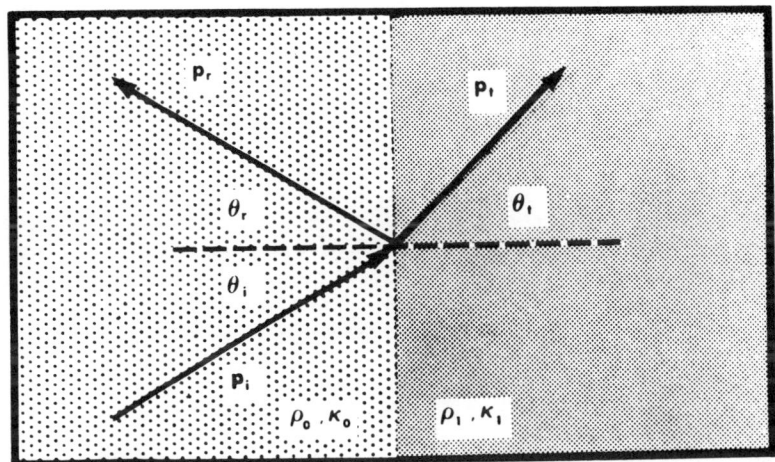

FIGURE 1: A ray diagram illustrating the incident (p_i), reflected (p_r), and penetrating (p_t) acoustic waves generated at the boundary between fluid media with different physical properties. Refracted wave shown for $\rho_o c_o < \rho_1 c_1$.

els in a direction θ_t which is determined by Snell's law (Equation 3) for acoustic waves.

$$(\cos \theta_i / c_o) = (\cos \theta_t / c_1) \tag{3}$$

If a reflection coefficient, C_r, is defined such that

$$p_r = C_r p_i, \tag{4}$$

the reflection coefficient can be written as

$$C_r = \frac{\rho_1 c_1 \cos \theta_i - \rho_o c_o \cos \theta_t}{\rho_1 c_1 \cos \theta_i + \rho_o c_o \cos \theta_t}. \tag{5}$$

Morse and Ingard (1968) present one approach to the development of Equation 5 (pp. 226-276). If we restrict our attention to the back-scattering, we can set $\theta_i = -\theta_r = 0^o$ and $\theta_t = 0^o$. The backscattering coefficient, C_b, can then be written as

$$C_b = \frac{\rho_1 c_1 - \rho_o c_o}{\rho_1 c_1 + \rho_o c_o}. \tag{6}$$

The magnitude of the backscattered wave is then dependent on the ratio of the difference of $\rho_1 c_1$ and $\rho_o c_o$ to their sum. The ρc product oc-

curs frequently in scattering problems and is called the impedance of
the boundary at normal incidence.

Substantial complications arise when one admits the more realistic
cases in which the impedance depends on angle of incidence and frequency. Additional complications evolve as the shape of the scattering object is introduced, refraction and scattering inside the object
are allowed and surface waves on the object's surface are included in
the calculation. Yet, after size and shape have been described for a
scatterer, specifying the density, ρ, and the sound speed, c, or the
density and the compressibility, κ, for the medium and the scatterer
is often sufficient to define a scattering problem. In practice, density contrast, g, is often used as is the sound speed contrast, h,
(Equations 7,8). Note that these three parameters are related through
Equation 1.

$$g = \rho_1/\rho_0 \tag{7}$$

$$h = c_1/c_0 \tag{8}$$

Dividing both the numerator and the denominator of Equation 6
by the quantity $\rho_0 c_0$ and substituting the relations in Equations 8
and 9 for the appropriate ratios yields

$$C_b = \frac{gh - 1}{gh + 1} . \tag{9}$$

In the limit of a large product of g and h, i.e., a rigid scatterer,
C_b approaches a value of one and most of the incident energy is reflected. For most marine organisms the values of g and of h are
very close to one. Thus, Equation 9 predicts a very small amount of
backscattering. For the fluid sphere model discussed in the next
section, Johnson (1977a) estimates that a 1% change in g or h will
cause a 40% change in theoretical backscattering.

Fluid Sphere Model

The scattering of sound from a fluid sphere was treated by Lax
and Feshback (1948), but the resulting form was not suitable for
numerical calculations. Anderson's (1950) formulation was more elegant, but also unsuitable for early computers and hand calculations
required two people with mechanical calculators some two months to
obtain 600 values of backscattering.

The Anderson model assumes that a zooplankter is homogeneous
with sound speed contrast, h, and density contrast, g. The animal
is approximated in shape by a sphere of radius, a, which has a
volume equivalent to the actual displacement volume. There is no
allowance for carapace, appendages or viscous and thermal effects
in this model. The backscattering predicted by the fluid sphere
model is a function of the sphere's radius and the acoustic wavelength. This ratio is usually expressed as the product of the wave-
number and the radius, ka. For small values of ka, i.e., long wave-

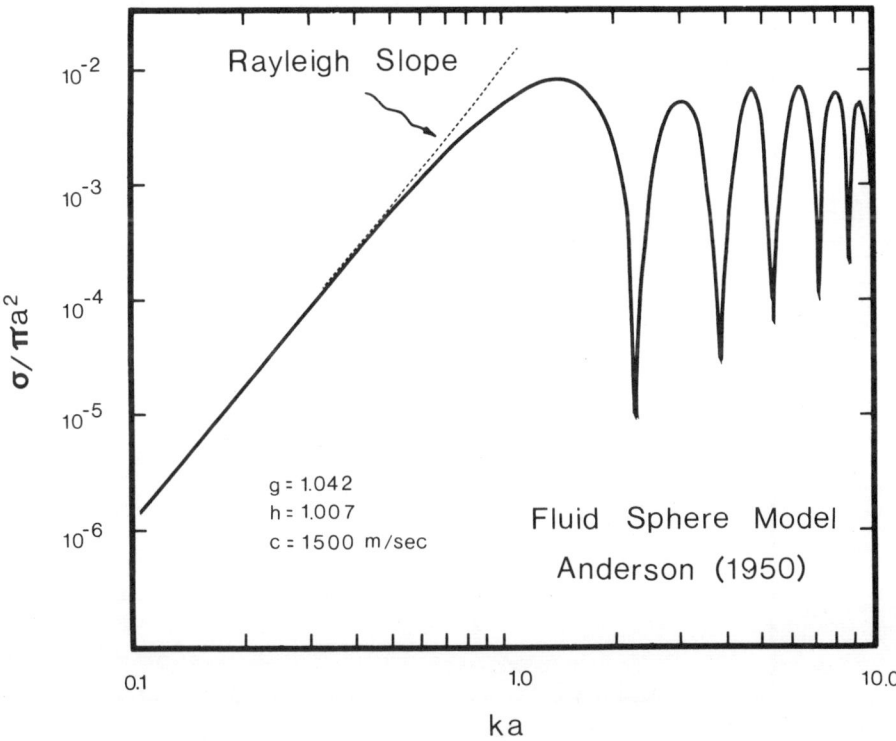

FIGURE 2: Fluid sphere model (Anderson, 1950).
Reflectivity for a fluid sphere with density
contrast, g = 1.042, speed contrast, h = 1.007
and a sound speed in sea water, c = 1500 m/sec,
plotted versus the sound product of the wave-
number and the sphere radius.

lengths relative to scatterer size, the fluid sphere model approxi-
mates the scattering predicted by Lord Rayleigh (1945). There is a
sixteen-fold increase in sound scattering intensity for each doubling
of the frequency in this region (Fig. 2). The reflectivity does not
continue to increase above ka $\simeq$ 1.4. Above this value the peaks in
scattering intensity approximate the level one would expect from a
rigid (impenetrable) sphere of radius a. A variety of resonance and
interferences between reflected and incident signals cause large deviations from the geometric scattering assymptote. The degree to
which these resonances and interferences reflect reality for common
zooplankton species is currently the subject of more speculation than
hard data. Our concern, however, is real, as a non-monotonic form
for scattering strength would be a mixed blessing. On the negative
side, one could not choose a minimum size class of interest in a particular experiment and then choose one frequency sufficiently high to
allow detection of all larger sizes. The problem of evaluating the
size distribution present in any open ocean sampling program would

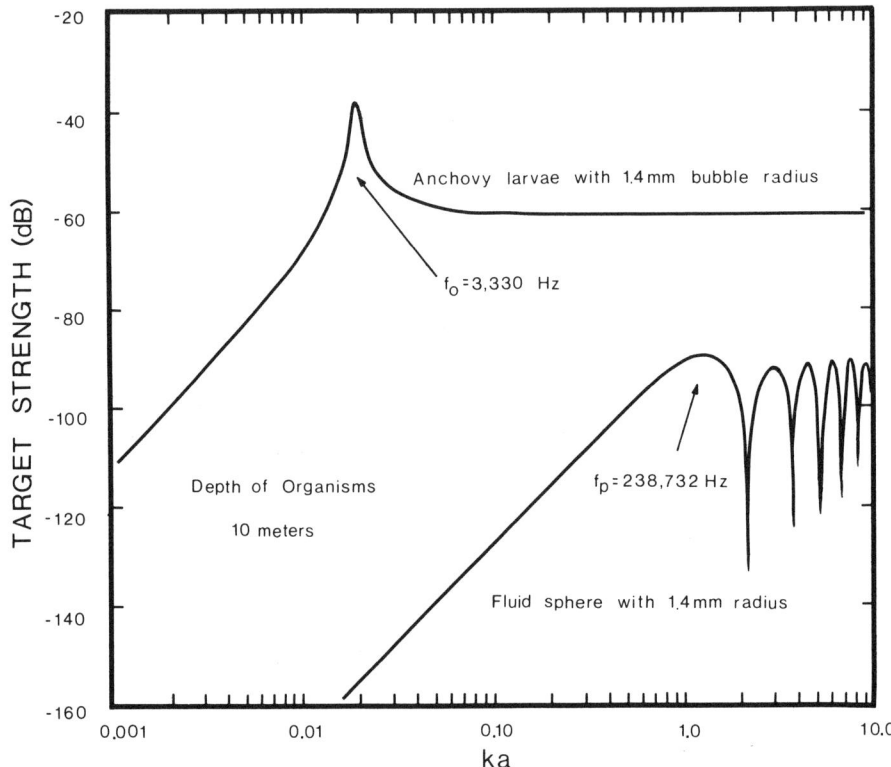

FIGURE 3: Target strength versus ka for a 1.4 mm
air bubble at 10 meters depth, typical of that in
a 45 mm (standard length) anchovy larvae and for
a fluid sphere of the same radius as the bubble.
The values of h and c were 1.042, 1.007 and 1500 m/
sec.

not be amenable to "back-of-the-envelope" calculations. On the
positive side, "character" in the curve admits the possibility of
sophisticated and potentially accurate estimates of zooplankton size
distributions through measurements and analysis of multi-frequency
backscattering data. Such calculations would require a medium size
computer in all but the most trivial cases. An approach to making
such calculations and several examples will be treated after a brief
discussion of the model most often used to describe acoustic scattering from gas-filled swimbladders.

Swimbladder Model

     Gas-filled swimbladders play an exceptionally important role in
marine bio-acoustics. Some fish are known to use these organs to
assist in the production of sound. Others are thought to make use of
the swimbladder as an aid in hearing (Tavolga, 1966). The acoustic
characteristics of the swimbladder in scattering are dependent on the

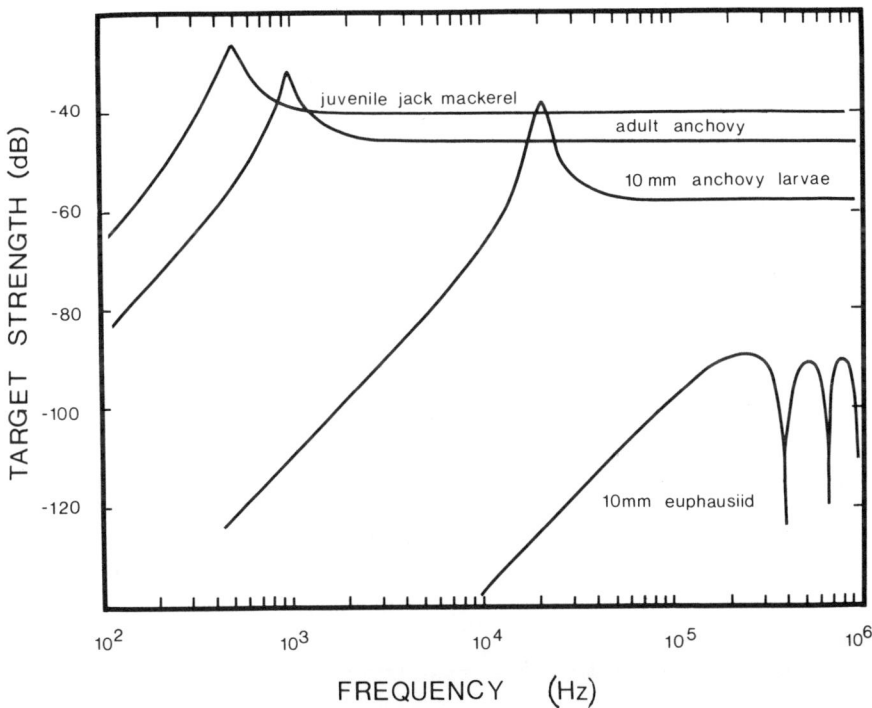

FIGURE 4: Estimated target strengths versus frequency for four typical marine organisms. The
jack mackerel and anchovy backscattering was assumed dominated by scattering from the air-bladder.
The euphausiid scattering is based on the Anderson
(1950) fluid sphere model. The value of Q was
taken as 5 for the jack mackerel and adult anchovy
while 10 was used for the anchovy larva, reflecting an assumption of a lesser effect of flesh and
swimbladder wall tissue.

static pressure at the depth of the fish, added pressure due to the
swimbladder wall and surrounding muscles, the shape and volume of the
included gas bubble and to a lesser degree on the type of gas in the
swimbladder.

The backscattering from a swimbaldder can be much larger than
one would assume based on its geometric cross section. A comparison
of the scattering from a fluid sphere and a swimbladder of the same
size (Fig. 3) reveals a 30 dB greater scattering strength for the
bubble at large values of ka. At resonance, the peak in scattering
cross section for the bubble, the difference is 118 dB. The onset
of the Rayleigh slope occurs at ka $\cong$ 0.01 for the bubble. The predicted acoustic target strength, $10 \log_{10}(\sigma/4\pi)$ where $\sigma/4\pi$ is the
backscattering cross section, was calculated for a 10 mm euphausiid
and for a 10 mm larva of the northern anchovy (*Engraulis mordax*).
The anchovy larva is by far the better sound scatterer (Fig. 4).

Authorities used for species names of fishes in this paper are given
in Bailey (1970). This is the case even though no allowance was made
for scattering by the flesh or bones of the anchovy. Only swimbladder scattering was used. The two target strength curves are
sufficiently different that a system could be easily built to detect
anchovy larvae in the presence of a large background of euphausiids.
For example, at 12 kHz, it would require nearly 1.6×10^7 euphausiids
in the same volume to create a level of scattering equal to that from
a single anchovy larva. Note however, that the reverse problem is
not easily solved. A single frequency system sufficiently sensitive
to detect euphausiids will also detect the anchovy. Scattering estimates for juvenile jack mackerel (*Trachurus symmetricus*) and adult
anchovies of about the same length (Fig. 4) reflect the change in
acoustic backscattering one expects from the swimbladder size difference in the two species.

RELATING ACOUSTIC AND BIOLOGICAL MEASUREMENTS

The emphasis in relating biological and acoustic measurements
has largely been in predicting acoustic scattering levels from biological samples. Almost without exception this work has been related to analyses of the deep scattering layer (DSL) constituents.
Although the basic properties of the DSL are thought to be reasonably
well understood, no remarkable quantitative successes in predicting
volume reverberation levels from net haul data stand out. One reason is the difficulty biologists have with quantitative sampling of
nekton (Pearcy, 1975). A contributing non-scientific reason is that
a significant amount of the federal support in this area has been
directed to collecting and archiving standard measures of volume reverberation on a world-wide basis.
One remarkable exception to this approach deserves species mention. An Office of Naval Research (ONR) sponsored conference was
held at Asilomar in Monterey, California in 1975. At that conference
scientists from four related oceanographic disciplines, oceanic
modeling, marine biology, marine chemistry and underwater acoustics
began the task of integrating observations and techniques from all
four disciplines in order to achieve a better understanding of ocean
processes and energy flow. The resulting symposium proceedings
(Andersen and Zaburanec, 1977) is recommended not necessarily for
its acoustics, but for its treatment of each of the disciplines in
the context of the larger problem.
Although the prediction of sound scattering levels due to biological organisms is of undeniable importance in the maintenance of
a viable anti-submarine warfare program, working the inverse problem
presents at least as great a technical challenge. In a short paper
prepared for the Asilomar Conference (Holliday, 1977a), the mathematical formulation for extracting biophysical properties of marine
organisms from acoustic signatures was documented. In the procedure
presented, the acoustic backscattering from an individual, layer,
aggregation, or school of marine life is measured at several independent acoustic frequencies. These data, when combined with a model for the scattering from individuals, can be used to extract such

information as presence or absence of gas-filled swimbladders and size distributions for the organisms present. While simple in concept, the mechanics of the procedure can become quite involved. The procedure has been demonstrated for mesopelagic fishes (Johnson, 1977b), for epipelagic fishes (Holliday, 1978) and for zooplankton (Greenlaw, 1978). Selected parts of the work on epipelagic fish will be discussed next in the context of one use of multi-frequency acoustic data for fisheries research and resource assessment.

ACOUSTIC MEASUREMENT OF SEASONAL CHANGES IN FISH SIZE DISTRIBUTIONS

The northern anchovy (*Engraulis mordax*) is the dominant fish found in the California Current between Pt. Conception, California and the U. S.-Mexican border. This area, extending several hundred miles offshore, is loosely called the Southern California Bight. The anchovy schools have a gas-filled (air) swimbladder and grow to a maximum length of about 170 mm.

Squire (1972), working with indices of relative abundance, estimated that 86% of the schooled biomass was northern anchovy. Jack mackerel, Pacific bonito (*Sarda chiliensis*) and Pacific mackerel (*Scomber japonicus*) account for the majority of the non-anchovy schooling biomass with 6.3%, 4.5% and 2%, respectively, of the total estimated tonnage sighted from the air during the survey period. A list of numerically abundant species known to occur in the area (Table I) includes some species not observed in aerial surveys. Examples are Pacific hake (*Merluccius productus*) and the deep sea smelt (*Leuroglossus stilbius*). Though seen occasionally from the air, squid abundance may be underestimated because of the probable vertical distribution of the organism. Squire also determined that just under one-half of the schools observed from the air were anchovy. About 20% were Pacific bonito and 17% were jack mackerel. While some deviations of the apparent abundance from absolute abundance are inevitable due to behavior and distribution of the fish in depth, egg and larvae surveys conducted in the same area generally support Squire's results for the principal nekton in this area.

The anchovy can be found at depths at least as great as 310 m (Davies, 1973), but are most frequently found in the vicinity of the seasonal thermocline (Holliday and Larsen, 1979). Packing densities in anchovy schools has been observed to range between 50 and 366 fish/m^3 (Graves, 1977). In some special instances, for example, under stress from a predator, compactions might momentarily exceed 12,000 fish/m^3 (Hewitt et al., 1976). These variations in packing densities make biomass estimates very difficult and subject to equivalent variations when based on school size measurements. This is the case whether the size data are acoustically derived (Smith, 1970) or photographically obtained (Squire, 1978). For schools of equal size, the variations in target strength are commonly 20 dB (Vent, 1978) and have been observed to fluctuate over as much as 36 dB (Larsen, 1974). While some of these fluctuations can be explained by packing density variations, there is evidence that target strength and biomass are not linearly related at "high" packing densities (Larsen, 1974). This might be expected in view of the potential for shadowing

TABLE 1: Common species found in the California
Current near the Southern California coast.

Species	Adult* Size Range (m)	Gas Filled Swimbladder	Comments
Northern anchovy (*Engraulis mordax*)	0.110 to 0.165	Yes	Abundant, upper mixed layer
Jack mackerel (*Trachurus symmetricus*)	0.25 to 0.81	Yes	Young exhibit dense packing, older fish sparse packing
Pacific mackerel (*Scomber japonicus*)	0.38 to 0.64 (fork length)	No	Medium packing density
Saury (*Cololabis saira*)	0.18 to 0.36	No	Loose schooling, near surface
Bonito (*Sarda chiliensis*)	0.50 to 1.01	No	Can be close packed
Hake (*Merluccius productus*)	0.15 to 1.00	Yes	Deep, slow, plumes, layers
Sardine (*Sardinops caerules/sagax*)	0.15 to 0.33	Yes	Rare, upper mixed layers
Pelagic red crabs (*Pleuroncodes planipes*)	0.06 to 0.10 (total length)	No	No gas bladder, lethargic
Skipjack (*Katsuwonus pelamis*)	0.43 to 0.90 (fork length)	?	Fast, strong swimmer
Yellowfin (*Thunnus albacares*)	0.60 to 1.50 (fork length)	?	Fast, loose-packed
Bluefin (*Thunnus thynnus*)	0.60 to 1.25	?	Loose schools
Albacore (*Thunnus alalunga*)	0.60 to 1.00	?	Dispersed or loose schools, fast
Rockfish (*Sebastes*)	0.15 to 0.90	Yes	Slow motions, plumes, frequently associated with bottom features

*Standard lengths except as marked.

and multiple scattering of sound in the school, but "high" has not
been defined either theoretically or experimentally with sufficient
precision. Thus, acquiring acoustic estimates of biomass for this
species is not within the current state-of-the-art.

While an absolute acoustic biomass estimate must still be
treated as a goal for densely schooling pelagic fish, another
measure, fish size distribution, may be of use in both fish behavior
studies and in population dynamics modeling of fisheries. In particular, even relative abundances in different size classes are thought
to be of value if one could cover the entire range of sizes for a
species and its entire geographic range as well. Such data could be
particularly valuable if ratios of adult fish to recruits could be
measured. For the faster nekton, this type of quantitative information is difficult and expensive to obtain with a series of calibrated
nets. Therefore, the development of an underway acoustic approach
could be a significant contribution.

In late 1976, the ability to estimate the mean size of the fish in an anchovy school and to detect size classes due to minority species in mixed schools by measuring the acoustic signature had been demonstrated (Holliday, 1972). A crude, but effective, technique for making swimbladder resonance reasurements underway had also been developed and demonstrated (Holliday, 1977b). The mathematical formulation for automated extraction of fish swimbladder size had been developed and documented (Holliday, 1977a), but not implemented for mass data processing. With this background, three acoustic resonance surveys of the fish schools in the Southern California Bight were conducted in March, May, and September of 1977. The average duration was about 10 days for each survey. The objective was to demonstrate that underway resonance techniques could be used to detect fish growth in the new year class. The months were chosen to span the peak in anchovy spawning (March) and to track subsequent fish growth into the fall. While space does not permit a comprehensive presentation of the results of these surveys, they are documented elsewhere (Holliday, 1978). Selected results from one survey, in March, will be presented here as an example of the use of the frequency domain in fisheries research.

Acoustic Method

Once the target species, anchovy, was selected, the range of swimbladder sizes and school depths of interest could be defined. These parameters were used to calculate the extremes of the frequency band that would contain the distribution information we wished to extract. The curves used were similar to those presented previously (Fig. 4). The result of this calculation was a requirement to cover a frequency band from about 100 to 25,000 Hz. One needs at least one measurement of backscattering at a different frequency for each size class in which an abundance estimate is desired because the number of measurements must equal or exceed the number of unknowns. A slight reduction in the frequency band to be covered was made in order to achieve a higher frequency resolution at frequencies sensitive to the presence of juveniles without increasing the data rate and subsequent computational load. This bandwidth reduction was made in such a way as to place more emphasis on juveniles and pre-recruits and slightly less on the smallest larvae. The band finally chosen was 80 Hz to 20 kHz. Our choice of swimbladder equivalent spherical radii at which relative abundance estimates were to be attempted were also chosen with a higher density of size bins in the large larvae to pre-recruit size range (Table 2).

Since details of the acoustic procedure are documented elsewhere (Holliday, 1972, 1977a, 1978) only an overview will be presented here. A school of fish can be considered a "black box" from a signal processing viewpoint. It is approximately linear in the sense that the word is used in electrical circuit theory (Thomas, 1969). As a consequence of the linear property, one can represent the frequency dependence of backscattering from a school in terms of an acoustic signature analogous to the transfer function of linear filter theory. The transfer function can be measured one frequency at a time, or, again in an analogy with linear circuit theory, at many frequencies

TABLE 2: Equivalent spherical swimbladder radii
in centimeters at which relative abundance estimates were made.

0.010	0.0170	0.024	0.0310	0.038	0.0450	0.052	0.0726
0.093	0.1135	0.134	0.1545	0.175	0.1955	0.216	0.2365
0.257	0.2775	0.298	0.3185	0.339	0.3595	0.380	0.4005
0.421	0.4415	0.462	0.8120	1.162	1.5120	1.862	2.2120
2.562	2.9120	3.262	3.6120	3.962	4.3120	4.662	

simultaneously. To make the measurement at many frequencies simultaneously involves producing a stimulus which is both loud and of short
duration, i.e., an impulse. It is this process borrowed from linear
circuit theory which we have adapted to the schooled fish problem.
In addition to linearity, an approximation of ergodicity over a few
echo ranging cycles was necessary and some ensemble averaging was
done to account for the random nature of the acoustic echoes characteristic of fish schools.

An adaptation of a geophysical survey instrument was used to
create a high amplitude, short acoustic impulse near the ocean surface. This instrument, an arcer, consisted of a large capacitor bank
repetitively charged to about 10,000 volts and then discharged directly into the sea through simple towed electrodes. The spark generated an intense, broadband acoustic signal which was used to echo-
range on schools of fish. The echoes were detected with a towed hydrophone array, partially processed in real-time, on-board, for quality control and recorded for detailed analysis in the laboratory.

The quality control instrumentation included spectrum analysis,
ensemble averaging, calculation and display of the transfer function,
and a variety of recording checks. The echo spectrum was displayed
for each echo, but only 2 or 3% of the schools were subjected to detailed analysis on-board as the computer used was chosen for low cost,
rather than for speed. The on-board spectrum analyzer was capable of
real-time processing of 500 independent frequencies in the 0-20 kHz
analysis band. The same instrument was used in the laboratory for detailed analysis of the data for one of the three surveys. The laboratory spectral analysis was done with a software FFT program (256
point transform) for two surveys.

Spectra for between 3 and 10 echoes were averaged for each school
for which data was processed. After normalization to account for the
spectral shape of the stimulus signal, the averages of echo spectra
approximate the transfer function for the school. This quantity is
proportional to the backscattering cross section. The values of the
transfer function were tabulated for 54 discrete frequency bands in
the range between 156 and 20,000 Hz. The frequencies were chosen to
exclude interference from other acoustic sensors in use to gather
ancillary data such as school size and depth. The transfer functions,
or acoustic backscattering functions, were submitted to a constrained

least squares analysis. This resulted in an optimal estimate of relative population abundance in each of 30 swimbladder size classes ranging from 0.01 to 4.7 cm equivalent spherical radius. This range includes fish sizes extending from 10 mm anchovy larvae to 35 cm jack mackerel.

Survey Strategy

The sampling strategy developed for the anchovy surveys consisted of four parts. They were the location of the school groups, definition of school boundaries, broadband acoustic sampling, and direct biological sampling. In March and May, the activities were carried out sequentially on each school group as it was encountered. In September, the procedure was changed, resulting in a significant improvement in efficiency and in area covered. The change amounted to locating the school groups and defining their boundaries first, for the entire region to be surveyed. This process took about four days; during this period six distinct school groups were located. Biological sampling was conducted at night from the start, but only to the extent that it did not seriously impede the traveling required to position the ship for the next day's survey effort.

Having completed the first two tasks, the remaining time for the cruise was allotted to the school groups according to size. Transits between groups were planned to make the best use of time between acoustic and biological sampling periods. Both biological and acoustic sampling were conducted for each group. Though progressively more effort was dedicated to school group relocation as the time between initial contact during the early survey and the desired detailed work increased, each school group was found to be within about 10 mi of its original location. Most were within 2 or 3 mi of their original position. One group was relocated after a period of 11 days at a position about 10 mi closer to the coast than it had been previously logged.

March Survey Results

The March survey was ten days long. During this period, approximately 61 hr were dedicated to resonance frequency operations. Eighteen midwater trawls, 29 bongo tows to 70 m and 26 manta neuston net tows were obtained during the same period in support of the resonance program. An additional 161 1-m net CalCOFI tows were made by other investigators.

The geographic area investigated was approximately bounded by a line from Pt. Dume south to the north end of San Clemente Island, along the eastern shore of San Clemente Island, and from the south end of the island to a point about 20 mi southwest of Oceanside, California, thence to Point Loma. This area contains about 3,400 square nautical miles.

Four school groups were studied (Fig. 5). Group A, located on the northside of Redondo Canyon in Santa Monica Bay, covered about 8.25 square nautical miles. During the sonar mapping pass, 0.31% of this surface area covered schooled fish. (Area of schools is $\pi(Dia/2)^2/1.72$: Squire's correction for school shape is 1.72 (Squire, 1978).)

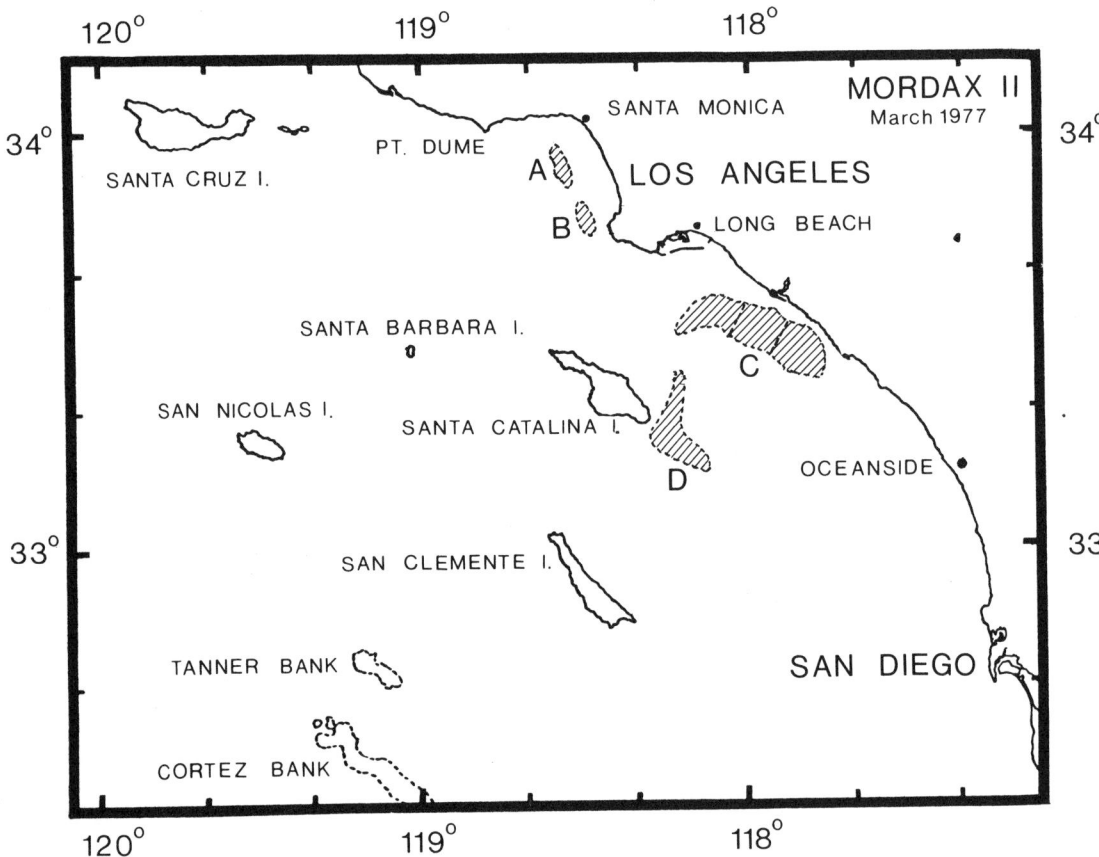

FIGURE 5: Geographic distribution of fish school
groups during the March 1977 MORDAX acoustic resonance survey.

Estimates of school characteristics were made by measuring the broadband acoustic signatures of 19 schools in this group. These data revealed that the dominant population in all 19 schools contained gas-
filled swimbladders. One school consisted essentially of larvae only,
but of two distinct sizes (Fig. 6). Two schools contained large larvae, juveniles and adult fish with swimbladder sizes consistent with
northern anchovy up to 155 mm standard length (Fig. 7). Acoustic
assessment of the remainder of the Group A schools indicated the presence of fish in a variety of sizes, ranging between larvae and adult
fish (Fig. 8).

School Group B was the largest and most extensively studied of
the four groups encountered in March. It was located between Dana
Point and Huntington Beach. Acoustic sampling and trawling was conducted in and near Group B for five days and nights. The acoustic
data were dominated by scattering from bubbles consistent in size
with larval anchovy. Significant components of juvenile fish and a
small (relative to numbers of larvae) fraction of adult fish, up to

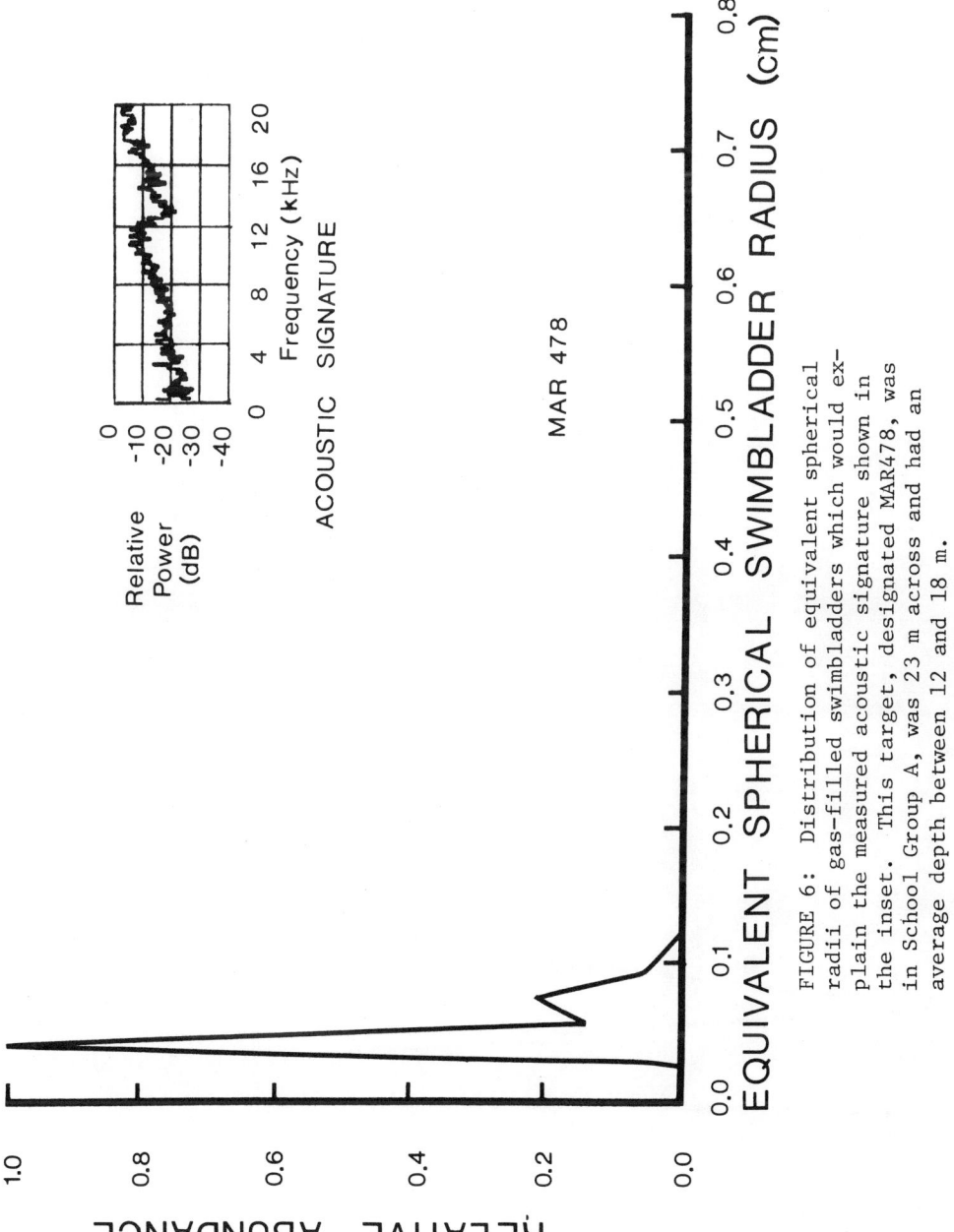

FIGURE 6: Distribution of equivalent spherical radii of gas-filled swimbladders which would explain the measured acoustic signature shown in the inset. This target, designated MAR478, was in School Group A, was 23 m across and had an average depth between 12 and 18 m.

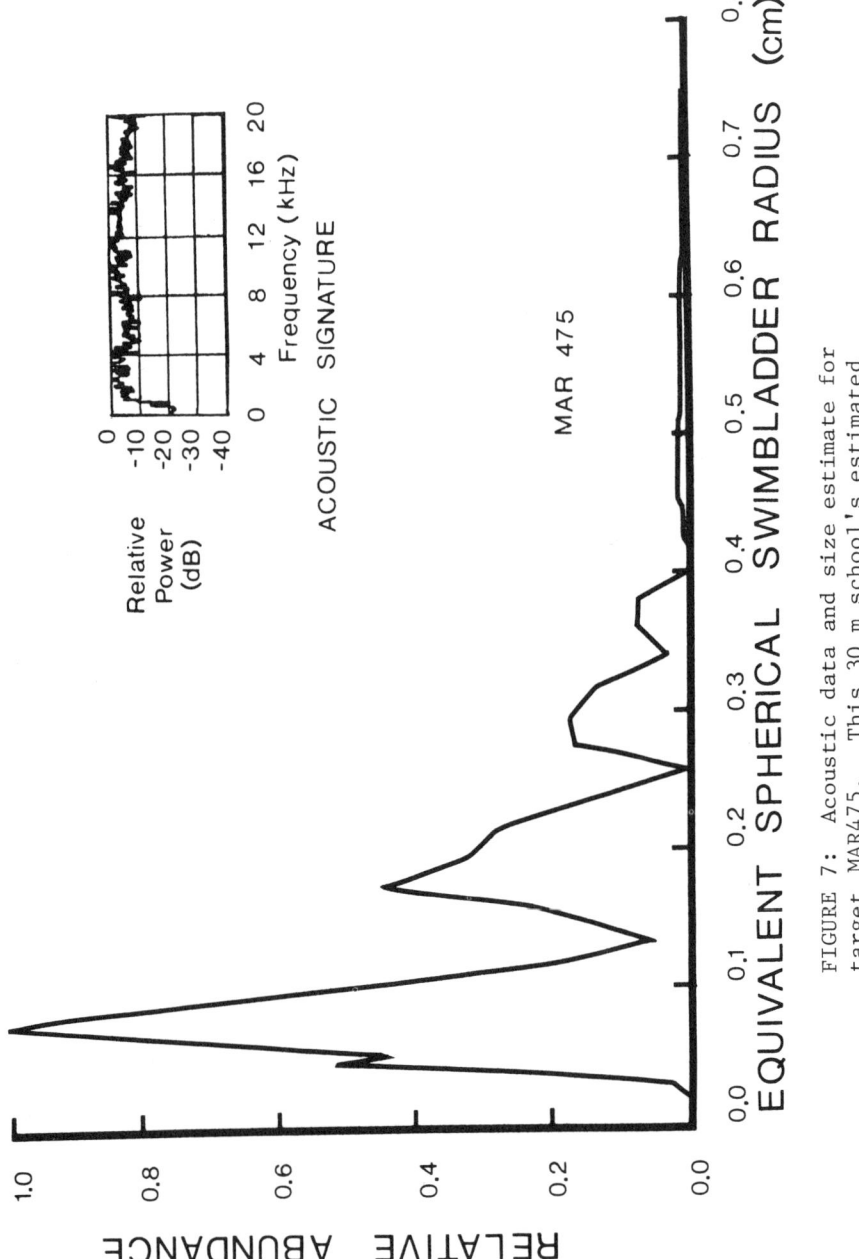

FIGURE 7: Acoustic data and size estimate for target MAR475. This 30 m school's estimated projected area on the surface was 411 square meters. The school's mean depth was between 12 and 18 m.

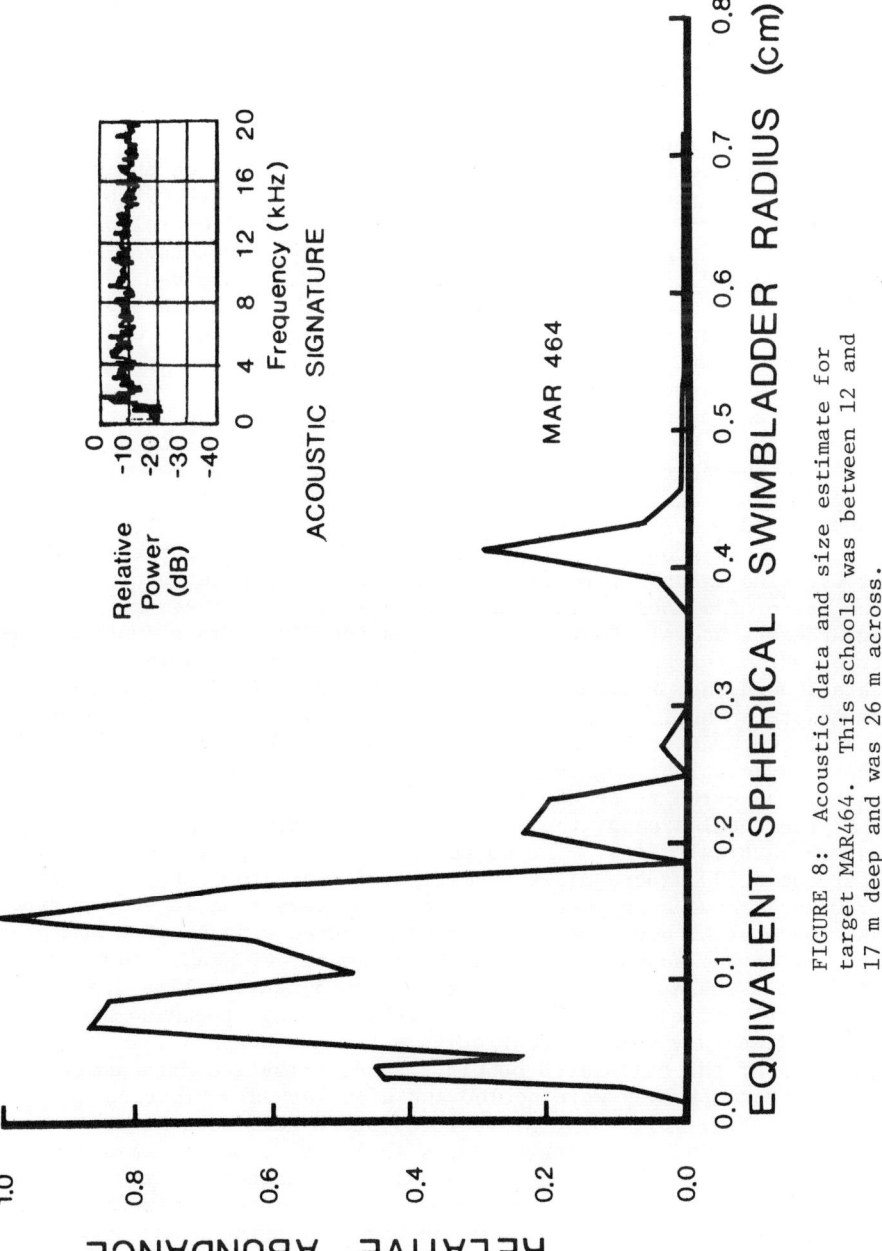

FIGURE 8: Acoustic data and size estimate for target MAR464. This schools was between 12 and 17 m deep and was 26 m across.

165 mm, were also indicated by the acoustic sampling. Direct sampling with the nets revealed the presence of larval anchovy in great abundance. Adult anchovy, 6 cm squid and one 30 cm Pacific mackerel were taken by the midwater trawl. A minor fraction (<10%) of the schools in Group B exhibited an acoustic signature thought to be characteristic of schools with a minor component of non-resonant (non-swimbladder bearing) individuals. One of the schools studied in Group B appeared to be dominated by squid or Pacific mackerel, based on the shape of the acoustic transfer functions. Taken as a whole, 0.09% of the area encompassed by Group B covered fish schools. The central part of the group was the densest concentration with 0.44% coverage. The western end averaged 0.02% and on the eastern end 0.10% coverage was measured.

A second schoolgroup was located over the southern edge of Redondo Canyon, near Pt. Vicente. This group, identified as Group C, covered 5.19 square nautical miles. The ratio of area over schools to the area surveyed was 0.12% for this group. Acoustic estimates of relative abundance in different size classes revealed a pattern for Group C (Fig. 9) which was very similar to that in Group A, a few miles to the north. The dominant population was larvae, with juvenile fish slightly less numerous and finally a low (relative) abundance of adults.

The last group, Group D, was located near the southeastern end of Catalina Island. Group D covered about 42 square nautical miles. The ratio of school horizontal area to surface area searched was 0.20% for these schools. Four schools, each over 100 m in diameter, contributed 64% of the school surface projection. Although the acoustic data showed the presence of larvae in the vicinity of school Group D, at least one-half of the schools were bottom-associated. These schools were detected on side-scan sonar due to severe ray bending caused by a strong negative thermal gradient. The remainder of the schools appeared to be non-resonant (flat transfer function).

The total area of the combined school groups was about 187 square nautical miles. The survey pattern was sufficiently dense over the 3,400 square miles investigated that groups the size of B or D would have been detected. Groups of less than 10 square miles, such as A or C, would only have been detected with about a 50% probability in the southern one-fourth of the survey area. In the northern three-quarters of the area (Fig. 5) survey line densities were sufficient to detect all school groups the size of A or C unless they were of particularly strange shape.

All of the calculated bubble size distribution data analyzed for the March survey were pooled and a summary distribution resulted (Fig. 10). Several features stand out. The first is that bubbles which are consistent with larval anchovy dominate the size distribution. Several peaks in abundance are resolved at bubble sizes below 0.5 cm radius. These could be associated with either peaks in spawning or in survival. The peaks near a swimbladder effective radius of 0.4 cm would be consistent with a fall (November, 1976) spawn. Evidence for this spawn has recently also been found in egg and larvae surveys, in otolith-based aging studies, and in commercial catch statistics (Smith, P. E., NMFS La Jolla, pers. comm.). There are also peaks, though at low relative abundances, for swimbladder sizes larger than can

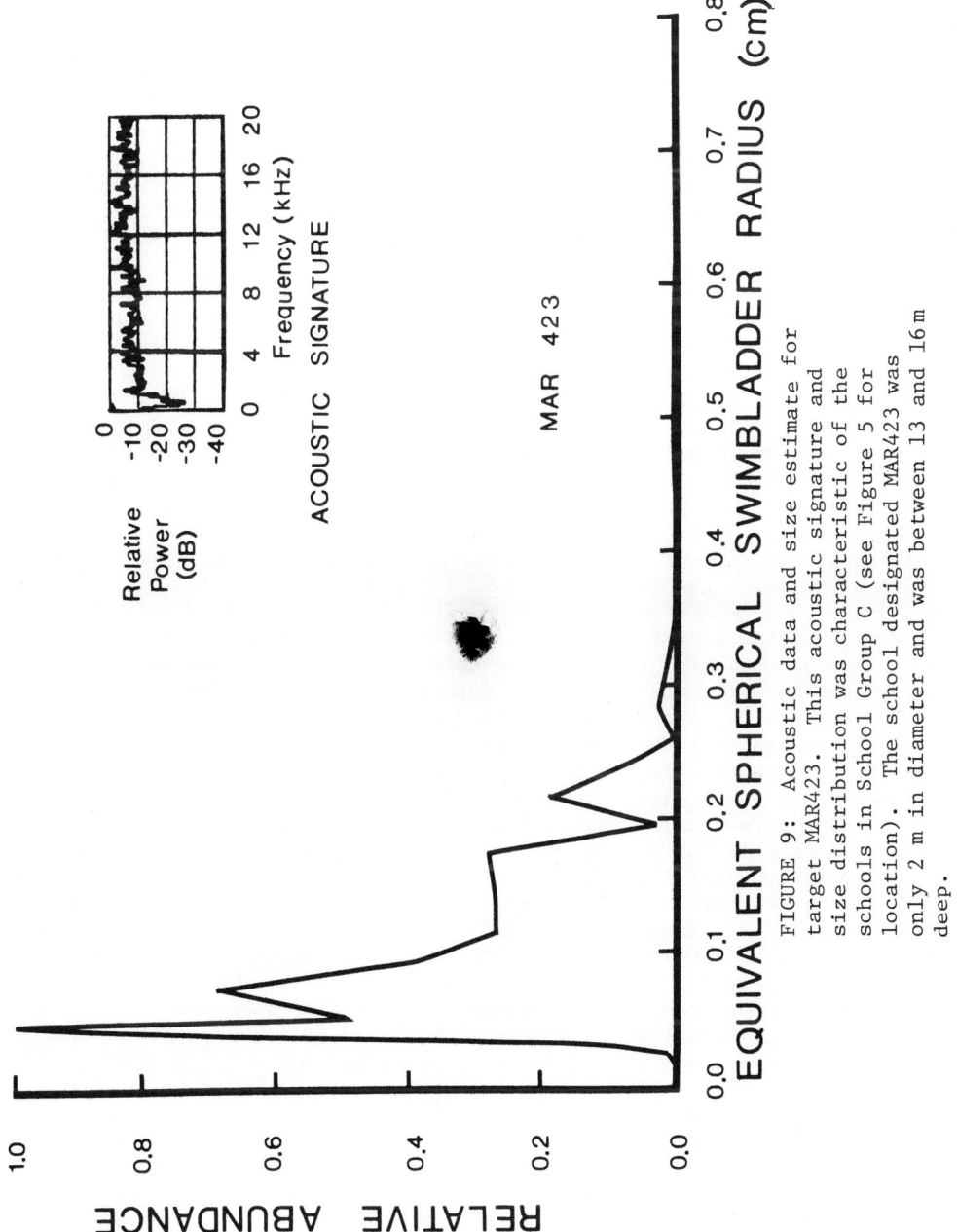

FIGURE 9: Acoustic data and size estimate for target MAR423. This acoustic signature and size distribution was characteristic of the schools in School Group C (see Figure 5 for location). The school designated MAR423 was only 2 m in diameter and was between 13 and 16 m deep.

be attributed to anchovy. Anchovy will appear at sizes less than
about 0.8 cm swimbladder radius. This infers at least three peaks
which must be explained by larger swimbladder-bearing fish. Finally, the slope of the curve in the anchovy size class range is suggestive of a mortality curve and should be the subject of further
investigation.

Regressions of anchovy swimbladder dimensions on fish length
are available for small larvae, for large juveniles, and for adults.
Few measurements are available between 30 and 60 mm (standard
length). As a consequence, the swimbladder size distribution was
transformed to fish size distribution only over the large juvenile
and adult range for School Group B. The dominant size class was
larvae for Group B, as in the summary for March. The data for
Group B were plotted relative to the calculated abundance at 70 mm
size class (Fig. 11). The 70 mm fish are potential recruits as the
minimum size for the fishery is 108 mm. The relative fish abundance indicated in the adult size class was about 1/100th that of
the juveniles. In order to compare biological data from trawls
conducted in the school group each night with daytime acoustic results, the data for Group B was plotted again, normalizing with
respect to the peak abundance in the adult anchovy range. When
the acoustic data and the trawl data were both normalized to one
at their respective peaks in abundance, the agreement supported a
correlation coefficient of 0.92 (Fig. 12).

In a different comparison, the size distribution derived from
commercial landings two weeks earlier in the vicinity of Group C
was plotted with the acoustic data from Group C. The peak in
abundance for the biological data occurred at the same size as
indicated by the acoustic data. When the values were normalized
to the same value in the adult size bin with the maximum abundance,
the shapes of the two curves (Fig. 13) are seen to be in good agreement above the legal commercial minimum, 108 mm.

There remains the question of the large bubbles whose presence
is indicated by the acoustic resonance data. These bubbles are
consistent qualitatively with the three year classes of jack mackerel
which are apparent in the size distributions developed from commercial fishery landings (Fig. 14). These data were kindly supplied by
John Sunada, California Department of Fish and Game. Unfortunately,
the regressions between swimbladder size in the jack mackerel and its
length have never been developed and our time and budget did not permit us to undertake this task. Thus, for the moment, the comparison
must remain only plausible, not quantitative.

SURVEY RESULTS

A detailed discussion of the anchovy resonance work is documented in Holliday (1978). Supporting biological and environmental
data are also documented in the same report series. Consequently,
only the major conclusions drawn at the project's completion will
be reported here.

   † For all three cruises, the acoustic estimates agreed excep-
     tionally well with all available biological samples.

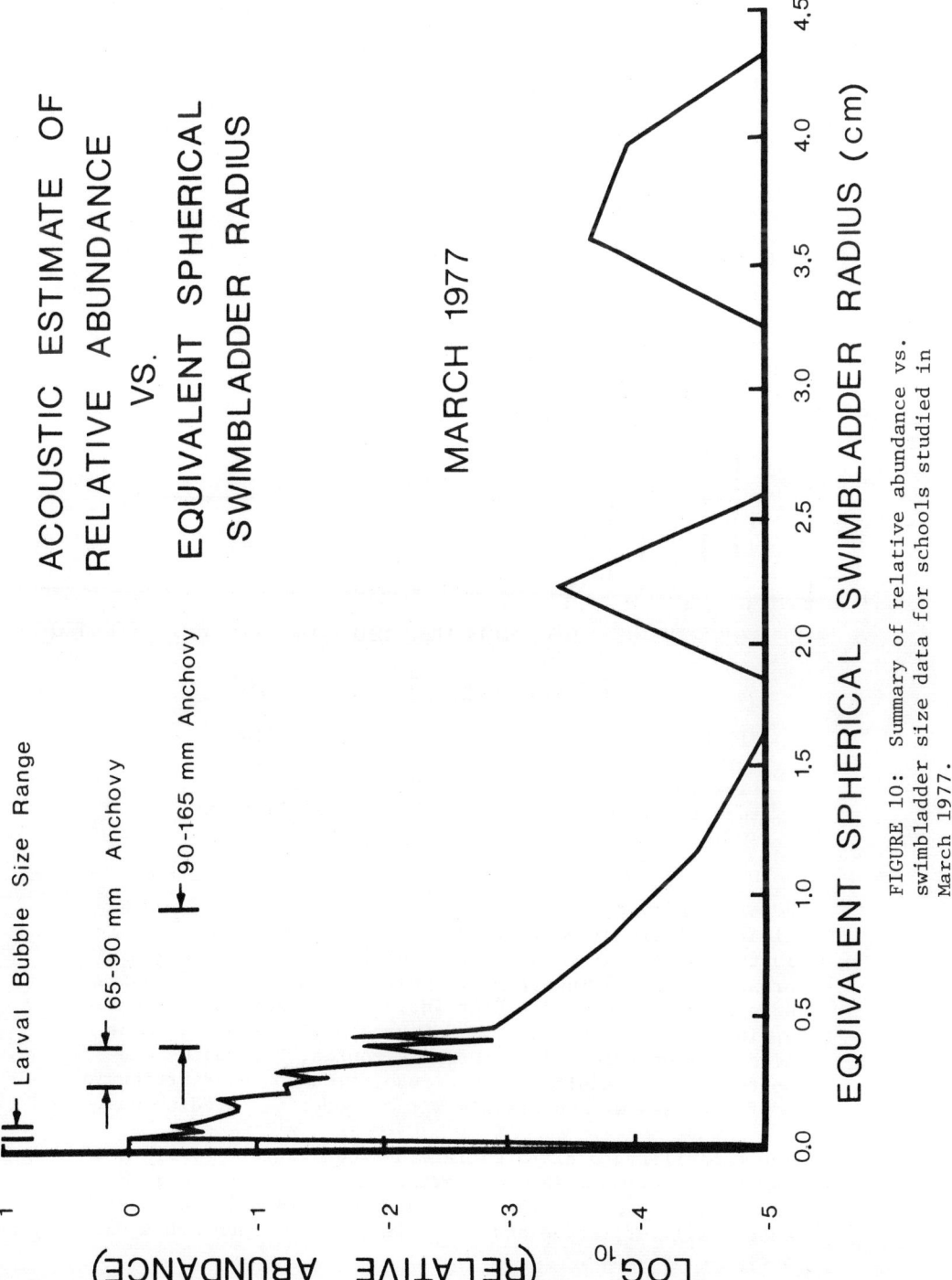

FIGURE 10: Summary of relative abundance vs. swimbladder size data for schools studied in March 1977.

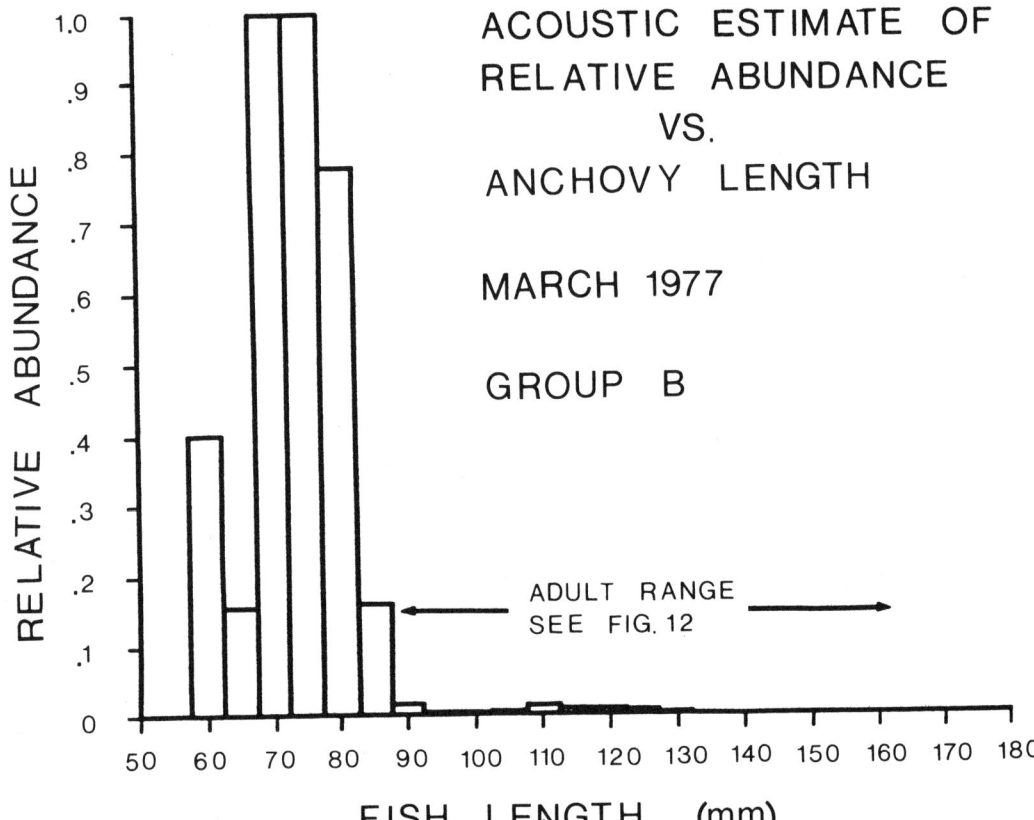

FIGURE 11: Relative abundance vs. fish length for Group B. The conversion from swimbladder size to fish size was not made for fish less than 60 mm long due to a lack of data relating swimbladder characteristics and fish size.

† Several individual size classes of northern anchovy were clearly distinguishable in the acoustic data. For example, in the March cruise data, six individual peaks in spawning or survival, ranging from 0.9 months of age to 11.5 months of age, were detected. Four individual peaks were detected during the September cruise. Corresponding ages of these size classes were 1.2 months, 1.9 months, 2.4 months and 3 months. This sensitivity and resolution to the distribution of fish sizes was comparable with theoretical predictions.

† The capability to differentiate between adult, juvenile and larval schooling fish was convincingly demonstrated in May when a 6,000 school group was shown acoustically to consist essentially of larvae and juvenile fish, with no adult anchovy in evidence.

† Neither the adult anchovy population nor the large larvae studied in March were apparent in the 5,700 square mile area surveyed during May. Larvae averaging 30 mm, apparently from

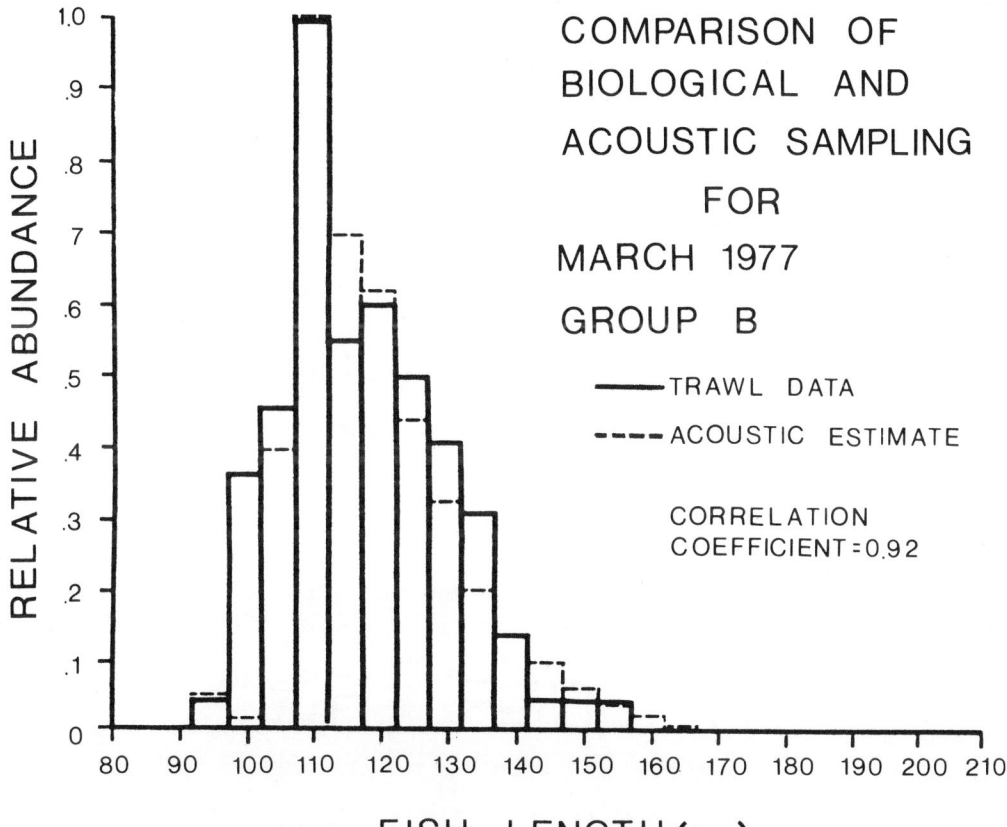

FIGURE 12: Comparison of trawl and acoustic
sampling for March 1977, School Group B. Both
data sets normalized to unit abundance at size
class exhibiting largest abundance in the adult
anchovy size range.

a spawn in March-April, were abundant. Unlike May, adult anchovy were relatively abundant in the survey area during
September.

† In all three surveys, fish with relatively large swimbladders
of a size likely to be consistent with jack mackerel were detected. The relative abundance of these fish was considerably
greater in September than during the other two surveys. In
many cases, these fish were associated with schools of smaller
or similar size fish, probably anchovy.

† In Santa Monica Bay, acoustic fish size estimates derived from
reverberation-like echoes indicated the presence of juvenile
anchovy. Such fish would not normally be sampled by nets or
measured with the sonar mapping techniques currently used.
Yet they may constitute an important proportion of the total
population. Directed sampling, based on on-board resonance frequency analyses, resulted in sizable catches of juvenile anchovy.

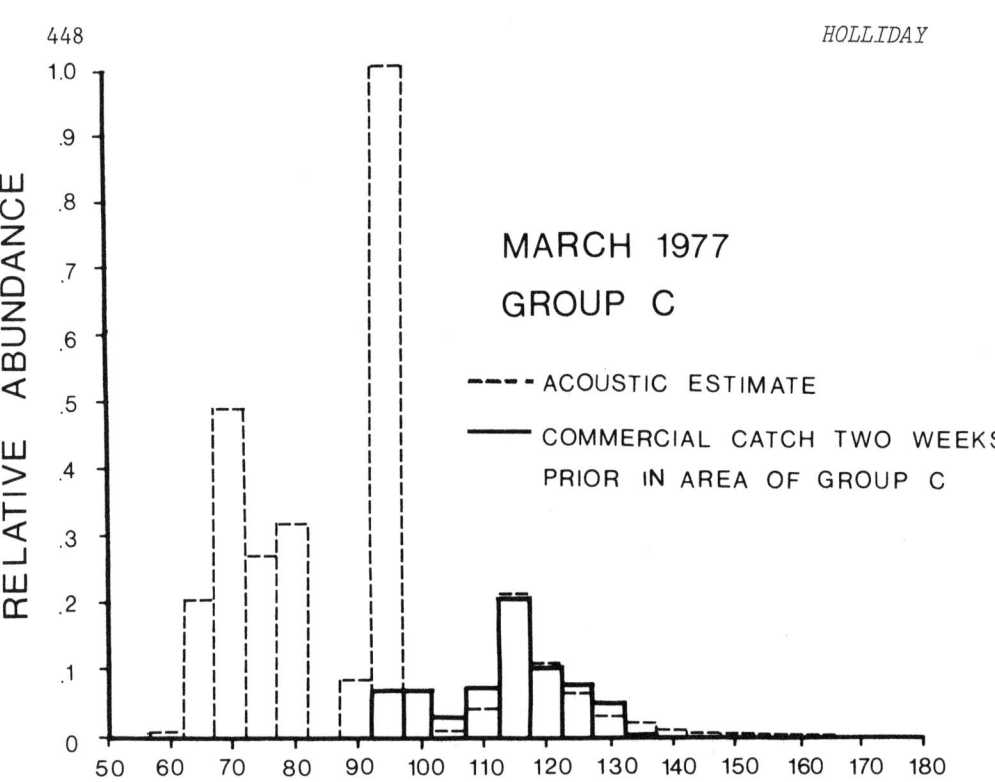

FIGURE 13: Relative abundance vs. fish length for Group C, March 1977. Size frequency distributions from the commercial landings two weeks earlier from the same area as was occupied by School Group C during the acoustic survey are compared with the distribution of fish sizes estimated by the acoustic resonance technique. The comparison of the curves below 100 mm is biased by the anchovy legal minimum size, 108 mm.

† The September results confirmed earlier indications that the broadband acoustic technique could separately assess adult fish and total schooling biomass for northern anchovy.

† The ability to assess swimbladder relative abundance over sizes from small larval anchovy to adult jack mackerel, sampling 2×10^6 m^3 of water each 5 sec, traveling at 5 to 9 knots has a significant cost advantage over other techniques for accomplishing the same work.

† Finally, a possibility exists that absolute assessment of larval stages of northern anchovy could be achieved by modifications of the techniques used in this program. It is also

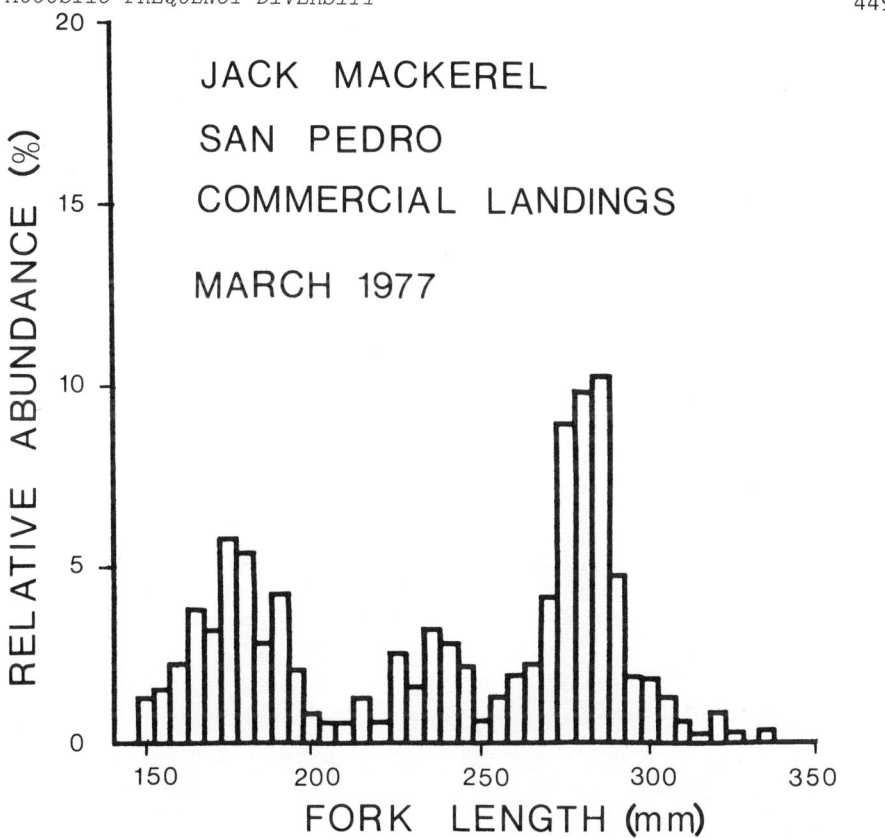

FIGURE 14: San Pedro, California commercial
landings of jack mackerel, length frequency
distribution for March 1977.

possible that the same techniques could lead to definition in
absolute terms of the vertical distribution of anchovy larvae
by size class.

ACOUSTIC MEASUREMENT OF ZOOPLANKTON DISTRIBUTIONS

With rare exceptions, acoustic volume backscattering in the
marine environment is associated with the population of biological
organisms in the water column (Farquhar, 1970; Andersen and Zahuranec,
1977). As indicated in previous sections of this paper, scattering
from gas-bubble or air-filled swimbladder bearing organisms has been
most intensively studied. Scattering from plankton has received a
great deal less attention.

In general, for non-bubble bearing organisms, acoustic frequencies must be used which are higher than those normally associated
with scattering from fishes. This is due in part to resonances in
gas-bubbles at lower frequencies, i.e., kHz and in part because of
the size differences between fishes and much of the important plankton.

The scattering strength for organisms not containing air increases rapidly with increasing frequency from low values to near a value of ka=1 (k=2πf, i.e., the wave number; a is the equivalent spherical radius of a sphere with the same displacement volume as the organisms; f is the acoustic frequency). Above ka=1, the exact form of the scattering is less well known, but tends to slow or even reverse the increase of scattering (Fig. 4). The details of various scattering models were discussed earlier. For the present discussion, it is sufficient to point out that higher frequencies are required to detect small plankton than are required for studying fish. For plankton with lengths on the order of 1 mm, frequencies in the megahertz range are necessary.

The absorption of sound in water increases dramatically as the acoustic frequency is increased. For example, at 3.08 mHz, one-half of the sound is lost to absorption when echo ranging is attempted on an object only 0.5 m distant from the sensor. This phenomenon makes it impractical to "echo sound" from the surface in the expectation of detecting small plankton. It is necessary to provide a sensor which can be lowered through the water column. Given such an instrument, one could look for structure in the vertical or horizontal dimensions by either a vertical profile or by maintaining the sensors at a constant depth while towing the device through the water.

Acoustic Instrumentation

In order to study the vertical and horizontal distribution of zooplankton of sizes about 1 mm, an acoustic instrument was designed to acquire backscattering strength data at four frequencies. Nominally, these frequencies were 0.5, 1.2, 1.8, and 3.0 mHz. Specific transducers differ slightly in their resonances and the set used during the most recent cruise were tuned to 0.54, 1.16, 1.8 and 3.08 mHz. Transducer beamwidths were between 2.5° and 5°, depending on the specific transducer in use. The final design for the underwater instrument array included four high-frequency acoustic transducers, a high-resolution thermistor, a depth sensor, a current meter, and the end (input) of a plankton pump hose (7.6 cm diameter). Vertical profiles (acoustic, biological, temperature) were made by slowly lowering the array through the water column to depths of 120 m. Most profiles were made with the main axes of the sound beams directed horizontally (Fig. 15). Measurements were taken to the side of the array at ranges between 2 and 10 meters. The vertical resolution, then, depended on the rate of lowering (or raising) of the array and varied from 10 cm sec^{-1} to 50 cm sec^{-1}. Pulse repetition rates were variable but were usually set at about one pulse each 100 msec. Most measurements were made with 50 μsec pulses. A variable width range gate was used to define a window between the transducer and the maximum range desired. Signals received within the range gate were integrated over the gate length after square law detection. Sixteen sequential echo ranging cycles were then summed and a single sample recorded. The process was then repeated. Recording was done in three modes, depending on the end analysis intended. In cases requiring the envelope of the echoes (simple statistical studies), direct recordings were made at 60 ips

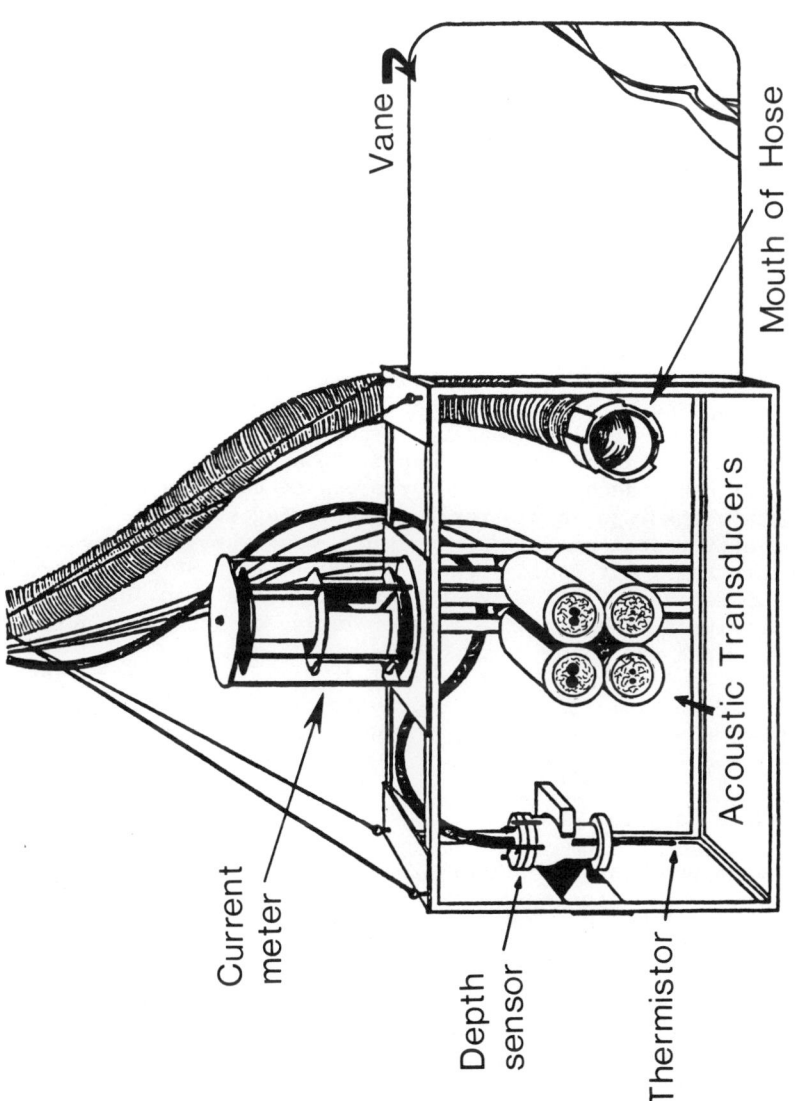

FIGURE 15: Underwater instrument array for high frequency acoustic studies of zooplankton distributions.

on an instrumentation recorder. For backscattering strength measurements, details of the individual echo waveforms are not required. Consequently, the integrator output was recorded. Frequency modulated recording at 1-7/8 ips was used, allowing several profiles to be placed on a single multichannel reel while gaining the benefit of improved fidelity with the FM mode. These data were used both for on-board processing and display as well as for later laboratory processing.

The on-board display was real-time and was principally intended for data quality control. Second order corrections, included in laboratory processing, were not used at sea in view of the speed of the computer. In addition to data quality control, the real-time display proved useful for directing biological sampling in real-time. Sampling depths could be effectively selected based on the detail in the structure of plankton distributions.

Calibration of the acoustic suite was accomplished by using the standard self-reciprocity technique (Bobber, 1970). Known signal levels were injected at the system input in order to assure accurate measurement of receiver/recorder gain. These calibration signals were applied before each sequence of profiles and at the beginning of each new reel of magnetic tape.

Laboratory processing included application of calibration data and transducer depth sensitivity measurements to the data base in order to obtain absolute values for backscattering strengths. For the display of vertical profile information, data were averaged in 2 m depth intervals, but would support increased resolution if desired. A limited number of horizontal profiles have been subjected to Fourier analysis with distance moved relative to the water defined as the independent variable.

Additional analyses in progress include calculating scattering strengths based on biological samples, evaluation of backscattering models and calculations of population densities in multiple size classes from the acoustic profiles.

Biological Sampling

A plankton pump was the primary zooplankton-sampling device for this study. The water was pumped at rates which varied from 500-700 liters min^{-1} into a filtering system (Fig. 16) where the plankton were filtered out with nets of 100 μm pore diameter. Sampling times for each sample ranged from 1 to 5 min depending on zooplankton concentrations. The volume of water sampled ranged from 0.5 to 3.5 m^3.

The biological samples were preserved in 10% buffered formalin in the field for later counting and identification. In the laboratory all large organisms (e.g., fish larvae, amphipods, euphausiids, etc.) were removed, counted and identified. Samples containing a large number of organisms were split up to a maximum of eight times using a plankton splitter. All organisms in the sub-sample were identified to species, counted, and the mean length determined. Calanoid copepods were separated into adult males, adult females, the different copepodite stages and nauplii. Other copepods were separated into adults, copepodites, and nauplii. Copepods and

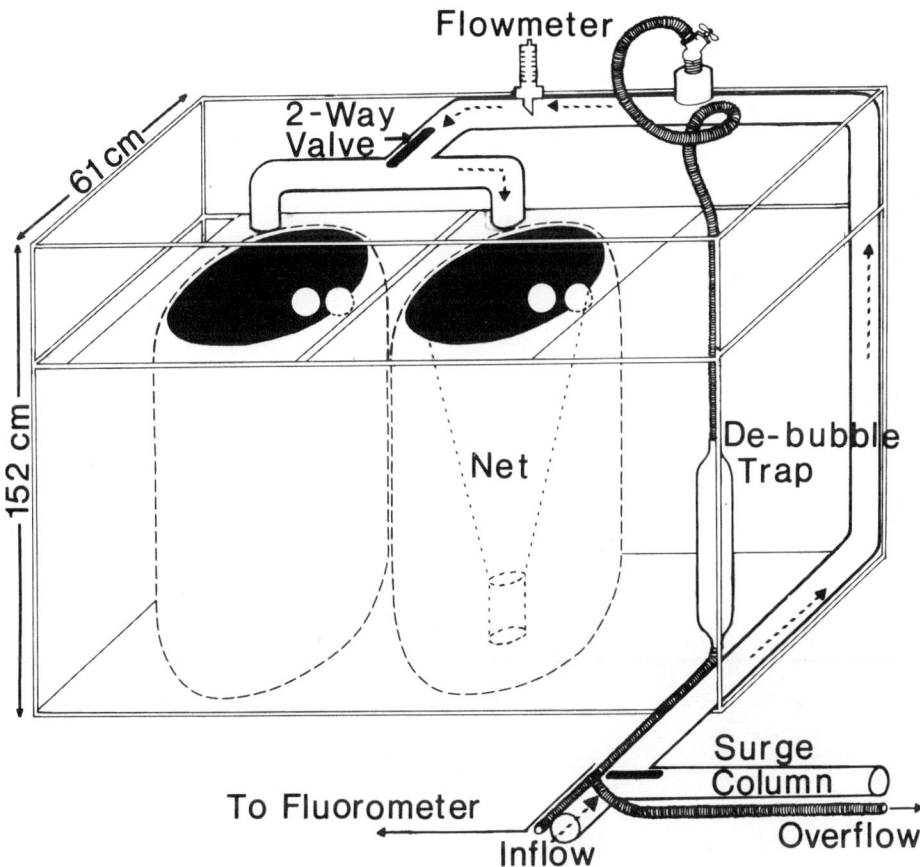

FIGURE 16: Filtration system for use with
plankton pump.

their developmental stages were the dominant organism collected,
with periodic high counts of cladocerans, larvaceans (oikopleura)
and eggs. Other organisms counted and identified included fish
larvae, decapod and euphausiid larvae, amphipods, mysids, ostracods, chaetognaths, pteropods, polychaetes, medusae, siphonophores,
salps, doliolids and developmental stages of benthic invertebrates.
The concentration of each species (or developmental stage) was converted to the number per m^3 of water filtered.

The mean lengths of all organisms have been converted into
values corresponding to the radius of an equivalent-volume sphere
(Anderson, 1950). These values, along with the concentrations will
be used to calculate the predicted scattering profiles at each frequency for comparison with measured values.

Chlorophyll profiles were determined for the last three cruises
either with a Turner Model 111 Fluorometer (*in situ*) measurements with
a flow-through door or from spot samples for later spectrophotometric
chlorophyll analyses. The chlorophyll samples were taken from the 3-

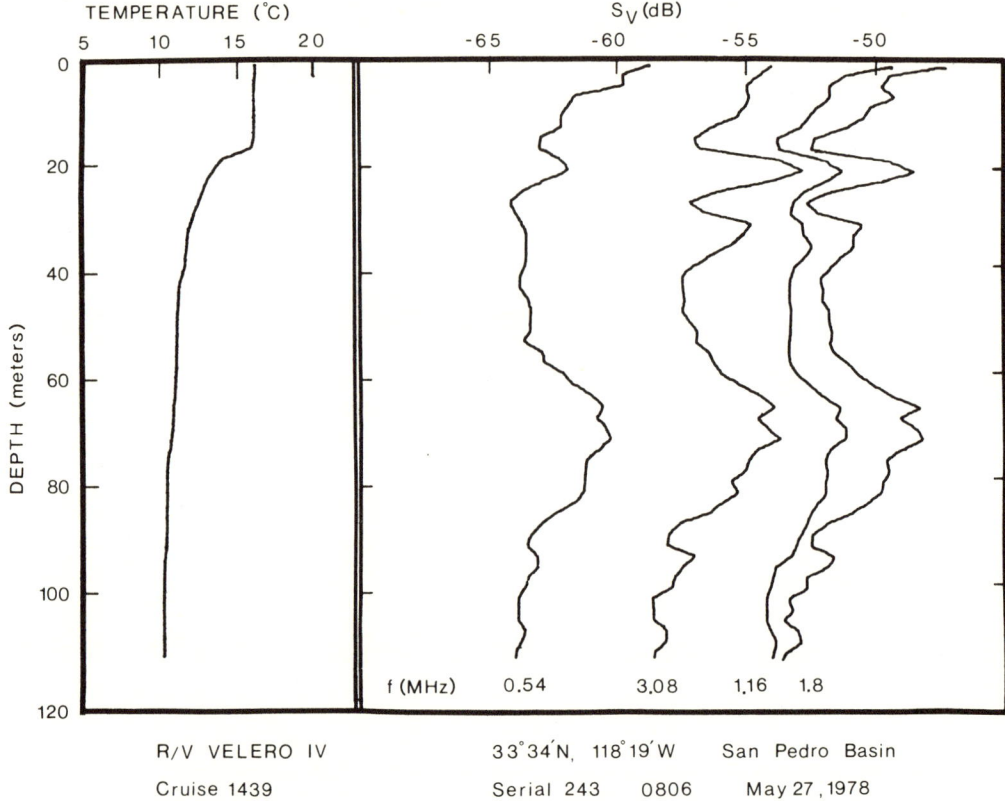

FIGURE 17: Scattering profiles at four frequencies to a depth of 112 m. Station 243,
0806, May 27, 1978.

inch plankton pump system.

Net samples were also taken on all cruises to collect the larger
organisms not sampled by the pump. Bongo nets (303 and 500 μm mesh)
were used on several cruises. An opening/closing, 1 m², multiple
plankton sampler (MPS) and an Isaacs–Kidd–Midwater Trawl (IKMT) with
attached MPS were used on one cruise by Dr. William Pearcy (Oregon
State University). One meter ring nets and a 6-ft opening/closing
Tucker trawl were used on two cruises. All samples were, again, preserved in 10% buffered formalin for later analysis in the laboratory.

Vertical Volume Scattering Strength Profiles

A vertical profile of acoustic backscattering was made in the
San Pedro Basin, near Santa Catalina Island in May 1978. The profile (Fig. 17) represents 9170 measurements of acoustic backscattering at each of four frequencies. It was made over a period of about
25 min. The data also included measurements of sensor depth and
water temperature. The depth information was used to average the
acoustical and temperature measurements into two meter depth intervals. The complex structure in the acoustic profiles is evident as

are differences between individual frequencies. At all depths in
this example, the scattering strength increased with frequency with
the exception of the 3.08 mHz signal. This apparent anomaly may be
due to the form of the backscattering curve in the geometric scattering region (e.g., see Fig. 4).

    A second notable feature is the relative scattering levels at
different depths. For example, the ratios of scattering strengths
in order of increasing frequency are 1 :8.7 :21.9 :11.5 at 20 m and
is 1 :4.6 :14.5 :10.2 at 50 m. Scattering strengths, in addition to
changing with acoustic frequency, are sensitive to the size distribution of the scattering zooplankters and to a lesser degree to the
density and compressibility contrasts of the animals. The changes
in ratios of scattering strength at the two depths are typical of
profiles of this type and are thought to be largely due to changes
in the distribution of zooplankter sizes at different depths.

    The scattering profiles also contain information on the absolute
numbers of organisms. In the least complex case, if all the organisms were one size, the number, N, of animals that it would take to
cause the observed scattering strength at a given depth could be
readily calculated from the relation

$$S_v(f) = TS(f,a) + 10 \log N. \tag{11}$$

Here $S_v(f)$ is the backscattering strength at frequency f. The term
$TS(f,a)$ is the target strength of a single animal with size a. For
example, assume we have reason to believe the equivalent spherical
radius of the animals causing the scattering at 20 m in the case of
Serial 243 (Fig. 17) was 0.2 mm. The backscattering strength at
0.54 mHz at 20 m is -62 dB. The target strength of a single animal
which would fill a sphere of radius 0.2 mm, assuming a specific density and a specific compressibility, might be -92 dB. Solution of
equation (11) results in N=1000. This example is too simple in view
of the "real world" distribution of sizes, but the principle remains the same even though the mathematics becomes significantly
more complex in attempting to apply this approach to real world data.

    A second and a third example of vertical scattering profiles
(Figs. 18, 19) exhibit a complex structure with several peaks in
scattering intensity as well as regions of relatively less scattering. Note that the order of frequencies is neither monotonic nor in
the same order as was the case in the first profile (Fig. 17).

Summary of Zooplankton Studies

    The analysis of the high frequency zooplankton data is only
partially complete. We have successfully measured volume scattering
strengths at four exceptionally high acoustic frequencies which were
selected to be sensitive to marine zooplankton in sizes between 0.1 mm
and 10 mm. Each of these profiles exhibits complex structures as a
function of depth. This structure is thought to reflect the zooplankton distribution on a scale not readily measured for this size
class of animals by any other means.

    Work is still in progress in analyzing the data base we have gathered during the past year. In summary, we have accomplished the following:

   † High depth resolution, quantitative acoustic backscattering measurements were made to a depth of 120 m at frequencies designed to detect small zooplankton.

   † Directed sampling was successfully accomplished for small zooplankton (copepods) acoustically, in real-time, at sea.

   † Zooplankton layering which is not associated with apparent thermal struture was observed.

   † A correlation was observed between thermal gradients, chlorophyll, zooplankton and in data from the pelagic fish studies discussed earlier, the thickness and depth distribution of fish schools.

SUMMARY AND CONCLUSION

     A simplified overview of the basic physics and the mathematical models which are used to describe sound scattering from biological organisms in the ocean has been presented. The importance of choice of acoustic frequency in instruments to be used for studying marine life has been stressed.
     We have attempted to illustrate the usefulness of frequency diversity by choosing examples from opposite ends of the frequency/ target size spectrum. In one example, broadband, low frequency sound has been shown to be of use in the rapid assessment of size distributions for swimbladder-bearing, schooling, pelagic zone fish. In another case, very high frequencies are currently in use for studying the distribution of small zooplankton, including copepods, in the 0.1 to 1 mm size range.
     Even in relating these two examples, we have only touched on two possible uses of multiple frequency acoustic systems. Uses of both the doppler effect and passive acoustics comes to mind immediately in the context of marine biological measurements. Others will undoubtedly surface soon in this still young, hopefully synergistic mix of well-established disciplines--marine biology and acoustics.

     FIGURE 18: Scattering profile for station 230.

     FIGURE 19: Scattering profile for station 272.

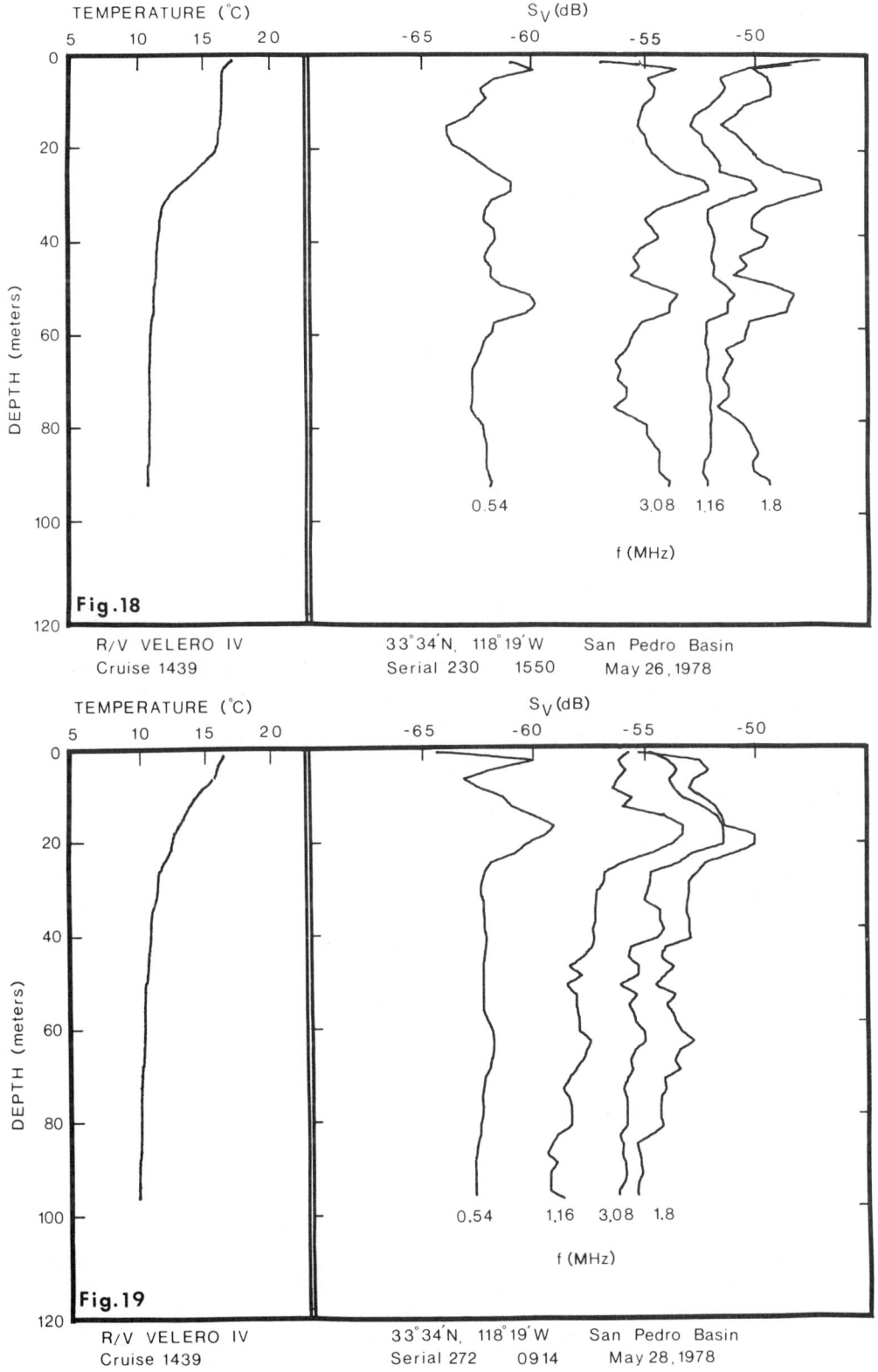

TEMPERATURE (°C)

S_V (dB)

Fig.18

R/V VELERO IV
Cruise 1439

33°34′N, 118°19′W   San Pedro Basin
Serial 230   1550   May 26,1978

f (MHz)

0.54   3.08 1.16   1.8

TEMPERATURE (°C)

S_V (dB)

Fig.19

R/V VELERO IV
Cruise 1439

33°34′N, 118°19′W   San Pedro Basin
Serial 272   0914   May 28,1978

f (MHz)

0.54   1.16 3.08 1.8

ACKNOWLEDGMENTS

The author is indebted to ONR, NSF, NOAA/NMFS and NAVSEA for
sponsorship of the several research projects from which the data
used as examples in this paper were taken. Two of these programs
have been joint efforts between Tracor's staff and scientists from
other institutions.

The ultra-high frequency zooplankton studies are jointly sponsored by NSF and ONR under a joint university/industry program
funded under NSF Grant OCE 76-13201 through the Institute of Marine
and Coastal Studies at the University of Southern California.
Richard Pieper of USC is our co-principal investigator in this work.
With the able assistance of his staff he conducted the biological
sampling efforts. Our particular thanks to Ms. H. Mumper for the
artwork in Figures 15 and 16 of the text. Dr. Pieper's contributions in planning and conducting the research and in the data analysis continues to be a key factor in insuring relevance of the
acoustic effort to the biological problem.

The work on resonance surveys of anchovy schools was sponsored
by the National Marine Fisheries Service, Southwest Fisheries Center, Coastal Fisheries Resource Division under Dr. Reuben Lasker.
Our co-principal investigator in this work has been Paul Smith of
that Division. The biological sampling was carried out for the anchovy program by Dr. Smith and his staff. His experience with the
anchovy population and support in planning, data acquisition and
analysis was invaluable in the conduct of the MORDAX acoustic surveys.

My sincere thanks to Hugh Larsen, David Horton, and Dr. David
Doan of Tracor, who did most of the difficult work in the acoustic
sampling programs. The work of others in our joint staffs, too
numerous to mention, is also gratefully acknowledged.

REFERENCES

Albers, V. M., 1972. *Benchmark Papers in Acoustics: Underwater
    Sound.* Dowden, Hutchison & Ross, Inc., Pennsylvania. 468 pp.
Andersen, N. R. and B. J. Zahuranec, 1977. *Oceanic Sound Scatter-
    ing Prediction.* Plenum Press, New York. 859 pp.
Anderson, V. C., 1950. Sound scattering from a fluid sphere. *J.
    Acoust. Soc. Amer.* 22: 426-431.
Andreeva, I. B., 1964. Scattering of sound by air bladders of
    fish in deep sound-scattering ocean layers. *Sov. Phys. Acoust.*
    10: 17-20.
Bailey, R. M., 1970. *A List of Common and Scientific Names of
    Fishes from the United States and Canada.* American Fisheries
    Soc. Spec. Pub. No. 6, Washington, D. C. 150 pp.
Bobber, R. J., 1970. *Underwater Electroacoustic Measurements.*
    Naval Research Laboratory, Washington, D. C. 333 pp.
Davies, I. E., 1973. Deep observations of anchovy and blue sharks
    from Deepstar 4000. *Fish. Bull., NOAA* 70: 510-511.
Devin, C., Jr., 1959. Survey of thermal, radiation and viscous
    damping of pulsating air bubbles in water. *J. Acoust. Soc.
    Amer.* 31: 1654-1667.

Duvall, G. E. and R. J. Christensen, 1946. Stratification of
sound scatterers in the ocean. *J. Acoust. Soc. Amer.* 20: 254.

Eyring, C. F., R. J. Christensen, and R. W. Raitt, 1948. Reverberation in the sea. *J. Acoust. Soc. Amer.* 20: 462-475.

Farquhar, G. B., ed., 1970. *Proceedings of an International Symposium on Biological Sound Scattering in the Ocean.* Maury Center
for Ocean Science, Washington, D. C. 629 pp.

Freedman, A., 1962a. A mechanism of acoustic echo formation. *Acustica* 12: 10-21.

Freedman, A., 1962b. The high frequency echo structure of some
simple body shapes. *Acustica* 12: 61-70.

Graves, J. E., 1977. Photographic method for measuring spacing
and density within pelagic fish schools at sea. *Fish. Bull.*
75: 230-234.

Greenlaw, C. F., 1979. Acoustical estimation of zooplankton populations. *Limnol. Oceanogr.* 24: 226-242.

Hersey, J. B. and R. H. Backus, 1954. New evidence that migrating
gas bubbles, probably the swimbladders of fish, are largely
responsible for scattering layers on the continental rise south
of New England. *Deep-Sea Res.* 1: 190-191.

Hewitt, R. P., P. E. Smith, and J. C. Brown, 1976. Development and
use of sonar mapping for pelagic stock assessment in the California current area. *U. S. Fish. Bull.* 74: 281-300.

Holliday, D. V., 1972. Resonance structure in echoes from schooled
pelagic fish. *J. Acoust. Soc. Amer.* 51: 1322-1332.

Holliday, D. V., 1977a. Extracting bio-physical information from
the acoustic signatures of marine organisms. In: *Oceanic
Sound Scattering,* (N. R. Andersen and B. J. Zahuranec, eds.),
pp. 619-624, Plenum Press, New York.

Holliday, D. V., 1977b. The use of swimbladder resonance in the
sizing of schooled pelagic fish. In: *Hydro-Acoustics in
Fisheries Research,* (A. R. Margetts, ed.), pp. 130-135, Denmark.

Holliday, D. V., 1978. *MORDAX II/III/IV.* Tracor Document No. T-
78-SD-002/3/4-U, San Diego, California. 2260 pp.

Holliday, D. V. and H. L. Larsen, 1979. Thickness and depth
distribution of epipelagic fish schools. *U. S. Fish. Bull.* 77.

Johnson, M. W., 1948. Sound as a tool in marine ecology, from data
on biological noises and the deep scattering layer. *J. Mar.
Res.* 7: 443-458.

Johnson, R. K., 1977a. Sound scattering from a fluid sphere re-
visited. *J. Acoust. Soc. Amer.* 61: 375-377.

Johnson, R. K., 1977b. Acoustic estimation of scattering-layer
composition. *J. Acoust. Soc. Amer.* 61: 1636-1639.

Larsen, H. L., 1974. *Distributions of Target Strengths and Horizontal Dimensions for Aggregations and School of Marine Organisms.* Tracor Document No. T-74-SD-1054-U, San Diego, California. 66 pp.

Lax, M. and H. Feshbach, 1948. Absorption and scattering for impedance boundary conditions on spheres and circular cylinders.
J. Acoust. Soc. Amer. 20: 108-124.

Mackenzie, K. V., 1971. A decade of experience with velocimeters.
J. Acoust. Soc. Amer. 50: 1321-1333.

Minnaert, M., 1933. On musical air bubbles and the sounds of running water. *Phil. Mag.* 16: 235-248.

Morse, P. M. and K. U. Ingard, 1968. *Theoretical Acoustics.* McGraw-Hill Book Company, New York. 927 pp.

Pearcy, W. G., ed., 1975. *Workshop on Problems of Assessing Populations of Nekton.* ONR Report ACR 211, Corvallis, Oregon. 30 pp.

Physics of Sound in the Sea, 1946. Nat. Defense Res. Comm. Div. 6 Sum. Tech., Rep. 8, Chapter 14: 284-288.

Pieper, R. E., 1977. Some comparisons between oceanographic measurements and high frequency scattering of underwater sound. In: *Oceanic Sound Scattering Prediction.* (N. R. Andersen and B. J. Zahuranec, eds.). pp. 667-675. Plenum Press, New York.

Rayleigh, Lord, 1945. *Theory of Sound.* Second Edition. Dover Publications, New York. 984 pp.

Skudrzyk, E., 1971. *The Foundation of Acoustics.* Springer-Verlag, New York. 790 pp.

Smith, P. E., 1970. The horizontal dimensions and abundance of fish schools in the upper mixed layer as measured by sonar. In: *Proceedings of an International Symposium on Biological Sound Scattering in the Ocean,* (G. Brooke Farquhar, ed.), pp. 563-600. Maury Center for Ocean Science, Washington, D. C.

Squire, J. L., Jr., 1972. Apparent abundance of some pelagic marine fishes off the southern and central California coast as surveyed by an airborne monitoring program. *U. S. Fish. Bull.* 70: 1005-1019.

Squire, J. L., Jr., 1978. Northern anchovy school shapes as related to problems in school size estimation. *U. S. Fish. Bull.* 76: 443-448.

Strasberg, M., 1953. The pulsation frequency of nonspherical gas bubbles in liquids. *J. Acoust. Soc. Amer.* 25: 536-537.

Tavolga, W. N., 1964. Sonic characteristics and mechanisms in marine fishes. In: *Marine Bio-Acoustics,* (W. N. Tavolga, ed.), pp. 195-211. Pergamon Press, New York.

Thomas, J. B., 1969. *An Introduction to Statistical Communication Theory.* John Wiley & Sons, New York. 670 pp.

Vent, R. J., 1978. Fish school target strength measurements off southern California. *J. Acoust. Soc. Amer.* 64: S96.

Weston, D. E., 1967. Sound propagation in the presence of bladder fish. In: *Underwater Acoustics,* Vol. 2, Chapter 5, (V. M. Albers, ed.), pp. 55-86, Plenum Press, New York.

Ocean Noise and the Behavior of Marine Animals: Relationships and Implications

Arthur A. Myrberg, Jr.

ABSTRACT: The present state of our knowledge
regarding underwater ambient noise and its
effects on selected marine biological systems
is reviewed. The report is limited to relevant
findings from marine fishes and selected marine
mammals since little or nothing is presently
known about the subject in other groups. Only
shallow, coastal regions of the Continental
Shelf are considered since they represent the
major habitats of those animals considered here.
A general synthesis of ambient noise conditions
for such regions is provided, including spectral
curves for traffic (industrial), wind, rain, and
selected sources of high level, biological noise.
Sound detection and localization by fishes and
marine mammals are considered in relation to
various types of noise with frequencies ranging
from less than 100 Hz to beyond 100 KHz. A
preponderance of evidence exists from both laboratory and field studies that ambient noise can
indeed affect audibility in such animals, especially
in the regions of greatest sensitivity. Various
simulations are provided in which auditory sensitiv-

ities of various fishes and mammals are placed into
the context of specific levels of noise for the
frequency ranges involved. The results, though
speculative, suggest that environmental noise does
affect audibility. The implications which arise
from such findings are covered in the final
discussion.

INTRODUCTION

There is general agreement among biologists that the acoustical
modality of aquatic animals probably constitutes their most important
distance receptor-system. This is particularly true among members of
two vertebrate lines, the fishes and the marine mammals, whose acoustical activities have been investigated at an ever-increasing rate of
sophistication during recent years. A major conclusion drawn from
these varied studies-inclusive of those using psychophysical, physiological and ethological methodologies-is that the acoustical system
can, and does, provide its owner appropriate information, readily and
rapidly, on a variety of functions related to food, competitors,
potential mates and predators.

Concomitant with the increasing sophistication of problem-queries,
experimental designs and available instrumentation, there is a growing
awareness that ambient noise itself can no longer be ignored in underwater bio-acoustics. Not only is it probable that such noise actually
affects, at least temporarily, the hearing abilities of the animals
concerned, it may also act to inhibit sound production as well. Such
a reduction in acoustic transmission and/or its reception can adversely affect the reproductive potential or even the survival of any given
species or population that is dependent on such a sensory process.
Additionally, available evidence indicates that excessive noise also
has other more direct, deleterious consequences on marine biological
systems (see below). This problem will become more widely recognized
as more is learned about ambient noise especially in the shallow,
coastal regions of the world's oceans.

The present state of our limited knowledge, regarding acoustical
noise and its effects on selected marine biological systems, will be
reviewed below. Also, I shall take this opportunity to speculate how
such knowledge may provide insight into problems that extend beyond
those restricted to bio-acoustics. This review has been limited to
relevant findings from marine fishes and selected marine mammals,
specifically the odontocete cetaceans and pinnipeds, since little or
nothing is presently known about the subject in other groups. Also,
only the shallow, coastal regions of the Continental Shelf are considered since they represent the major habitats of those animals
considered here.

SHALLOW-WATER AMBIENT NOISE

Measurements of shallow-water, ambient noise from widely differing locations (e.g., Heindsmann et al., 1955; Dietz et al., 1960;
Wenz, 1962; Piggott, 1964; Arase and Arase, 1966; Widener, 1967;

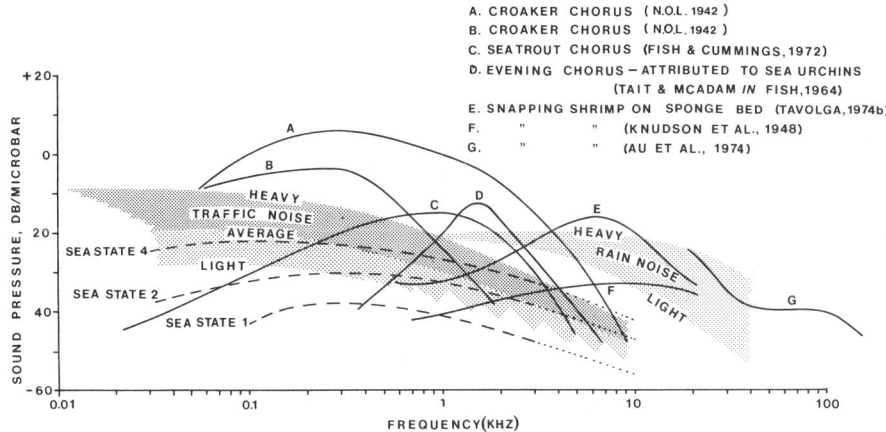

A. CROAKER CHORUS (N.O.L. 1942)
B. CROAKER CHORUS (N.O.L. 1942)
C. SEATROUT CHORUS (FISH & CUMMINGS, 1972)
D. EVENING CHORUS—ATTRIBUTED TO SEA URCHINS
   (TAIT & MCADAM *IN* FISH, 1964)
E. SNAPPING SHRIMP ON SPONGE BED (TAVOLGA, 1974b)
F.   "   "   (KNUDSON ET AL., 1948)
G.   "   "   (AU ET AL., 1974)

FIGURE 1:   Shallow water (<70 m), ambient noise
(spectrum level).   Data have been extracted from
numerous sources and redrawn.   Major references
include Albers, 1965, Urick, 1975 (traffic
noise); Piggott, 1964 (sea state); Heindsmann et
al., 1955 (rain noise).

Banner, 1968; Banner, 1972; Myrberg et al., 1972; Wenz, 1972;
Tavolga, 1974a; Myrberg et al., 1976) have shown that sound-levels
in coastal waters--including bays and harbors--vary greatly in
time and place (Wenz, 1964; Albers, 1965; Urick, 1975).   Also at
frequencies below 1000 Hz, such noise often constitutes a mixture
of different types of noise, each originating from different sources,
e.g., traffic (and industry), wind, and soniferous marine animals.
At frequencies higher than 1000 Hz, rain noise replaces the predominantly low frequency traffic (or industrial) noise as a contributing
factor.   Despite considerable variation in existing noise at given
locations from time to time, if conditions remain stable for reasonable periods, the resulting curves do show some predictive similarity
and constancy.
     Figure 1 provides a general synthesis of the ambient noise
conditions that prevail for shallow, coastal regions whose depths
are generally less than 70 m.   The levels attributed to respective
sea states have been taken from Piggott (1964), their slopes being
similar to those determined off the coast of Bimini, Bahamas, at
comparable states and recording depth (see Myrberg et al., 1969).
Such levels, being strongly dependent upon the wind, are generally
5 to 7 dB higher than corresponding levels for the same sea state
in deep waters (Knudsen et al., 1948; Wenz, 1962) especially at
frequencies above 500 Hz.   Below that point, the corresponding
deflections of the curves representing specific sea states are not
parallel.
     Corresponding levels for traffic (and industrial) noise across
the spectrum were taken from Albers (1965) and Urick (1975) with
the relatively flat, high-level, low-frequency extension between 10
and 50 Hz taken from Myrberg et al. (1976).   The nature of the

curves in this particular region has been recently reviewed by
Wenz (1972). The slopes are in reasonable agreement with those
obtained by Maniwa (1971) and Olsen (1971) from noise measurements
made in the vicinity of large fishing vessels. There can be
little doubt that under quiet or reasonable sea states, traffic
and/or industrial activity constitute in certain locations a major
source of noise in the two decades between 10 and 1000 Hz (see
Urick, 1975, for an insightful review).

Levels of rain noise were provided from Heindsmann et al.
(1955). Within the spectral region of influence, there is likewise no question that rain is a significant source of background
noise, at least near the surface.

The most difficult type of ambient noise to characterize
through time and space is that produced by soniferous animals.
At times, it can characterize, however, a given region. Biological sounds in the sea are many and varied; a few examples are
provided in Figure 1. These were chosen since their levels were
consistently high (as much as 50 dB above normal) for long
periods of time. Kaneohe Bay, Hawaii, and Bimini Bay, Bahamas,
are, for example, characterized by high-level ambient-noise above
2000 Hz, this resulting from enormous populations of snapping
shrimp. Such sounds, be they from fish, shrimp or sea urchins,
often span a broad frequency range with peaks of intensity
dependent on the animals involved. Finally, many such sources
are associated with particular seasons of the year, phases of
the moon, or times of the day (or night) (for review, see Cummings
et al., 1964 and Steinberg et al., 1965).

SOUND DETECTION AND LOCALIZATION BY FISHES AND MARINE MAMMALS

Recent reviews of our knowledge concerning the detection and
localization of sounds by fishes (Tavolga, 1971; Chapman, 1973;
Hawkins, 1973; Popper and Fay, 1973; Fay, 1974; Sand and Enger,
1974; Schuijf, 1974; Tavolga, 1974b; Hawkins and Chapman, 1975)
preclude the necessity of reviewing this broad subject here. I
shall therefore only briefly discuss a few topics that appear
relevant to the subject at hand, i.e., ambient noise. This will
also be the case regarding those mammals of interest--the
odontocete (toothed) cetaceans and pinnipeds (for recent reviews,
see Norris, 1969; Schusterman et al., 1972; and Evans, 1973).

Selected audiograms of three marine fishes and three marine
mammals are provided in Figure 2, the examples chosen in each
group possessing different levels of sensitivity. These allow
direct comparison of thresholds relative to spectral range and
sensitivity. An aerial audiogram for man is also included to
show that directly comparable levels of sensitivity are attained
by at least some species of fishes at lower frequencies and by
various species of marine mammals at a higher spectral range.

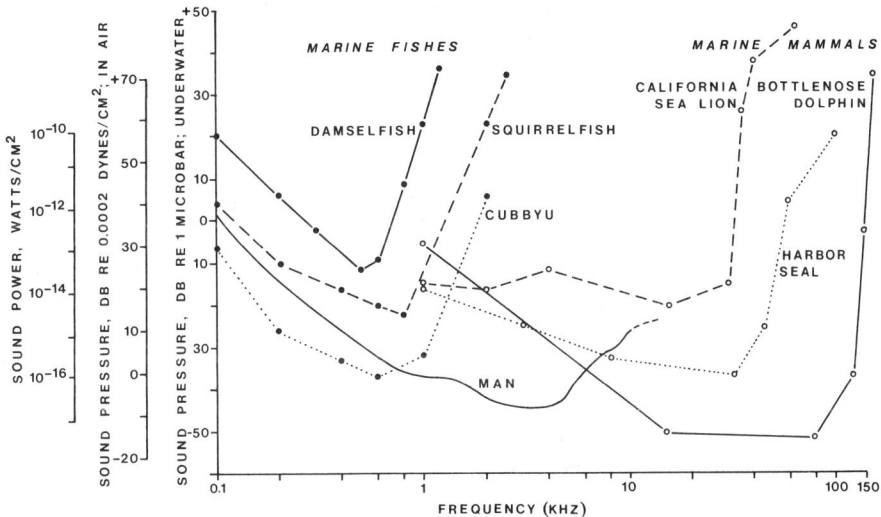

FIGURE 2:   A comparison of hearing curves among
selected marine fishes, marine mammals (authors
cited in text), and man (Sivian and White, 1933).

FISHES

    Hearing studies, carried out on many species found in widely
differing habitats, show that although certain marine fishes can
detect quite high frequencies (e.g., herrings--5 to 10 KHz) the
great majority are sensitive only to frequencies below 2000 Hz.
    It is difficult to form meaningful generalizations from the
audiograms presently available for more than thirty species of
marine fishes (many are provided in the review of Popper and Fay,
1973).   There are, nevertheless, certain patterns of relative
consistency.   Besides the dichotomy that is often made between
the auditory specialists (i.e., the cypriniform fishes and other
selected species) and the so-called non-specialists (i.e., those
showing a reduced range of spectral sensitivity along with a
general decreased sensitivity across that range, relative to the
other group), the latter can also be divided into two groups,
based largely on speculative candor and convenience.   The members
of Group 1 (Fig. 3) have a rather broad region of greatest sensitivity at frequencies generally between 75 and 300 Hz, while
those of Group 2 (Fig. 4) have a narrower range of greatest sensitivity around 400 to 800 Hz.   Also, although there are exceptions,
the ascending slopes of decreasing sensitivity above the regions of
peak sensitivity are rather steep (35 to 40 dB/octave) and similar
among many species from both groups regardless of relationship.
This function appears to be independent of environmental noise and
it is not unreasonable to suggest that a similar physiological
mechanism(s) is controlling the function in most species.   The
corresponding slopes below the regions of peak sensitivity in these

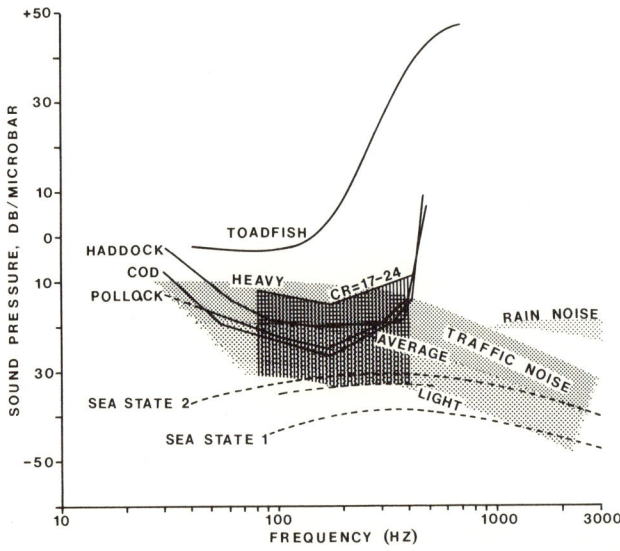

FIGURE 3: Low frequency ambient noise and its
probable masking effect on the hearing abilities
of selected marine fishes (Group 1 - see text),
whose peak sensitivities are found within that
spectrum. The four audiograms shown were determined either totally, or partially, in the
field. The hatched area is the region chosen to
show the amount of masking that would occur above
the arbitrarily chosen spectrum levels of sea
state (<2) and traffic noise (light, see Fig. 1)
for those species possessing the critical ratios
(CR) as given.

same fishes appear, on the other hand, to vary considerably at
times and there are strong suggestions that this function was
dependent upon the magnitude of ambient noise in that spectrum at
the time of testing. When noise is maintained at some constant
level at a given frequency, audiograms from different species
having similar regions of peak sensitivities show remarkably similar
slopes and associated sensitivities (Chapman, 1973; Myrberg and
Spires, 1977).

A contrasting study has been conducted by Tavolga and Wodinsky
(1963) on nine species of fishes common in the waters off Bimini,
Bahamas. These authors reported an extraordinary 60 dB variation
in sensitivity within the lowest frequency ranges tested (100 to
200 Hz) and yet many, if not all, of the species examined were from
the same general habitat. Large differentials were even present
among those species possessing similar peak sensitivities at the
same or similar frequencies. It is possible that the fishes
examined in that study were responding to the particle displacement
component of the acoustical stimulus at certain times and to the

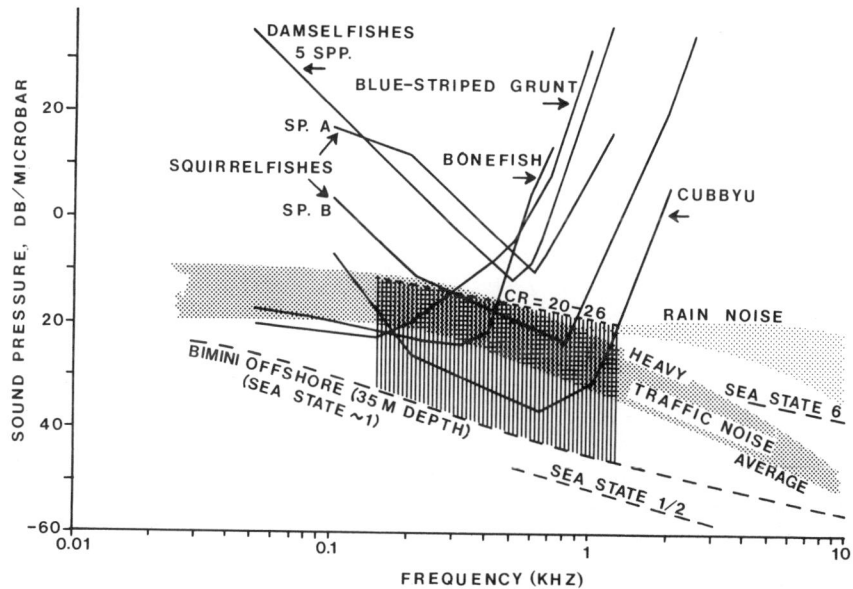

FIGURE 4: Low frequency ambient noise and its
probable masking effect on the hearing abilities
of selected marine fishes (Group 2 - see text),
whose peak sensitivities are found within that
spectrum. The hatched area is the region chosen
to show the amount of masking that would extend
above the arbitrarily chosen level of ambient
noise (in spectrum level). The ambient chosen
was offshore the island of North Bimini, Bahamas
where all the species reside. Audiograms were
selected from various authors (cited in text).

pressure component at other times. Shifting of the modality used
to determine threshold sensitivity would most likely occur at those
low frequencies and this could explain at least some of the variability noted. A somewhat more parsimonius explanation for such variation may be that differences in ambient noise were present in test
tanks or possibly even holding tanks, resulting in temporary threshold shifts (Ha, 1968).

Closely related species show similar hearing abilities, so long
as they are residing in a similar habitat (e.g., the cod-like fishes
of the family, Gadidae - Chapman, 1973; damselfishes of the genus
Eupomacentrus - Myrberg and Spires, 1977). Apparent exceptions are
the squirrelfishes of the genus *Holocentrus* examined by Tavolga and
Wodinsky (1963).

Another trend noted among the audiograms of fishes is one
suggesting that the less sensitive a species is to sound (i.e., the
more distant its sensitivity is from probable influence by ambient
noise), the steeper is its low frequency slope. Although there

again appear to be exceptions, this can be noted in the figures
supplied in various reviews, especially that by Popper and Fay
(1973). This, in turn, suggests that the low frequency slopes of
those species possessing good hearing abilities are being affected
by some factor which is causing a 'flattened' range of sensitivity
in that portion of their audible spectrum. The thesis developed
below is that ambient noise is that factor.

The long-standing question of sound localization in the acoustic far-field by fishes appears recently to have been answered at
least in part for some species of teleosts and elasmobranchs.
Strong evidence from a number of recent studies has favored such
localization (Myrberg et al., 1969; Nelson, 1969; Olsen, 1969;
Myrberg et al., 1972; Nelson and Johnson, 1972; Schuijf et al.,
1972; Chapman, 1973; Enger et al., 1973; Schuijf and Siemelink,
1974; Myrberg et al., 1976). Present evidence also indicates that
for fishes with gas bladders (e.g., the cod), sound-detection
thresholds are below those required for sound-localization (Chapman,
1973). This relationship apparently differs, however, in fishes
without gas bladders (e.g., the lemon shark) where one threshold
served both functions (Banner, 1972).

ODONTOCETE CETACEANS AND PINNIPEDS

Relatively little quantitative information is available on the
auditory sensitivity of marine mammals. Sound-detection thresholds
have been obtained over a wide range of frequencies for only five
species of odontocetes: four marine species--the bottlenose dolphin
Tursiops truncatus (Johnson, 1966, 1967), the harbor porpoise
Phocoena phocoena (Andersen, 1970), the common porpoise *Delphinus
delphis* (Belkovich and Solntseva, 1970, in Hall and Johnson, 1971)
and the killer whale *Orcinus orca* (Hall and Johnson, 1971) and one
freshwater species--the Amazon River dolphin *Inia geoffrensis*
(Jacobs and Hall, 1972). In each case only single animals were
studied. Morozov has, however, recently confirmed Johnson's results
on the bottlenose dolphin (W. Evans, personal communication). Five
species of pinnipeds have been studied in like-manner. Schusterman
(1974) and Schusterman et al. (1972) have established air and underwater audiograms for the single otariid studied, the California sea
lion *Zalophus californianus* while Terhune and Ronald (1971, 1972)
and Møhl (1968) have carried out similar studies on two phocids, the
harp seal *Pagophilus groenlandicus* and the harbor seal *Phoca
vitulina*, respectively. Additionally, underwater audiograms have
been determined for the ringed seal *Pusa hispida* (Terhune and Ronald,
1975a) and the gray seal *Halichoerus grypus* (Ridgway and Joyce, 1975).
As in the case of the toothed whales, only one, or at most, two
individuals of each pinniped species were examined. Nonetheless, the
results, to date, show a number of clear correlates within the
respective taxonomic groupings, as well as among the group as a whole,
despite differences in methodologies.

Although the various odontocetes have shown different low
frequency limits, all have demonstrated extremely high sensitivity
to a wide range of high frequency sounds. This is exemplified by

the harbor porpoise whose range extends below 8 KHz to 140-150 KHz.
All species appear to have an extremely rapid cut-off in sensitivity
at their upper hearing limit (Møhl, 1968; Andersen, 1970).

The regions of peak sensitivity fit well the animals' own
signal characteristics (e.g., Diercks et al., 1973, but see Møhl
and Andersen, 1973 and Watkins, 1974), this fact correlating well
with that noted in fishes and other animal groups where similar
data are available. It is well known that marine mammals possess
a broad repertoire of widely differing sounds; but the most cele-
brated ones, at least in the case of the odontocetes, are the
echolocation-clicks. There is little doubt that the high frequency character of these brief, high energy pulses is reflected
in the high frequency sensitivity of these animals.

When considering pinniped audibility, the remarkable similarity shown by the four phocid species (i.e., no difference in
sensitivity greater than 20 dB at any frequency) suggests that
one may refer to a general phocid audiogram. Although Schusterman
et al. (1972) considered that differences in the sensitivities
existing between their otariid curve and the only phocid curve
available to them at the time might actually be due to methodo-
logical differences and individual variation, some investigators
presently believe otherwise (e.g., Terhune and Ronald, 1975a).
The underwater audiogram of the sea lion shows an extremely broad
region of highest sensitivity ($\sim$ -15 dB/µbar) between 1 and about
28 KHz. Between the latter point and 36 KHz, decreasing sensitivity attains a slope of about 60 dB/octave. This same rate of
loss in sensitivity was also noted in the phocid seals between 45-
50 and 64 KHz. An interesting leveling-off of this slope to about
12-14 dB/octave occurred, however, in both groups near their upper
limit of hearing (the sea lion, near 32 KHz; the phocids, around 64
KHz). This phenomenon has been considered by Møhl (1968) and
Schusterman et al. (1972) to be 'pseudo-hearing' of ultrasonic
sound through bone conduction ('pseudo-hearing' meaning that only
intensity differences are perceived among given sounds).

Although the differences between phocid and otariid hearing
abilities may well be real, there appears to be sufficient similarity to justify a general comparison of audibility between them and
the odontocetes. The odontocetes appear to have a 20 to 30 dB
superiority over the pinnipeds in their respective regions of
greatest sensitivity and, except for the killer whale (high frequency cut-off similar to that of the sea lion, i.e., about 30 KHz),
their audibility extends 1 to 2.5 octaves beyond that of the
pinnipeds (ignoring here the region of pseudo-hearing). These
differences are no doubt due in large part to the echolocating
capability of at least certain odontocetes and as Schusterman
et al. (1972) have appropriately stated: "...It does not seem
unreasonable to assume that echolocation was a major source of
selective pressure for excellent hearing sensitivity at frequencies
above 60 KHz..." The development of high frequency audibility in
seals may have been confined by their limited use of echolocation
as well as their amphibious nature which requires that they possess
relatively good hearing both underwater and in air (for review, see
Schusterman, 1974).

AUDIBILITY AND ENVIRONMENTAL NOISE: THE RELATIONSHIPS

When considering the effects of ocean noise upon audibility
among both the marine fishes and mammals, the frequency range
involved is extensive. The most important region of sound detection
in most fishes rests between about 40 and 1000 Hz, while in the
pinnipeds and odontocetes, it is located between 500 Hz and 30 to
45 KHz and between 8 KHz and 120 to 145 KHz, respectively.

I have divided this portion of the report into three sections.
The first deals with those fishes whose hearing sensitivity rests
in what may be termed the extremely low register, i.e., between 10
and 500 Hz. These fishes, including the cod and its relatives (e.g.,
the haddock, pollack and ling), the toadfish and the sharks, all
appear keenly adapted to this particular range of frequencies, the
teleosts because of their own respective signal characteristics
(Brawn, 1961; Gray and Winn, 1961; Winn, 1967; Fish and Offutt, 1972)
and the elasmobranchs because of the various sounds produced by their
prey (Nelson and Gruber, 1963; Myrberg et al., 1976; Nelson and
Johnson, 1976; Myrberg, 1978).

The second group is made up of those marine fishes whose peak
sensitivities extend from about 200 to 1000 Hz. The majority of
teleost fishes appears to belong to this group and many of these in
turn belong to the advanced and highly radiated taxonomic order, the
Perciformes. Although there certainly are glaring exceptions, this
rather generalized group constitutes the most speciose and populous
fishes in shallow water. Their domain includes rocky coastlines,
mangrove flats, grassy areas and the coral reefs.

Based on our present knowledge, the third group--marine
mammals--can be divided into three sub-groups: the otariid and
phocid pinnipeds and the odontocetes (the delphinids and the single
phocoenid studied appear to possess more similarities than differences
in their auditory capabilities). All three sub-groups have audio-
sensitivities that extend beyond 20 KHz. Their habitats include
generally the coastal and littoral regions, although in the case of
two delphinids--the common porpoise and the killer whale, the preferred habitats appear to be over deep, oceanic waters (former) and
either over deep water or the littoral region (latter) (Evans, 1973).

GROUP 1

Figure 3 depicts various ambient noise effects in relation to
auditory thresholds from four selected species whose sensitivities
span most of the frequency spectrum shown. It is clear that the
threshold sensitivities obtained by Chapman (1973) for the cod,
the haddock, and the pollack (all shown in the Figure) are close
to, and could easily be exceeded by, the ambient noise. These
three thresholds (a fourth, not shown, was from the ling), are
especially informative since they were all obtained directly in
the field off the Scottish coast and their associated ambient
levels fitted well those recorded by Piggott (1964) also off the
Scottish coast. Chapman mentioned that there was a direct correlation between thresholds for the haddock and the spectrum level
of the noise. This same correlation was also reported for the

cod (Chapman and Hawkins, 1973) and based on many determinations,
sea state appeared to be the causative factor. Their results
showed unquestionably that masking of threshold by ambient noise,
although negligible in calm sea conditions, invariably occurred
at higher sea states. The frequencies not affected by such noise
levels were invariably those at, or near, the respective ends of
the hearing curves where the animals' sensitivities were sufficiently low to preclude further impairment or masking by noise. The
remaining threshold curve shown in Figure 3 is from the toadfish.
It was initially determined by Fish and Offutt (1972) in the laboratory with specific points later confirmed in the field. The species
has low sensitivity compared to that of the gadoid fishes at similar
frequencies; but it is likely that reasonable levels of ambient
noise would also mask its most sensitive region.

In the above studies, the investigators also found that a
specific relationship apparently existed between a given frequency
threshold and the spectrum level noise at that same point. This
was subsequently investigated in the cod by Hawkins and Chapman
(1975), following the leads by Tavolga (1967) and Buerkle (1968,
1969) that ambient noise could mask the detection of sounds by
fishes.

An equally illuminating study by Banner (1972) also demonstrated that the hearing capability of lemon sharks *Negaprion
brevirostris* depends on a remarkably consistent relationship
between signal and ambient noise. Signal to noise (spectrum
level) capability measured in the laboratory at two different
frequencies was shown to be statistically identical to the
signal to noise response levels shown by free-ranging lemon
sharks in the field. Since he had previously shown (Banner,
1967) that hearing thresholds for the species were dependent
on particle motion rather than pressure, it is interesting that
identical signal to noise capabilities were found despite the
fact that the signals and associated noise in the field apparently
were directed vertically (shallow water) whereas those in the
laboratory were predominantly horizontal.

The above studies, plus others by Fay (1974), Ha (1968) and
Tavolga (1974b) have established beyond question that, in fishes,
a specific relationship does exist between the spectrum level
noise at a given frequency and the detectability of a signal at,
or near, that frequency. This relationship does not hold, of
course, when detection is not impaired by noise, e.g., when the
level of background noise is decreased after absolute threshold
is reached.

The relationship between a masked threshold determination
and its corresponding spectrum level noise has been variously
termed: signal/noise ratio (Tavolga, 1974b), threshold: noise
ratio (Hawkins and Chapman, 1975), and the critical ratio : CR
(Albers in Ha, 1968; Ha, 1968; Johnson, 1968; Fay, 1974; Terhune
and Ronald, 1975b). Although CRs (in Hz) were interpreted by
Hawkins and Chapman (1975) to be calculated values for the critical
band rather than empirically derived values, it is apparent that
others use the term to refer specifically to the relationship (in
dB) described above.

Perusal of the literature regarding the constancy of the

above ratios resulted in Table 1. These specific threshold:
spectrum level noise values, or critical ratios (CRs), ranged from
14 to 29 for all fishes, except the goldfish (the single specialist
listed). A clearly suggestive trend is also apparent regarding the
values for a given species, i.e., a general (but not universal)
increase as frequency increases, e.g., cod, (Buerkle, 1968). This
same directional trend is also seen among the values for man and
the marine mammals. Finally, as Fay (1974) has pointed out, since
the CRs are indirect estimates of the frequency range over which
two stimuli interfere with one another (Licklider, 1959), a growth
in the size of the CR with frequency indicates that the resolving
power of the system is declining.

Based on these data, specific values were assigned various
frequencies across the best hearing range of the cod. These were
then added to the spectrum level noise of an arbitrary sea state
(slightly less than sea state 2) at the same frequencies. A light
level of traffic (industrial) noise coinciding with the selected
sea state was then added. The resulting simulation is shown in
Figure 3. From this simulation, it was apparent that if the
thresholds had been determined under the simulated conditions, a
line of masked thresholds would have resulted (=line drawn below
CR=17 24). The unique shape of the simulated hearing curve retains
almost precisely the curves of the actual thresholds obtained for
the cod and pollack by Chapman (1973). This suggested that thresholds in that region may well be affected not only by wind, but also
by traffic (or industrial) noise as well. The particular simulation
used is a reasonable one from the standpoint of ambient noise and it
clearly substantiates the statement made by Chapman and Hawkins
(1973) that only a calm sea produced reliable thresholds. Even at
sea state 1, with extremely light traffic noise, the cod's threshold
would probably have been masked. It is also clear that any increase
in ambient noise at such low frequencies should have a greater
effect on audibility than a similar increase at higher frequencies.
This probably accounts for the wider and flatter region of good
sensitivity in this portion of the spectrum compared to that seen
in higher regions. The above implies that fishes do experience
hearing loss. This particular point, recently examined in a few
species, will be briefly discussed later in this report.

Before leaving this section, it should be pointed out that in
those animals sensitive to particle-motion (e.g., sharks), direction-
ality of ambient noise may also be an important factor. A fish with
directionally sensitive hearing might actually discriminate a sound
source from background noise by taking advantage of differences in
the orientation of source and the ambient noise particle-motion. If
so, one must account for near-field effects of both types of sound
(see Banner, 1968). Under such circumstances, a displacement-sensi-
tive receptor might be at a disadvantage unless the direction of the
particle-motion was not the same as that of the source. In any case,
it is apparent that both signal and noise must be measured in the
same manner and under the same conditions as experienced by the test
animal.

TABLE 1: Auditory threshold: spectrum-level noise (or critical ratio) (dB) in the presence of masking noise.

	$\times 10^1$ Hz				$\times 10^2$ Hz								$\times 10^3$ Hz						$\times 10^4$ Hz			
	4	5	6	7	1	1.5	2	3	4	5	7	8	1	1.2	2	4	6	8	1.6	3.2	6	8
Man in Air (Hawkins & Stevens '50)					19	18		16			17		18		20	23		27				
Marine Mammals																						
Ringed seal (Terhune & Ronald '75b)																			32	34	35	
Bottlenose dolphin (Johnson '68)																30	22	25	32	32	37	39
Marine Fishes																						
Cod (Hawkins & Chapman '75)	18	17	16		16		19	21	21													
Cod (Buerkle '68)				21		24			21													
Ling (Chapman '73)	17		24		24		25	25	26													
Haddock (Chapman '73)	21		25		21		23	22														
Pollack (Chapman '73)	21				22			26														
Blue-striped grunt (Tavolga '74b)								20		24												
Pinfish (Tavolga '74b)										29												
Longspine squirrelfish (Tavolga '74b)										14												
Lemon shark (Banner '72)	21											25										
Freshwater Fishes																						
Goldfish (Tavolga '74b)		8			14	22		23		23	23											
Goldfish (Fay '74)					13	17		19	23	23	22			25								
Black-chinned mouth-brooder (Tavolga '74b)								23		24												

GROUP 2

Audiograms of ten species of marine fishes (Tavolga and Wodinsky 1963, 1965; Tavolga, 1974a; Myrberg and Spires, 1977) are shown in Figure 4, along with ambient noise effects within the selected spectrum. The spectrum level noise of a relatively calm sea state (approaching 1 over a depth of 35 m; Myrberg et al., 1969) at Bimini, Bahamas is also added since all of the included species are found in the waters off that island. It is apparent that wind (sea state) and traffic noise are again the major determinants of sea noise in the spectrum being considered. Biological noise would, of course, add to any effect.

Those species having greatest sensitivity in the region designated for this group have shown, with only one exception to date, that their critical ratios fall between 19 and 26 dB (Table 1). These values are somewhat greater than those obtained from the few species that were examined in Group 1. Although the groupings are based largely upon convenience and many more values are needed to confirm the indicative trend, species of Group 2 apparently do possess a somewhat reduced resolving power compared to those of Group 1. Based on the reasoning by Fletcher (1940) and subsequent authors dealing with mammalian hearing, the maximum band of frequencies which can effectively mask a given tone, i.e., the critical band, recently has been examined in fishes by various investigators. Although controversy and speculation presently abound regarding the usefulness of the concept in fishes, I agree with Hawkins and Chapman (1975) that there are advantages to be gained by describing the masking function in terms of an effective band of finite width.

Since the level of noise within a band width Δf_c is equal to the spectrum level plus $10 \log_{10} \Delta f_c$, then Δf_c equals the critical band. By using the CRs provided in Table 1, it can easily be determined that such bands, if present, would be 40 to 160 Hz wide for frequencies below 100 Hz and from 80 to 400 Hz wide for frequencies between 100 and 1000 Hz.

When a simulation, such as that shown in Figure 3 is applied to the data provided in Figure 4, using as its base the noise level for a relatively calm sea state off Bimini, it is again apparent that the hearing abilities of the most sensitive species will be impaired to a level similar to those species initially far less sensitive than they. Also any increase in ambient noise above that of a calm sea probably would affect all the thresholds shown in Figure 4. The simulation points out that statements concerning the auditory capacities of fishes must be related to background noise.

A comparison of the respective shapes of the audiograms shown in Figure 4 shows that the more sensitive a species is to sound, the less steep is its ascending slope of sensitivity in the lower frequencies. This indicates that even under quiet conditions of the laboratory, thresholds in that region are still being influenced by extraneous noise.

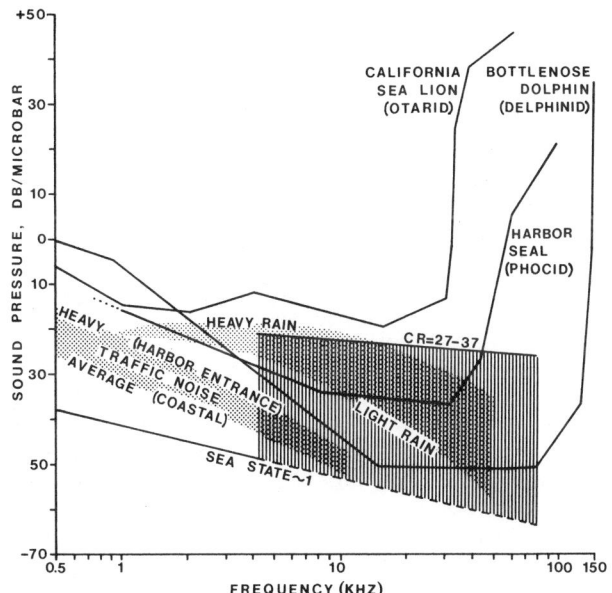

FIGURE 5: High frequency ambient noise and its
probable masking effect on the hearing abilities
of selected marine mammals whose peak sensitivities are found within that spectrum. Audiograms
were redrawn from several authors (cited in text).
The hatched area is the region chosen to show the
amount of masking that would extend above the arbitrarily chosen level of ambient noise (in spectrum level) with the critical ratios (CR) provided.

GROUP 3

Evidence shows that all marine mammals produce a variety of
sounds. Hearing capability has been examined, however, only among
certain odontocetes and pinnipeds. Representative audiograms are
shown in Figure 5 and their divergencies reflect the taxonomic
divergence of the groups. Although these mammals certainly perceive
underwater signals below 500 Hz as well as a variety of air-borne
sounds, discussion will be limited to their higher frequency, underwater capability.

The audiogram of the California sea lion is the only one available for the family Otariidae (Schusterman et al., 1972). It is
essentially flat between 1 and 30 KHz. This stands in contrast to
the available audiograms of the harbor seal, representing the Phoci-
dae and that of the bottlenose dolphin, representing the odontocetes
(Johnson 1966, 1967). It is probably not coincidental that the slope
of decreasing sensitivity for the harbor seal parallels reasonably
well the slope of ambient noise in the same region. Although the

equivalent slope for the dolphin departs somewhat from that normally associated with noise, it nevertheless indicates that some type of traffic (or industrial) noise may well have affected sensitivity during testing.

The 15 dB increase in peak sensitivity of the phocid threshold over that of the otariid suggests a real difference in capability between members of these two families. This is borne out especially well by the great similarity in sensitivity recorded from all the phocids studied thus far. When considering the effects of ambient noise on hearing in these two groups, taxonomic or physiological differences may be masked by the reality of the underwater world in which these animals live. Schusterman (personal communication) points out, however, that data being obtained on the northern fur seal, *Callorhinus ursinus*, suggests no difference in absolute sensitivity from *Phoca*.

Critical ratios have been determined for only one pinniped. Terhune and Ronald's elegant study (1975b) of the ringed seal provides the unique opportunity to do a bit more than merely speculate about the possible effects of noise upon the hearing abilities of these animals and their near relatives. Results showed that the CRs varied from 30 to 35 dB between 4 and 32 KHz (Table 1). When these values (increasing with higher frequencies) are added to a relatively low ambient level such as that provided by a sea state 1, the actual thresholds for the ringed seal, as well as those of the harbor seal and the remaining phocids, become masked, with the resulting sensitivities approaching that of the less sensitive sea lion. This can be seen in the simulation provided in Figure 5 (a similar interpretation is made also by Terhune and Ronald, 1975b). Slightly different CRs were used so that the simulation might also be applicable to the odontocetes.

If masking takes place under a calm sea, the effects of any traffic or industrial noise above that level should add to a further decrease in sensitivity. Those sources plus that of sea state are reduced to extremely low levels, however, when frequencies beyond 100 KHz are reached. Rain noise, though an intermittent variable, may, however, play havoc with hearing during certain times of the day or during certain seasons of the year. When critical ratios are taken into account during periods of even moderate rain, hearing in phocids could decrease to levels of +5 to +10 dB/μbar over a broad range of frequencies. And since the CRs of the odontocetes are not greatly dissimilar from those of the phocids, it is reasonable to assume that those of the otariids will also be similar.

Sound emissions by the otariid and phocid pinnipeds have been described by various authors (Schevill et al., 1963; Evans, 1967; Poulter, 1968; Schusterman, 1968; Schusterman and Balliet, 1969; Schusterman et al., 1970). Many of the sounds emitted in air are also emitted underwater. Although there is much energy below 2 KHz, a reasonable amount exists also in the 8 to 12 KHz range (e.g., barks--Otariidae: roars, moos, yodels, hums and clicks--Phocidae). Within the latter range, ambient noise will probably regulate sensitivity.

The excellent sensitivity of various odontocetes at frequencies between 50 and 145 KHz correlates well with the known echolocating

emission by these animals (see Evans, 1973). Although rain noise
(Fig. 5) and certain biological sounds (e.g., shrimp, see Fig. 1) do
constitute major components of noise at frequencies between 20 and
100 KHz (and even higher, see Au et al., 1974), the major sustaining
factor affecting sensitivity in these animals should be their critical
ratios. Fortunately, we again have a single, but important study that
deals specifically with these values in the bottlenose dolphin
(Johnson, 1968). The CRs reported in that study are also provided in
Table 1 and they were used in part for the simulation provided in
Figure 5. Based on the high values of these CRs, it is apparent that
the actual thresholds of these animals will also be masked, with
resulting sensitivity being only slightly below those of the pinnipeds.
Although the degree of apparent masking is considerable, the reduced
sensitivity still appears fully sufficient for the task of echolocating
(see below). At frequencies below 15 KHz, background noise increases,
but this is compensated by decreasing values of the CR. This interesting interplay between an animal's sensitivity and specific factors
in its environment may well have implications relative to the numerous
sounds, other than those of echolocation, produced by these highly
soniferous animals.

AUDIBILITY AND ENVIRONMENTAL NOISE: THE IMPLICATIONS

Marine Fishes

It is reasonable to assume that in the shallow, coastal regions
of the oceans, sound reception by the majority of fishes is limited
by environmental noise. Freedom from such an influence is probably
enjoyed by members of a given species only at the extremes of their
hearing range, where physiological restraints then set the limits.
Although various studies in the past referred to a rather
specific relationship existing between threshold sensitivities and
spectrum level noise, i.e., a 20 dB (+ a few dB) difference at
frequencies of a few hundred hertz (Banner, 1967; Nelson, 1967;
Buerkle, 1969; Cahn et al., 1969; Myrberg et al., 1969), consideration of ambient noise and particularly its spectrum levels was,
until recently, often relegated to a non-empirical factor that had
only to be discussed away. Now that this specific relationship has
been firmly established by a number of studies, it is clear that
this critical ratio is a clear indicator of masking. If its value
at any given frequency remains relatively constant, sensitivity is
most probably being masked; if its value exceeds the point of
constancy, masking is unlikely. The reason for the older studies
often referring to a rather similar value is now obvious--they
were dealing with masked sensitivities.
While discussing masking by ambient noise in fishes, Tavolga
(1967) mentioned that its effects in nature were probably minor,
except under highly unusual conditions for a few species with unusually good hearing. I plead here the contrary: the effects of
ambient noise are paramount to hearing abilities except under the
often unusual condition of a very calm sea with less than a 'whisper'
of mechanical (i.e., traffic or industrial) noise. When seas of

state 1 or higher are noted, when traffic or industrial noise is
apparent, or when other low frequency sound sources such as
shifting bottoms, tidal currents, or noisy animals are encountered, sound reception in all but the deafest fishes is probably impaired, the degree of such impairment being directly dependent upon the noise-level attained. This particular point
has rather far ranging implications.

Underwater ambient noise is characterized by a Gaussian amplitude spectrum, while sounds of most, if not all, marine vertebrates
often possess recognizable frequency spectra, along with appropriate
modulations. Therefore, those animals which rely on acoustic reception probably utilize some form of signal processing so as to perceive specific (biologically important) signals even at low amplitudes. The spectral content of given signals, their redundancy and
relative stability all provide ways and means to accomplish such
feats. Nevertheless, in the following section, I shall speculate
upon findings from pure tone sensitivities with inferences made
about actual sounds and their reception by the animals concerned.
Some readers will probably consider such actions unwarranted because
of the absence of confirming evidence. This point is especially
justified in the case of the marine mammals since their hearing
abilities cover such a broad frequency range that differential
masking may actually pose little or no problem to them. Yet, with
our present knowledge, I do not think it is unreasonable to suggest
that the possibility that ambient noise can affect biological signals appears to be changing to probabilities, the levels of which
must be determined in the future by concerned scientists. By using
an extremely simplified and conservative approach, the findings
provided below would represent actuality only if signal processing
was at a minimum. If, on the other hand, complexity characterizes
such processing, sound-detection limits would clearly exceed those
given below. For purposes of this symposium, I am going far out on
a limb of speculation and the limb will most assuredly be cut off
behind me; how close the cut is to the trunk (or to my position)
remains to be seen.

Although we know that the vast majority of fishes makes sounds,
detailed knowledge about the biological significance of sound production is scant except in a few instances. Four such instances
are provided in Table 2, along with certain facts and some rather
conservative estimates about sound production and reception. Simple
calculations have, accordingly, revealed information relative not only
to acoustical biology but also to other problems seemingly unrelated
to the modality of sound.

The major bio-acoustical function in the case of the first three
species listed in Table 2 is intraspecific communication. Sounds
inform conspecifics that the sender is physiologically ready for
specific types of activities, be they reproductive, agonistic or
whatever. Among nearby conspecifics, there will be receivers who, in
turn, are also physiologically ready for the same types of activities
and it is they who respond. Since it is imperative in such situations
that the message from the sender reach a receiver, the actual distance
between such individuals is important. If ambient noise adversely
affects the level of the signal (message), communication is precluded

TABLE 2: Estimated sound-detection-distances under different ocean-noise conditions for selected species of marine fishes. Conspecific source levels used in all calculations, except for those involving the lemon shark (see text); audio frequencies selected from regions of peak energy for the respective sound sources.

Species	Sound-Source Level (dB/μbar re 1 m)	Selected Audio-Frequency (Hz)	Audio Threshold; Spectrum Level Noise Ratio (dB)	At Sea State	Most Sensitive Audio Threshold (dB/μbar)	Estimated Maximum Detection Distance (Meters)	At Traffic Level (Sea State 1)	Most Sensitive Audio Threshold (dB/μbar)	Estimated Maximum Detection Distance (Meters)
Eupomacentrus partitus Bicolor damselfish	+ 7	500	23*	1 2	-12 - 7	9 5	Light Average Heavy	-12 - 3 + 6	9.0 3.5 1.0
Holocentrus rufus Longspine squirrelfish	+13	600	23	1 2	-16 - 8	30 12	Light Average Heavy	-13 - 4 + 5	20.0 7.5 1.5
Opsanus tau Toadfish	+35	100	17*	1 2	- 2 - 2	75 75	Light Average Heavy	- 2 - 2 + 7	75.0 75.0 30.0
Negaprion brevirostris Lemon shark	+30*	300	20	1 2	-13 -10	150 105	Light Average Heavy	-12 - 3 + 6	130.0 50.0 20.0

*Assumes values (see text)

or, at least, interrupted. When such a signal is involved in initiat-
ing and maintaining the courtship ritual, as in the bicolor damselfish
(see Myrberg, 1972; Myrberg and Spires, 1972), reproductive activity
may well be curtailed.

The fourth example, a shark, being a predator, would find it to
his advantage to hear and interpret the sounds of his prey. Numerous
studies have, in fact, shown that sharks clearly use their sense of
hearing for that purpose (e.g., Nelson and Gruber, 1963; Banner, 1972;
Myrberg et al., 1972; Nelson and Johnson, 1972; Myrberg et al., 1976;
Nelson and Johnson, 1976; Myrberg, in press). Again, such information transfer being distance-related can be affected by the level of
ambient noise. If prey detection and localization is dependent on
the hearing modality, the stake has become the severest of games--
survival.

All values within the second through the fourth columns of
Table 2 are published except where asterisks are used. Although an
audiogram is available for the bicolor damselfish (Ha, 1973), a study
dealing with proven masked thresholds has not been done. Therefore,
an estimated CR was supplied, its value being the mean of those
which were obtained on other teleosts at 500 Hz. This procedure was
also used to supply an appropriate CR to the toadfish. There is no
published sound-source level for the longspine squirrelfish and so
an arbitrary level was also assigned in that instance, it being
almost twice the level of the sound produced by the bicolor damselfish (personal experience with both sounds in the field indicates
that the level is reasonable). The sound-source level used for the
lemon shark is that of a natural prey-sound ("stampeding" bonefish,
Albula vulpes) measured and used by Nelson and Johnson (1970) in
attraction studies on sharks.

The calculated values in the tables are based on established
audiograms for each species, and by applying the appropriate CRs
to levels of the various types of noise, maximum-detection distances (in meters) were determined. Such determinations are the
first attempts to provide a most conservative idea of how far an
individual must send its message so as to reach an appropriate
receiver. Since fishes are apparently unable to vary appreciably
the sound-level of their signals, the distance over which a
sender transmits its signal to an appropriate receiver should
remain relatively constant (i.e., all else being equal).

In the case of the bicolor damselfish, large, reproductively
active males within a colony maintain territories of such a size
that the nearest large males are between 3 and 5 m away. Since
the courtship sounds of one such male in its territory cause nearby males to begin competitive courtship rituals, it is noteworthy
that such distances also happen to be the maximum distances of
detection for their respective sounds under conditions that are
normal for our region (i.e., sea state 2 with a moderate level of
traffic).

The distance over which the sounds of the longspine squirrelfish travel exceeds that normally defended by such animals. It is
reasonable to assume, however, that their sounds are used for
functions that demand either operative distances of 8 to 12 m (for
conditions normal to our region) or high levels close to their

residences. Toadfish produce some of the loudest sounds made by
fishes. Since males often reside near one another, indications
are that the corresponding levels are used for the purpose of
attracting females over a wide area. Except under conditions of
heavy traffic, that area should have a radius of approximately
75 m under reasonably quiet sea states.

In the case of attracting a shark to the source of a relatively low, biologically interesting sound, both detection and
localization (see Banner, 1972) should occur between 100 and
150 m away under reasonable sea states and with light traffic.
It is noteworthy that under a calm sea, the maximum proven distance over which carcharhinid sharks have been attracted to
similar sounds is between 300 and 400 m.

Thus, based only on these four examples, social function
and even predation may well be dependent for their success on
the level of environmental noise. Signals are enhanced as
noise is reduced and vice versa; slight changes in environmental
states result in clear changes in receptive fields.

Although sound reception has been emphasized in this section some evidence has accumulated that sound transmission itself is adversely affected by noise. Winn (1967) obtained, for
example, clear evidence during a field study of vocal facilitation in the toadfish that members of the species significantly
reduced their calling rate when background noise, normal to the
area, was played back at levels far less than that of the
species' own sounds (which increased calling rates). This
finding indicates that fishes can perceive differences in sea
noise. The fact that fishes are limited in their hearing by the
level of ambient noise also suggests that they are able to detect differences in that noise.

Still another implication of the results, summarized here,
has to do with commercial fisheries. The sounds of fishing
vessels and their associated gear can probably be heard by
their potential catch when levels (now 'signals') reach some
point greater than the background noise (see Maniwa, 1971; Olsen,
1971). Since response thresholds and those of detection need
not necessarily be the same, further investigations regarding
this important factor in coastal and high seas fisheries are
clearly warranted.

Similarly, important questions to be answered in the future
are whether marine animals suffer temporary or even permanent
hearing loss when subjected to unusually noisy environments for
long periods of time or whether they avoid or leave such locali-
ties. Although almost no information is presently available
regarding these questions, Popper and Clarke (1976) have recently shown that goldfish experience temporary hearing loss (about
24 hours) after being exposed to intense signals of approximately
+49 dB/μbar (equivalent to +123 dB/0.0002 μbar) for four hours.
Such losses are manifested by appropriate temporary threshold
shifts. Similar findings, using less intense signals, have also
been shown for the lane snapper, *Lutjanus synagris* by Ha (1968).

The final implication to be mentioned is that based on
Banner and Hyatt's study (1973) of acoustical noise and its

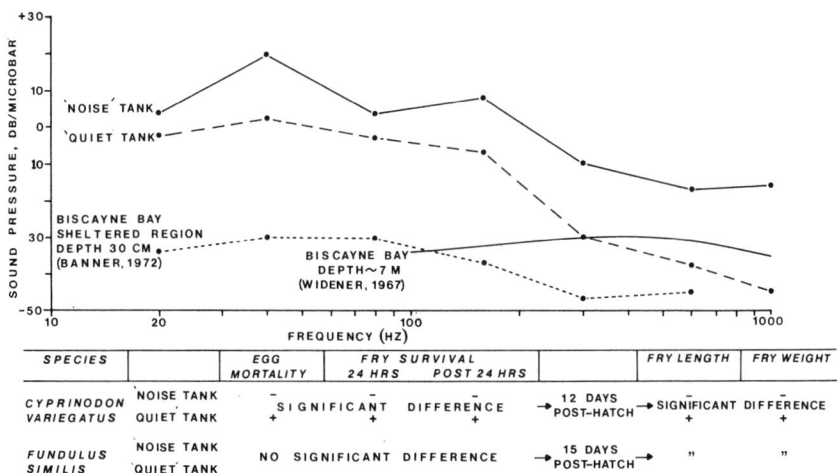

SPECIES		EGG MORTALITY	FRY SURVIVAL 24 HRS	POST 24 HRS		FRY LENGTH	FRY WEIGHT
CYPRINODON VARIEGATUS	NOISE TANK	–	– SIGNIFICANT	DIFFERENCE –	→ 12 DAYS POST-HATCH →	SIGNIFICANT DIFFERENCE	–
	QUIET TANK	+	+	+		+	+
FUNDULUS SIMILIS	NOISE TANK		NO SIGNIFICANT DIFFERENCE		→ 15 DAYS POST-HATCH →	"	"
	QUIET TANK						

FIGURE 6: Effects of sustained noise on egg mortality and fry survival and growth in two species of estuarine fishes (*Cyprinodon variegatus* and *Fundulus similis*). Data redrawn from Banner and Hyatt, 1973. Ambient noise portrayed in spectrum levels.

effects on the development of two common estuarine fishes (Fig. 6). These authors noted that under controlled testing viability of the eggs of *Cyprinodon variegatus* was significantly reduced in aquaria when a low frequency (40 to 1000 Hz) noise level, approximately 40 to 50 dB above that normally encountered in the natural habitat, was maintained over a number of consecutive days. Additionally, growth rates of fry in that same species, as well as in *Fundulus similis*, were significantly less than those noted when noise levels were reduced by about 20 dB during the same time period. These results are reminiscent of the pathological effects of excessive noise which have been observed in various mammals, including man, during recent years (e.g., Athey, 1970; Fry et al., 1970; Miller et al., 1971). This area of interest will unquestionably expand in importance as economic development continues to increase within, or adjacent to, aquatic regions.

The Marine Mammals

The following simulations are provided only to emphasize concern for the problems of sound detection by these animals and to touch on the complexity of the total problem that might face these animals in the acoustical domain.

Tables 3 and 4 follow basically the same format as that used in Table 2. To standardize sound source levels, all values have been expressed in peak to peak values (to obtain rms values, subtract 20 $[\log_{10} 2(1/\sqrt{2})]$).

TABLE 3: Estimated sound-detection distances
under different ocean-noise conditions for the
harbor seal, *Phoca vitulina*. Selected audio-
frequency, 9 KHz (see text); audio-threshold:
spectrum level noise ratio = 30 dB. Sound-
source level (p-p) in dB/μbar re 1 m (e.g.,
conspecific) = +38.

	At Sea-State	
	1	2
Most sensitive threshold (dB/μbar)	-22	-15
Estimated maximum detection distance (meters)	1000	500

	At Sea-State 1			
	Traffic level		Rain level	
	Average	Heavy	Light	Heavy
Most sensitive threshold (dB/μbar)	-19	-14	-3	+9
Estimated maximum detection distance (meters)	750	425	120	30

The audio-frequency selected for the harbor seal (Table 3),
falls within the region of peak energy for its 'buzz-like' call.
Those used for the bottlenose dolphin (Table 4) are, likewise,
within the regions of peak energy for the respective echolocation-
clicks at different sound-source levels. Two-way transmission was
considered in all calculations regarding the dolphin. An arbitrary
CR value was assigned to the harbor seal, it being slightly below
that of its relative, the ringed seal. The absorption coefficient
used in the calculations of Table 4 was 0.05 dB/m (Urick, 1975); Au
et al., (1974) used a similar value which agreed with theoretical
considerations.

The simulation, shown in Table 3, involves the detection of a
signal by a conspecific at a distance under two rather calm sea
states and under differing levels of traffic and rain noise. Low
sea states were used to demonstrate capability, it being reduced
considerably at higher states. Threshold values were based on the
established audiogram of the species (Møhl, 1968).

It is apparent that signal detection by the harbor seal (and
probably by its near relatives) may also be dependent on background noise. Maximum detection-distance for reception in a relatively calm sea (state 1) is estimated at 1000 m. Although traffic
(or industrial) noise affects reception, it is small compared to
the loss in sensitivity during periods of rain. When such conditions become severe, communication is probably reduced to a few 10s
of meters.

In the case of the bottlenose dolphin (Table 4), difficulties
were initially encountered when attempting to produce a simple
table. Results from the literature provided quite different sound-
source levels for the species. Based on the expertise of the observers, choosing a single 'correct' level was impossible. Rather,
it was obvious that members of the species do not maintain a certain level of sound production, but actually vary the magnitude of
their signals so as to conform to the situation that confronts
them. It was also noted that a corresponding change in the level
of the echolocation-clicks resulted in a change in the spectrum of

peak energy. Based on the comments by Au et al., (1974), the
effects of sustained noise by snapping shrimp were also considered
(sound-pressure level based on measurements in Kaneohe Bay).

It is clear that a 40 dB increase in sound-source level would
increase effective echolocating distances by 3 to 4 times. This
is a reasonable factor based on spreading and absorption. At the
quietest source level, detection distances should be reduced considerably, especially in the presence of high levels of biological
noise. At the two highest levels used in the simulation, the
effects of traffic (or industrial) noise upon sensitivity appear
essentially nil.

The maximum distance for echolocation under a relatively calm
sea was estimated to be somewhere around 650 m. Under the same
sea conditions, but in regions of high-level biological sources,
such as snapping shrimp, that distance dropped almost one-half.

Thus, hearing by odontocetes may well be affected by environmental noise, and if so, the effective spectrum is quite different
from that which affects hearing in marine fishes. It is also clear
that extremely high frequencies are little affected by environmental noise; but the animals which are sensitive to that spectrum
will probably have to deal with their own 'noise', such as the
limits that are imposed by underlying physiological mechanisms.

Before ending this final section, it should be mentioned that
the great whales of the order Mysticeti produce a wide variety of
sounds, ranging from the extremely complex 'songs' of the humpback
(Winn et al., 1970) to the '20 cycle-pulse' of the blue whale
(Cummings and Thompson, 1971). Since the sound levels produced by
these mammoth creatures are truly astounding, it is reasonable to
assume that at least certain of the sounds are used for communication. If such is the case, their extreme low frequency vocaliza-
tions may well be affected by traffic noise, whether in coastal
regions or on the high seas. Also, with increasing industrial and
related noise, it is probable that the same processes that affect
the tiniest of fishes may well affect the largest animals on this
planet.

ACKNOWLEDGMENTS

I am extremely grateful to many individuals for their insight
and critical advice to this project. These include Arnold Banner,
Harry A. De Ferrari, William E. Evans, Charles R. Gordon, Joseph D.
Richard, Ronald J. Schusterman, Juanita Y. Spires, William N.
Tavolga, William F. Watkins, and Gordon M. Wenz. I have incorporated most recommendations of these and other scientists and all
have strengthened the final story. Certain recommendations were
not followed, however; the responsibility for any factual errors
or controversial subjects that might appear in the report rests
solely with the author. This work relates to Department of the
Navy Contract N00014-76-C-0142 issued by the Office of Naval
Research. The United States Government has a royalty-free li-
cense throughout the world in all copyrightable material contained
herein.

TABLE 4: Estimated echolocation-limits, in meters, under different ocean-noise conditions for the bottlenose dolphin, *Tursiops truncatus*. Capability of varying sound-source level assumed; audio-frequencies selected from regions of peak energy for the respective sound-source levels (see text); absorption coefficient – 0.05 dB/ 1 m (m.d.d. = maximum detection distance).

Source Level (average p–p) (dB/ μbar re 1 m)	Selected Audio-Frequency (KHz)	Audio-Threshold: Spectrum Level Noise Ratio (db)	At Sea State	Audio-Threshold (dB/μbar)	Est. m.d.d. (m.)	At Traffic Level (Sea State 1)	Audio-threshold (dB/μbar)	Est. m.d.d. (m.)	At Rain Level (Sea State 1)	Audio-Threshold (dB/μbar)	Est. m.d.d. (m.)
+ 45	30	32	1	-27	60	Average	-25	55	Light	-17	40
			2	-23	55	Heavy	-22	45	Heavy	+ 3	15
			Sustained noise by snapping shrimp	+ 2	10	N.A.			N.A.		
+ 85	50	37	1	-26	260	N.A.			N.A.		
			2	-22	215	N.A.			Light	-10	160
			Sustained noise by snapping shrimp	- 3	105	N.A.			Heavy	+ 1	90
+120	110	39	1	-29	640	N.A.			N.A.		
			2	-25	600				Heavy	-15	520
			Sustained noise by snapping shrimp	- 1	340				N.A.		

REFERENCES

Albers, V. M., 1965. *Underwater Acoustics Handbook II*. Pennsylvania State University Press, University Park. 356 pp.

Andersen, S., 1970. Auditory sensitivity of the harbour porpoise, *Phocoena phocoena*. In: *Investigations on Cetacea*, (G. Pilleri, ed.), Vol. 2, pp. 255-258. Hirnanat. Inst., Berne.

Arase, E. M. and T. Arase, 1966. Correlation of ambient sea noise. *J. Acoust. Soc. Amer.* 40: 205.

Athey, S. W., 1970. Acoustics technology-a survey. National Aeronautics and Space Administration, Washington, D. C. 139 pp.

Au, W. W. L., R. W. Floyd, R. H. Penner, and A. E. Murchison, 1974. Measurement of echolocation signals of the Atlantic bottlenose dolphin, *Tursiops truncatus* Montagu, in open waters. *J. Acoust. Soc. Amer.* 56: 1280-1290.

Banner, A., 1967. Evidence of sensitivity to acoustic displacements in the lemon shark, *Negaprion brevirostris* (Poey). In: *Lateral Line Detectors*, (P. Cahn, ed.), pp. 265-273. Indiana University Press, Bloomington.

Banner, A., 1968. Measurements of the particle velocity and pressure of the ambient noise in a shallow bay. *J. Acoust. Soc. Amer.* 44: 1741-1742.

Banner, A., 1972. Use of sound in predation by young lemon sharks, *Negaprion brevirostris* (Poey). *Bull. Mar. Sci.* 22: 251-283.

Banner, A. and M. Hyatt, 1973. Effects of noise on eggs and larvae of two estuarine fishes. *Trans. Amer. Fish. Soc.* 102: 134-136.

Brawn, V. M., 1961. Sound production by the cod (*Gadus callarias* L.). *Behaviour* 18: 239-255.

Buerkle, U., 1968. Relation of pure tone tresholds to background noise level in the Atlantic cod (*Gadus morhua*). *J. Fish. Res. Bd. Canada* 25: 1155-1160.

Buerkle, U., 1969. Auditory masking and the critical band in the Atlantic cod, *Gadus morhua*. *J. Fish. Res. Bd. Canada* 26: 1113-1119.

Cahn, P. H., W. Siler, and J. Wodinsky, 1969. Acoustico-lateralis system of fishes: tests of pressure and particle-velocity sensitivity in grants, *Haemulon sciurus* and *Haemulon parrai*. *J. Acoust. Soc. Amer.* 46: 1572-1578.

Chapman, C. J., 1973. Field studies of hearing in teleost fish. *Helgol. wiss. Meeresunters.* 24: 371-390.

Chapman, C. J. and A. D. Hawkins, 1973. A field study of hearing in the cod, *Gadus morhua* L. *J. Comp. Physiol.* 85: 147-167.

Cummings, W. C. and P. O. Thompson, 1971. Underwater sounds from the blue whale, *Balaenoptera musculus*. *J. Acoust. Soc. Amer.* 50: 1193-1198.

Cummings, W. C., B. D. Brahy, and W. J. Herrnkind, 1964. The occurrence of underwater sounds of biological origin off the west coast of Bimini, Bahamas. In: *Marine Bio-Acoustics*, (W. Tavolga, ed.), pp. 27-43. Pergamon Press, New York.

Diercks, K. J., T. Trochta, and W. E. Evans, 1973. Delphinid sonar: measurement and analysis. *J. Acoust. Soc. Amer.* 54: 200-204.

Dietz, F. T., J. S. Kahn, and W. B. Birch, 1960. Nonrandom associations between shallow water ambient noise and tidal phase. *J. Acoust. Soc. Amer.* 32: 915.

Enger, P. S., A. D. Hawkins, O. Sand, and C. J. Chapman, 1973. Directional sensitivity of saccular microphonic potentials in the haddock. *J. Exp. Biol.* 59: 524-533.

Evans, W. E., 1967. Vocalization among marine mammals. In: *Marine Bio-Acoustics*, Vol. 2, (W. N. Tavolga, ed.), pp. 159-186. Pergamon Press, New York.

Evans, W. E., 1973. Echolocation by marine delphinids and one species of fresh-water dolphin. *J. Acoust. Soc. Amer.* 54: 191-199.

Fay, R. R., 1974. Masking of tones by noise for the goldfish (*Carassius auratus*). *J. Comp. Physiol. Psychol.* 87: 708-716.

Fish, J. F. and W. C. Cummings, 1972. A 50-dB increase in sustained ambient noise from fish (*Cynoscion xanthulus*). *J. Acoust. Soc. Amer.* 52: 1266-1270.

Fish, J. F. and G. C. Offutt, 1972. Hearing thresholds from toadfish, *Opsanus tau*, measured in the laboratory and field. *J. Acoust. Soc. Amer.* 51: 1318-1321.

Fish, M. P., 1964. Biological sources of sustained ambient sea noise. In: *Marine Bio-Acoustics*, (W. N. Tavolga, ed.), pp. 175-194. Pergamon Press, New York.

Fletcher, H., 1940. Auditory patterns. *Rev. Mod. Phys.* 12: 47-65.

Fry, F. J., G. Kossoff, R. C. Eggleton, and F. Dunn, 1970. Threshold ultrasonic dosages for structural changes in the mammalian brain. *J. Acoust. Soc. Amer.* 48: 1413-1417.

Gray, G. A. and H. E. Winn, 1961. Reproductive ecology and sound production of the toadfish, *Opsanus tau*. *Ecol.* 42: 274-282.

Ha, S. J., 1968. Masking effects on the hearing of the lane snapper, *Lutjanus synagris* (Linnaeus). Thesis, University of Miami. 51 pp.

Ha, S. J., 1973. Aspects of sound communication in the damselfish, *Eupomacentrus partitus*. Ph.D. Dissertation, University of Miami. 78 pp.

Hall, J. D. and C. S. Johnson, 1971. Auditory thresholds of a killer whale *Orcinus orca* Linnaeus. *J. Acoust. Soc. Amer.* 51: 515-517.

Hawkins, A. D., 1973. The sensitivity of fish to sounds. *Oceanogr. Mar. Biol. Ann. Rev.* 11: 291-340.

Hawkins, A. D. and C. J. Chapman, 1975. Masked auditory thresholds in the cod, *Gadus morhua* L. *J. Comp. Physiol.* 103: 209-226.

Hawkins, J. E. and S. S. Stevens, 1950. The masking of pure tones and of speech by white noise. *J. Acoust. Soc. Amer.* 22: 6-13.

Heindsmann, T. E., R. H. Smith, and A. D. Arneson, 1955. Effects of rain upon underwater noise levels. *J. Acoust. Soc. Amer.* 27: 378-379.

Jacobs, D. W. and J. D. Hall, 1972. Thresholds of a freshwater dolphin, *Inia geoffrensis* Blainville. *J. Acoust. Soc. Amer.* 51: 530-533.

Johnson, C. S., 1966. Auditory thresholds of the bottlenose porpoise (*Tursiops truncatus*, Montagu). *U. S. Naval Ordinance Test Station Rep.* (NOTS TP 4178). 22 pp.

Johnson, C. S., 1967. Sound detection thresholds in marine mammals.

In: *Marine Bio-Acoustics,* (W. N. Tavolga, ed.), pp. 247-255. Pergamon Press, New York.

Johnson, C. S., 1968. Masked tonal thresholds in the bottle-nosed porpoise. *J. Acoust. Soc. Amer.* 44: 956-967.

Knudsen, V. O., R. S. Alford, and W. Emling, 1948. Underwater ambient noise. *J. Mar. Res.* 7: 410-429.

Licklider, J., 1959. Three auditory theories. In: *Psychology: A Study of a Science,* Vol. 1, (S. Koch, ed.), McGraw-Hill, New York.

Maniwa, Y. 1971. Effect of vessel noise in purse seining. In: *Modern Fishing Gear of the World,* (H. Kristjonsson, ed.), pp. 294-296. Fishing News (Books) Ltd., London.

Miller, J. D., S. J. Rothenberg, and D. H. Eldredge, 1971. Preliminary observations on the effects of exposure to noise for seven days on the hearing and inner ear of the chinchilla. *J. Acoust. Soc. Amer.* 50: 1199-1203.

Møhl, B., 1968. Auditory sensitivity of the common seal in air and water. *J. Auditory Res.* 8: 27-38.

Møhl, B. and S. Andersen, 1973. Echolocation: high frequency components in the click of the harbour porpoise (*Phocoena* ph. L. (sic)). *J. Acoust. Soc. Amer.* 54: 1368-1372.

Myrberg, A. A., Jr., 1972. Ethology of the bicolor damselfish, *Eupomacentrus partitus* (Pisces: Pomacentridae). A comparative analysis of laboratory and field behavior. *Animal Behav. Monogr.* 5: 197-283.

Myrberg, A. A., Jr., 1978. Underwater sound: its effect on the behavior of sharks. In: *Sensory Biology of Elasmobranch Fishes,* (R. Mathewson and T. Hodgson, eds.), pp. 391-417. U. S. Government Printing Office.

Myrberg, A. A., Jr. and J. Y. Spires, 1972. Sound discrimination by the bicolor damselfish, *Eupomacentrus partitus.* *J. Exp. Biol.* 57: 727-735.

Myrberg, A. A., Jr. and J. Y. Spires, 1977. Comparative analysis of hearing among damselfishes of the genus, *Eupomacentrus.* Abstract. *57th Ann. Meeting, Amer. Soc. Ichthyologists and Herpetologists,* Gainesville.

Myrberg, A. A., Jr., A. Banner, and J. D. Richard, 1969. Shark attraction using a video-acoustic system. *Mar. Biol.* 2: 264-276.

Myrberg, A. A., Jr., C. R. Gordon, and A. Peter Klimley, 1976. Attraction of free ranging sharks by low frequency sound, with comments on its biological significance. In: *Sound Reception in Fish,* (A. Schuijf and A. D. Hawkins, eds.), pp. 205-228. Elsevier, Amsterdam.

Myrberg, A. A., Jr., E. Spanier, and S. J. Ha, 1978. Temporal patterning in acoustical communication. In: *Contrasts in Behavior,* (E. S. Reese and F. J. Lighter, eds.), pp. 137-179. John Wiley & Sons, New York.

Myrberg, A. A., Jr., S. J. Ha, S. Walewski, and J. C. Banbury, 1972. Effectiveness of acoustic signals in attracting epipelagic sharks to an underwater sound source. *Bull. Mar. Sci.* 22: 926-949.

Naval Ordinance Laboratory, 1942. Measurement of background noise in the water at Cape Henry, Virginia, due to surf and marine life. *N. O. L. Rept. No. 594.*

Nelson, D. R., 1967. Hearing thresholds, frequency discrimination
    and acoustic orientation in the lemon shark, *Negaprion brevirostris*
    (Poey). *Bull. Mar. Sci.* 17: 741-768.
Nelson, D. R. and S. H. Gruber, 1963. Sharks: attraction by low-fre-
    quency sounds. *Science* 142: 975-977.
Nelson, D. R. and R. H. Johnson, 1970. Acoustic studies on sharks.
    Rangiroa Atoll, July, 1969. *Tech. Rept. 2*, Office of Naval
    Research Contr. No. N00014-68-C-0318. 15 pp.
Nelson, D. R. and R. H. Johnson, 1972. Acoustic attraction of Pacific
    reef sharks: effect of pulse intermittency and variability. *J.
    Comp. Biochem. Physiol.* 42A: 85-95.
Nelson, D. R. and R. H. Johnson, 1976. Some recent observations on
    acoustic attraction of Pacific reef sharks. In: *Sound Reception
    in Fish*, (A. Schuijf and A. D. Hawkins, eds.), pp. 229-237. El-
    sevier, Amsterdam.
Nelson, D. R., R. H. Johnson, and L. G. Waldrop, 1969. Responses in
    Bahamian sharks and groupers to low frequency, pulsed sounds.
    Bull. So. Calif. Acad. Sci. 68: 131-137.
Norris, K. S., 1969. The echolocation of marine mammals. In: *The
    Biology of Marine Mammals*, (S. Andersen, ed.), pp. 391-423. Aca-
    demic Press, New York.
Olsen, K., 1971. Influence of vessel noise on behaviour of herring.
    In: *Modern Fishing Gear of the World*, (H. Kristjonsson, ed.),
    pp. 291-293. Fishing News (Books) Ltd., London.
Piggott, C. L., 1964. Ambient sea noise at low frequencies in
    shallow water of the Scotian shelf. *J. Acoust. Soc. Amer.* 36:
    2152-2163.
Popper, A. N. and N. L. Clarke, 1976. The auditory system of the
    goldfish (*Carassius auratus*): effects of intense acoustic
    stimulation. *J. Comp. Biochem. Physiol.* 53A: 11-18.
Popper, A. N. and R. R. Fay, 1973. Sound detection and processing
    by teleost fishes: a critical review. *J. Acoust. Soc. Amer.*
    53: 1515-1529.
Poulter, T. C., 1968. Marine mammals. In: *Animal Communications*,
    (T. Sebeok, ed.), pp. 405-456. Indiana University Press,
    Bloomington.
Ridgway, S. H. and P. L. Joyce, 1975. Studies on seal brain by
    radiotelemetry. *Rapp. P.-v. Reun. Cons. int. Explor. Mer.*.
    169: 81-91.
Sand, O., 1974. Directional sensitivity of microphonic potentials
    from the perch ear. *J. Exp. Biol.* 60: 881-889.
Sand, O. and P. S. Enger, 1974. Possible mechanisms for di-
    rectional hearing and pitch discrimination in fish. *Rheinisch.
    Westfälische Akad. L. Wiss.* 53: 223-242.
Schevill, W. E., W. F. Watkins, and C. Ray, 1963. Underwater
    sounds of pinnipeds. *Science* 141: 50-53.
Schuijf, A., 1974. Field Studies of Directional Hearing in Marine
    Teleosts. Ph.D. Dissertation, University of Utrecht. 119 pp.
Schuijf, A. and M. E. Siemelink, 1974. The ability of cod (*Gadus
    morhua*) to orient towards a sound source. *Experientia* 30:
    773-774.
Schuijf, A., J. W. Baretta, and J. T. Wildschut, 1972. A field
    investigation on the description of sound direction in *Labrus*

bergylta (Pisces: Perciformes). *Neth. J. Zool.* 22: 81-104.

Schusterman, R. J., 1968. Experimental laboratory studies of pinniped behavior. In: *The Behavior and Physiology of Pinnipeds*, (R. J. Harrison, R. C. Hubbard, R. S. Peterson, C. E. Rice, and R. J. Schusterman, eds.), pp. 87-171. Appleton-Century-Crofts, New York.

Schusterman, R. J., 1974. Auditory sensitivity of a California sea lion to airborne sounds. *J. Acoust. Soc. Amer.* 56: 1248-1251.

Schusterman, R. J. and R. F. Balliet, 1969. Underwater barking by male sea lions (*Zalophus californianus*). *Nature* 222: 1179-1181.

Schusterman, R. J., R. F. Balliet, and J. Nixon, 1972. Underwater audiogram of the California sea lion by the conditioned vocalization technique. *J. Exp. Anal. Behav.* 17: 339-350.

Schusterman, R. J., R. F. Balliet, and S. St. John, 1970. Vocal displays underwater by the gray seal, the harbor seal, and the stellar sea lion. *Psychon. Sci.* 18: 303-305.

Sivian, L. J. and S. D. White, 1933. On minimum audibility fields. *J. Acoust. Soc. Amer.* 4: 288-321.

Steinberg, J. C., W. C. Cummings, B. D. Brahy, and J. Y. MacBain (Spires), 1965. Further bio-acoustic studies off the west coast of North Bimini, Bahamas. *Bull. Mar. Sci.* 15: 942-963.

Tavolga, W. N., 1967. Masked auditory thresholds in teleost fishes. In: *Marine Bio-Acoustics*, Vol. 2, (W. N. Tavolga, ed.), pp. 233-245. Pergamon Press, New York.

Tavolga, W. N., 1971. Sound production and detection. In: *Fish Physiology*, Vol. 5, (W. S. Hoar and D. J. Randall, eds.), pp. 135-205. Academic Press, New York.

Tavolga, W. N., 1974a. Sensory parameters in communication among coral reef fishes. *Mt. Sinai J. Med.* 41: 324-340.

Tavolga, W. N., 1974b. Signal/noise ratio and the critical band in fishes. *J. Acoust. Soc. Amer.* 55: 1323-1333.

Tavolga, W. N. and J. Wodinsky, 1963. Auditory capabilities in fishes. *Bull. Amer. Mus. Nat. Hist.* 126: 177-240.

Tavolga, W. N. and J. Wodinsky, 1965. Auditory capacities in fishes: threshold variability in the blue-striped grunt, *Haemulon sciurus. Anim. Behav.* 13: 301-311.

Terhune, J. M. and K. Ronald, 1971. The harp seal, *Pagophilus groenlandicus* (Erxleben, 1777). 10. The air audiogram. *Can. J. Zool.* 49: 385-390.

Terhune, J. M. and K. Ronald, 1972. The harp seal, *Pagophilus groenlandicus* (Erxleben, 1777). 3. The underwater audiogram. *Can. J. Zool.* 50: 565-569.

Terhune, J. M. and K. Ronald, 1975a. Underwater hearing sensitivity of two ringed seals (*Pusa hispida*). *Can. J. Zool.* 53: 227-231.

Terhune, J. M. and K. Ronald, 1975b. Masked hearing thresholds of ringed seals. *J. Acoust. Soc. Amer.* 58: 515-516.

Urick, R. J., 1975. *Principals of Underwater Sound*. McGraw Hill, New York. 384 pp.

Watkins, W. A., 1974. Bandwidth limitations and analysis of

cetacean sounds, with comments on "delphinid sonar:
measurements and analysis. [K. J. Diercks, R. T. Trochta,
and W. E. Evans. *J. Acoust. Soc. Amer.* 54: 200-204 (1973)]
J. Acoust. Soc. Amer. 55: 849-853.

Wenz, G. M., 1962. Acoustic ambient noise in the ocean: spectra
and sources. *J. Acoust. Soc. Amer.* 34: 1936-1956.

Wenz, G. M., 1964. Curious noises and the sonic environment in the
ocean. In: *Marine Bio-Acoustics*, (W. N. Tavolga, ed.), pp. 101-119. Pergamon Press, New York.

Wenz, G. M., 1972. Review of underwater acoustic research: noise.
J. Acoust. Soc. Amer. 51: 1010-1024.

Widener, M. W., 1967. Ambient-noise levels in selected shallow water
off Miami, Florida. *J. Acoust. Soc. Amer.* 42: 904-905.

Winn, H. E., 1967. Vocal facilitation and the biological significance of toadfish sounds. In: *Marine Bio-Acoustics*, Volume 2,
(W. N. Tavolga, ed.), pp. 283-303. Pergamon Press, New York.

Winn, H. E., P. J. Perkins, and T. C. Poulter, 1970. Sounds of
the humpback whale. In: *Proceedings of the Seventh Annual
Conference on Biological Sonar and Diving Animals*. Stanford
Research Institute, Menlo Park.

Transect Sampling Methods and Their Applications to the Deep-Sea Red Crab

G. P. Patil
C. Taillie
R. L. Wigley

ABSTRACT: A line transect is a survey technique
where the number of organisms of interest are recorded along a path, usually a straight line,
through the survey area. The National Marine
Fisheries Service applied this technique to a
quantitative photographic survey of the deep-sea
red crab, *Geryon quinquedens*, off the northeastern United States. The purpose of the present
report is to review the statistical aspects of
line transect theory as it pertains to this survey.

INTRODUCTION

A line transect is a survey technique wherein the number of
organisms of a given species are recorded along a path, usually a
straight line, through the survey area. The National Marine
Fisheries Service applied this technique to a quantitative photographic survey of the deep-sea red crab, *Geryon quinquedens* Smith,
off the northeastern United States, June–July 1974. Survey operations were conducted in continental slope waters at depths between
229 and 1,646 m in the area extending from offshore Maryland (38°N.,

74°W.) northeastward to the eastern end of Georges Bank (41.5°N., 66°W.). The primary purpose of the survey was to obtain quantitative estimates of the number and biomass of red crabs in that region. Secondary purposes were to assess the size composition of this species, and to obtain additional information on its distribution, life history, and ecology. General information resulting from this survey was reported by Wigley et. al. (1975). Emphasis in that report was on the quantitative distribution of the red crab in terms of density and biomass; ecological and statistical aspects were secondary. The purpose of the present report is to review the statistical aspects of line transect theory as it pertains to this survey.

MATERIALS AND METHODS

The survey was conducted from aboard the research vessel *Albatross IV*, a 57 m vessel operated for the Northeast Fisheries Center by the National Ocean Survey, Office of Fleet Operations, NOAA. Sampling gear of two types was used on this survey: (1) an underwater photographic system (Figure 1), which took *in situ* photographs of the sea bottom and constituent epibenthic fauna; and (2) an otter trawl used to catch red crabs.

Photographs of the sea bottom were obtained to determine the density of red crabs. The photographic system consisted of a 70 mm camera and stroboscopic light mounted on a large steel sled. The sled was 2.7 m long, 2.1 m wide, and 1.9 m high, and was constructed of heavy-guage 6.4 cm diameter steel pipe, and runners 25.4 cm broad and 2.5 cm thick. The sled weighed 1,225 kg. The camera was a Hydro-Products Deep Sea Photographic Camera, Model PC-705; the electronic flash unit was a Hydro-Products, Deep Sea Stroboscopic, Model PF-730. Film used was Kodak Ektachrome EF daylight (color) and Kodak Tri-X Pan (black and white).

Stations were pre-selected according to a stratified random design. Previous information indicated that crab densities could be expected to be highest at water depths between approximately 250 and 500 m. Thus, a higher proportion of sampling stations were designated for that bathymetric zone. At each station where the sea floor was suitable, the sled was towed at a speed of 1 to 2 knots. Suitability of the bottom was evaluated by means of echo-sounding prior to launching the sled. At depths less than 585 m, an ELAC fathometer was used, while in deeper water an EDO instrument was employed. The duration of the tow at each station ranged from 30 to 75 min, depending on local conditions (bottom roughness, anchored fishing gear, etc.). The camera was programmed to obtain a photograph every 10 sec; thus the maximum number of photographs during one tow was approximately 400. Upon completion of the tow, the film was removed, and a short strip (1 to 2 m) was developed to monitor focus, strobe light position, exposure, and other parameters. The remainder of the film was commercially processed. A total of 18,000 photographs was obtained at 33 stations. Of this total, 8,262 photographs representing the best quality for counting crabs were selected for quantitative analyses.

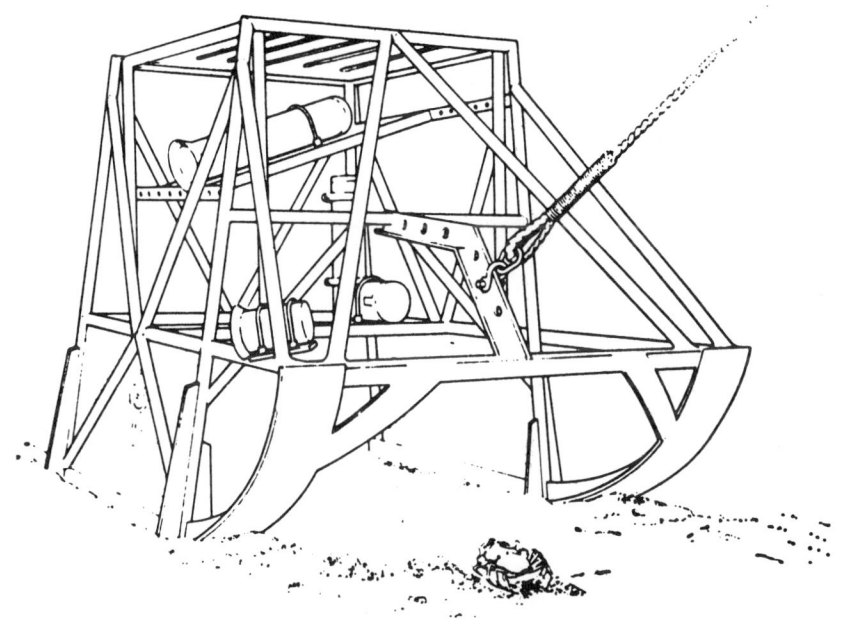

FIGURE 1: Front-oblique view of the sled-mounted
photographic system. Camera is mounted at upper
right-center of sled, the electronic flash unit
is in the forward corner, the power pack is fastened horizontally at the left center of the sled,
and the orientation pinger is mounted vertically
in the rear corner.

STATISTICAL METHODOLOGY

     In the usual line transect procedure, the observer moves along
a transect and records the number of organisms sighted. However,
not all organisms present in the study region may be observed;
those farther away from the transect are less likely to be sighted.
Thus, observed density tends to be an underestimate of the true
density of the organisms. In order to remove this visibility bias,
the investigator may either inflate the recorded count of the organisms or replace the area of the region by something smaller,
called effective area. The line transect theory provides a method
to determine the effective area by using right angle distances to
the organisms sighted, for examples, see Gates (1968), Seber (1973),
Burnham and Anderson (1976), Eberhardt (1978), and Gates (1979).
The effective area depends on the visibility function g(x) defined
as the probability of sighting an organism present at right angle
distance x from the transect. The shape of the visibility function
g(x) is usually reflected in the empirical graph of the number of
sighted organisms against their right angle distances. Generally,
these graphs show the number sighted decreasing with distance.

The counts of red crabs in this study first increase and then decrease with the right angle distance. This unusual feature can be explained, however, by decreasing visibility coupled with the expanding field of view of the camera with distance from the sled. This composite effect can be represented by a weighted visibility function, which we use to develop methods for estimating population density.

Weighted Visibility Function and Effective Area

What appears to be a rectangle in a photograph is in fact a trapezoid on the ocean floor as displayed in Figure 2. Let a and H be the base and the height of the trapezoid. The length of the trapezoid at distance x is of the form a+bx, where b is a dimensionless constant. For our setup, the values of a, b, and H are 2.868m, 0.961, and 5.49m respectively.

Initially, when a crab was detected on the photograph, its right angle distance x was measured by superimposing a grid and counting the number of squares to the crab. This procedure was tedious and time-consuming, however. So, instead, the field of view was divided into five zones as indicated in Figure 2, and the number of crabs was counted for each zone. Originally all the five zones were of equal width. But sediment clouds obscured the visibility near the sled, and, therefore, only the upper half of the first zone has been analyzed.

If $g(x)$ is the visibility function, and $n(x)$ is the number of crabs sighted in the trapezoid strip $(x, x + \Delta x)$,

$$E[n(x)] = sD[a+b(x+\tfrac{1}{2}\Delta x)]\, \Delta x g(x), \tag{1}$$

where s is the number of frames (photographs) and D is the density of crabs per square meter. This implies that the probability density function $f(x)$ of recorded x is proportional to the weighted visibility function, $(a+bx)g(x)$, giving

$$f(x) = \frac{(a+bx)g(x)}{w} \quad 0 \le x \le H \tag{2}$$

where

$$w = \int_0^H (a+bx)g(x)\,dx. \tag{3}$$

Further, from equation (1) we can estimate density D by

$$\frac{n(x)}{s(a+bx)g(x)\Delta x}.$$

The minimum variance pooled estimate is then given by

$$\hat{D} = \frac{n}{s\int_0^H (a+bx)g(x)\,dx} = \frac{n}{sw}, \tag{4}$$

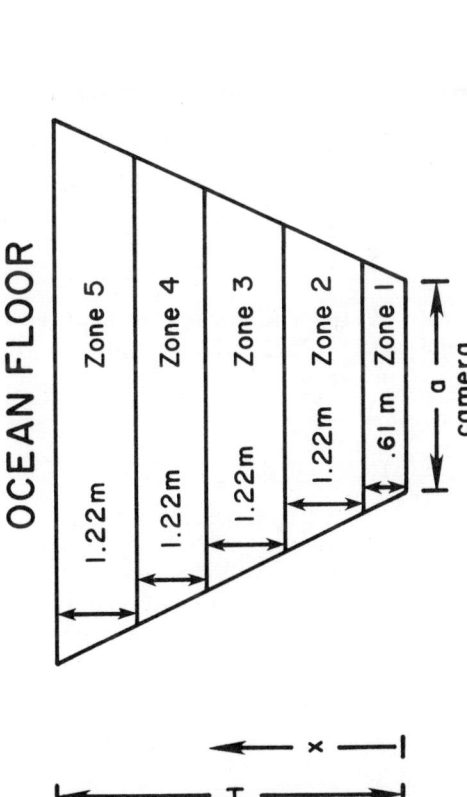

FIGURE 2: Schematic diagrams of the actual field of view on the ocean floor and its photographic image.

where n is the total number of crabs recorded on s frames. Because of the nature of equation (4), w is called the effective area per frame.

In order to utilize (4), we need to know w, which appears as a parameter in the probability density function of the recorded right angle distances as given in equation (2). The parameter w can be estimated in principle from these distances. We discuss two possible methods in the next two sections.

Exponential Estimators

The exponential sighting function has received considerable attention in the line transect literature. In the present study we use

$$g(x) = e^{-\lambda x}, \quad 0 \le x \le H. \tag{5}$$

The graph of the corresponding weighted visibility function $(a+bx)e^{-\lambda x}$ increases for all small x and decreases for large x as needed in our present study. While the appropriateness of the exponential sighting function for our data is discussed later, we provide the necessary results for estimating the density here. The effective area is given by

$$w = \frac{a}{\lambda}(1 - e^{-\lambda H}) + \frac{b}{\lambda^2}(1 - e^{-\lambda H} - \lambda H e^{-\lambda H}). \tag{6}$$

The maximum likelihood estimate of λ is the solution of

$$\bar{x} = -\frac{1}{w}\left(\frac{dw}{d\lambda}\right) = \frac{2}{\lambda} - \frac{1}{w}\left[\frac{a}{\lambda^2}(1 - e^{-\lambda H}) + \frac{(a+bH)H}{\lambda}e^{-\lambda H}\right] \tag{7}$$

where $\bar{x}$ is the average x. Using the method of statistical differentials (see Johnson and Kotz, 1969), the asymptotic variance of $\hat{D}$ in (4) is given by

$$\text{Var}(\hat{D}) = \frac{D}{sw}\left[\frac{\text{Var}(n)}{E[n]} + \frac{(\frac{dw}{d\lambda})^2}{\frac{d^2w}{d\lambda^2}w - (\frac{dw}{d\lambda})^2}\right]. \tag{8}$$

The variance to mean ratio is known to be related to spatial pattern. It is one for Poisson pattern and exceeds one for aggregated pattern. Although there is indication of some aggregation in our data, in the analysis section we have set the variance to mean ratio to be one, thus giving an underestimate for the asymptotic variance of $\hat{D}$. We feel, however, that the bias is not serious.

Cox Estimators

Cox Estimators   As Eberhardt (1978) has discussed, the Cox estimators provide estimates of the density when no parametric form is assumed for the sighting function. Putting $x = 0$ in equation (2) gives

$$f(0) = \frac{a}{w} \qquad (9)$$

provided $g(0)$ can be assumed to be unity. The Cox method obtains the estimate of w by estimating $f(0)$ as follows. Consider the first two zones with widths Δ and 2Δ respectively. Let n_1 and n_2 be the observed counts of the crabs in these zones. Then $f(\frac{1}{2}\Delta)$ and $f(2\Delta)$ are estimated by $\frac{n_1}{n\Delta}$ and $\frac{n_2}{2n\Delta}$.. The linear extrapolation back to the origin gives the estimate

$$\hat{f}(0) = \frac{8n_1 - n_2}{6n\Delta} . \qquad (10)$$

Combining (4), (9) and (10) gives the Cox estimator

$$\hat{D} = \frac{1}{sa\Delta} \left(\frac{8n_1 - n_2}{6}\right) . \qquad (11)$$

If the spatial pattern is random, n_1 and n_2 are independent and follow Poisson distributions. It follows that $Var(\hat{D})$ can be estimated by

$$\hat{Var}(\hat{D}) = \frac{1}{(sa\Delta)^2} \left(\frac{64 \, n_1 + n_2}{36}\right) \qquad (12)$$

DATA ANALYSIS

Density Estimation by Zone Counts

We illustrate the above methodology using zone counts from three stations (16, 21, 67). Sighting distances by station are given in Table 1. Computations and estimates are summarized in Table 2. Note that $\bar{x}$ is computed from grouped data in the form of zone counts. The effect of grouping on the estimates is not sizable, as discussed in a later section.

We have assumed the exponential visibility function for a parametric approach. This assumption appears reasonable, at least according to the usual chi-square test. The Cox estimates are based on non-parametric procedures that do not assume a form for the visibility function. Comparing the exponential estimates with Cox estimates, the Cox estimates have larger standard errors as expected of non-parametric procedures. The two density estimates are similar for Station 16, but for Station 21, the Cox estimate is smaller than the exponential estimate, whereas the reverse is true for Station 67. For this reason, we will examine more closely the validity of an exponential visibility function.

TABLE 1:   Sighting distances by station.

Station 16:   51 Crabs

```
 56    10    38    36    18    51
 86   163   122   142   142   142   102   117   173   127    61   173   147   168   112
173    61
254   183   295   224   183   239   254   244   213   254   193   213   244
305   396   335   366   335   396   325   361   335   325
498   503   488   498   498
```

Station 21:   93 Crabs

```
 10     5    25    20     5    20    51    25    31
 81   142   132   147   112    61    76    91   152   165   163   142    71   168    61   152
132    76    61   142   112   142   152   132   132   158   122   158   158
274   305   244   183   239   183   239   264   198   244   290   224   302   234   254
254   254   198   193   264   290   284   183   295   254   183   239   254   193
335   386   386   356   386   335   396   340   356   305   305   345   356   315   330
371   305   356
427   457   427   518   457   457   427   488
```

Station 67:   52 Crabs

```
 46    41    41    30    41
112   163   107   132    81   122    61    61   168    81    81   112   152   173   163
 81
203   185   244   279   274   193   213   193   224   213   193   183   239   284
305   335   305   351   396   366   366   422   386   416   366
427
```

Appropriateness of Exponential Visibility Function

For each zone, the observed and expected number of crabs were divided by the zonal area. These densities are plotted in Figure 3 against the zone midpoints. The observed and expected points closely match for Station 16. For Station 21, the observed plot

Table 2.: Density estimation using zone counts and exponential sighting function (estimates are ± one standard deviation).

Station	Photo Zone	Zone Mid-point (Meters)	Number of Crabs Observed	Number of Crabs Expected	Estimates
16	1	0.305	11	10.7	Exponential Estimates
Depth: 530–530 m	2	1.220	22	22.5	$\hat{\lambda} = .207$ m^{-1}; $\hat{w} = 16.47$ m^2
	3	2.440	22	22.6	$\hat{D} = 149 \pm 30$ crabs/ha
394 frames	4	3.660	23	21.5	Cox Estimates
	5	4.880	19	19.8	$\hat{w} = 15.42$ m^2
					$\hat{D} = 160 \pm 65$ crabs/ha
		$\bar{x} = 2.69$	97	$x^2_3 = .17$	
21	1	0.305	17	20.5	Exponential Estimates
Depth: 393–412 m	2	1.220	40	35.9	$\hat{\lambda} = .414$ m^{-1}; $\hat{w} = 9.92$ m^2
	3	2.440	34	28.0	$\hat{D} = 299 \pm 48$ crabs/ha
404 frames	4	3.660	12	20.7	Cox Estimates
	5	4.880	17	14.8	$\hat{w} = 13.12$ m^2
					$\hat{D} = 226 \pm 79$ crabs/ha
		$\bar{x} = 2.20$	120	$x^2_3 = 6.3$	
67	1	0.305	18	14.0	Exponential Estimates
Depth: 412–960 m	2	1.220	26	26.5	$\hat{\lambda} = .323$ m^{-1}; $\hat{w} = 12.24$ m^2
	3	2.440	18	23.1	$\hat{D} = 189 \pm 36$ crabs/ha
423 frames	4	3.660	17	19.1	Cox Estimates
	5	4.880	19	15.3	$\hat{w} = 8.72$ m^2
					$\hat{D} = 266 \pm 77$ crabs/ha
		$\bar{x} = 2.41$	98	$x^2_3 = 3.5$	

appears to be more nearly half-normal than exponential. The use of exponential sighting function in such a case tends to over-estimate the population density. This may explain the smaller value of the non-parametric Cox estimate for Station 21. At Station 67, the observed points fall off more rapidly and are more convex that the expected points. In this case, the exponential estimate is likely to be an underestimate. This may explain the larger value of the non-parametric Cox estimate for Station 67.

In view of the above observations, a richer and more flexible family of visibility functions which includes a wide variety of shapes is desirable. A promising possibility lies in the exponential-power-family defined by

$$g(x) = \exp(-\lambda x^\gamma) \ .$$

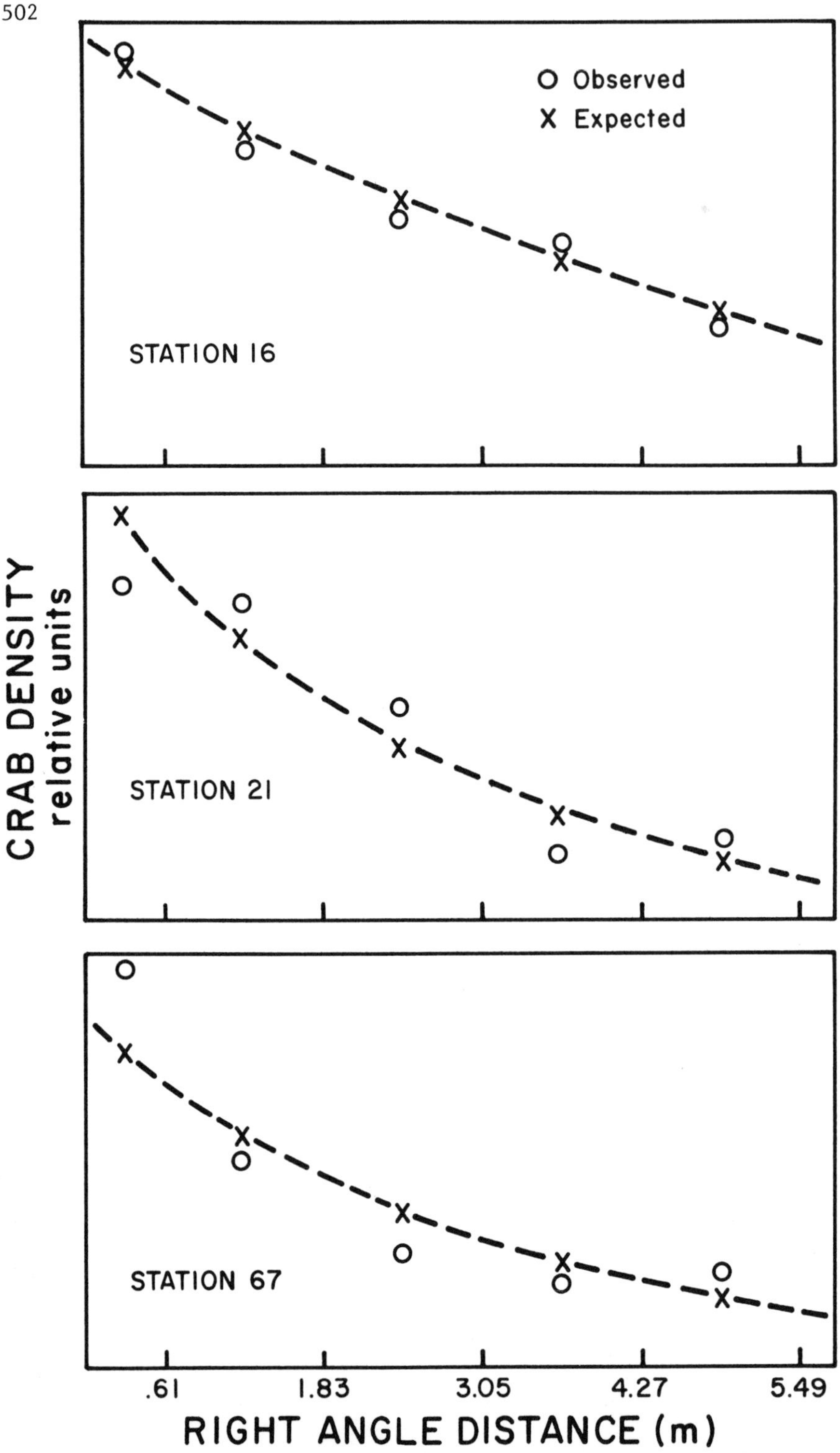

502

O Observed
X Expected

STATION 16

CRAB DENSITY
relative units

STATION 21

STATION 67

.61   1.83   3.05   4.27   5.49
RIGHT ANGLE DISTANCE (m)

Table 3 : Grouping effect on exponential estimates.

Station 16: 51 Crabs	
Ungrouped Estimates	**Grouped Estimates**
$\bar{x}$ = 2.23 m	$\bar{x}$ = 2.26 m
$\hat{\lambda}$ = .400 m^{-1}	$\hat{\lambda}$ = .387 m^{-1}
$\hat{w}$ = 10.24 m^{2}	$\hat{w}$ = 10.55 m^{2}
$\hat{D}$ = 240 ± 55 crabs/ha	$\hat{D}$ = 233 ± 54 crabs/ha

Station 21: 93 Crabs	
Ungrouped Estimates	**Grouped Estimates**
$\bar{x}$ = 2.21 m	$\bar{x}$ = 2.30 m
$\hat{\lambda}$ = .410 m^{-1}	$\hat{\lambda}$ = .370 m^{-1}
$\hat{w}$ = 10.01 m^{2}	$\hat{w}$ = 10.96 m^{2}
$\hat{D}$ = 297 ± 53 crabs/ha	$\hat{D}$ = 271 ± 49 crabs/ha

Station 67: 52 Crabs	
Ungrouped Estimates	**Grouped Estimates**
$\bar{x}$ = 1.99 m	$\bar{x}$ = 2.07 m
$\hat{\lambda}$ = .510 m^{-1}	$\hat{\lambda}$ = .472 m^{-1}
$\hat{w}$ = 8.12 m^{2}	$\hat{w}$ = 8.77 m^{2}
$\hat{D}$ = 285 ± 62 crabs/ha	$\hat{D}$ = 264 ± 58 crabs/ha

This family includes the exponential ($\gamma=1$), half-normal shapes ($\gamma>1$), and shapes more convex than the exponential ($0<\gamma<1$). Estimation procedures for this family are presently under investigation.

Effect of Grouping

For the three stations under consideration, some data was also available in the form of measurements of right angle distances. This has enabled us to study the effect of grouping referred to earlier. This study is summarized in Table 3. It shows that grouping tends to underestimate the density estimates and their standard errors. The bias introduced in the density estimates seems, however, small for the sizes of the standard errors. Thus, one may choose to work with zonal count data and not expend any effort in measuring right angle distances. In this context, the question of 'how many zones?' should be examined further.

FIGURE 3: Observed and expected crab density versus right angle distance at three stations.

REFERENCES

Burnham, K. P. and D. R. Anderson, 1976. Mathematical models for
    nonparametric inferences from line transect data. *Biometrics*
    32: 325-336.
Eberhardt, L. L., 1978. Transect methods for population studies.
    Journal of Wildlife Management 42: 1-31.
Gates, C. E., 1968. Line transect method of estimating grouse
    population densities. *Biometrics* 24: 135-145.
Gates, C. E., 1979. Line transects and related issues. In:
    Sampling Biological Populations (R. Cormack, G. P. Patil, and
    D. S. Robson, eds.), pp. 71-154, Satellite Program in Statisti-
    cal Ecology, International Co-operative Publishing House,
    Fairland, Maryland.
Johnson, N. L. and S. Kotz, 1969. *Discrete Distributions*. Wiley,
    New York. p. 29.
Seber, G. A. F., 1973. *The Estimation of Animal Abundance*. Hafner
    Press, New York. p. 28.
Wigley, R. L., R. B. Theroux, and H. E. Murray, 1975. Deep-sea
    crab, *Geryon quinquedens*, survey off northeastern United States.
    Marine Fisheries Review 37: 1-21.

Survey Design in Marine Environment: Three Examples

Woollcott Smith

ABSTRACT: Three examples are used to illustrate
the range of statistical design problems encountered in marine research.   The first example
uses nonlinear design methods to find the optimum
survey pattern for accurately locating the positions of acoustic beacons used in precision navigation.   In the second example, time series methods are used to evaluate environmental survey
designs and to determine the number of times to
repeat a survey over time.   In the third example,
I show that a variation of double sampling procedures can be implemented quickly in response to
an oil spill, providing greater flexibility in
evaluating the effect of the oil spill.

INTRODUCTION

The investigation of physical and biological processes in the
ocean is expensive and time-consuming.   It is important, then, to
evaluate marine surveys and experiments before they are initiated.
Classical statistical theory can provide us with some tools for
comparing the effectiveness of alternate survey programs.   Simple

Contribution Number 4390 from the Woods Hole Oceanographic Institution.

statistical models cannot reflect the complexity and variety of
oceanic processes. However, carefully chosen models can capture
the fundamental nature of the survey and provide guidelines, but
not exact recipes, for efficient designs.

Survey design methods can be as varied and complex as the
processes under investigation. Rather than give a general review
of survey design techniques, this paper addresses the range of
problems by discussing three specific examples from my experience
in marine research: acoustic transponder surveys; environmental
surveys of time varying processes; and oil spill surveys. These
examples cover a wide range of marine survey problems and use
different mathematical methods to solve them. However, each
example contains elements that are common to all survey design
problems.

To define a survey design problem, one needs three basic
ingredients: first, a statistical model of the system under investigation; second, a specific statement about the goals of the
survey (i.e., which parameters of the system do we wish to estimate); third, a definition of the constraints under which the survey is to be conducted (i.e., so many hours of ship time will be
used, so many samples will be taken, total cost of the survey
shall not exceed so many dollars, specific measurement techniques
which will be used, etc.). These ingredients are listed in the
order of importance according to classical statistics. We will
discuss them, however, in order of practical importance, that is,
in reverse order.

In a volume on marine measurement systems, it may seem re-
dundant to emphasize that measurement systems form an important
constraint in marine surveys. For instance, plankton density
estimates can be obtained by a wide variety of techniques (see
the 1968 UNESCO volume on zooplankton sampling). New developments in measurement techniques are often discussed by engineers
and scientists with little attention to the implications of
these methods for survey design. In turn both theoretical and
applied statisticians discuss the design problem in the context
of outdated technology. For instance, this paper discusses samples as if they are taken at discrete points in space and time,
whereas in fact the technology exists to take nearly continuous
measurements in time (i.e., pumped samples), with continuous
oxygen, salinity, temperature, and other measurements. Constraints used by the statistician in solving the design problem
should be set by the limits of technology and the value of the
answers that the survey will provide. Too often, the constraints
are defined by conventional measurement methods and an uncertain
understanding of the ultimate goal of the survey.

Before one can evaluate a survey design, the goal of the survey must be carefully stated. A survey designed to answer one set
of questions may be of little use in answering another set of
questions. Failure to address this point has been one of the main
problems in the design of surveys for environmental impact studies.

Finally, one must specify the statistical properties of the
system under investigation. In general, the more detailed information one has about the system, the better and more efficient the

survey. The examples given here will illustrate this problem. In the acoustic navigation problem we are dealing with a well-defined nonlinear system, and a careful survey design can improve the efficiency by nearly an order of magnitude. In the oil spill example, little is known before the study is initiated, and one can give only general guidelines to survey strategy.

Since the detailed mathematics of these examples have been discussed elsewhere, we will now present the methods and philosophy involved in survey design in each case (Smith, 1978, 1979; Smith et al., 1975).

ACOUSTIC TRANSPONDER SURVEY

We begin with the most complex of the examples we will discuss. It is also the example in which the details of the system are known well in advance of the survey.

Many oceanographic experiments depend on the precise navigation of ships, instrument packages, or submersibles. In the deep ocean, acoustic transponders are now widely used to determine position. The transponders are moored near the ocean floor and the acoustic travel times between the fixed transponders and the vehicle are used to estimate the ranges. If the positions of the transponders are known, simple geometric calculations will give the position of the vehicle relative to the transponders.

The first and most critical step in this navigation method is to determine the relative position of the transponders located at fixed, but unknown, positions near the ocean floor. To do this, a ship operating at a known fixed depth measures the acoustic slant ranges to the transponders at a series of survey points. The transponder positions can then be estimated by finding the best fit of the observed slant ranges to the theoretical ranges. This least squares estimation procedure has been described by Lowenstein (1965) and others.

The Goal

The system parameters to be estimated from the acoustic slant ranges are the relative positions of the beacons located on or near the bottom. Our ability to estimate these positions depends not only on how well we measure the acoustic slant ranges, but on where the ship on the surface measures the slant ranges. The survey design problem is to find the set of ship survey positions that gives us the most information about the transponder positions.

The Model

To begin to answer the survey design problem, we must specify the geometric and statistical model. The mathematical notation in this section is of necessity rather complex. For the general reader, the numerical results given in the survey design section might best summarize the error analysis and design methods developed here.

For clarity we will consider here only the 3-beacon survey prob-

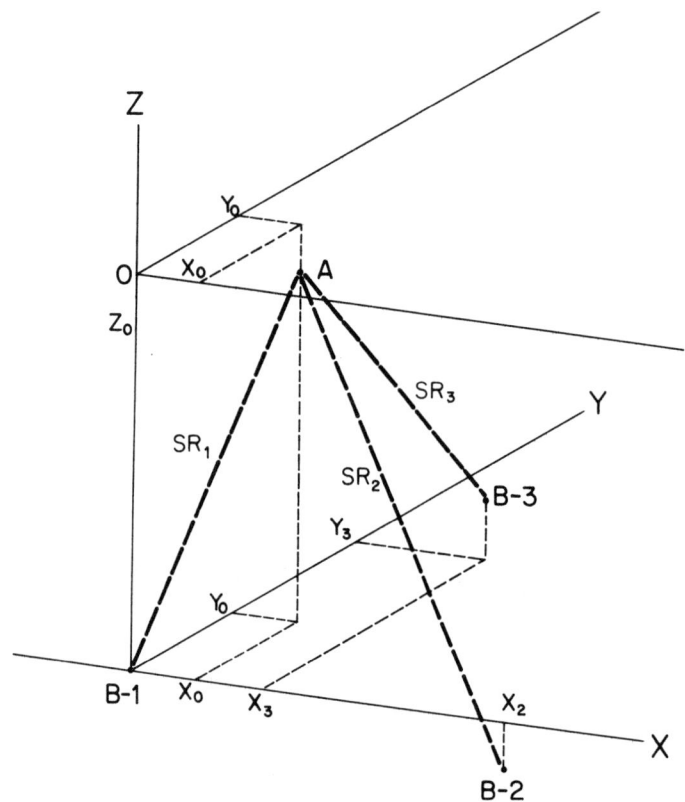

3 BEACON GEOMETRY

FIGURE 1:   Geometry for the 3 Transponder Survey

lem, although this method can easily be extended to more than three
beacons.   Figure 1 describes the 3-beacon situation; the beacons are
positioned at $(0, 0, z_1)$, $(x_2, 0, z_2)$, and (x_3, y_3, z_3).   The ship's
transducer at constant depth, zs, near the surface, measures the
slant ranges to the beacon positions at a set of n survey points,
(xs_i, ys_i, zs), $i = 1,2,\ldots,n$.   Let $\underline{S_1} = (S_{i,1}, S_{i,2}, S_{i,3})$, $i = 1.2$,
$\ldots n$, denote the vector of observed slant ranges at point $(xs_i, ys_i,$
$zs)$.   Let $sr_{i,j}$, $j = 1\ldots3$, denote the exact geometrical slant range
to the jth beacon at survey point i,

$$sr_{i,j} = [(xs_i - x_j)^2 + (ys_i - y_j)^2 + (zs - z_j)^2]^{\frac{1}{2}},$$

where $x_1 = y_1 = y_2 = 0$.

For convenience we will place all coordinates of the survey in
a single vector.   Let θ_1 denote the column vector of survey coordinates for the n survey points.   Its transpose, $\underline{\theta_1}$, is the row vector
$(xs_1, ys_1, xs_2, ys_2,\ldots xs_n, ys_n)$.   Similarly let $\underline{\theta_2}$ denote the column

vector of beacon coordinates, $\underline{\theta}_2 = (x_2, x_3, y_3, z_1, z_2, z_3)$. Finally let $\underline{\theta}$ denote the column vector of the 2n+6 parameters of the system, $\underline{\theta}' = (\underline{\theta}_1', \underline{\theta}_2')$. We denote the vector of slant ranges from a survey point i to the three beacons by $\underline{sr}_i'(\underline{\theta}) = (sr_{i,1}(\underline{\theta}), sr_{i,2}(\underline{\theta}), sr_{i,3}(\underline{\theta}))$ and the slant range vector for all n survey points by

$$\underline{sr}'(\underline{\theta}) = (\underline{sr}_1'(\underline{\theta}), \ldots, \underline{sr}_n'(\underline{\theta})) .$$

This completes the description of the geometry of the survey. The next step is to specify the statistical properties of slant range observations.

We assume that the observed slant range observations, $S_{i,j}$, can be defined by,

$$S_{i,j} = sr_{i,j}(\underline{\theta}) + \varepsilon_{i,j} ,$$

where the $\varepsilon_{i,j}$ are independent identically distributed normal random variables with mean 0 and variance, σ^2. Under these assumptions the maximum likelihood estimate is the $\underline{\hat{\theta}}$ that minimizes the sum of squared errors,

$$SS(\underline{\hat{\theta}}) = \min_{\underline{\theta}} \sum_{i=1}^{n} \sum_{j=1}^{3} (S_{ij} - sr_{ij}(\underline{\theta}))^2 . \tag{1}$$

The problem of finding the $\underline{\theta}$ that minimizes (1) is discussed in Lowenstein (1965) and Smith et al. (1975).

If we can find the covariance matrix of the beacon position estimates, then the survey design problem is reduced to finding the ship position that minimizes the magnitude of the diagonal elements of the covariance matrix. The covariance matrix for the parameter estimates can be approximated by finding a Taylor series expansion for the slant ranges in terms of the parameters, $\underline{\theta}$. For small changes in $\underline{\theta}$ denoted by $\Delta\underline{\theta}$ we can represent $\underline{sr}(\underline{\theta}+\Delta\underline{\theta})$ by a Taylor series expansion

$$\underline{sr}(\underline{\theta} + \Delta\underline{\theta}) \approx \underline{sr}(\underline{\theta}) + X \Delta\underline{\theta} , \tag{2}$$

where X is the 3n by 2n+6 matrix of the derivatives of the slant ranges with respect to the parameters $\underline{\theta}$.

With the Taylor series representation for the slant ranges, one can apply standard linear regression theory, Draper and Smith (1966), to determine an approximate error covariance matrix, Σ, for $(\underline{\hat{\theta}}_1', \underline{\hat{\theta}}_2')$, yielding,

$$\Sigma = \begin{bmatrix} \Sigma_{11} & \Sigma_{12} \\ \Sigma_{21} & \Sigma_{22} \end{bmatrix} \approx \sigma^2 [X^t X]^{-1} \tag{3}$$

where σ^2 is the variance of the slant range errors.

Since the goal of the survey is to estimate beacon position we

are only interested in reducing the diagonal elements of the $\sum_{22}$
matrix. Smith et al. (1975) give a computational form for $\sum_{22}$ that
does not require the inversion of the 2n+6 by 2n+6 X'X matrix.
Equation (3) shows clearly that there are two nearly independent
ways of reducing the size of $\sum_{22}$, part of the error covariance matrix. One can reduce σ^2; that is, one can make more accurate slant
range measurements by improving instrumentation. Or one can reduce
the magnitude of the relevant diagonal elements in the $(X'X)^{-1}$ matrix; since the X matrix depends only on beacon and ship positions,
this matrix can only be reduced by altering the survey pattern. We
have now completed the specification of the survey model.

Constraints

There are two kinds of constraints imposed in this survey design problem. The first set of constraints is imposed by the technology we are using. These constraints include the fact that we are
using three beacons, that slant range measurements can be made only
from the ship at the surface to the beacons, etc. This set of constraints is included in the development of the model, but the constraints can be and have been altered by using different instrument
designs, Spindel et al. (1976).

The second set of constraints is imposed by the time allocated
for the ship to make the survey. For simplicity in the example
given, we have assumed that the survey will include only six survey
points. This is a rather artificial constraint; we impose it in
order to make the survey design results easier to follow.

Survey Design

The results given in the preceding sections can be used both
to evaluate actual survey data and to plan surveys to minimize the
overall survey error. The position of the survey points relative
to the beacons is important in determining the size of errors in
the beacon position estimates. By evaluating $\sum_{22}$ defined in equation (3) for different sets of survey points, one can determine a
survey design that minimizes these errors. Of course, both the
positions at the beacons and the survey points are unknown; however, in practice one can obtain approximate positions by using
other navigation systems.

The approximate covariance matrix given in (3) is a function
of two independent effects: slant range measurement error, and
the position of the survey points. The standard error in estimating a beacon parameter, denoted by σ_i, is composed of two
independent terms,

$$\sigma_i = \sigma \cdot M_i \cdot ,$$

where σ is the standard error of the slant ranges and depends only
on the accuracy of the slant range measurement. M_i is the error
magnification term for the ith parameter. M_i depends on the number
and positions of the survey points. We have defined a measure of

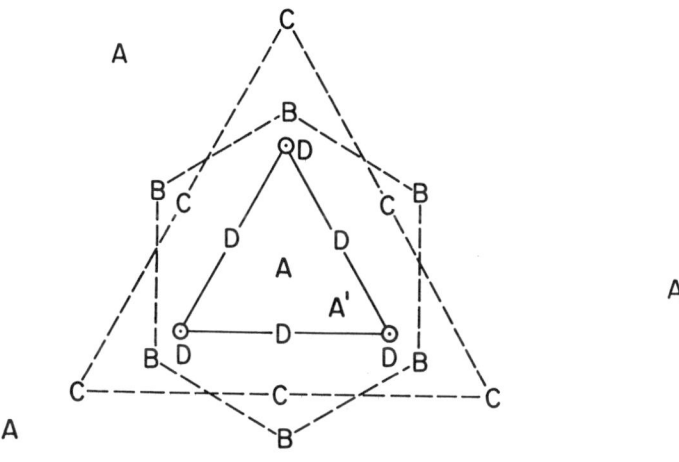

FIGURE 2: Beacon positions - . - and survey
points for surveys A, A', B, C, and D. Surveys
A and A' are identical except for the survey
point in the center of the array.

TABLE 1: Error magnification terms and survey
error for the survey points in Figure 2.

Survey	X_2	X_3	Y_3	Z_1	Z_2	Z_3	Survey Error C
A	1.35	1.74	1.20	1.15	1.11	1.09	10.05
A'	1.37	1.74	1.26	1.34	.89	1.40	11.07
B	23.81	13.73	18.83	4.01	5.30	4.39	1173.58
C	1.94	2.41	1.61	.98	.98	.98	15.01
D	4.24	6.04	3.59	1.00	1.00	1.00	70.34

the efficiency of the survey, the survey error, C, which is the sum
of the squares of the six error magnification terms. This is proportional to the sum of the diagonal elements of the $\sum_{22}$ matrix. In
evaluating survey design we will use the error magnification terms,
M_i, and the survey error C, since these terms are both independent
of scale and measurement error.

In Figure 2 and Table 1 we give the results for a series of surveys labeled A, A', B, C, and D, each consisting of six survey points.
The beacons in this example form an equilateral triangle 1,000 meters

on a side and the depth of each beacon is 1,000 meters. The error
magnification terms, M_i, are the square roots of the diagonal ele-
ments of the $[\underline{X}'\underline{X}]^{-1}$ matrix defined in equations (2) and (3). Figure
2 gives the positions of the survey points and the beacons, and Table
1 gives the individual error magnification terms and the survey error,
C, for each of the surveys.

The important lesson to learn from Figure 2 and Table 1 is that
survey error, that is, the accuracy of the survey, can vary by an order of magnitude or more. Without the design tools developed here,
it would be difficult to judge which of these surveys would yield the
most information about the beacon positions.

The survey points in design A we found by minimizing the survey
error, C, with respect to the six survey coordinates. These theoretical calculations are only a guide to designing good survey patterns. The theoretical model used to calculate the error magnification terms does not allow for the limited range of the sonar signal.
Thus in many situations the optimal survey points, A, are impractical. Surveys A and A' are the same except that there is a different
point in the center of the array in the two surveys. The results for
survey A and A' indicate that the efficiency of a survey is relatively insensitive to small changes in the position of the survey points.
Survey C will use less ship time, but there is a slight increase in
the survey error. Survey D takes even less ship time, but this time
there is a substantial increase in the survey error. Survey B is a
good example of how large the survey error can become. In survey B
the pattern of points falls nearly on a conic section (i.e., a circle) where Vanderkulk (1961) has shown that the transponder positions
are not uniquely determined.

Using this analysis, Hunt et al. (1974) have developed some
guidelines for survey design in deep ocean acoustic navigation. The
same techniques have been used by Spindel et al. (1976) for a more
complex acoustic navigation system.

Both the statistical models and the constraints are a simplification of the actual conditions encountered in acoustic surveys. The
most critical assumption made in formulating the model is that the
slant range measurement errors were uncorrelated. In fact, the differences between the true slant range and the measured slant range
is a complex function of uncorrelated random measurement error and
bias, largely due to initial errors in measuring the sound velocity
profile which is used to convert acoustic travel time to slant range.
Because of this measurement bias, our simplified analysis will yield
over-optimistic results when many more than six survey points are
used. For this reason we have limited the analysis to six-point
surveys.

Also, it is more realistic to limit time taken to complete the
survey rather than the number of points used in the survey. The
ship's travel time between survey points is an important factor,
since the distance between survey points could be as much as 10 km.
Thus the more accurate surveys, A and C, are at least twice as costly in ship time as survey D. The prudent designer might well choose
the less time consuming survey. Again, these results, derived from
a simplified model, should be thought of as an aid in evaluating surveys rather than as specifying the final design in detail.

The survey problem discussed here is an example of the design of non-linear experiments (Chernoff, 1953). Box and Hunter (1965) give an excellent review of this general area.

SURVEY DESIGN FOR TIME VARYING PROCESSES

We now consider a widely encountered problem in marine surveys: how to construct an efficient survey when the parameter that we are interested in is changing over time. For instance, concentration of a chemical pollutant or plankton density in an estuary is changing in response to water temperature, river run-off, etc. One must sample at different time points to understand the behavior of the system over a season or a year.

The Goal

If the parameter varies over time, then a survey might have a number of possible goals. One could use survey data to estimate the maximum value of the parameter in some time interval, for instance, the maximum concentration of a chemical pollutant in a year. One might also want to estimate the rate of change of the pollutant concentration from year to year. Each goal will imply a different optimum survey design. In this paper we will assume that the goal of the survey is to estimate the mean value of the parameter in some time interval. More formally, let $\theta(t)$ denote the value of the parameter of interest at time t, then the goal of the survey is to estimate

$$\bar{\theta} = \frac{1}{T} \int_0^T \theta(t) \, dt \ . \tag{4}$$

The design question is how many times in the interval $(0,T)$ should we conduct the survey, and how much sampling effort should we spend at each time point? To answer these questions we must first specify the sampling model and the constraints of the sampling program.

The Model

We will develop here a special case of the more general sampling model discussed in Smith (1978). There are two steps in specifying the sampling model. The first step is to develop a model for the fluctuations in the $\theta(t)$ process over time. The second step is to define the sampling model at each time point.

From the point of view of survey design, the process can be thought of as random, since its behavior cannot be predicted at the time the survey is planned. In this paper we will assume that the process is a stationary Markov process. Under these assumptions the process is defined by its mean and by its covariance function,

$$C(t) = Cov(\theta(s),\theta(s+t)) = C_o e^{-\beta|t|} \ ,$$

Jenkins and Watts (1966). Intuitively this implies that the process is continuously being displaced from its mean, so that the variance about the mean is C_0. However, the effect of any particular displacement decays at rate β.

In chemical and biological sampling, samples must be taken at discrete points in time, often at considerable expense. At each survey time point, we have only an estimate of the true value of $\theta(t)$; that is, at discrete time points t_1, t_2, ..., t_n we observe

$$\hat{\theta}_i = \theta(t_i) + \varepsilon_i, \quad i = 1,2,\ldots,n \tag{5}$$

The ε_i's denotes independent random errors due to sampling and measurement errors at time t_i. We assume that ε_i has mean 0 and sampling variance, σ_s^2. The size of σ_s^2 will depend on the sampling effort expended at each time point.

If we know little about the time series properties of the process $\theta(t)$, a relatively robust sampling and estimation procedure for $\bar{\theta}$ is the following. Let N denote the number of time points when sampling will occur. Place the sampling time at the midpoints of each of the N equal intervals that divide the time interval (0,T). That is

$$t_i = \frac{1}{2}\frac{T}{N} + \frac{T}{N}(i-1), \quad i = 1,2,\ldots N \tag{6}$$

The estimate of $\bar{\theta}$ is then

$$\hat{\bar{\theta}} = \frac{1}{N}\sum_{i=1}^{N}\hat{\theta}_i . \tag{7}$$

The survey design problem is to allocate the sampling effort between sampling at more time points (increasing N) and obtaining more accurate estimates of $\theta(t)$ at specified time points (decreasing σ_s^2). To evaluate a survey design one needs to find the variance of $\hat{\bar{\theta}}$. Smith (1978), using the results of Tubilla (1975) and Cochran (1963), found the variance of $\hat{\bar{\theta}}$ in this situation. The variance of the $\hat{\bar{\theta}}$ estimator is

$$\sigma_{\bar{\theta}}^2(N) = \sigma^2(N) + \frac{\sigma_s^2}{N} . \tag{8}$$

$\sigma^2(N)$, the first term of the right hand side of this equation, is the mean squared error due to sampling a time varying process at N discrete time points. The second term on the right hand side is the mean squared error due to sampling and measurement error at single time points. The expression for $\sigma^2(N)$ has the form

$$\sigma^2(N) = C_0\left[\frac{1}{N} - \frac{2}{\beta T}(1 + U/\beta T) + \frac{2}{V}(e^{-\beta T/N}(1-\frac{U}{V}) + \right.$$

$$\left. \frac{2}{\beta T}e^{-\beta T/2N})\right] \tag{9}$$

where the constants U and V are defined by

$$U = 1-e^{-\beta T}$$

and

$$V = N(1-e^{-\beta T/N})$$

The single time point sampling variance σ_s^2 depends on sampling strategy at each time point. In the special case where $\theta(t)$ is the mean of a population and a random sampling program is conducted, at each sampling time the variance of the single time point estimate is

$$\sigma_s^2 = \frac{\sigma^2}{m}, \tag{10}$$

where σ^2 denotes the population variance and m denotes the number of samples taken at each time point.

Constraints

The constraints imposed on the sampling program are again of two kinds. The first constraints are instrument and sampling limitations that we have incorporated into the model. For instance, we have assumed in our sampling model that samples are taken at discrete points rather than continuously over time. The second set of constraints is imposed by the limits put on the cost of a survey program. These costs include the cost of using a boat and crew to sample at a single time point, denoted by g, and the cost of analyzing each sample, denoted by h. The cost of the survey is then a function of the number of time points sampled, N, and the number of samples taken at each time point, m. In this simplified situation, the total cost of the survey has the following form:

$$S = Ng + Nmh$$

Let S_0 be the maximum cost of the survey. The optimum survey design minimizes equation (8) subject to the constraint that $S < S_0$.

Survey Design

To develop a survey design for a particular marine application, one needs at least rough estimates for the parameters in the model. These parameters include the variance of the natural process C_0 and the decay rate β, as well as the sampling variance σ^2. These can be obtained from historical data, field trials, and an understanding of the circulation processes in the area. Typical results are illustrated in Figure 2, from Smith (1978), for an ichthyoplankton survey conducted over a 26-week period. The parameters of this sampling model were $\sigma^2 = .59$, $C_0 = .27$, and $\beta = .86$, and the cost constraints of the survey were described by g = 750, h = 100, and $S_0 = 50,000$ dollars. The ratios of σ^2 to C_0 and h to g are more important for survey design than their absolute values.

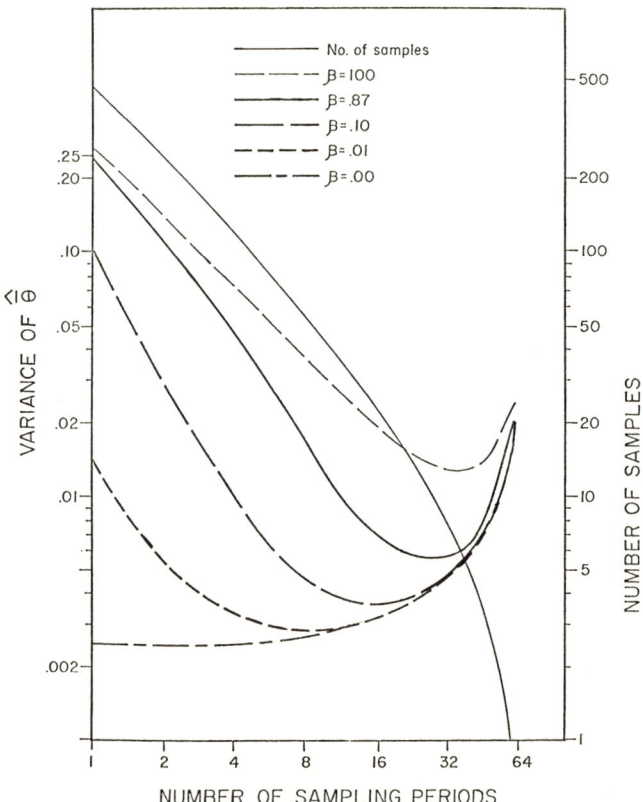

FIGURE 3: The expected mean squared error for
different sampling plans with a fixed cost of
$50,000. The fixed cost determines the maximum
number of samples (solid line) given the number
of sampling periods.

The trade-off between the number of time points and the number
of samples within a time point can be investigated by tabulating the
possible combinations of N and m that produce surveys with a total
cost less than $50,000. Figure 3 gives the results of these compu-
tations on a log-log plot. The number of sampling periods are plotted
against the variance of $\hat{\theta}$, equations (8), (9), and (10). With weeks
as the time unit, β is varied to illustrate how the sampling design
will change depending on the rate at which changes in the $\theta(t)$ occur.
For large β ($\beta \gg 10$) the process varies rapidly over time, and results
will be improved by more frequent sampling. In our example, the
minimum is about 40 times in the 26-week period with four samples at
each time point. With β = .1 and β = .01, we see that the optimum
number of time points decreases because with a slowly varying process
it becomes less important to sample frequently. When $\beta \ll .001$ or
nearly 0, the optimum strategy is to sample only at one time.
An important feature of these results is that they are relative-

ly insensitive to small changes in β and that there is a fairly wide
choice of sampling frequencies which yield mean squared errors that
are close to the optimum. This is fortunate since at the design
stage of an environmental survey, we have only a rough knowledge of
the stochastic properties of the θ(t) process. For practical reasons
the sampling period must be in even multiples of days or weeks. This
will usually not be at the theoretical minimum, but since the function
$\sigma^2_{\bar{\theta}}$ (N) is nearly level in the neighborhood of the minimum, one will not
lose much efficiency by choosing a sampling period that conforms with
the calendar.

The sampling theory discussed here does not include the effect of
aliasing. Aliasing occurs when processes with a natural periodic component due to tidal cycles, daily fluctuations, etc., are sampled.
The Markov processes discussed here do not exhibit such periodic
variation. If there is a large periodic component in the process
sampled, the error in estimating the time averaged mean will be dependent on the relationship between the sampling frequency and the
frequency of the periodic component.

There are other approaches to the design of surveys over time,
depending on the goals and sampling model assumed, Moore (1973),
Scott and Smith (1974), and Glass et al. (1975).

AN OIL SPILL SURVEY STRATEGY

In the preceding examples, we have assumed that the investigator
has a good deal of knowledge about the processes he is investigating.
The goals of the survey are stated and the survey constraints are well
defined. Thus, before designing a survey, the investigator has a
clear quantitative definition of the three ingredients of survey design.

An oil spill is an unplanned event. It may occur in an area of
the world where there is little knowledge about biology, sediment
type, or water circulation patterns. Even if this information were
available, it would be difficult on short notice to assemble the
baseline data to develop an adequate sampling model.

At the time an oil spill survey program is planned there may
be no consensus about the important questions that data from a survey should answer. These questions may be developed in legal proceedings long after the relevant data should have been collected.
Finally, at the initiation of the spill study, the constraints,
such as the total cost of the spill study, cannot be specified.

Even under these circumstances, classical statistical theory
can give at least some qualitative rules for survey design. We
begin again with a statement of the three parts of survey design.

The Goals

Even though the exact nature of the scientific and legal
questions cannot be stated, one can say something about the general
nature of the questions. Rather than detailed information about a
single location, the legal question will most certainly center

around the total effect of the spill: the total area of the benthos taken out of production, the number of bushels of shell fish
destroyed, and total fish mortality, among other parameters.

The Model

It is difficult to specify a detailed sampling model without
a thorough knowledge of the area under investigation. However, it
is reasonable to assume that most of the properties of the spill
area are highly correlated over space. That is, if the sediment
were muddy in one location it is likely to be muddy in a location
100 meters removed from the first location. In the same way the
spill effect will be correlated over space. A sampling strategy
should take advantage of these general spatial characteristics.

Constraints

Although the funds allocated for the survey are unknown, we
know something about the relative cost of each step in the survey.
First, the cost of collecting and preserving the samples is an
order of magnitude less than the cost of detailed hydrocarbon
analysis, or the analysis of the species composition of a sample.
However, there are also a number of other measurements that one
can make on a sample that are relatively inexpensive. For benthic
samples these measurements would include sediment type, whether
there is visible oil in the sample, depth of the sample, and so
forth. These easy to measure variables could be used to refine
the sampling model for the area.

Survey Design

This general description of the goals, sampling model, and
cost of an oil spill survey in turn yield general non-quantitative
guidelines for survey design.

The combination of the goals of the survey, estimating the
total effect of the spill, and the spatial characteristics of
the effect implies that a systematic survey should be performed
(Cochran, 1963). In two dimensions, this implies a grid survey
(Tubilla, 1975). Figure 4 gives a hypothetical oil spill event;
superimposed is a possible grid sampling scheme.

A quick glance at such a sampling plan reveals a fundamental
flaw. The cost of analyzing all the samples collected is prohib-
itive. However, the cost of collecting and preserving field
samples is small in comparison with the cost of analyzing these
samples. Thus, a desirable sampling procedure would be to over-
sample the area, using a grid sampling scheme, and then use the
information on sediment type, presence of oil, etc. as a guide to
subsampling the original samples for the more costly chemical and
biological analysis. This well-studied sampling method is called
double sampling (Cochran, 1963).

Finally, a sampling plan like this has two practical advantages
to the oil spill investigator. The first is that the field part of
the sampling program is easy to implement on short notice. Second,

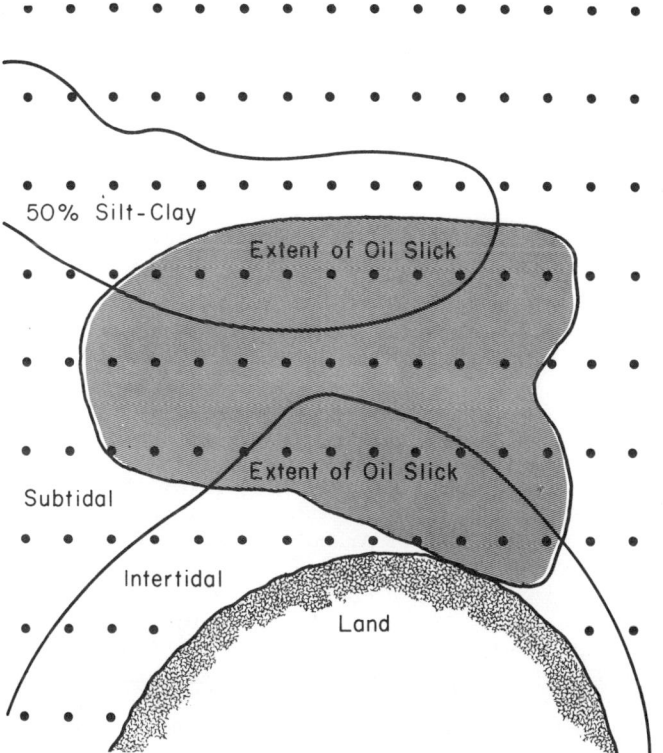

FIGURE 4:   Hypothetical oil spill and grid sampling.

the difficult part of the plan, subsampling the original samples,
can be postponed until the goals of the particular study are better
defined.   The sampling plan outlined is based on well-known classical sampling procedures and would be relatively easy to defend in
an administrative or legal proceeding.   Katz (1975) discusses the
problems faced by a statistician in presenting survey results in a
legal setting.

CONCLUSIONS

     The examples given here illustrate the complex nature of sta-
tistical design problems in marine science.   In this area the
casual application of standard survey methods often provides little
help.   It is only by analyzing in detail the properties of the system,
the nature of the constraints, and the questions asked, that one
can begin to improve the efficiency of the survey.   The time and
energy spent in analysis before a large survey or experiment is
initiated is seldom wasted.

ACKNOWLEDGMENTS

    This work was supported by NOAA Sea Grant 04-8-M01-149, and
The Ocean Industries Program.

REFERENCES

Box, G. E. P. and W. G. Hunter, 1965. The experimental study of
    physical mechanisms. *Technometrics* 7: 23-41.
Chernoff, H., 1953. Locally optimal designs for estimating para-
    meters. *Annals of Mathematical Statistics* 24: 586-600.
Cochran, W. G., 1963. *Sampling Techniques*, Second Edition. Wiley,
    New York. 413 pp.
Draper, N. R. and H. Smith, 1966. *Applied Regression Analysis*.
    Wiley, New York. 350 pp.
Glass, F. V., V. L. Wilson, and J. M. Gottman, 1975. *Design and
    Analysis of Time-Series Experiments*. Colorado Associated
    University Press, Boulder, Colorado. 241 pp.
Hunt, M. M., W. M. Marquet, D. A. Moller, K. R. Peal, W. K. Smith,
    and R. C. Spindel, 1974. An acoustic navigation system. *Woods
    Hole Oceanographic Institution, Technical Report 74-6*. 67 pp.
Jenkins, G. M. and D. G. Watts, 1968. *Spectral Analysis and Its
    Applications*. Holden-Day, San Francisco. 525 pp.
Katz, L., 1975. Presentation of a confidence interval estimate as
    evidence in a legal proceeding. *Amer. Stat*. 29: 138-142.
Lowenstein, D. C., 1965. Computations for transponder navigation.
    Scripps Institute of Oceanography, Report No. MPL-U-2/65. 6 pp.
Moore, S. F., 1973. Estimation theory application to design of
    water quality monitoring systems. *Journal of Hydrology, ASCE*
    815-831.
Scott, A. J. and T. M. F. Smith, 1974. Analysis of repeated surveys
    using time series methods. *J. Amer. Statistical Association* 69:
    674-678.
Smith, W. K., 1978. Environmental survey design: a time series
    approach. *Estuarine and Coastal Marine Science* 6: 217-224.
Smith, W. K., in press. An oil spill sampling strategy. In:
    Sampling Biological Populations, (R. M. Cormack, G. P. Patil,
    and D. S. Robson, eds.). International Statistical Ecology
    Program.
Smith, W. K., W. M. Marquet, and M. M. Hunt, 1975. Navigation
    transponder survey: design and analysis. *IEEE Ocean 75
    Conference Record* 563-576.
Spindel, R. C., R. P. Porter, W. M. Marquet, and J. L. Durham,
    1976. A high-resolution pulse-doppler, underwater acoustic
    navigation system. *IEEE Journal of Ocean Eng*. 1: 6-16.
Tubilla, A., 1975. Error convergence rates for estimates of
    multidimensional integrals of random functions. *Department
    of Statistics, Stanford University Technical Report NO. 72*.
    75 pp.
UNESCO, 1968. Zooplankton sampling. *UNESCO*, Paris. 174.
Vanderkulk, W., 1961. Remarks on a hydrophone location method.
    U. S. Navy Journal of Underwater Acoustics 11: 241-250.

WORKING GROUP REPORTS

Problems in Marine Biological Measurements

D. V. Holliday, Chairman
Robert L. Edwards
A. A. Yayanos
Howard L. Weinert
G. Daniel Hickman

INTRODUCTION AND BACKGROUND

Marine environmental research ranges from broad scale, long term monitoring programs to (in time and space) process-oriented studies highly restricted in time and space. For convenience, programs may be categorized as macro-, meso-, or microscale. Broadly speaking, the macroscale programs are logistically complex and scientifically conservative (more applied than basic) while the microscale projects are logistically simple and often highly theoretical. While there are many exceptions, the above generalization does tend to suggest where government agencies should and do concentrate their efforts and where one is more often apt to find the academic scientist and the private research institution. It is often overlooked that each end of this spectrum of activity depends upon the opposite and that if the ocean research program is to be credible and progressive and to serve all the varied needs of society, an appropriate mixture of all types of work is mandatory. Finding this balance is not easy.

Monitoring programs requires a commitment of logistic support and time that in many, if not most cases is well beyond that of private institutions. Since such programs usually service the immediate needs of society, they must be scientifically conservative so that the results are easily acceptable by non-specialists. At the same time they must also be sufficiently sophisticated that they provide a useful frame of reference so that the meso- and mi-

croscale research carried out within their limits may be extrapolated beyond their limits.

Continued appropriate basic research provides the protocols
for future monitoring programs. The process is continuous and evolutionary and over a period of time will tend to reduce the need of
logistic commitment to monitoring per se. Inasmuch as the ocean
ecosystems are relatively three dimensional and without the geographic stability of the terrestrial ecosystems, site specific studies are less useful than on land. For example, a specific site off
the coast of Virginia will have relevance to most nektonic species
for only a relatively short period each year as each species mi-
grates through the area. Without comprehensive monitoring programs,
studies on nektonic species have no frame of reference.

REGIONAL PROGRAMS

There have been many successful large interdisciplinary programs, particularly international programs. Such programs involve
only a relatively small number of marine biologists. Marine biologists generally tend to work in specialized areas and tend also not
to be involved in substantive interdisciplinary efforts. Further,
there appear to be a significant number of factors working against
interdisciplinary programs. These include funding problems and the
significantly different rates at which problems in specific disciplines are dealt with, beginning with the initial collection of data and extending to the analytic techniques involved and the preparation of the final reports.

It is suggested that encouragement can be provided for interdisciplinary studies by giving a regional focus to problems. Examples are "Pollution Impacts in Chesapeake Bay" and "Fish and
Mammal Interactions in the Aleution Islands." To provide a useful
focus for such studies, the present-day confusion of varied federal
and state agency purviews in such areas needs to be dealt with. To
some extent NOAA's MESA program in the New York Bight was reasonably successful in providing such a focus. It did achieve a considerable degree of governmental interaction in that substantial
progress was achieved without confusing agency purview. While MESA
was an experimental step in the right direction, that experiment
should be extended to situations where there is a clear mandate for
integrated, interdisciplinary research. Among the options suggested
for management of such programs was that of an interagency and in-
stitutional 'board of directors' with funds provided from a central
fund. There are other options. Clearly the smaller the geographic
focus, the less complex the management problem.

STATE-OF-THE-ART AND RECOMMENDATIONS

There is an obvious divergence between the physical sciences
and marine biology with respect to their sampling and analysis capabilities. Differences in the state-of-the-art have also developed between disciplines within marine biology itself, e.g., bio-

chemistry and fishery or zooplankton assessment programs.

Measurements techniques, instruments and data analysis tools
for major disciplines within marine biology have shortcomings in
several areas. These include the following.

† On a per sample basis, standard sampling procedures are
   more expensive in marine biology than in the physical
   sciences.

† Standard sampling procedures are not generally synoptic,
   i.e., characterized by timely, wide area coverage.

† Standard sampling procedures and many marine biological
   measurements are not traceable to an acceptable standard; therefore, intercalibration is much more difficult
   or even impossible for most investigators.

† Standard sampling procedures cannot resolve spatial or
   temporal variance on scales less than twice their extent
   or duration.

† Methods of rapid sample analysis, i.e., sorting, counting, sizing, chemical, physical property and nutrient
   analyses, etc., are labor intensive, expensive and slow.

† For a wide range of analyses on live organisms, the
   technology for making measurements over the range of
   parameters (temperature, salinity, pressure, etc.) they
   encounter in their natural environment is poorly developed.

The overall impact of these perceived problem areas is that
often, the framework of marine biological premises, theories and
models are developed from and tested against limited numbers of
marginally adequate samples. We recommend that the attention of
research funding agencies, the management of research institutions
and the individual researchers be directed to three general areas
in which conscious efforts could bring rapid (i.e., five years) advancement to sampling and data analysis in the marine biological
disciplines.

Interdisciplinary and International Investigations

Marine biological studies are perhaps unique in the extraordinary amount of interdisciplinary effort required. For example, the
development of an acoustic study of a population or community could
demand contributions from physicists, chemists, biologists, and engineers. Straightforward mechanisms do not exist for funding such
efforts because of the compartmentalization of funds among traditional disciplines. Exceptions to this situation may often be
found where the solution of an applied problem is desired, but seldom occur where basic research is the motive.

The world-wide nature of marine research could be advantageous in allowing one to avoid duplication, to share unusual opportunities and facilities including data banks and to share new technologies. The mechanisms for achieving these lie in increased communication and in international expeditions. Topics of meaningful workshops should be indentified and funded on a regular basis. The identification and establishment of standards and standard methods could be the topics of one such workshop.

Regional, national and international workshops should be fostered in an interdisciplinary fashion. Attendees should include scientists from both the public and private sectors. In the same vein, current government policies tend to discourage interested scientists from attending professional meetings outside their specialty even when there is reasonable cause to believe an interdisciplinary exchange could be of long term benefit to their research areas. Budgetary and management action to encourage interdisciplinary exchanges is one step in the process of effecting a sorely needed technology transfer and of developing symbiotic relationships between biologists and physical scientists. Such relationships should be encouraged, both for two individuals from different fields in the specific common interests and for small interdisciplinary groups with a defined common goal.

Technology Transfer

New technology from diverse sources is more readily absorbed into some scientific disciplines than others. Physical oceanographers have little apparent trouble in assimilating microprocessor technology and new developments in telemetry or in adapting advances in mathematical sciences to requirements and problems. Marine biologists have not had comparable success on the whole in technology transfer. There are three actions which may improve the rate of assimilation for new technology from diverse sources.

The first such area involves providing an opportunity and funding for structuring interdisciplinary study programs for marine biologists. Advances in understanding and our technical ability to deal with research efforts in the marine environment are occurring at an ever increasing rate. Technological development may lag a decade or more behind implementation by the researcher. Advances in understanding may lag even more significantly in areas peripheral to a scientist's principal interest. As the need rapidly grows for more integrated and interdisciplinary efforts, the need for each scientist to expand his education becomes more critical.

To some extent the needs for continuing education can be met by special short courses while on the job, by visiting lecturers, and by sabbatical leaves. Depending upon the situation and/or the individual, any one or more of those options will often serve the purpose. Recent advances, however, are occurring on all fronts simultaneously and the rate at which individuals are tending to become anachronisms is becoming a matter of serious concern.

Sabbatical leave has traditionally been a period of intellectual refreshment more or less left to the discretion of the individual insofar as the planning of his activities is concerned. To

some extent the individual can exercise sufficient options to adequately catch up on all relevant areas of particular interest. Often, however, at any one time or place, these options do not allow
for a comprehensive upgrading.

It has been suggested that there may be merit in a structured
sabbatical period, with the curriculum approved by the institution.
Such a curriculum would be planned to include all relevant subject
areas of interest tailored to general classes of marine scientists.
The material presented would particularly include recent advances
in supportive sciences and technologies, as for example, statistical theory or satellite technology as well as material in the field
of the scientist. Formal structuring of this sort would make it
easier for government and industry scientists to continue their
educations.

Second, there are many new instruments and advances that have
been developed outside of the field of marine biology that could be
of great use in current research. A lag of five to ten years appears to exist in transferring much of this new technology into
marine biological research programs. Such delays could be due to
one or more of the following reasons: (a) funding constraints do
not allow for the procurement of these instruments; (b) manufacturers perceive a small market among marine biologists and so do not
make their instruments adaptable to or suited for marine research,
which results in additional development costs for marine biologists,
retarding further adaptation of such instruments in their research;
and (c) there may not be an appreciation at the policy levels of
government that marine biological research is in urgent need of acquiring new technological developments. The anticipated increased
use of the resources of the oceans places a demand on marine biologists to produce extensive quantitative data regarding these resources. These scientists now require rapid automated and remote
measurement techniques.

The third area of concern is related to the relationship of
industry, government laboratories and academic institutions. While
we cannot speak to the intent which led to the NSF policy of en-
couraging transfer of technology by sponsoring meritorous joint
industry-university programs, in practice, there has been more emphasis on the transfer of new ideas and technology from the academic community to application or product in industry than in the
reverse direction. We would like to point out that the potential
transfer of specific high level interdisciplinary applications
technology in use in industry and in government laboratories may be
of benefit to the sampling and analysis problems of more academic
biological science investigators. Current policies are not effective in developing this reverse flow of information and technology.

Standardization

Results from similar studies done by different laboratories
often cannot be compared. This lack of agreement may possibly be
reduced by the adoption of traceable standards and standard methods.
A repository and authoritative body for standards in marine biological research must be identified. Standard methods for universal-

ly performed measurements or analyses must be identified and accepted. The replacement of one standard method by a new one should
be based on error analysis and statistical criteria. This proposal
is not intended to inhibit the development of new methods or instrumentation, but is intended to provide the basis for evaluation of
such developments and to aid in the calibration and comparison of
such methods and techniques. This philosophy of traceable standards has been successfully used in the physical sciences for many
years and has played an important role in the interchange of information, intercalibration of systems, and comparison and confirma-
tion of results on both a local and international basis. Few, if
any, analogous standards exist in the field of marine biology.

    We suggest that a management decision be made at a level
which can provide adequate international funding and support to develop traceable standards and methods for marine biological measurements. To reiterate, this international forum would also serve
to transfer technology among nations and to encourage cooperative
international programs.

MISCELLANEOUS SPECIFIC PROBLEM AREAS IN ADVANCING MARINE SAMPLING
TECHNOLOGY

    Several areas which deserve special attention were identified
during the workshop. These areas are specific and involve problems
which, if solved, can individually have a significant impact on the
state-of-the-art in marine biological sampling and analysis. This
is in contrast to the programs suggested above.

    Sample Processing

    The time required for sorting, counting, identifying, sizing,
measuring chemical and biophysical properties and the costs of
these activities force biologists to work with small sample sizes.
In many cases this results in inadequate data in time and in space
(horizontal and vertical). This leads to aliased data. Aliasing
is a systematic error. Replicate sampling will not eliminate this
type of error. The potential for errors in conclusions based on
aliased data is obvious.

    Technology is currently available which, if adapted to the
problem of sorting, etc., can be used in a phased research and development program to first, provide a tool or aid to speed up the
manual process significantly and within five years to provide some
reasonable level of automatic processing. Adequate funds should be
supplied for both hardware prototype development, mathematical
analysis, image processing, pattern recognition and software development. The possibilities of optical signal processing should not
be overlooked. In view of the potential of this work in the near
term, research and development funds and support should be provided
a high level of priority.

Archiving of Data and Samples

Some samples are subject to further analysis after they have been collected and analyzed for some immediate purpose. Deep-sea photographs are in this categroy. Regional data collections for these films should be established. Publicly funded principal investigators should be required to deposit their films (after obtaining working copies) into regional collection within 30 days after they are taken. The deposited film would be kept proprietary for a reasonable period (e.g., one year) and thus would allow the principal investigator who obtained the data sufficient time to analyze and publish his findings. At the end of the period, the film would become available to any interested person.

Samples of whole organisms and tissues constitute another set of important archivable samples. The availability of such samples, frozen or otherwise preserved, taken as a function of time from aquatic environments could provide the capability in the future of determining the time course in the accumulation of some presently unknown pollutant.

Funding Cycles and Sampling/Monitoring Programs

Programs for sampling or monitoring marine biological populations or environmental parameters will be effective only if the program can be conducted long enough to determine the frequency with which the parameter or population varies and on a scale commensurate with the patchiness and extent of the distribution.

For oceanic processes and populations the time and space scales may be long and large while patchiness may force closely spaced samples. Beginning programs without the necessary dedication and funding potential to carry them out properly could lead to a situation where no information would be better than bad information. These factors must be carefully considered in the management decision process for such surveys.

Marine Biological Data and Measurements Needed to Advance Remote and Hydroacoustic Sensing

Remote sensing (electromagnetic) and hydroacoustics both offer unique opportunities to marine biologists for gaining near-synoptic information on parameters related to the presence of marine organisms, their numbers, sizes, biophysical characteristics and distributions.

Behavior of organisms, if known, can be used to design specific sensors, processing strategies and displays to enhance the detection, localization, and classification of one species of animal in the presence of others. Such behavior includes distribution in time and space, schooling or aggregation, swimming motions, etc. Programs designed to provide biological information in areas subject to remote monitoring should be encouraged as should subsequent or parallel development of sensors, processors and survey techniques and their applications.

Similarly, the ability of both electromagnetic and acoustic sensor systems to detect, localize and classify organisms depend on *a priori* knowledge of an appropriate signature of the organism. In the electromagnetic domain, examples are water color due to the organism's presence, fluorescence and radiance. Association of particular patterns of distribution for temperature, salinity, near-surface chlorophyll, etc., with animal distributions may also be important as indirect indicators. In the acoustic domain, examples of important biophysical properties include presence or absence of an included gas bubble, the physical properties of the surrounding membrane, the density and compressibility of the tissue, carapace, included oil, etc., for parts of organisms and whole organisms.

We recommend that studies in basic marine biology which will provide these data be funded and encouraged. Joint investigations involving biologists and specialists from the physical disciplines should also be encouraged. Cooperative efforts will also be necessary in "ground truth" efforts and in operational evaluations of new systems and approaches to this means of improving biological sampling and data.

Some of the measurements, cataloging and archiving of basic properties data, may be eventually reduced to routine, but labor-intensive operations. These measurements may be needed for a large number of species over ranges of environmental parameters commensurate with their natural environment. In most cases, it is essential to work with live organisms. We urge consideration of the development of international programs to provide such measurements in order to use skilled people, and potentially complex equipment efficiently and to aid in obtaining a high degree of data quality control via standard methods with traceability to physical and chemical standards.

Survey Design

We recommend the evaluation of the methods and mathematics of communications theory or signal physics for problems in survey design as a supplement to conventional statistics.

Statistical and Mathematical Methodology

Woolcott Smith, Chairman
Arthur Baggeroer
Bradley W. Dickinson
G. P. Patil
J. B. Suomala

Statistical and mathematical methods play an important role in the development, testing, and application of marine measurement techniques. In this report specific areas are outlined where quantitative methods could make a significant contribution to marine biological measurement. Several steps are then recommended to bring about progress in these areas. In general, these steps involve bringing together scientists, mathematicians, and statisticians to work on solutions to specific problems and to disseminate those solutions to the general marine community.

QUANTITATIVE METHODS IN MARINE BIOLOGICAL MEASUREMENT

To give some idea of the scope of quantitative methods in marine biology three important problem areas are briefly outlined: data handling; statistical design; and general theory. Where these general areas have impact on marine measurement techniques is indicated.

Data exchange and data analysis methods, important in any science, are essential parts of large multidisciplinary programs in marine science. Incompatibility of data formats and other standardization problems make data sharing more difficult than it should be in marine science. This problem is complicated by the diverse

531

kinds of data archived from the wide variety of biological measurement programs. It is also difficult to assess data quality for sets of observations taken under varying conditions and with varying methods.

When measurement programs are costly it is important that full use of data be facilitated by good data base software. A specific need exists for a standard data base management system for marine biological data. This data base system should be machine independent. Lack of a general data base system hinders full exploitation of the data available from many programs. Existing data base systems could be modified for marine data.

Also, many of the data analysis programs used in marine science suffer from lack of standardization and validation. Since accuracy of results depends crucially on statistical software, it is important that reliability and standardization of software receive as much attention as calibration of the sensors that collected the data in the first place. It is also important that the software for new computer-based techniques and methods be widely distributed within the marine community; this distribution would make more efficient the expensive process of software development. An example of computer-based methods is the automatic sorting and pattern recognition techniques for sorting of marine plankton samples. Another example is the data display software used to analyze satellite and aircraft remote sensing data.

Sound statistical inference requires clear problem identification followed by the execution of an experiment designed for the efficient gathering of data relevant to the problems posed. Research in experimental and survey design in the following areas could make the marine measurement programs more efficient:

†   The use of remote sensing as an aid in the design of stratified oceanographic surveys;

†   Optimal design of sensor evaluation experiments;

†   Development of sampling schemes designed to improve the efficiency and accuracy of classical sorting methods for plankton samples; and

†   Optimal design of experiments and sensors for the detection of rare events. For example, in studies of slumps, slides, or biogenic events in the deep sea and their effects on marine life.

Mathematical and statistical theory has played an important role in ecological theory and practice. Examples of such theories applied to ecological problems include: spatial and temporal analysis, diversity measures, optimization methods in resource management, study of the properties of ecosystems through the use of numerical models, and multivariate methods for the study of community structure. Many of these techniques have already proved useful in marine biology; further applications of these ideas to marine measurement should be explored. For example, good quantitative models for many ecological systems now exist, but their application to efficient design and evaluation of marine biological measurement systems is not well developed. As a consequence, a measurement system often contributes little to the understanding of the biological process it was built to investigate. Properly used, ecological models can aid in the design of measurement systems to answer more fully important biological questions.

RECOMMENDATIONS

In comparison with other problem areas in marine biological measurement, the steps identified to improve the use of quantitative methods are straightforward and relatively inexpensive. Since research activity involving quantitative methods in marine biology is diffused throughout marine institutions, our recommendations require mechanisms to bring together researchers in these areas to increase communication and to solve outstanding problems. In many ways this organizational problem is similar to the problems faced by the marine instrument engineering community, but the latter has already developed a number of mechanisms to solve these problems. Quantitative marine biologists, mathematicians, and statisticians must follow the engineers' example in this area.

An important step in promoting the advance of statistical and modeling techniques in marine biological research is to bring researchers in applied mathematics and statistics into the appropriate environment. One-or two-year positions for visiting academic statisticians and applied mathematicians could be arranged at institutions involved in marine biological research. Postdoctoral positions for newly graduated mathematical scientists would prove beneficial both to the participants and to the institution of residence. We are convinced that these visitors will encounter challenging and interesting problems, and that the academic community's interest in these areas of research will increase.

There is an immediate need to develop research workshops to focus efforts in quantitative methods in marine science. These workshops, one to two weeks in length, could bring together marine scientists, statisticians, and mathematicians to discuss current research of importance to marine measurement programs.

On a more general level, both graduate students and active scientists need to appreciate the full scope of quantitative methods in marine science. Individual research institutions often cannot provide appropriate exposure for their scientists. An effective means of promoting such an appreciation would be to hold a short-term advanced research workshop. The course could include instructional courses, graduate seminars, and presentations of current research problems.

Finally, a need exists for a prototype center of research in marine biometry to coordinate activities, to publish technical manuals, and to develop training programs in marine biometry.

There are a number of ways that data archiving methods and software systems could be improved. Standardized data formats can

be developed, perhaps enlarging the approach of the Society of
Exploration Geophysics. Also, a thoroughly tested statistical
software, compiled as a publication of reliable special purpose
algorithms for marine biological data analysis, would be invaluable. This would be similar to a project already undertaken by
researchers in digital signal processing. Useful data base systems have already been developed by a number of scientific disciplines. The development of a standardized, transportable software
system is a more important and more easily achieved goal than the
standardization of computer hardware systems.

Instrumentation Concepts and Measurements

Earl E. Hays, Chairman
Bruce C. Coull
Franklin H. Farmer
Michael Karweit
T. T. Packard
P. S. Tabor

One of the major problems in marine biology programs is the lag time required for separation and classification of samples. This creates an appreciable delay between the field sampling phase of the program, and the time analysis and data interpretation phase. Further delay occurs in correlating the biological data with the environmental data, which is usually available much sooner. Sample complexity varies greatly. Samples from the deep ocean sediments present a different separation problem from those of estuaries and from those of mid-water trawls. The varieties to be separated and classified are enormous; therefore, a completely automatic sorting scheme seems impractical to expect at this time, even though pattern recognition techniques are beginning to be well understood and used in many instances. As the technology of signal processing improves and as large scale integrated circuits develop, the cost and time for data processing will be reduced, and the realization of an automatic system will soon be possible. Meanwhile, sizeable reductions in the lag time can be made by the application of video viewing and large image processing. Counting and sizing techniques for one or two component samples are available today. Truly automatic sorting and classification will benefit by future hardware developments but at this time considerable work on the criteria for sorting should be done.

While sensors for measuring many of the variables of impor-

tance to marine biology exist (O_2, pH, EH, chlorophyll, etc.),
their sensitivity is not sufficient to be useful in many oceanographic experiments. These experiments, which often are dealing
with the functional aspects of a community rather than its structure, can be influenced by small but long term changes in concentrations of chemical and biological constituents. A sudden increase
in one of the constituents may have a pronounced effect. The time
series monitoring of the ambient conditions and a sampling program
(or counting by whatever method works) seems like the logical approach to the experiment. Many of the sensors do need improvement
in sensitivity and stability for this kind of experiment.

The availability of microprocessors and high density digital
tape recorders now makes it possible to build self-contained instruments that can sample low frequency phenomena (tides, storms,
seasons). The data collection and storage (i.e., from the time-
dependent sensor outputs to digital tape), are available as well-
defined engineering practice. Therefore, a development of the
sensors is warranted. Existing sensors require improved sensitivity,
while others require improved reliability (fouling drift, etc.).
While some problems can be solved by the use of microprocessors
monitoring the sensors, other problems will require innovation. We
may not be able to make improvements in all of them, but upgrading
existing sensors to be compatible with existing data processing
practices should be examined. One example of equipment which requires improvements is the fluorometer. A survey of marine biologists showed that this instrument was considered one of the most
important in observing phytoplankton biomass and that presently,
continuous (in a sampled sense) remote measurements were not practical. The development costs and marketing possibilities of a
self-contained fluorometer should be analyzed.

The potentially high data acquisition rate in marine biology
will require efficient data assessment and analysis systems. This
has been done to some degree in biological oceanography, but not to
the extent it has been done in the physical and chemical counter-
parts. This is partially an education problem and partially an
instrumentation/hardware problem (not development but availability).
The education problem is not obviously solved. Perhaps the best
place to start is in the funding agencies. If these personnel can
be informed about the possibilities existent in present day technology for processing marine biological data, and can see the benefits for their programs, it would not be long before the "technology
transfer" would take place; those fortunate few who do not depend
upon funding agencies would soon find out they would have to follow
suit or be left behind. As instrumentation/hardware is an efficiency improvement, the funding agencies should be sympathetic to
requests for such improvements.

Marine biological experimentation requires sampling of physical, chemical, and biological variables. Some of these can be
effectively done by point measurements with time series, others
are inherently density measurements (units/volume). The usual
means for obtaining the latter include nets, grabs, and dredges,
for example, and have been the source for most biological data.
Photography has been used in some fields and has proved useful, but

is not the final solution. The counting and identification of objects in the water column or in the sediment layers by other means of collection and sorting, etc. (resulting in a major time-consuming process), would be a major step forward in marine biology. The range of the space scales of the volumes to be scanned and of the size of objects to be detected is very large and no single system will give all answers.

Again, the development of large scale integrated circuits of low power consumption mean that systems can be now considered to perform measurements that a few years ago would have been impractical. Each volume/size regime may require different physical sensing, optics, acoustics, electric field, e.g., but the assembling, sorting and recording of the data is within the capability of standard engineering practice.

Marine Populations: Processes and Interactions

Kenneth Sherman, Chairman
J. W. Deming
J. Frederick Grassle
Richard V. Lynch
Eric Shulenberger
C. Taillie

INTRODUCTION

Studies in biological oceanography have, with few exceptions, been conducted over limited geographic areas for short periods of time. The result has been a series of snapshots of marine species and their environments in time and space with limited coherence among the observations. Significant progress has been made in describing the more abundant species and communities. Some important groups of organisms, nevertheless, remain poorly understood including fast-moving nekton (e.g., squids, tunas, sharks), soft-bodied gelatinous plankton, and bacteria, particularly with respect to the role they play in the energy flux of the ocean. Large gaps remain in understanding the linkages between primary production and secondary production, and the production of fish and other large predators.

Some important advances have been made in developing quantitative methods for estimating changes in the abundances of marine populations. However, much of the literature describes changes in population levels of single species, often with no consideration of the impact of these changes on other species.

More comprehensive assessments of the impact of changes in population levels on the marine ecosystem will require long-term observations of population-environmental interrelationships at specific sites in the intertidal area, on the continental shelf, and in the deep ocean. As these population studies are developed, attention should be given to sampling design and the use of statistical analyses in determining the appropriate numbers of samples for analyses.

The experimental approach in studying population interactions using large-scale enclosures (e.g., CEPEX, MERL), and *in situ* "patch-type" studies should be encouraged. Shallow-water field experiments have led to new developments in fundamental ecological theory. Efforts to extend these techniques to the continental shelf and deep sea should be encouraged. In the context of advancing ecological theory, fisheries studies have already provided an opportunity for conducting large-scale field experiments on the effects of the removals of large biomasses of predators and prey on the energy flux of continental shelf ecosystems.

Population processes

Significant problems exist in understanding the growth and mortality processes that occur within populations, among populations, and between populations and the environment. The effects of population density on predator-prey interactions and of physical and chemical factors on age-specific rates of birth, growth, and mortality are largely unknown. For example, testing the effect of a given pollutant at a single stage of an organism's life is almost worthless for predicting the effects that substance will have on a population of that species unless all life stages are considered. Elucidation of the relationships of population processes to physical and biological variables requires controlled experimentation, using either laboratory culture methods or field manipulation. Currently, both approaches are very labor intensive. Also, organisms that are easy to culture generally have certain types of life histories which may not be representative of all, or even the important, species. Improved culture techniques for a variety of fish and invertebrates, particularly automated culture techniques, would greatly facilitate future laboratory population studies. In this regard, the marine analog of the Wistar rat, as a biological indicator of physiological condition in relation to pollution levels is a promising area for further culture research. Rewarding field population experiments on the continental shelf, now utilize benthic species as indicators of environmental health because the problems of motility are much less severe than with free-swimming species. These studies should be encouraged and enlarged to encompass strategic areas of both impacted and relatively "clean" marine environments.

Major Problem

Studies of the interrelationships among ecosystem components can best be approached by posing key questions dealing with the basic processes of recruitment, growth, and mortality of marine populations. To predict the responses of ocean ecosystems to manmade and natural perturbations, we ask how do the individual lifecycles of populations and interrelationships among species relate to patterns of environmental variation.

PLANKTON

The state of knowledge in plankton biology varies by region.
Most advances in understanding the processes of reproduction, growth,
mortality, and distribution have been attained in accessible, relatively nearshore environments. There, some regions have been intensively described. There is a need to expand descriptive efforts,
particularly in the open ocean, especially on scales smaller than
ocean-wide. While both long-term and short-term descriptive efforts must continue, insights into processes and rates are best obtained by experimental manipulation of defined systems. In the
nearshore environment, various intentional or unintentional perturbations (e.g., formal experimental manipulations, pollution events,
fisheries) have given limited insight into ecosystem processes and
functioning. No methodology exists for doing such experiments in
the open ocean; even selection of appropriate scales is a difficult
problem that can only be solved by detailed descriptive efforts.
Considerable progress has been made in understanding some ecological proceses through laboratory experiments. We need to develop
appropriately-scaled experimental techniques tailored to both nearshore and open oceanic environments, so that the laboratory results
may be verified in natural systems. Aspects needing such large
scale treatment include very basic processes such as rates of photosynthesis, productivity, second-order and higher level productivities, and reproduction. Other problems needing detailed investigation on realistic scales include questions of competition, predation, and community structure, stability, and evolution.

BENTHOS

Studies of benthic ecosystems have progressed to the extent
that the most rapid advances are the result of experimentation involving alteration of single variables in field situations. In
shallow water it has been possible to work out the functional role
of individual species in maintaining community structure. The role
of predators, larval dispersal and recruitment, and networks of
competitive relationships between species have all been active
areas of investigation. Concepts are being developed that relate
the factors influencing survival of individuals to more broadscale
patterns of community structure.

The technology is now available to extend these experimental
studies into the deep sea. Here the relative uniformity of conditions over substantial areas allows us to generalize from experiments on spatial scales of meters. It is possible now to measure
how the basic biological rate processes of individuals and populations influence the major chemical and biochemical fluxes in the
benthic boundary layer. Deep-sea experiments of the future will
involve detailed studies of the movements, dispersal and mortality
of organisms, fluxes and dissolution rates of chemical constituents
and movements of particles above and below the sediment-water interface. These processes are so intimately related that one cannot
be studied without the other. The deep-sea experiments to study

the benthic-boundary layer processes must be done in areas where
the physical regime and influx of particles from the upper layers
of the ocean are being studied simultaneously.

BACTERIA

Marine bacteria are ubiquitous in the ocean environment, distributed as free-living planktonic species, associated with particulates, or in commensal or symbiotic relationships with other life
forms. A major role of marine bacteria in the sea is the regeneration of nutrients and mineralization of allochthonous materials, a
critical role in nutrient cycles. Techniques for determining rates
of activity and degradative potential of free-living microbial
populations have been applied to the water column, but still require considerable effort and increased field application. Deep-
sea microbiology, still in the developmental stage, has the technology available to measure *in situ* metabolic rates of deep-sea
microbial populations that exist under extreme conditions of low
temperature and nutrient levels and elevated hydrostatic pressure.
Initial studies indicate that metabolic rates for free-living bacteria in the deep-sea trenches and basins are markedly reduced
relative to activity under atmospheric pressure and temperature.
Recent studies of bacteria associated with intestinal tracts of
deep-sea amphipods suggest the nutrient-rich guts of macroorganisms
are sites of considerable microbial activity even under deep-sea
conditions. Studies of the importance of physical associations of
bacteria with particulates and with other organisms should be expanded in order to understand periphytic and commensal or symbiotic
activities of microorganisms and their role in food webs of the sea.
It is increasingly evident that rapid evaluation of man-made
and natural perturbations in the ocean can best be accomplished by
examining impact at the microbial level. Bacterial community structure will reflect the environment from which the organisms are isolated. In the open ocean, where the nutrient supply is limited,
functional oliogotrophs predominate. Studies of deep-sea sediment,
water, and intestinal gut flora have shown that taxonomically distinct groups of bacterial types are associated with a specific habitat. Environmental changes will be accompanied by changes in community structure and/or microbial biomass. Although the last decade
has seen rapid advances in assessment of microbial biomass, the
various methods, including epifluorescence microscopy, ATP and/or
GTP analysis, heterotrophic uptake, and *Limulus* lysate assay, remain to be standardized, inter-calibrated and automated for large-
scale studies of marine bacteria. In addition to further development
and application of these techniques to marine studies, sampling
problems, in particular, equipment interference in the integrity of
samples collected for microbiological analyses must be overcome.
To date, the microbiologist does not have available equipment or
technology for aseptic collection of deep sea sediments. Future
studies in marine microbiology should stress a comprehensive approach, including taxonomy, ecology, and physiology, for improving
the definition of autochthonous bacterial populations, as well as

providing a basis for quantitating the sensitivity of the microbial communities to impingements of man and his activities on the ocean environment.

FISH

Studies of fish are moving rapidly from the more traditional assessments of single species populations to investigations of fish as principal predators in complex marine ecosystems. The newly emerging fisheries ecology studies focus on the importance of primary production, secondary production, and predation on the growth, recruitment, and mortality of fish stocks. Considerable effort is directed to the partitioning of mortality into natural mortality, mortality due to fishing, and mortality that may be related to increasing ocean pollution.

One of the major problems in fisheries studies is the present lack of understanding of the relationship between the size of the parent stock and the numbers of larvae and juveniles they produce as new recruits to the fishery. A better understanding of the stock-recruitment relationship is required if predictions of future levels of abundance are to be improved. Programs have been developed and are now underway to deal with this problem using a combination of laboratory experimentation on the growth and survival of eggs and larvae and large-scale systematic assessments of changing levels of larval, juvenile, and adult fish abundance. These studies should be expanded to sample adequately the newly established fisheries management zone of the U. S. Present survey methods could be improved with the refinement and application of acoustic systems for estimating population biomass.

Technology Needs

In the laboratory, two time-consuming procedures could be improved with the application of image scanning systems: ageing of scales and otoliths, and the sorting, sizing, and counting of phytoplankton, zooplankton, fish eggs, and larvae. Development of hydroacoustic and other pertinent advanced technology for increasing the efficiency of present trawling methods used to monitor changes in the abundance of fish stocks should be accelerated.

Expanding Fishery Ecosystem Studies

Time-series measurement of changes in the abundance levels of fish stocks and their prey over space and time should be expanded to include all of the area now under the jurisdiction of the U. S. as described in the recently enacted Fisheries Conservation and Management Act of 1977.

Fisheries studies provide a unique opportunity to view the changing abundance levels of fish as large-scale field experiments in ecosystem dynamics. For example, off the northeast coast during the past decade there has been a decrease of about 50% in the biomass of fish. Much of the decline is attributed to heavy fishing

mortality. Local environmental conditions and longer-term climatic
changes may also have contributed to the decline. Significant
questions remain unanswered. Does the reduction of principal species (e.g., herring, cod, haddock, yellowtail flounder) release
secondary production to be consumed by short-lived, fast growing,
smaller, less desirable species? What are the probabilities associated with the return of over-exploited species to former abundance levels? Present studies focus on the critical linkages among
the principal sources of the food of fish and the recruitment, survival, and productivity of the fish stocks from the Gulf of Maine
to Cape Hatteras. The availability of these stocks to U. S. fishermen is the end product of a complex series of interactions among
the fish stocks, benthos, plankton, and hydrography of the region.
Studies of single species have not provided the kind of scientific
information required for effective management of multispecies fisheries operating at different trophic levels. While it is important
to continue these studies, they must be conducted within a broader
matrix that measures changing abundances of key species in the
ecosystem. Fisheries ecosystem studies will lead to improved understanding of the factors controlling fish production, and this
information in turn will allow for greatly improved forecasts of
abundance levels of fish stocks.

Synoptic Measurements and Remote Sensing

Robert W. Johnson, Chairman
Thomas M. Fitzgerald
Anil K. Jain
V. Klemas
Roland E. Nagle

Advances in measurement and analysis techniques indicate that remote sensing and marine biology have many areas of mutual interest. Some of these are listed below along with suggestions to aid in their implementation.

EXPERIMENTAL OPPORTUNITIES

A current list of experimental opportunities should be developed and distributed to principal investigators in multidisciplinary experiments in coastal zone and oceanographic environments. The chief scientist may contact or be contacted by investigators to define objectives, measurements, analysis techniques, and other experimental factors. An initial effort is being made by the University of Rhode Island to meet this need through a bi-monthly publication.

REMOTE SENSING DATA AVAILABILITY

The utilization of remotely sensed data requires access to the basic observation. There are a number of methods already established that facilitate access to these data. For example, NOAA/

NESS publishes monthly catalogues of mapped polar stereographic imagery acquired from the NOAA/TIROS -N satellites; the NASA LANDSAT Program has an extensive archival/distribution system to allow browsing through available data and to submit orders for specific data from a centralized facility. It is recommended that a list of existing facilities be prepared from which remotely sensed data may be acquired; the list should contain points of contact and references to the documents detailing data request procedures. Making this information available to marine biological researchers would provide the basic data by which coincident remote sensing data can be acquired.

ESTABLISH A LINK BETWEEN REMOTELY SENSED DATA AND CRITICAL MARINE BIOLOGY PARAMETERS

In order to aid application of synoptic remote sensing to the marine environment the marine biology community should identify critical parameters that may be measured or inferred from remote sensing measurements. Approaches to this include:

> † Identification of parameters and associated radiance characteristics.

> † Identification of radiance measurement that may be interpreted to provide information to marine scientists.

> † Use of data to construct synoptic maps of existing conditions in four dimensions (x, y, z, t) by permitting extrapolations of ship data in space and time.

Temperature and chlorophyll may be two such universal parameters.

AUTOMATED SAMPLE ANALYSIS

Advanced pattern recognition and spectral analysis techniques should be applied to processing and analysis of large numbers of marine biological samples. Such analysis systems were developed for medical laboratory analyses over a decade ago. The processing system used at the National Marine Fisheries Service Narragansett Laboratory is an initial step toward this technology transfer and should be critically evaluated by marine biologists.

RESEARCH VS. APPLICATION REQUIREMENTS

The increasing concern for ocean resources has elevated marine biology to a worldwide interest level. Concomitant with this change in emphasis, marine biologists must broaden their requirements to coordinate them with the information requirements of management decisionmakers. The type and precision of observations required for the pursuit of scientific problems may vastly exceed what is required to allow critical management decisions. In this context, marine biological scientific requirements may be difficult if not impossible to achieve by remote sensing techniques whereas remote sensing might satisfy the management requirements. A clear definition of both management and scientific requirements should be formulated.

REMOTELY SENSED DATA AND MODELING

Synoptic wide-area remotely sensed data may be used for verification and updating analytical and predictive needs. Analytical models include parameters that by direct measurement or inference provide needed information to marine biologists or ecosystem managers. Predictive models may incorporate spatial and temporal measurements to increase the understanding of the past, current, and future state of an ecosystem.

DATA BASE APPLICATIONS OF OCEANOGRAPHIC EXPERIMENTS

Encourage oceanographic investigators to consider current or future use of remote sensing by including in their experiment design and reports:

† Location of the experiment on regional or global coordinate system,

† an attempt to make data collections when satellite overpasses occur, and

† the determination of representative characteristics of area sampled.

The Future of Hydroacoustics as an Aquatic Biological Tool

F. R. Harden Jones, Chairman
Charles L. Brown
Arthur A. Myrberg, Jr.
H. L. Warner
William A. Watkins

The aquatic environment is well suited to the use of acoustic techniques which may be applicable to a broad range of research problems ranging, for example, from the study of the behavior of individuals to the determination of the abundance of populations, and from studies on bottom sediments and topography to the shape and performance of fishing gear.

Two areas of aquatic biology have been identified where the application of acoustics could lead to significant advances in the next five years.

   † Determination of the distribution and abundance of populations and their identification to species level should be accomplished if feasible.

   † The behavior of individual animals should be studied in order to predict the behavior and distribution of the wild population and thus add to our basic knowledge and understanding. Such studies would add significantly to the understanding of the organism as "a going concern" and would be particularly relevant to studies of conservation, management, and environmental protection.

In the aquatic environment there are many special topics and fields of study where acoustic techniques can be applied. For ex-

ample, the following were of particular interest to members of this
group; (a) the use of specific frequencies in attracting or repel-
ling fish or whales; (b) training techniques for special studies;
(c) a study of the movements of fish through passes or estuaries;
(d) the small-scale structure of plankton distributions; (e) a variety of hydrographic applications; and (f) special studies of
near-field effects ahead of sampling gear.

An attempt has been made to identify those constraints which
would appear to limit the successful application of acoustic techniques to problems in the aquatic environment. The most serious
constraint is imposed by the failure of biologists to specify their
requirements in quantitative terms that would provide the basis for
a feasibility study by the physicist and acoustic engineer. A typical example is provided by the application of acoustic techniques
to the estimation of population abundance in the fisheries field.
Fisheries require stability, requiring managers to control and limit fishing mortality. This control might well be achieved in terms
of a total allowable catch (TAC) calculated to give some form of
maximum or optimum yield. TAC's are traditionally calculated from
catch per unit effort. To be useful the estimated TAC should be
within ± 10% of the value corresponding to the optimum yield. This
requirement sets a constraint for the physicist and the acoustic
engineer: can this requirement be met by counting and sizing individual fish or by estimating the biomass by echo integration?

When the feasibility of any project has been determined, the
next consideration is that of cost. No research project is free
from this 'chain', but the 'links' are particularly heavy in acoustics, where the cost of equipment may be beyond that which could be
funded through normal channels. This difficulty will probably not
be resolved until it is recognized and treated accordingly at the
appropriate funding level.

There is little merit in burdening this report with a catalog
of problems approaches: specific problems must be identified and
clearly defined quantitatively before appropriate acoustic techniques can be tailored to achieve a particular objective.

Recommendations

To facilitate significant advancement in sampling and data analysis in marine biological disciplines within the next five years, three general areas need specific attention by funding agencies:

Interdisciplinary and International Investigations

†   Measurements of parameters important to marine biologists require a multidisciplinary effort. For example, an acoustic study of a population of fish could use the expertise of physicists, chemists, biologists, and engineers.

†   International workshops using an multidisciplinary approach would enhance research productivity and reduce the potential for duplication of effort, saving time and money.

Technology Transfer

Advances in understanding technological development may lag a decade behind the scientist's understanding of data derived from

observational programs. This is especially true in the case of marine biologists who do not have a strong background in the physical
sciences. Three actions are recommended to haste technology transfer:

   † Multidisciplinary study programs, such as continuing
     education, short courses, and sabbaticals.

   † The development of an awareness in the biological com-
     munity of the observation potential of new instruments.

   † Ready access of technological developments of industrial
     and governmental laboratories to academia.

Standardization of Techniques

   At the level of national and international sponsoring agencies,
support is needed to develop traceable standards and methods for
marine biological measurements.
   Progress in the solution of specific measurement problems is
necessary. These problems include sample processing and archiving
of biological samples and data, and measurement methods applicable
to marine biological hydroacoustics and remote sensing.
   Research opportunities in marine-related areas must be made
known to potential principal investigators. A current list of such
opportunities should be developed and distributed. An initial effort to meet this need is being made by the University of Rhode
Island through a bi-monthly publication.
   Funding agencies must be made aware of the need for programs
for sampling or monitoring marine biological populations or environmental parameters which allow sufficient time to determine the frequency with which the parameter or population varies, and which are
on a scale commensurate with the patchiness and extent of the distribution.

MARINE POPULATIONS: PROCESS AND INTERACTIONS

Broad Requirements

   † Comprehensive assessments of the impact of changes in
     population levels on the marine ecosystem require long-
     term observations of population-environmental interrela-
     tionships at specific sites. Particular attention should
     be given to sampling design and statistical determination
     of appropriate sample sizes.

   † The experimental approach in studying population inter-
     actions using large-scale enclosures, e.g., CEPEX, MERL,
     and *in situ* "patch-type" studies should be encouraged.

   † Improved culture techniques for a variety of fish and
     invertebrates, particularly automated culture techniques,
     would greatly facilitate laboratory population studies.

† Field population studies on the continental shelf now utilize benthic species as indicators of environmental health. These types of studies should be encouraged and expanded.

Plankton Research

† Descriptive efforts, particularly in the open ocean, should be expanded.

† Appropriately-scaled experimental techniques tailored to both nearshore and open oceanic environments should be designed so that laboratory verification is possible. Aspects needing large-scale treatment include rates of photosynthesis, productivity, second-order and higher level productivities, and reproduction. Additional problems requiring detailed investigation on realistic scales include questions of competition, predation, and community structure, stability, and evolution.

Benthic Research

Deep-sea experiments to study the benthic-boundary layer processes must be done in areas where the physical regime and influx of particles from the upper layers of the ocean are simultaneously examined.

Bacterial Research

† Techniques for determining rates of activity and degradative potential of free-living microbial populations have been applied to the water column, but still require increased field application.

† Studies of the importance of physical associations of bacteria with particulates and with other organisms should be expanded in order to understand periphytic and commensal or symbiotic activities of microorganisms and their role in food webs of the sea.

† Various methods for assessing microbial biomass, including epifluorescence microscopy, ATP and/or GTP analysis, heterotrophic uptake and *Limulus* lysate assay, remain to be standardized, intercalibrated and automated for large-scale studies of marine bacteria.

† Future studies should stress a comprehensive approach, including taxonomy, ecology, and physiology to improve the definition of autochthonous bacterial populations, as well as to provide a basis for quantifying the sensitivity of the microbial communities to anthropogenic activities.

Fish Research

† A better understanding of the stock-recruitment relationship is required if predictions of future levels of
abundance are to be improved. Studies using a combination of laboratory experimentation on the growth and
survival of eggs and larvae and large-scale systematic
assessments of changing levels of larval, juvenile and
adult fish abundance should be expanded.

† In the laboratory, two time-consuming procedures could
be improved with the application of image scanning systems: (1) ageing of scales and otoliths and (2) sorting,
sizing, and counting of phytoplankton, zooplankton, fish
eggs, and larvae.

† The development of hydroacoustic and other pertinent advanced technology must be accelerated to increase the
efficiency of present trawling methods used to monitor
changes in the abundance of fish stocks.

† Time-series measurement of changes in the abundance
levels of fish stocks and their prey over space and time
should be expanded to include all United States jurisdiction fisheries.

INSTRUMENTATION CONCEPTS AND MEASUREMENTS

Several instrumentation approaches could alleviate the lag
time required for separation and classification of marine biological samples.

† Video-viewing and large-image processing might be applied.

† Improved counting, sizing, and sorting criteria and
techniques are required.

† Sensors for time-series monitoring of ambient conditions
need improvement.

A specific example of an existing instrument which needs further development for use in routine marine biological sampling is
the fluorometer. Development costs and marketing possibilities of
a self-contained fluorometer should be analyzed, as marine biologists have indicated it is one of the most important tools in determining phytoplankton biomass.

ACOUSTIC TECHNIQUES

Acoustic techniques have broad application to measurements of
parameters important to marine biology, especially the distribution,
abundance, and behavior of populations. The following problems
need specific attention in order to accelerate significant advances.

† Biologists need better training in physical acoustics that would provide a basis for feasibility studies by physicists and acoustic engineers.

† Acoustic studies are expensive and require special funding considerations by granting agencies.

SYNOPTIC MEASUREMENTS AND REMOTE SENSING

Remote Sensing Data Archiving and Utilization

There are a number of methods already established that facilitate access to remotely sensed data. For example, NOAA/NESS publishes monthly catalogues of mapped polar stereographic imagery acquired from the NOAA/TIROS -N satellites; the NASA LANDSAT Program has an extensive archival distribution system to allow users to scan the available data and to submit orders for specific data from a centralized facility. It is recommended that such specific capabilities be identified and appropriate lists be disseminated.

Remotely Sensed Data and Critical Marine Biology Parameters

The marine biology community should identify critical parameters that may be measured or inferred from remote sensing measurement with particular emphasis on the synoptic aspects of remote sensing data. Approaches include:

† Identification of parameters and associated radiance characteristics.

† Identification of radiance measurements applicable to marine biological observations.

† Use of data to construct synoptic maps of existing conditions in four dimensions (x, y, z, t) from extrapolations of ship data.

It appears that temperature and chlorophyll may be two such universal parameters.

Automated Sample Analysis

There is a need to apply advanced pattern recognition and spectral analysis techniques to process and analyze the large numbers of samples gathered by marine biologists. Such systems were developed for medical laboratory analyses over a decade ago. The processing system used at the National Marine Fisheries Service, Narragansett Laboratory, provides the initial step for marine biological analysis.

Research vs. Application Requirements

The increasing concern for ocean resources has elevated marine biology to a worldwide level of interest. Concomitant with this increased emphasis, the marine biologists must broaden their research requirements to coordinate with the information requirements of management decisionmakers. The type and precision of observations required for the pursuit of scientific solutions vastly exceed what is required for critical management decision. In this context, it may be some time before a full set of research needs may be applicable to management requirements. A clear definition of both management and scientific requirements should be formulated.

Remotely Sensed Data and Modeling

Synoptic wide-area remotely sensed data may be used for verification and updating analytical and predictive needs.

Data Base Applications of Oceanographic Experiments

Encourage oceanographic investigators to consider current or future use of remote sensing by including the following in their experiment design and reports:

   † Locate experiment on regional or global coordinate system.

   † Attempt to make data collections when satellite overpasses occur.

   † Determine representative environmental characteristics of area sampled.

STATISTICAL AND MATHEMATICAL METHODOLOGY

A specific need exists for a marine biological reference data base. Existing data base systems, like the Environmental Protection Agency's BIOSTORET system could be modified for marine data.
Data analysis programs require standardization and validation. It is also important that the software for new computer-based techniques and methods be widely distributed within the marine community to improve the efficiency of the expensive process of software development. Examples of computer-based methods are the automatic sorting and pattern recognition techniques used for analyzing marine plankton samples. Another example is the algorithyms software which transforms remotely sensed signals into a variety of display patterns; their interpretation provides clearer understanding of the displayed data.
Research in experimental and survey design in the following areas could make the marine measurement programs more efficient:

† The use of remote sensing as an aid in the design of stratified oceanographic surveys.

† Optimal design of sensor evaluation experiments.

† Development of sampling schemes designed to improve the efficiency and accuracy of classical sorting methods for plankton samples.

† Optimal design of experiments and sensors for the detection of rare events.

† Further development of ecological models of marine systems.

Data archiving methods and software systems require improvement. Standardized data formats should be developed, and a thoroughly tested statistical software package, compiled as a publication of reliable special purpose algorithms for marine biological data analysis, should be developed

A need exists for a prototype center of research in marine biometry to coordinate activities, to publish technical manuals, and to develop training programs in marine biometry.

F.P.D.
F.J.V.
D.Z.M.

PARTICIPANTS

ARTHUR BAGGEROER*, *Department of Electrical Engineering, Massachusetts Institute of Technology, Cambridge, Massachusetts.*

D. S. BARTLETT, *College of Marine Studies, University of Delaware, Newark, Delaware.*

CHARLES L. BROWN, *Naval Underwater Systems Center, New London Laboratory, New London, Connecticut.*

CLARENCE A. BROWN, JR., *National Aeronautics and Space Administration, Langley Research Center, Hampton, Virginia.*

JANET W. CAMPBELL, *National Aeronautics and Space Administration, Langley Research Center, Hampton, Virginia.*

R. R. COLWELL, *Department of Microbiology, University of Maryland, College Park, Maryland.*

BRUCE C. COULL, *Belle W. Baruch Institute for Marine Biology and Coastal Research, University of South Carolina, Columbia, South Carolina.*

*Paper not available at time of publication.

J. W. DEMING, *Department of Microbiology, University of Maryland, College Park, Maryland.*

BRADLEY W. DICKINSON, *Department of Electrical Engineering and Computer Science, Princeton University, Princeton, New Jersey.*

FERDINAND P. DIEMER, *Office of Naval Research, Arlington, Virginia.*

ROBERT L. EDWARDS, *National Oceanic and Atmospheric Administration, National Marine Fisheries Service, Northeast Fisheries Center, Woods Hole, Massachusetts.*

FRANKLIN H. FARMER, *National Aeronautics and Space Administration, Langley Research Center, Hampton, Virginia.*

THOMAS M. FITZGERALD, *Naval Underwater Systems Center, Newport Laboratory, Newport, Rhode Island.*

J. FREDERICK GRASSLE, *Woods Hole Oceanographic Institution, Woods Hole, Massachusetts*

F. R. HARDEN JONES, *Fisheries Laboratory, Lowestoft, Suffolk, United Kingdom.*

EARL E. HAYS, *Woods Hole Oceanographic Institution, Woods Hole, Massachusetts.*

G. DANIEL HICKMAN, *Applied Science Technology, Inc., Arlington, Virginia.*

D. V. HOLLIDAY, *Tracor, Inc., San Diego, California.*

ANIL K. JAIN, *Department of Electrical Engineering, University of California, Davis, California.*

OLIN JARRETT, JR., *National Aeronautics and Space Administration, Langley Research Center, Hampton, Virginia.*

ROBERT W. JOHNSON, *National Aeronautics and Space Administration, Langley Research Center, Hampton, Virginia.*

MICHAEL KARWEIT, *Chesapeake Bay Institute, The Johns Hopkins University, Baltimore, Maryland.*

V. KLEMAS, *College of Marine Studies, University of Delaware, Newark, Delaware.*

J. B. LOZOW, *The Charles Stark Draper Laboratory, Inc., Cambridge, Massachusetts.*

RICHARD V. LYNCH, *Marine Biology and Biochemistry Branch, Ocean Sciences Division, Naval Research Laboratory, Washington, D. C.*

ROMAINE R. MAIEFSKI, *Engineering Consultant, 1035 Golden Road, Encinitas, California.*

HUGH B. MARTIN III, *Ocean Applied Research Corporation, San Diego, California.*

DONNA Z. MIRKES, *Belle W. Baruch Institute for Marine Biology and Coastal Research, University of South Carolina, Columbia, South Carolina.*

ARTHUR A. MYRBERG, JR., *Rosenstiel School of Marine and Atmospheric Science, University of Miami, Miami, Florida.*

ROLAND E. NAGLE, *Naval Environmental Prediction Research Facility, Monterey, California.*

T. T. PACKARD, *Bigelow Laboratory for Ocean Sciences, West Boothbay Harbor, Maine.*

G. P. PATIL, *Department of Statistics, The Pennsylvania State University, University Park, Pennsylvania.*

W. D. PHILPOT, *College of Marine Studies, University of Delaware, Newark, Delaware.*

KENNETH SHERMAN, *National Marine Fisheries Service, Northeast Fisheries Center, Narragansett, Rhode Island.*

ERIC SHULENBERGER, *San Diego Natural History Museum, Balboa Park, San Diego, California.*

WOOLLCOTT SMITH, *Woods Hole Oceanographic Institution, Woods Hole, Massachusetts.*

WELDON L. STATON, *National Aeronautics and Space Administration, Langley Research Center, Hampton, Virginia.*

J. B. SUOMALA, JR., *The Charles Stark Draper Laboratory, Inc., Cambridge, Massachusetts.*

P. S. TABOR, *Department of Microbiology, University of Maryland, College Park, Maryland.*

C. TAILLIE, *Department of Statistics, The Pennsylvania State University, University Park, Pennsylvania.*

F. JOHN VERNBERG, *Belle W. Baruch Institute for Marine Biology and Coastal Research, University of South Carolina, Columbia, South Carolina.*

H. L. WARNER, *Naval Coastal Systems Center, Panama City, Florida.*

DOUGLAS WARTZOK, *The Johns Hopkins University, Baltimore, Maryland.*

WILLIAM A. WATKINS, *Woods Hole Oceanographic Institution, Woods Hole, Massachusetts.*

HOWARD L. WEINERT, *Department of Electrical Engineering, The Johns Hopkins University, Baltimore, Maryland.*

R. L. WIGLEY, *National Marine Fisheries Service, Northeast Fisheries Center, Woods Hole, Massachusetts.*

A. A. YAYANOS, *Scripps Institute of Oceanography, University of California San Diego, La Jolla, California.*

INDEX

564